AF248660

Fundamental and Applied Aspects of Chemically Modified Surfaces

Fundamental and Applied Aspects of Chemically Modified Surfaces

Edited by

Jonathan P. Blitz
Eastern Illinois University, Charleston, USA

Charles B. Little
Supelco Inc., Bellefonte, USA

The proceedings of the 7th International Symposium on Chemically Modified Surfaces held at the
Northwestern University, Evanston, Illinois, USA on 24–28 June 1998

The front cover illustration is taken from the contribution by
A. Singh, M.A. Markowitz, P.E. Schoen and C. Costellanos, p. 14.

Special Publication No. 235

ISBN 0-85404-714-X

A catalogue record for this book is available from the British Library

Published by The Royal Society of Chemistry,
Thomas Graham House, Science Park, Milton Road,
Cambridge CB4 0WF, UK

For further information see our web site at www.rsc.org

Printed and bound by Bookcraft (Bath) Ltd.

Preface

Many researchers have an interest in the chemical modification of particle surfaces. The reason for this is twofold: (1) a fundamental understanding of surface chemistry is elusive because of its inherent complexity; and (2) chemically modified surfaces exhibit many technologically useful properties in a broad range of fields. When workers get together to discuss the latest developments in the field, it is not surprising then that a mix of academic and industrial researchers is attracted. Furthermore, given the interdisciplinary nature of this work, one finds a mix of chemists, materials scientists, engineers and physicists all working on similar types of problems but thinking about them from unique perspectives. This is exactly what occurred in June, 1998, when many leaders in the field met at Northwestern University to discuss their most recent work. Approximately 60 different studies were reported by scientists from 12 countries at the 7th International Symposium on Chemically Modified Surfaces. This volume is a compilation of approximately two-thirds of the papers presented at this unique meeting.

This proceedings volume is structured such that similar papers are grouped together. The book itself has not been divided into parts, however, because the editors feel that such delineations are unnecessarily artificial. All of the work presented in this volume is in a sense related. Reading the perspectives of workers from a variety of backgrounds working on similar problems is beneficial, and contributions from various parts of the book may have many similarities. The inevitable bias of the editors is minimized by not dividing this volume into sections of papers so that readers will not overlook contributions of potential interest.

With that said, it is incumbent upon us to provide some rationale for the structure of this volume. The papers are grouped in broad categories starting with those relating to self-assembled monolayers, followed by polymer grafting and composite materials, chromatography and organosilane-modified silicas, plasma polymerizations, heterogeneous catalysis, electrochemical applications, general surface treament, metal binding and biological applications.

This book provides an excellent snapshot of the latest concepts and tools applied to the field of chemically modified particle surfaces. Some relevant trends include the large growth in the study of self-assembled monolayers, and new tools such as combinatorial synthetic and other methodologies for the synthesis of improved materials. On the analytical side, tools which were just a few years ago new and exciting are now routine. Newer methods such as scanning probe microscopies are exhibiting their power. In concert with the more established techniques, enhanced understanding of these scientifically interesting and technologically useful materials is now obtainable.

The field of chemically modified particle surfaces is exciting and dynamic. This volume contains the latest and best work in the field in a timely manner. We believe that any worker in this field will find at least some of the information contained herein useful to their own work.

Contents

SILOXANE-ANCHORED MONOLAYERS AS TEMPLATES FOR OXIDE FILM DEPOSITION

Hyunjung Shin, Yuhu Wang, Sitthisuntorn Supothina, Rochael J. Collins, Monika Agarwal, Mark R. De Guire, Arthur H. Heuer and Chaim N. Sukenik*

Departments of Chemistry and Materials Science and Engineering
Case Western Reserve University, Cleveland, OH 44106, USA
and Bar Ilan University, Ramat Gan, 52100, Israel

1 INTRODUCTION

Modern materials research continues to seek new processing paradigms for the creation of adherent, uniform oxide thin films. For example, by avoiding high temperature or line-of-sight limitations, films can be applied to thermally sensitive and/or irregularly shaped substrates. This has led to a number of novel approaches, among them the biomimetic deposition of inorganic thin films onto ordered organic templates.[1] Among the substitutes for the membranes and/or proteins used by living organisms as templates are Langmuir monolayers at an air–water interface and Langmuir Blodgett films or self-assembled monolayers (SAMs) on solid substrates.[2]

The use of self-assembled monolayers to achieve stable, uniform surface modification has been extensively developed over the past 20 years.[3] While much elegant work has been done using various monolayer anchoring strategies, the present discussion will concentrate on siloxane-anchored films created by the self-assembly of alkyltrihalo- or trialkoxy-silanes from homogeneous solution.[3b] Such films have been deposited onto metal, semiconductor, oxide, and/or polymeric substrates. Under optimal conditions, the SAM array provides a powerful tool for surface modification. They are very thin (typically 15–30 Å) and very uniform. They are comprised of close-packed hydrocarbon chains aligned nearly perpendicular to the substrate surface. This packing both presents a uniform interface and provides a barrier layer for the protection and further stabilization of the anchoring siloxane network. Using siloxane anchored SAMs as templates takes advantage of their relatively robust nature and of the ease with which they can be chemically manipulated.

In addition to the use of functionalized SAMs to achieve uniformly modified surfaces, a number of strategies have used SAMs to achieve patterned surfaces. Microcontact printing allows for application of the SAMs to only selected regions of a surface,[4] while patterned photo-ablation allows for removal of SAM-forming material from selected regions of the surface.[5]

There are also many reports of the *in situ* modification of film functionality. These surface functional groups control the chemical and physical properties manifested by the monolayer array. Functionalized SAMs provide a vehicle for construction of thin-film multilayer assemblies.[6] They have been used for controlling the surface interactions with, and for the surface anchoring of, both biomolecules (*e.g.* proteins)[7] and intact biological cells.[5c,8] They have also found application in controlling the attachment of inorganic (metallic/ceramic) overlayers created by CVD,[9] and by other processing methods (including sol-gel and catalytic electroless plating).[5d,10] Functional group interconversion also provides another approach to surface patterning (below).

Work in our group has focussed on using SAM templates as part of a processing paradigm for ceramic thin-film materials that promotes the low-temperature formation of uniform, adherent, pore-free oxide layers with control over film thickness and morphology.[11] We will illustrate our approach to SAM-assisted oxide deposition by discussing the formation of metal oxides on both planar and particulate SAM-coated substrates. We will also consider the role of the SAM as a tool for controlling the lateral patterning of the oxide film. Finally, we will discuss the issue of SAM stability and uniformity and the integrity of the oxide–substrate interface in the face of the thermal and/or chemical challenge presented by the oxide deposition medium and/or by post-deposition processing.

2 RESULTS AND DISCUSSION

In each case, the work described herein begins with the creation of the template. This involves the cleaning and/or activation of a substrate surface, coating it with SAM-forming organic compounds, characterizing the SAM film, and (often) an *in situ* transformation of its surface functional groups. The work reported for planar substrates was done on single-crystal silicon wafers that were cleaned and activated by H_2SO_4–H_2O_2 (*piranha*) solution treatment. The coating of particles was done on commercial TiO_2 powder (rutile; 50–500 nm particle size) which was cleaned by sonication in an organic solvent. The alkyl trichlorosilanes were either purchased (*e.g.* octadecyltrichlorosilane, OTS) or were prepared in a few steps from commercially available 10-undecenyl alcohol. A schematic representation of the deposition and *in situ* modification of the siloxane-anchored SAM is shown below. The use of this scheme to install polar/ionic functionality (specifically hydroxyls[12] and sulfonates[13]) that are useful as oxide template surfaces is embodied in systems, for example, where $X = ONO_2$, $Y=OH$ and where $X = SCOCH_3$, $Y = SO_3H$.

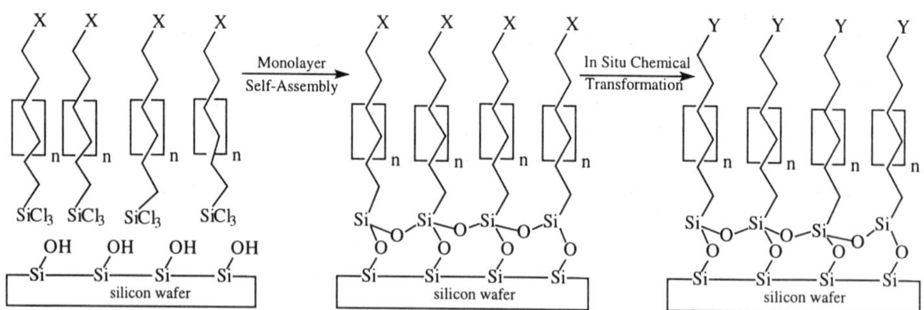

Figure 1   *Deposition and in situ transformation of a siloxane-anchored self-assembled monolayer (SAM) on a silicon wafer.*

The two sets of surface functionality alluded to above can be each transformed using a solution reagent (nitrate reduced to alcohol using $LiAlH_4$ in ether; thioacetate oxidized to sulfonate using Oxone in water), but these transformations can also be done photochemically. The photochemical route allows for the patterned installation of the target functionality when the irradiation is done through a standard lithographic mask. In this way, the creation of well-defined, 2-dimensional, surface domains that are intended to selectively favor or disfavor oxide deposition is possible.

Siloxane-based SAMs have also been used to coat particles. Our attachment of both OTS and thioacetate (TA) functionalized SAMs onto titania powder[14] compares well with an earlier report of OTS coatings on silica particles.[15] In addition to the change in particle wettability (as assessed by flotation studies) that accompanies the coating of the hydrophilic oxide particle (silica or titania) with a hydrophobic OTS coating, we can follow the progress of functional group interconversion on the particle surface as well. Specifically, XPS analysis of both flat and particulate substrates coated with TA-functionalized alkyltrichlorosilane, clearly shows the divalent sulfur signal (*ca.* 164 eV) which is converted into an oxidized sulfur signal (*ca.* 168 eV) on treatment with an aqueous Oxone solution. Figure 2 shows the progress of such an oxidation on a silicon wafer[13] and Figure 3 illustrates the same processing and analysis on particles.[14] Figure 4 demonstrates how this same analysis can be applied to the partial surface oxidation achieved by photo-patterning.[16]

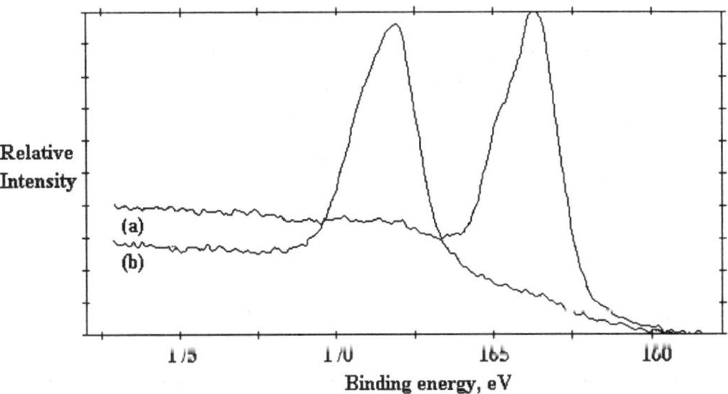

Figure 2 *S2p region of the XPS analysis of a silicon wafer with (a) as-deposited thioacetate-terminated monolayer and (b) oxidized monolayer (sulfonic acid obtained by treatment with aqueous Oxone).*

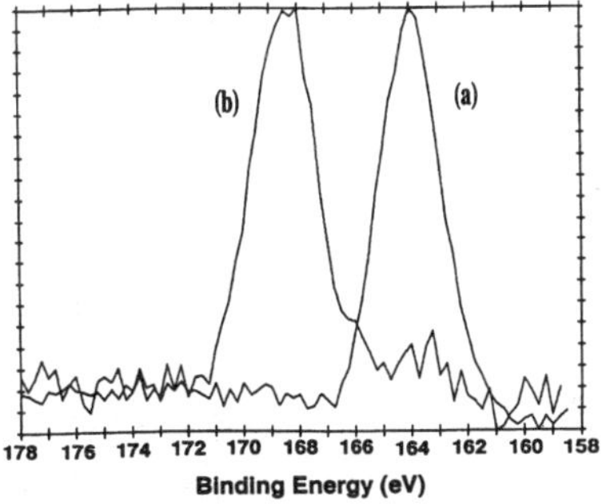

Figure 3 *S2p region of the XPS analysis of commercial titania powder with (a) as-deposited thioacetate-terminated SAM and (b) oxidized SAM (sulfonic acid obtained by treatment with aqueous Oxone).[17]*

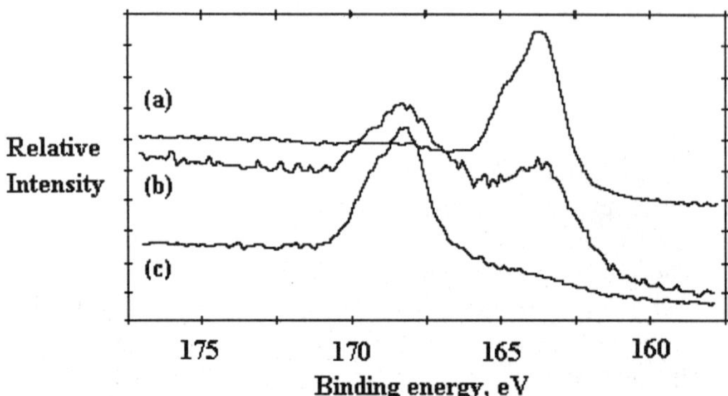

Figure 4   *S2p region of the XPS analysis of: (a) as-deposited thioacetate-terminated SAM; (b) photo-patterned SAM, bearing both thioacetate and sulfonate functionalities; (c) fully photo-oxidized, sulfonate-functionalized SAM.*[17]

The oxide thin films were all deposited by precipitation from a solution of a hydrolysable metal salt at temperatures between 30 and 80 °C. Aqueous solutions were preferred and in most cases control of pH and temperature was useful in controlling both the rate of film deposition and its final thickness. In some cases (TiO_2, ZrO_2, FeOOH, SnO_2), the as-deposited films consisted largely of randomly oriented oxide nanocrystals, usually 2–10 nm in size. In other cases (Fe_2O_3, Fe_3O_4, Y_2O_3, ZnO), obtaining the oxide (which was crystalline in all cases except for ZnO) required heat treatments, typically for 2 h at temperatures that ranged from 300 to 600 °C.[11,18] Figures 5 and 6 highlight the uniform, continuous, pore-free nature of the conformal film obtained for ZrO_2[19] and for SnO_2. Later in this paper, examples of comparable data for titanium and aluminum oxides will be presented. We emphasize both the range of metal oxides for which unequivocal evidence of conformal pore-free film attachment has been achieved, as well as the versatility of the methodology in achieving comparably compact films (*e.g.* for the aluminum basic sulfate deposition) on a particulate substrate[14] (Figure 7).

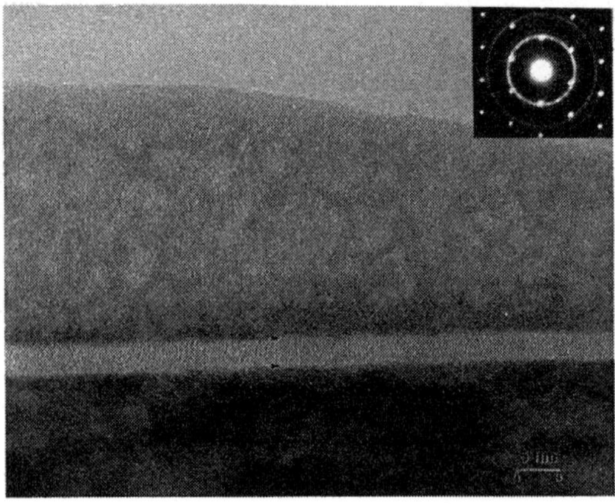

Figure 5   *Cross-sectional TEM image of a nanocrystalline tetragonal ZrO_2 film as deposited from aqueous solution onto a SAM-modified Si substrate.*[17]

Figure 6   *Cross-sectional TEM image of a nanocrystalline SnO$_2$ film as deposited from aqueous solution onto a SAM-modified Si substrate.*

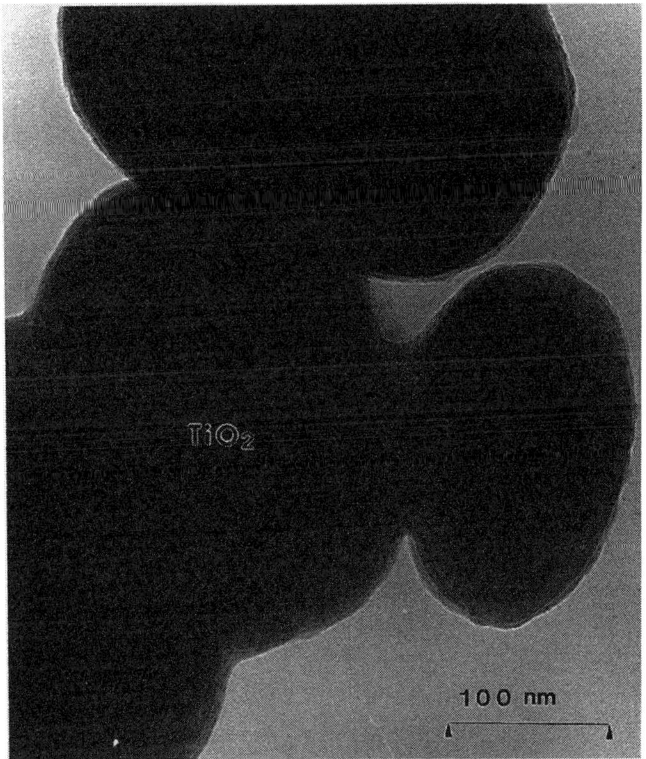

Figure 7   *Cross-sectional TEM image of an amorphous aluminum basic sulfate, as deposited from aqueous solution onto SAM-modified TiO$_2$ powder.[17]*

An additional demonstration of the power and versatility of this approach is the use of patterned SAM substrates to create patterned oxide films.[16] Figure 8 shows an SEM image of a patterned sulfonate surface obtained using photo-oxidation, while Figure 9, a Ti map made using a scanning Auger microprobe, reveals that the TiO_2 layer deposited on the photo-patterned surface reflects the same patterning.

Figure 8   *SEM of photo-patterned thioacetate SAM.[17]*

Figure 9   *Auger elemental (Ti) map of deposited TiO_2.[17]*

Finally, it is important to consider the stability of both the SAM itself and of the oxide-on-SAM composite. When dealing with functionalized SAMs, there is always the possibility that the oxide precipitation conditions will affect the SAM functionality. That is, does the oxide deposition medium in any way alter the pendant surface functionality? A case in point is the deposition of TiO_2 onto TA- and sulfonate-bearing SAMs. The patterning described above relied on selective attachment of the TiO_2 onto sulfonate functionality. It is interesting to note that the strong acid conditions (6N HCl, 80 °C) of the deposition medium do not in any way affect the sulfonate functionality. It is likely that this surface maintains the negative charge of the sulfonate anions even in 6N HCl. Interestingly, we find that any remaining TA functionality (CH_2SCOCH_3) is hydrolysed to thiol (CH_2SH) under these conditions. Thus, the observed selective precipitation attributable to different surface functionalities is in fact a comparison of sulfonate versus thiol and not versus TA.

In the more general issue of SAM stability and the integrity of the anchoring siloxane network, the robustness of siloxane-anchored SAMs in the face of a wide range of organic and aqueous solvents is well established. However, it is also known that solvents like hot DMSO or warm aqueous base can readily undermine SAM integrity. In this regard, solution precipitation conditions must be chosen which avoid such problems.

An interesting example of a case where siloxane-anchored monolayer instability is delicately balanced with oxide precipitation is in the formation of Zn oxide/hydroxide films.[18] The most continuous of such films were deposited from solutions with [Zn] = 0.056 M, pH = 11, and temperature of 35 °C. Deposition times of 2–4 h yielded films with the highest Zn content and with a thickness (based on RBS measurements) of 42 nm. The Zn content of the films reached a maximum value after 2–4 h of substrate immersion and then decreased gradually thereafter. Since depletion of Zn from the source solution upon extended immersion would be negligible when forming such thin films, it is reasonable to suggest that the basic deposition solution (pH 11, 35 °C) may have adversely affected either the surface functionality or the siloxane anchoring network connecting the SAM to the native SiO_2 on the substrate, leading to erosion of the Zn-containing overlayer. The basic hydrolysis of TA functionality to thiol, which might be a less favorable surface for the deposition of the film, could contribute to the observed loss of Zn. However, since a similar loss of Zn was observed from films that formed on sulfonate-functionalized SAMs under identical conditions, the apparent degradation of the zinc-containing overlayer is more likely due to a common cause for both the TA- and sulfonate-SAM surfaces, *i.e.* it is caused by attack of the siloxane anchoring network by the hot alkaline deposition medium.

The adherence of the inorganic film to its substrate is another issue related to the overall stability and durability of these materials. To date, this adherence has been tested by a simple tape peel test. Adhesive tape is firmly applied to the deposited film, then peeled off. XPS analysis is then performed on both the tape (to check for traces of inorganic material removed from the film) and the film (to check for exposed substrate, which likewise would indicate removal of material from the film). In general we found that films attached through SAM templates exhibited no loss of adherence in such tests. Only the zinc oxy-hydroxide films described above showed less than complete adherence on a SAM substrate, and ZrO_2 films were distinctly more adherent on the sulfonate SAM than on OTS surfaces. In several cases (basic Al sulfate, Y hydroxy-carbonate, and SnO_2), the films on bare Si adhered as well as on SAMs.

Another issue related to stability is whether the oxide films survive various kinds of post-deposition processing. In this regard, we cite two important examples. As-deposited films of basic aluminum sulfate were exposed to water at pH = 7 to remove sulfate ion.[14] The micrographs in Figures 10 and 11 clearly show that the compact, adherent, pore-free nature of the basic aluminum sulfate film was present both before and after the sulfate-leaching procedure.

A similar concern is whether the adherent oxide film will remain adherent and crack-free after heat treatment. Such treatments might be used, for example, to crystallize an initially amorphous film, to coarsen the grain size, or to convert a non-oxidic (*e.g.* chloride, sulfate, or carbonate) film to a simple oxide. Because sol-gel-derived films frequently crack or delaminate during such treatments, and because the temperatures of such treatments can be expected to exceed the window of thermal stability for the organic SAMs, it is by no means obvious that the present films would survive such heating. Nevertheless, in every system we have studied, the films on SAMs maintain their adherence and their structural uniformity after heating, despite pyrolysis of

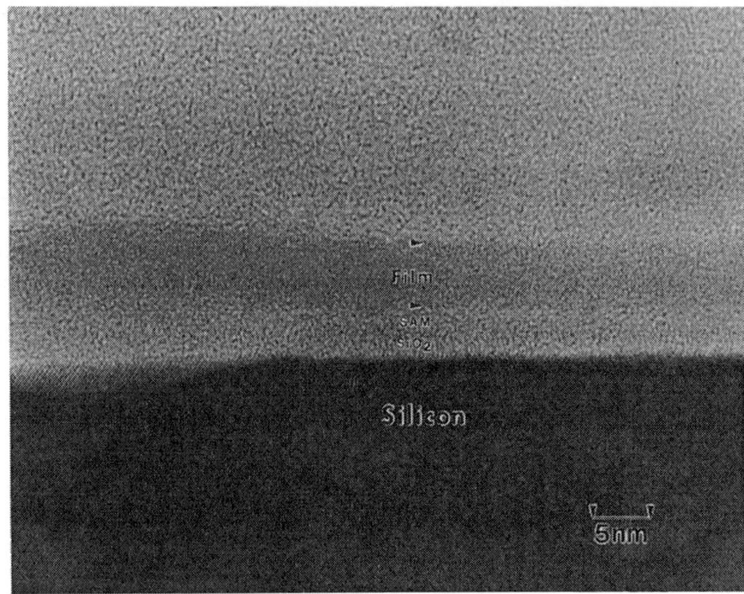

Figure 10   *HRTEM micrograph of as-deposited basic Al sulfate on sulfonate SAM on Si.[17]*

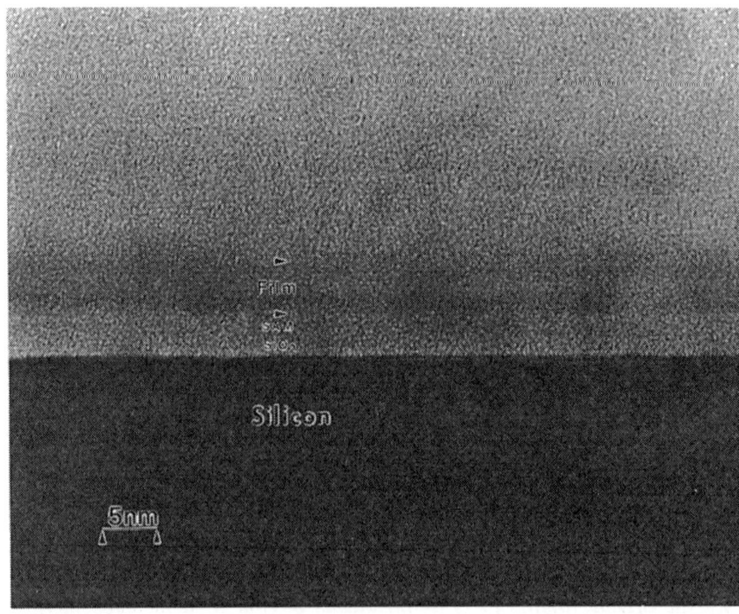

Figure 11   *HRTEM micrograph of leached basic Al sulfate film on sulfonate SAM on Si.[17]*

the SAM interlayer. The results of detailed studies of the thermal breakdown of the SAM are summarized as follows. TGA of alkylsilane SAMs on silica[15] and titania[20] powders shows significant weight loss beginning at just over 200 °C in air and at 300 °C under nitrogen. XPS was used to monitor the thermolysis of TA-functionalized SAMs on both powder and flat substrates.[20] SAM-coated titania powders were heated at 200, 400, 600, and 800 °C in air and showed some loss of carbon from the powder surface after heating at 200 °C, a much greater loss on heating at 400 °C, and essentially no further loss on heating at 600 and 800 °C. Sulfur (from the TA groups on the SAM) steadily decreased with increasing heat treatment temperature; it was barely detectable after heating at 600 °C and undetectable after heating at 800 °C. Similar studies were carried out on SAMs on Si wafers. These results mirror those of the heat-treated TiO_2 powders: no significant loss of carbon was observed until 200 °C, with much greater loss occurring at 250–300 °C, and carbon levels reaching background amounts after heating at either 400 or 600 °C. The smaller temperature increments of these experiments showed that sulfur loss began at 250 °C and was essentially complete by 300 °C. Heat treatments in vacuum showed that after 30 min at 195 °C or after 1 min at 250 °C, the levels of all elements were essentially unchanged from their as-deposited values. The only significant change detected using XPS was the partial oxidation of thioacetate sulfur to sulfonate sulfur in these specimens. Thirty minutes at 250 °C in vacuum caused essentially complete loss of the sulfur, while the carbon level remained unaffected. Additional heating in vacuum at 400 °C for 2 h led to loss of most of the carbon, though significantly more carbon remained than after pyrolysis in air at 400 °C.

These experiments provide a useful backdrop for a direct study of the thermal fate of an oxide-on-SAM multilayer.[20] The decomposition of SAMs incorporated into a multi-layer oxide structure was monitored using cross-sectional TEM images of planar specimens. Figure 12 shows a TEM micrograph of the interface between an as-deposited TiO_2 film and the Si substrate. The amorphous layer between the TiO_2 and the Si consists of the sulfonated SAM layer (2.5 nm thick) and the native SiO_2 layer (1.5 nm thick) on the Si. The dark fringe near the middle of the amorphous layer is caused by the difference in mean inner potential (which is simply related to the mean atomic number per unit volume) between the SAM and the native oxide layer.

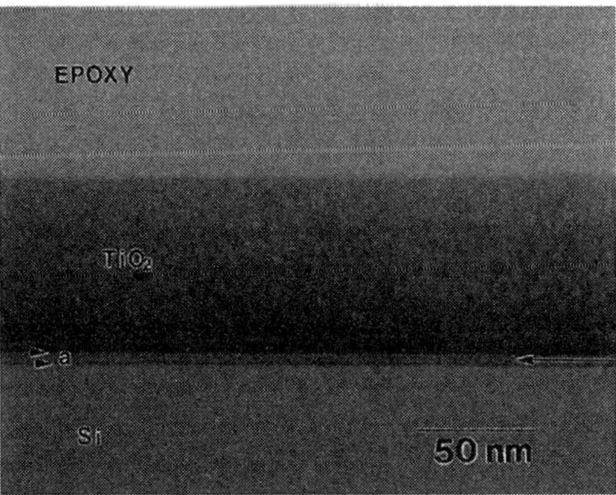

Figure 12 *Cross-sectional TEM micrograph of an as-deposited TiO_2 film. "a" indicates an amorphous region containing the SiO_2 and the SAM. The dark fringe between the layers (indicated by an arrow) shows the boundary between the SiO_2 native oxide and the SAM.*[17]

Figure 13 shows cross-sectional TEM micrographs of planar samples on Si after annealing at 300 and 400 °C under flowing nitrogen for 2 h. (Note: The TEM analysis of samples heated at 200 or 300 °C were identical.) In Figure 13a, the dark fringe in the middle of the amorphous layer can still be seen and the total thickness of the amorphous layer is *ca.* 4 nm, the same as in the as-deposited sample. After 2 h at 400 °C in nitrogen, the thickness of the amorphous layer had decreased to *ca.* 2 nm and the fringe between the SAM and the native oxide layer was no longer present (Figure 13b). This indicates that the decomposition of the SAM layer was complete after 2 h at 400 °C under nitrogen. A comparable cross-sectional TEM analysis done on TiO_2 layers on sulfonate-terminated SAMs on silicon wafers heated in air instead of nitrogen[21] likewise did not reveal any delamination or disruption within the ceramic–wafer interface in samples heated at temperatures up to 600 °C. However, whereas the amorphous layer shrank from 4 nm to 2 nm when the sample was heated under nitrogen at 400°C, heating in air at 400 °C resulted in an amorphous layer still 4 nm thick but lacking the inner fringe indicative of a $SAM–SiO_2$ interface. Heating at 600 °C caused the amorphous inner layer to grow to a thickness of 10 nm. These TEM results indicate that the TiO_2 overlayer enhances the thermal stability of the underlying SAM at 300 °C in nitrogen, relative to the above-cited TGA and XPS work. Also, the only consequence of whether the heat treatment of such samples is done in air or in an inert atmosphere is the growth of the interfacial silicon oxide layer when working in air. Under no circumstances is any delamination or disruption of the interface seen.

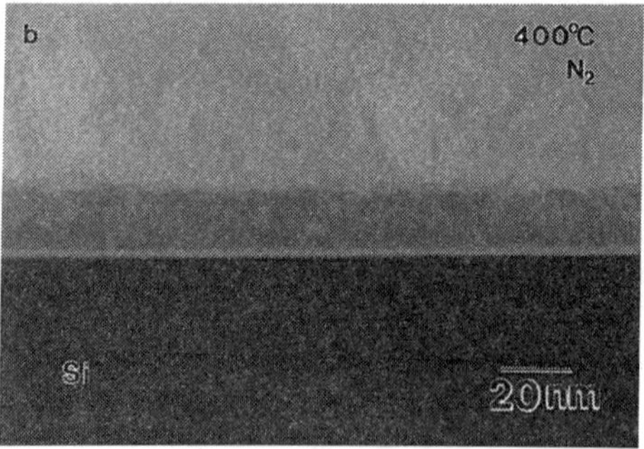

Figure 13   *Cross-sectional TEM micrographs of the TiO_2 films annealed at (a) 300 and (b) 400 °C for 2 h under flowing N_2.*

3 CONCLUSION

The chemical robustness and thermal stability of siloxane-anchored, self-assembled monolayer films, along with the relatively uniform interfacial layer that they present, suggests the potential application of these nanoscale assemblies as versatile underlayers capable of directing the attachment of subsequent organic and/or inorganic overlayers. We have reported herein on the use of functionalized siloxane-anchored films to direct the attachment of mineral oxide thin films. Our overall strategy includes: using organic monolayer films as templates for the formation of uniform, pore-free, oxide layers; the subsequent manipulation (typically by heat treatment or by solution leaching) of these multilayer assemblies; and, the use of a patterned organic layer to create a patterned mineral oxide thin film. The applicability of this method to variously shaped substrates (including both flat substrates and particulate materials) and its ability to produce highly adherent and robust films are among its key features.

4 ACKNOWLEDGEMENTS

The authors gratefully acknowledge the following agencies for their support of various portions of the work reported herein: the U.S. Air Force Office of Scientific Research, the Israeli Ministry of Science, the Max-Planck-Gesellschaft, the U.S. Basic Missile Defense Office, and the Government of Thailand. Current addresses are: for RJC – Avery Dennison Corp., Painesville, OH, USA; for HS – Samsung Adv. Inst. Of Tech., Microsystems Lab., Kihung, KOREA; for YW – CeraMem Corporation, Waltham, MA, USA; and for MA – Quester Technology, Fremont, CA, USA.

References

1. a) P. Calvert and S. Mann, *J. Mater. Sci.,* 1988, **23**, 3801;   b) P. C. Rieke, 'Nucleation and Growth of Calcium Carbonate Crystals on Surface-Modified Polyethylene and Polystyrene,' *Atomic and Molecular Processing of Electronic and Ceramic Materials*, I. A. Aksay, G. L. McVay, T. G. Stoebe and J. F. Wager, eds., 1988, Materials Research Society, Pittsburgh, PA, pp. 109–114; c) P. Calvert, *Mater. Res. Soc. Symp. Proc.* 1990, **180**, 619; d) B. J. Tarasevich and P. C. Rieke, 'Ceramic Oxide Thin Film Formation Utilizing Biological Processes,' *Materials Synthesis Utilizing Biological Processes*, P. C. Rieke, P. D. Calvert and M. Alper, eds., *Mater. Res. Soc. Symp. Proc.*, 1990, **174**, 51; e) P. C. Rieke, B. J. Tarasevich, S. B. Bentjen, G. E. Fryxell and A. A. Campbell, 'Biomimetic Thin-Film Synthesis,' *Supramolecular Architecture — Synthetic Control in Thin Films and Solids*, T. Bein, ed., *ACS Symposium Series*, 1992, **499**, 61; f) P. Calvert, 'Biomimetic Processing of Ceramics and Composites,' *Ultrastructure Processing of Advanced Materials*, D. R. Uhlmann and D. R. Ulrich, eds., John Wiley & Sons, New York, 1992, pp. 149–157; g) B. J. Tarasevich, P. C. Rieke, G. L. McVay, G. E. Fryxell and A. A. Campbell, 'Synthesis of Ceramic Ultrastructures Utilizing Biologic Processing,' *Chemical Processing of Advanced Materials*, L. L. Hench and J. K. West, eds., Wiley Interscience, New York, 1992, pp. 529–542.
2. a) J. Kuther, R. Seshadri, W. Knoll, and W. Tremel, *J. Mater. Chem.*, 1998, **8**, 641; b) A. A. Campbell, G. E. Fryxell, G. L. Graff, P. C. Rieke and B. J. Tarasevich, *Scanning Microscopy*, 1993, **7**, 423;   c) B. C. Bunker, P. C. Rieke, B. J.

Tarasevich, A. A. Campbell, G. E. Fryxell, G. L. Graff, L. Song, J. Liu, J. W. Virden and G. L. McVay, *Science*, 1994, **264**, 48; d) P. C. Rieke, B. J. Tarasevich, L. L. Wood, M. H. Engelhard, D. R. Baer, G. E. Fryxell, C. M. John, D. A. Laken and M. C. Jaehnig, *Langmuir,* 1994, **10**, 619; e) P. C. Rieke, B. D. Marsh, L. L. Wood, B. J. Tarasevich, J. Liu, L. Song and G. E. Fryxell, *Langmuir,* 1995, **11**, 318.

3. a) A. Ulman, 'An Introduction to Ultrathin Organic Films from Langmuir–Blodgett to Self-Assembly,' Academic Press, New York, 1991; b) A. Ulman, *Adv. Mater.*, 1990, **2**, 573.

4. J. Tien, X. Younan and G. M. Whitesides, *Thin Films*, 1998, **24**, 227.

5. For diverse recent examples see: a) L. M. Tender, R. L. Worley, H. Fan and G. P. Lopez, *Langmuir*, 1996, **12,** 5515; b) C. S. Dulcey, J. H. Georger, Jr., M. S. Chen, S. W. McElvany, C. E. O'Ferrall, V. L. Benezra and J. M. Calvert, *Langmuir*, 1996, **12**, 1638; c) M. Matsuzawa, K. Umemura, D. Beyer, K. Sugioka and W. Knoll, *Thin Solid Films*, 1997, **305,** 74; d) C. S. Dulcey, S. L. Brandow, J. M. Calvert, M. S. Chen, W. J. Dressick, S. Mcelvany and H. H. Nelson, *Polym. Mater. Sci. Eng.*, 1997, **77**, 414.

6. a) L. Netzer and J. Sagiv, *J. Am. Chem. Soc.*, 1983 **105**, 674; b) N. Tillman, A. Ulman and T. L. Penner, *Langmuir,* 1989, **5**, 101; and reference 12 (below).

7. a) M. Mrksich and G. M. Whitesides, 'Using Self-Assembled Monolayers that Present Oligo(ethylene glycol) Groups to Control the Interactions of Proteins with Surfaces,' *ACS Symp. Ser.*, 1997, **680**, 361; b) M. Lestelius, B. Liedberg and P. Tengvall, *Langmuir*, 1997, **13**, 5900; c). S. S. Cheng, K. K. Chittur, C. N. Sukenik, L. A. Culp and K. Lewandowska, *J. Colloid Interface Sci.*, 1994, **162**, 135; d) S. Margel, E. A. Vogler, L. Firment, T. Watt, S. Haynie and D. Y. Sogah, *J. Biomed. Mater. Res.*, 1993, **27**, 1463; e) K. L. Prime and G. M. Whitesides, *Science,* 1991, **252**, 1164.

8. a) M. Mrksich, *Cell. Mol. Life Sci.*, 1998, **54**, 653; b) M. Matsuzawa, K. Umemura, D. Beyer, K. Sugioka and W. Knoll, *Thin Solid Films*, 1997, **305**, 74; c) C. D. Tidwell, S. L. Ertel, B. D. Ratner, B. Tarasevich, S. Atre and D. L. Allara, *Langmuir,* 1997 **13**, 3404; d) E. Cooper, R. Wiggs, D. A. Hutt, L. Parker, G. J. Leggett and T. L. Parker, *J. Mater. Chem.*, 1997, **7**, 435; e) D. A. Hutt, E. Cooper, L. Parker, G. J. Leggett and T. L. Parker, *Langmuir*, 1996, **12**, 5494; f) M. Mrksich, C. S. Chen, Y. Xia, L. E. Dike, D. E. Ingber and G. Whitesides, *Proc. Natl. Acad. Sci. U. S. A.*, 1996, **93**, 10775; g) R. S. Potember, M. Matsuzawa and P. Liesi, *Synth. Meth.*, 1995, **71**, 1997; h) K. M. Wiencek and M. Fletcher, *J. Bacteriol.*, 1995, **177**, 1959; i) G. P. Lopez, M. W. Albers, S. L. Schreiber, R. Carroll, E. Peralta and G. M. Whitesides, *J. Am. Chem. Soc.*, 1993, **115**, 5877; j) C. N. Sukenik, N. Balachander, L. A. Culp, K. Lewandowska and K. Merritt, *J. Biomed. Mater. Res.,* 1990, **24**, 1307.

9. a) J. Weiss, H. J. Himmel, R. A. Fischer and C. Woell, *Chem. Vap. Deposition,* 1998, **4**, 17; b) N. L. Jeon, R. G. Nuzzo, Y. Xia, M. Mrksich and G. M. Whitesides, *Langmuir*, 1995, **11**, 3024; c) S. J. Potochnik, D. S. Y. Hsu, J. M. Calvert and P. E. Pehrsson, 'Selective Copper CVD on Diamond Surfaces Using Self-Assembled Monolayers,' in *Advanced Metalization for Devices and Circuits – Science, Technology and Manufacturability*, S. P. Murarka, A. Katz, K. N. Tu and K. Maex, eds., Mater. Res. Soc. Symp. Proc., 1994, Vol. 337, pp. 429–434.

10. a) P. G. Clem, N. L. Jeon, R. G. Nuzzo and D. A. Payne, *J. Am. Ceram. Soc.,* 1997, **80**, 2821; b) G. C. Herdt, D. E. King and A. W. Czanderna, *Z. Phys. Chem.* (Munich), 1997, **202**, 163; c) R. Rizza, D. Fitzmaurice, S. Hearne, G. Hughes, G.

Spoto, E. Ciliberto, H. Kerp and R. Schropp, *Chem. Mater.*, 1997, **9**, 2969; d) K. Bandyopadhyay, V. Patil, K. Vijayamohanan and M. Sastry, *Langmuir*, 1997, **13**, 5244; e) N. L. Jeon, P. Clem, D. Y. Jung, W. Lin, G. S. Girolami, D. A. Payne and R. G. Nuzzo, *Adv. Mater.*, 1997, **9**, 891.

11.　M. R. DeGuire, H. Shin, R. Collins, M. Agarwal, C.N. Sukenik and A.H. Heuer, *Integrated Optics and Microstructures III*, M. Tabib-Azar, ed., *Proc. SPIE*, 1996, **2686**, pp. 88–99.

12.　R. J. Collins, I. T. Bae, D. A. Scherson and C. N. Sukenik, *Langmuir*, 1996, **12**, 5509.

13.　R. J. Collins and C. N. Sukenik, *Langmuir*, 1995, **11**, 2322.

14.　Y. Wang, S. Supothina, M. R. DeGuire, A. H. Heuer, R. Collins and C. N. Sukenik, *Chem. Mater.*, 1998, **10**, 2135.

15.　S. Brandriss and S. Margel, *Langmuir*, 1993, **9**, 1232.

16.　R. J. Collins, H. Shin, M. R. De Guire, A. H. Heuer and C. N. Sukenik, *Appl. Phys. Lett.*, 1996, **69**, 860.

17.　Figures 3, 7, 10 and 11 are reprinted with permission from *Chem. Mater.*, 1998, **10**, 2135, Copyright © 1998 American Chemical Society. Figures 5 and 12 are reprinted with permission from 'Deposition of Oxide Thin Films on Silicon Using Organic Self-Assembled Monolayers,' in *Integrated Optics and Microstructures III*, Massood Tabib-Azar, ed., *Proc. SPIE*, 1996, **2686,** pp. 88–99. Figures 4, 8 and 9 are reprinted with permission from *Appl. Phys. Lett.*, 1996, **69**, 860.

18.　M. R. De Guire, T. P. Niesen, S. Supothina, J. Wolff, J. Bill, C. N. Sukenik, F. Aldinger, A. H. Heuer and M. Rühle, *Z. Metallkd*, 1998, **89**, 758.

19.　M. Agarwal, M. R. De Guire and A. H. Heuer, *J. Am. Ceram. Soc.*, 1997, **80** 2967.

20.　H. Shin, Y. Wang, U. Sampathkumaran, M. R. DeGuire, A. H. Heuer and C. N. Sukenik, 'Pyrolysis of Self-Assembled Organic Monolayers on Oxide Substrates,' *J. Mater. Res.*, submitted for publication

21.　H. Shin, R. J. Collins, M. R. DeGuire, A. H. Heuer and C. N. Sukenik, *J. Mater. Res.*, 1995, **10**, 692.

SYNTHESIS AND CHARACTERIZATION OF SELF-ORGANIZED MICROSTRUCTURES WITH CHEMICALLY ACTIVE SURFACES AND EVALUATION OF THEIR TECHNICAL UTILITY

Alok Singh,* Michael A. Markowitz, Paul E. Schoen, and Carolina Costellanos

Center for Bio/Molecular Science &'Engineering
Code 6930 Laboratory for Molecular Interfacial Interactions
Naval Research Laboratory
Washington, DC 20375-5348 USA

1 INTRODUCTION

The self-assembly process provides an efficient means to make morphologically different microstructures in a rational manner. The morphology of these microstructures can be tailored by altering the structures of molecular building blocks. The resulting microstructures can then be utilized as templates or scaffolds for the formation of more intricate or robust structures.[1-4] Both free standing and surface grafted molecular assemblies have been reported. Literature reports indicate substantial efforts have been focused on the synthesis of molecular-assemblies supported on static surfaces. For example, ordered chemi- or physisorbed mono- and bilayer assemblies as well as size-selective self-assembled patterns have been formed on surfaces.[5,6]

Free-standing bilayer membrane structures include vesicles, tubules, ribbons, or helices. Our studies have focused on exploring the technical usefulness of free-standing structures, tubules and helices shown in Figure 1. The chemistry, formation properties, and applications of these structures have been discussed in a number of research papers.[3,7-11] Functionalization of these structures to explore interfacial interactions or chemical reactions at the micro- structure surface, as well as the processing of these structures, often poses interesting chemical and physical challenges, retention of structural integrity being the prime concern.

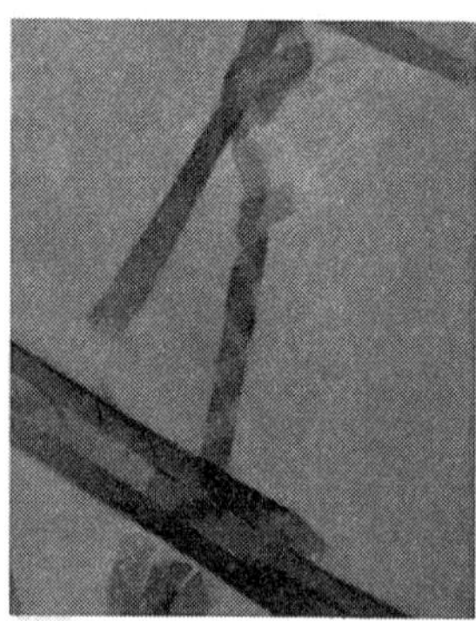

Figure 1 *Electron micrograph of tubules and helices made from diacetylenic phosphocholine.*

Applications of these structures are dependent on their morphologies and therefore physical stabilization must be achieved without affecting their respective morphologies.[12] Consequently our research efforts have been directed towards the formation of free-standing molecular assemblies in various morphologies, their subsequent stabilization, and the exploration of their technical utility.[12–18] Our approach relies on the design and synthesis of diacetylenic phospholipids. We chose diacetylenic groups because of their ability to modulate and stabilize the microstructures at the same time.

In this article, we report several strategies for differentially manipulating the reactive surfaces of microstructures by utilizing synthetic headgroup-modified phospholipids (Figure 2). The surface reactive microstructures have been used in nanoparticle synthesis, noncovalent enzyme immobilization, and for noncovalent tethering of microstructures to form ordered supramolecular assemblies.

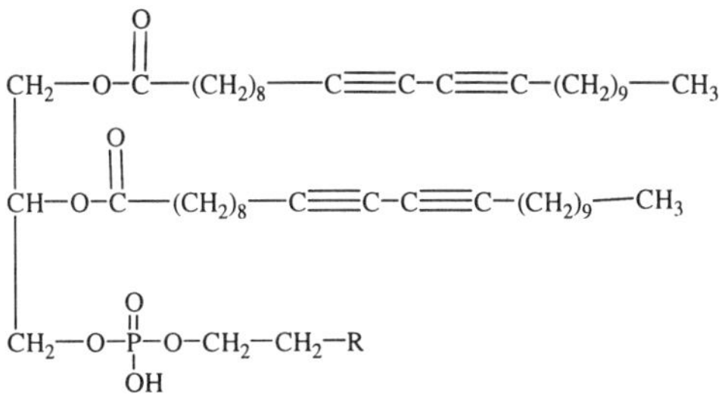

(1)   R = $-N^+Me_3$
(2)   R = $-OH$
(3)   R = $-N(CH_2COOH)_2$
(4)   Compound **3** with dipalmitoyl chain

Figure 2   *Phospholipids utilized to form surface-active microstructures.*

2 MATERIALS AND METHODS

2.1 General

All solvents used in this study were glass distilled. Bovine carbonic anhydrase II (EC 4.2.1.1) and *p*-nitrophenyl acetate were obtained from Sigma Chemical Company. An 80 µg/mL enzyme stock solution was prepared in 100 mM Tris -HCl (pH 8.5) buffer and stored refrigerated. 1,2-Bis(tricosa-10,12-diynoyl) -*sn*- glycero-3-phosphocholine (**1**) and 1,2-bis (tricosa-10,12-diynoyl) -*sn*- glycero-3-phosphohydroxyethanol (**2**) were prepared following a published procedure depicted in Scheme 1.[19,20] For column chromatography 70–230 mesh silica gel and for flash chromatography 230–400 mesh silica gel 60 were used. Cationic impurities were removed by using Biorad resin AG 50W-X8, 200–400 mesh. Purity of all the compounds was monitored by thin layer chromatography on silica gel 60F-254 (E.M. Merck) using chloroform : methanol : water (65:25:4) solvent system. FT-IR and FT-NMR (Bruker MSL-360) spectrometry was used in the characterization of lipids and synthetic intermediates. Enzyme assays were performed using a Durrum stopped-flow spectrophotometer.

Lipid assemblies were visualized by transmission electron microscopy (TEM) using a Zeiss EM-10 or JEOL JEM200CX microscope operated at 60 KV or 200 KV, respectively. High resolution transmission electron microscopy (HRTEM) was performed with a Hitachi 9000 UHR microscope operated at 300 kV. Crystallographic information was obtained by using selected area electron diffraction technique as well as HRTEM lattice imaging.

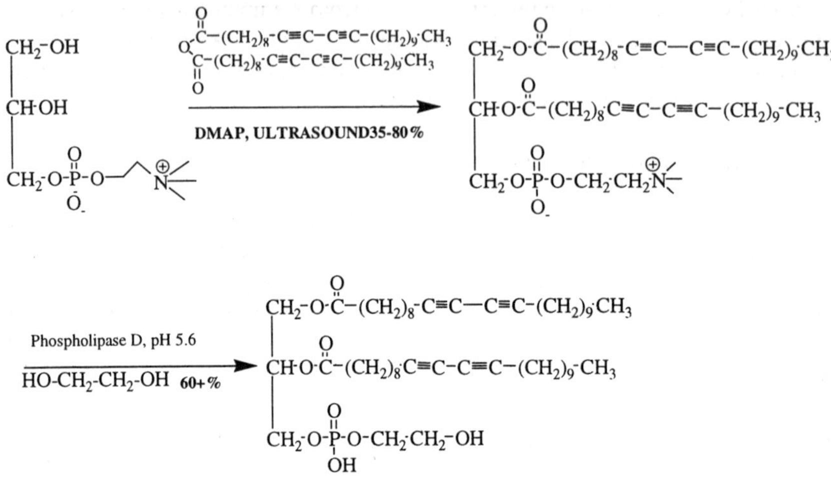

Scheme 1   *Synthesis of Phospholipids* **1** *and* **2**

2.2 Synthesis of Phospholipids

Metal chelating phospholipid 1,2-bis(tricosa-10,12-diynoyl)-*rac*-glycero-3-phospho-*N*-(2-ethyl)-iminodiacetic acid **3**, its dipalmitoyl analogue **4**, and their intermediates were synthesized following a reported procedure[23] with some modifications as depicted in Scheme 2. In general, *rac*-glycero-3-phospho-*N*-(2-ethyl)-iminodiacetic acid was reacted with tricosa-10,12-diynoic anhydride in the presence of 4-dimethylaminopyridine with the aid of ultrasound agitation for 2 hours followed by overnight magnetic stirring. After the reaction was complete, chloroform was evaporated under reduced pressure and the residue was dissolved in chloroform : methanol (1:1) and passed through a cation exchange column

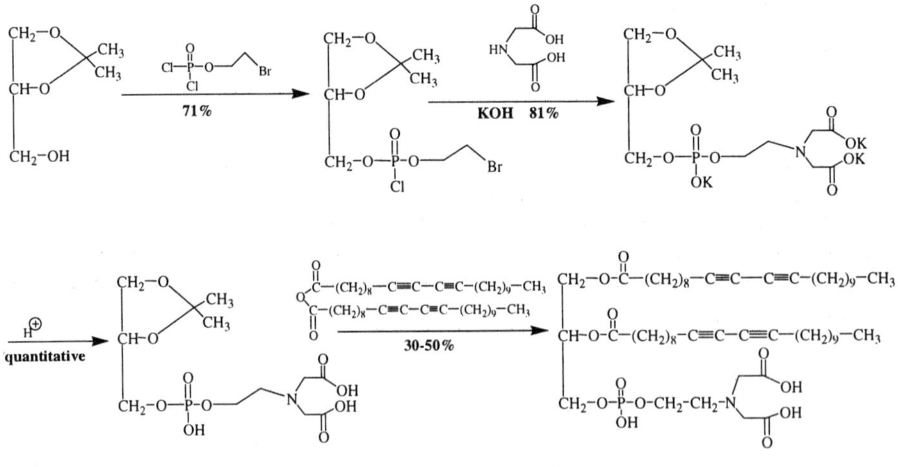

Scheme 2   *Synthesis of Metal Chelating Phospholipid* **3**

to remove DMAP and cationic impurities. Lipid from the mixture was separated by flash chromatography using as eluants chloroform, 5% methanol/chloroform, and then 10% methanol/chloroform. The phosphoIDA lipids have R_f values between 0.45 and 0.50 in a chloroform : methanol : water (65:25:4) solvent system.

2.3 Protocol for Nanoparticle Synthesis

The particles were synthesized following a literature procedure.[16] A thin film of a mixture of **1** and **2** was hydrated at 70 °C for 30 min in the presence of an equimolar amount of $Pd(NH_3)_4Cl_2$. The total concentration of lipid in each sample ranged from 2 mg/mL to 20 mg/mL. The mixture was vortex mixed and then sonicated (Branson sonifier, Model 450) at 60 °C for 5 min. The dispersion was then allowed to cool to room temperature and polymerized by exposure to UV radiation (254 nm, Rayonet Photochemical Reactor) at 20 °C for 15 min. An amount of EDTA (tetrasodium salt) equimolar to the amount of $Pd(NH_3)_4Cl_2$ present was added and the vesicles were then immediately gel filtered.

Metal particle deposition or formation of vesicles was accomplished with a gold plating bath containing chloroauric acid and sodium hypophosphite as major components. The vesicle dispersions (2 mL) were diluted with an equal volume of the plating bath. Plating was allowed to continue for 6 h and then the dispersions were dialysed against water (5 L) for 24 h to remove excess plating bath.

2.4 Preparation of Enzyme Immobilized Vesicles and Enzyme Assay

Enzyme immobilization was accomplished following a literature procedure.[22] Polymerizable lipid **1** was mixed with polymerizable metal chelating lipid **3** in 90:10 (mole/mole) ratio using chloroform as solvent. The solvent was removed to form a thin film of lipid on the wall of the glass tube. The resulting film was dried under vacuum (4 h), hydrated in 2 mL 0.05 M Tris-HCl (pH 8.5) by incubating at 55 °C for 2 h, and dispersed by sonication at 50°C for 12 min in a sonicator bath. The vesicle dispersions were then treated with 0.01mM $CuCl_2$ in Tris buffer. The unbound copper salt was removed by gel filtration on a Sephadex G 75-125 column. Copper bound vesicles were then divided into two equal volumes. One portion was polymerized at 4 °C by irradiating with 254 nm light for 10 minutes. Both polymerized and unpolymerized samples were incubated with enzyme bovine carbonic anhydrase II (EC 4.2.1.1) at room temperature for 3–4 h, and gel filtered to remove unbound enzyme from enzyme-bound vesicles.

One syringe of the stopped-flow spectrophotometer was loaded with substrate (5 mM *p*-nitrophenyl acetate in 30% acetone/water), and the other syringe was loaded with the preparation to be tested in 100mM Tris (pH 8.5). Samples were simultaneously mixed and the data was collected for 120 seconds at a rate of 18 points per second. Absorbance was monitored at 402 nm with a slit width set to 1.0 nm. Data was analysed using the "DATAFIT" module of the stopped-flow operating system. All data analysed were fit with a linear regression.

2.5 Tethering the Structures

Imidazole linker *N,N*-(*m* xylene)bis-imidazole was prepared by following a literature procedure.[21] A 5 mg vacuum dried film from lipid **1** and **4** mixture (9:1 mole/mole) was dissolved in 2 mL of EtOH/water (75:25) mixture. The solution was then heated to 55 °C for 60 min and vortex mixed to give a homogeneous dispersion. The sample was then placed in an insulated beaker and allowed to cool slowly to room temperature. The tubule solution was dialysed against water to exchange ethanol with water. A 1 mL aliquot of a 12 mM copper sulfate solution was then added to the tubule dispersion, and the solution was dialysed to remove excess copper. Bis-imidazole linker was then added to the dispersion and tubule tethering was monitored by optical microscopy and characterized by transmission electron microscopy.

3 RESULTS AND DISCUSSION

One of our research goals has been to investigate the microstructure formation properties of headgroup-modified diacetylenic phospholipids and to evaluate their technological utility. Utilizing diacetylenic functionality, both in morphology modulation as well as morphology stabilization, we prepared stabilized vesicles and tubules.[15,17,23] Tubules formed from diacetylenic phosphocholine are hollow, cylindrical structures of fixed 0.5 μm internal diameter. These structures are stable in the laboratory environment. Structures with one bilayer thick walls have also been reported.[8] In order to extend the technological capability of vesicles and tubules, microstructure formation properties and surface reactivity of vesicles and tubules formed from headgroup-modified negatively charged diacetylenic phospholipids have been investigated.[24] The headgroup geometry, surface charge density, and spatial arrangement of the negatively charged phospholipids influence the morphologies, surface behavior, and reactivity of the lipid membranes. Therefore, in the experiments reported in this article structures were stabilized before carrying out any chemistry on membrane reactive sites.

One criterion for making an affordable technology is the availability of large amounts of lipids on a cost effective basis. We have developed synthetic procedures suitable for large-scale headgroup modified lipid preparation utilizing an enzymatic approach.[20] Crude enzyme extract from savoy cabbage provided high yields of **2**. In addition, ultrasound-assisted agitation of reaction mixtures has been utilized to synthesize diacetylenic phospholipids.[19]

The phosphoiminodiacetic acid lipids **3** and **4** were synthesized following a simple and straightforward synthetic route illustrated in Scheme 2. Acylation of the 1,2-hydroxyl groups of glycerol backbone after linking the headgroup phosphoIDA to the glycerol 3 position provides a problem free route without any possibility of side reaction. Carboxylic acid groups of IDA moiety were not protected because anhydride/dimethylaminopyridine (DMAP) will acylate the hydroxyl groups of the glycerol backbone.[25] Protection of acid functionality is neither practical nor cost effective in the current synthetic scheme because the conditions needed for deprotection may cause hydrolysis of acyl chains. Acylation of the colorless to off-white waxy *rac*-glycero-3-phospho-*N*-(2-ethyl)-iminodiacetic acid produced metal chelating lipids as off-white waxy solids. The results discussed in the following sections were obtained using mixed lipid systems involving charge neutral 1,2-bis(tricosa-10,12-diynoyl)-*sn*-glycero-3-phosphocholine (**1**), negatively charged lipid (**2**), and metal chelating lipids **3** or **4** (Figure 2).

3.1 Differentiation of Reactive Surfaces

Spherical bilayer microstructures have two membrane surfaces, which may be selectively utilized as reactive or functionalized surfaces. The inner membrane surface defines the aqueous lumen of bilayer microstructures. Ions or other molecules bound to this surface would be available for reactions within the lumen. The products of such reactions will depend on the nature of any encapsulated molecules, the diffusion of molecules through the membrane, and the effect of confined space of the lumen on the phase behavior of the reaction products. The outer membrane is continually exposed to the aqueous environment of the colloidal dispersion. Changes in that environment will have an immediate effect on ions or molecules bound to that surface. Because it can accommodate macromolecules easily, the outer bilayer membrane has been exploited for covalent binding of large biomolecules. We have selectively exploited the inner membrane as a catalytic surface for metal nanoparticle synthesis, and we have utilized the outer membrane to develop a noncovalent means of binding biomolecules.

3.1.1 Reactions at the Inner Bilayer Membrane Surface. Vesicle stabilization has been achieved by polymerization by crosslinking acyl chains. Because of the vesicle curvature and topotectic polymerization of diacetylenes, polymer domains are formed in the membrane. The formation of the polymer domains produced channels of free volume in the membrane through which cations were able to diffuse. If catalytic metal ions are bound to one or both of the vesicle membrane surfaces, the ions can then be used as catalytic centers for electroless metal deposition. This property allowed polymerized vesicles, formed from lipids **1** and **2,** to be used as reaction vials. Synthesis of metal nanoparticles was carried out.[16] Since it was desirable to form the metal particles inside the vesicles in order to prevent their irreversible agglomeration, palladium metal ions were selectively bound to the negatively charged lipids **2** in the presence of the zwitterionic lipids **1**. This was accomplished by forming the vesicles in the presence of $Pd(NH_3)_4Cl_2$ and then stabilizing the vesicles by exposure to UV radiation. Palladium ions were removed from the external vesicle surface by ion exchange chromatography or by addition of tetrasodium EDTA followed immediately by gel filtration, leaving only the inner vesicle membrane with bound palladium ions. After addition of the plating bath, the resulting solution was allowed to stand for 6 to 24 h. No differences in the degree of vesicle metallization occurred over this time period. TEM reveals gold nanoparticles formed within the polymerized vesicles (Figure 3). Nucleation of gold occurred both from single nucleation and multiple nucleation sites for different vesicles. In the former case, this resulted in the formation of single crystals. By selectively binding the palladium ions to the inner vesicle membrane and directing the metallization process to take place within the polymerized vesicles, unagglomerated gold particles were synthesized.

3.1.2 Noncovalent Linkage of an Enzyme to the Outer Bilayer Membrane Surface. Enzymes and antibodies have been tethered to surfaces via covalent linkages and several applications have resulted particularly in the biosensor area.[26,27] However, one drawback of using covalent bonds to link proteins to surfaces is the significant loss of protein activity that sometimes occurs. Randomly distributed IDA-phospholipids in lipid membranes offer an opportunity for non-covalent binding of proteins containing multiple surface-available histidine sites due to the affinity of histidine towards Cu^{2+}-iminodiacetate.[28] With this in mind, two sets of vesicles were prepared by separately mixing polymerizable lipids **1** and **3** in a 9:1 molar ratio. Sonication for about 10 minutes provided dispersion with constant turbidity at 400 nm. The copper-bound vesicles came in the void volume during gel-filtration while the free copper eluted later. The copper-bound vesicles were divided into two parts, one of which was polymerized, and then the enzyme carbonic anhydrase was bound to the surface of both polymerized and non-polymerized vesicles. The vesicle/protein mixtures were eluted on a gel-filtration column. The fractions were monitored at 400 nm for the presence of vesicles, which eluted in the void volume. Carbonic anhydrase was retained on the column

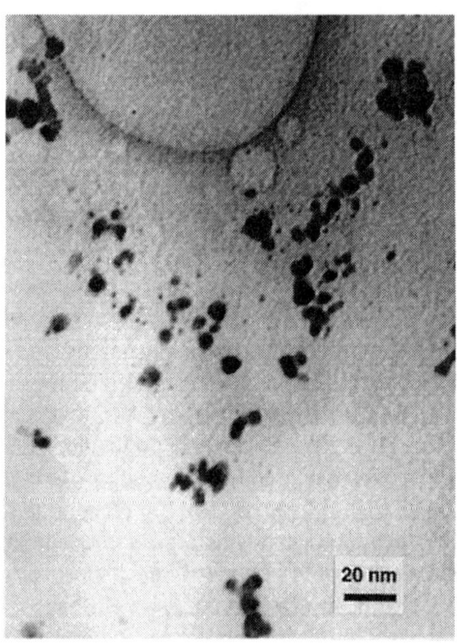

Figure 3 *Transmission electron micrograph of gold filled vesicles.*

longer than the vesicles. To examine the stability and activity of the vesicle-bound enzyme, two sets of experiments were performed employing (a) unpolymerized vesicles and (b) polymerized vesicles made from diacetylenic PC and PIDA. The results are summarized in Figure 4. Maximum activity was observed for enzyme-bound polymerized vesicles containing diacetylenic phosphocholine and metal chelating lipids. The graph depicted in Figure 4 illustrates that the estimated activity is equivalent to an enzyme concentration between 1.6 and 3.2 µg/mL carbonic anhydrase. No enzyme activity was observed for the enzyme-bound unpolymerized vesicles (Figure 4, graph on left). The relative activity of immobilized carbonic anhydrase on unpolymerized and fully polymerized vesicles demonstrates that

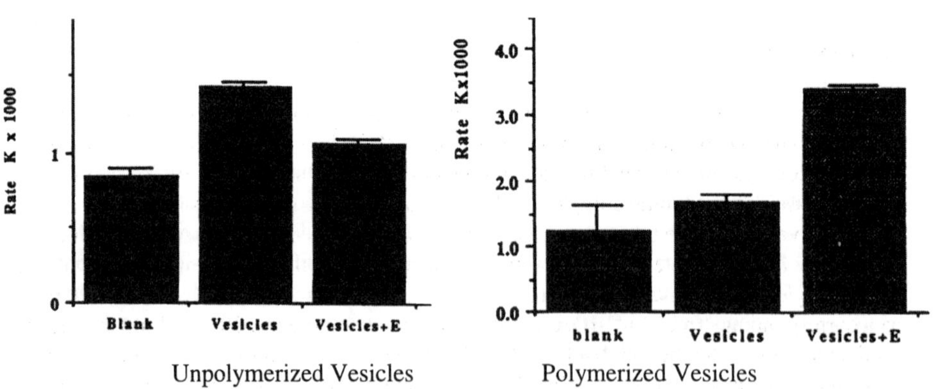

Figure 4 *Activity of carbonic anhydrase non-covalently bound to vesicles from lipid 1 and 3.*

cross-linking of the bilayers is a necessary step for the noncovalent immobilization of enzymes on vesicle surfaces. The results demonstrate that this method of noncovalent attachment of proteins to surfactant membranes provides an attractive alternative for immobilization of macromolecules to surfaces.

3.2 Tethering of Tubules

In an effort to extend the applicability of tubules, we have attempted to create an array of tubules using complementary linkers. Making an array of lipid tubules is like putting a puzzle together with the following pieces: lipid tubules, the linkage sites, and the complementary linker (or strap molecule). Polymers containing copper (II) iminodiacetate sites recognize and bond with imidazole substrates.[28] With our previous experience linking biomolecules to Cu^{2+}-IDA headgroups, we decided to utilize a bis-imidazole to connect tubules having surface Cu^{2+}-IDA headgroups. In this study, tubules were formed from mixtures of diacetylenic PC **1** and dipalmitoyl phosphoIDA **4** in a 75:25 EtOH/H$_2$O solvent system. The phosphoIDA headgroup was selected to provide binding sites on the tubule surface, and *N,N-(m*-xylene)bis-imidazole was used as a linker or strap molecule to align the tubules (Figure 5). Tubule alignment is initiated by the addition of Cu^{2+}, which binds to the phosphoIDA headgroups. The resulting Cu^{2+}-IDA groups in turn serve as linkage sites for the strap molecule.

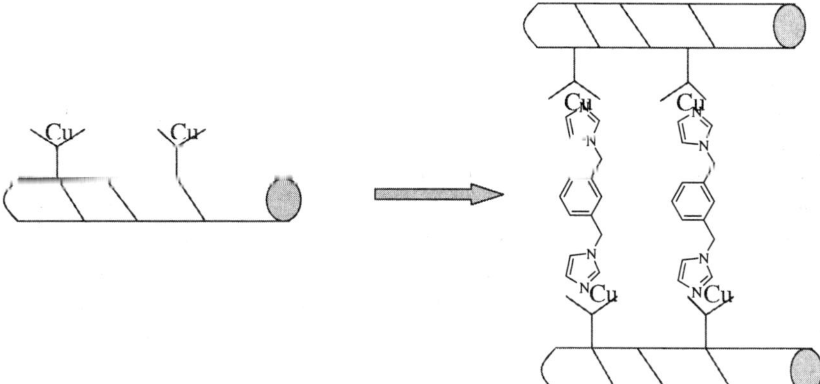

Figure 5 *Schematic of tubule tethering via strap molecules.*

There are a number of kinetic and thermodynamic obstacles to be overcome in order to align tubules via bis-imidazole molecules. Given the relative diffusion rates of dissolved bis-imidazole and precipitated tubules, the positioning of a second tubule for connection with a bis-imidazole–tubule complex will be the rate limiting step in this process. Another important variable to be taken into consideration is the ratio of Cu^{2+}-IDA linkage sites on tubules to bis-imidazole strap molecules. Once one of the imidazole groups of the *N,N'-(m*-xylene)bis-imidazole binds to a Cu^{2+}-IDA surface group on one tubule, there must be a corresponding "free" Cu^{2+}-IDA group on a separate tubule available to bind the second imidazole group. As the ratio of Cu^{2+}-IDA linkage sites on tubules to bis-imidazole strap molecules decreases, so does the likelihood of tubule alignment via the bis-imidazole.

Nevertheless, TEM revealed the existence of aligned tubules after the addition of CuSO$_4$ to an ethanol–water mixture of *N,N'-(m*-xylene)bis-imidazole and tubules formed from a 9:1 (wt/wt) mixture of **1** and **4**. In order to diminish the possibility of saturating tubule Cu^{2+}-IDA sites with bis-imidazole, the concentration of bis-imidazole relative to the concentration of **4** was kept very low. Figure 6 shows a transmission electron micrograph of a pair of tubules, which are apparently linked. An alternative explanation, which has been previously observed by TEM, would be that the tubules have aligned due to evaporation of the solvent. However, in those instances when such tubule alignment has been previously observed, the tubules retain their normally rigid cylindrical morphology. In Figure 6, tubule morphology is distorted and the tubules each followed the contour of the other. The linkage of the tubules may induce deformations in the lipid bilayers of the tubules resulting in the observed curvature.

Figure 6   *Transmission electron micrograph of a tubule tethered by bisimidazole linker.*

4 ACKNOWLEDGEMENT

We gratefully acknowledge funding for this program from the office of Naval Research. An NRL/NSF Summer Research Internship program supported Carolina Costellanos.

References

1.   J. M. Lehn, 'Supramolecular Chemistry: Concepts and Perspectives', VCH Weinheim, Germany 1995, Chapter 9, p. 139.
2.   G. M. Whitesides, E. E. Simanck, J. P. Mathias, C. T. Seto, D. N. Chin, M. Mammen and D. M. Gordon, *Acc. Chem. Res.*, 1995, **28**, 37.
3.   J. M. Schnur, *Science*, 1993, **262**, 1669.
4.   A. Ulman, 'An Introduction to Ultrathin Organic Films', Academic Press, NY, 1991.

5.   W. J. Dressick and J. M. Calvert, *Jpn. J. Appl. Phys.*, 1993, **32**, 5829.

6.   J. M. Calvert, *Lithographically Patterned Self-Assembled Films* in 'Organic Thin Films and Surfaces', A. Ulman, ed., Academic Press, Boston, 1995, p. 109.

7.   T. Kunitake, *Angew. Chem. Intl. Ed. Engl.*, 1992, **31**, 709.

8.   B. R. Ratna, S. Baral-Tosch, B. Kahn, A. S. Rudolph and J. M. Schnur, *Chem. Phys. Lipids*, 1992, **63**, 47.

9.   J. H. Georger, A. Singh, R. R Price, J. M. Schnur, P. Yager and P. E. Schoen, *J. Am. Chem. Soc.*, 1987, **109**, 6169.

10.   H. Ringsdorf, B. Schlarb and J. Venzmer, *Angew. Chem. Int. Ed. Engl.*, 1988, **27**, 113.

11.   T. Gulik-Krzywicki, C. Fouquey and J. M. Lehn, *Proc. Natl. Acad. Sci. U. S. A.*, 1993, **90**, 163.

12.   A. Singh and J. M. Schnur, 'Phospholipid Handbook', G. Cevc, ed., Marcel Dekker, New York, 1993, p. 233.

13.   M. A. Markowitz and A. Singh, *Chem. Phys. Lipids*, 1996, **84**, 65.

14.   M. A. Markowitz, A. Singh and E. L. Chang, *Biochem. Biophys. Res. Commun.*, 1994, **203**, 296.

15.   A. Singh and M. A. Markowitz, *New J. Chem.*, 1994, **18**, 377.

16.   M. A. Markowitz, G-M. Chow and A. Singh, *Langmuir*, 1994, **10**, 4095.

17.   A. Singh, M. Markowitz and L-I. Tsao, *Chem. Phys. Lipids*, 1992, **63**, 191.

18.   A. Singh, M. A. Markowitz, L-I. Tsao and J. Deschamps, 'Polymeric Materials in Diagnostics and Biosensors', A. Usmani, ed., ACS, Washington, DC, 1994, p. 252.

19.   A. Singh, *J. Lipid Res.*, 1990, **31**, 1522.

20.   A. Singh, M. A. Markowitz and L-I. Tsao, *Syn. Commun.*, 1992, **22**, 2293.

21.   P. K. Dhal and F. H. Arnold, *Macromolecules*, 1992, **25**, 7050.

22.   A. Singh, L. Tsao and D. Puranik, *Synth. Commun.*, 1995, **25**, 573.

23.   M. Markowitz, J. Schnur and A. Singh, *Chem. Phys. Lipids*, 1992, **62**, 193.

24.   C. M. Gupta, R. Radhakrishnan and H. G. Khorana, *Proc. Natl. Acad. Sci. (USA)*, 1977, **74**, 4315.

25.   D. R Shnek, D. W. Pack, D. Y. Sasaki and F. H. Arnold, *Langmuir*, 1994, **10**, 2382.

26.   S. K. Bhatia, M. J. Cooney, L. C. Shriver-Lake, T. L. Fare and F. L. Ligler, *Sensors and Actuators*, 1991, **B 3**, 311.

27.   J. D. Hua, L. Liu and Y-M. Qiu, *J. Macromol. Sci. – Chem.* , 1998, **A26**, 45.

28.   E. Sulkowski in 'Protein Purification: Micro to Macro', R. Burgess, ed., Alan R. Liss Inc., 1987, pp. 149–162.

THIOL-MODIFIED PHTHALOCYANINES AND THEIR SELF-ASSEMBLED MONOLAYERS ON GOLD SURFACES

Zhiyong Li and Marya Lieberman*

Department of Chemistry and Biochemistry
University of Notre Dame
Notre Dame, IN 46556 USA

1 INTRODUCTION

Molecules can be useful components in electronic devices; one example of a practical molecular electronic device is the liquid-crystal display in a pocket calculator. In our group, we view the chemical modification of surfaces with molecules that can perform some desired task as an enabling technology for new kinds of molecular electronics.

We are interested in using the self-assembled monolayer (SAM) technique[1] developed in the groups of Whitesides, Ulman, Nuzzo, and others to form ultra thin films of well-oriented silicon phthalocyanine derivatives on metal surfaces. Thin films of phthalocyanines have been used in organic photovoltaic cells, xerographic reproduction, and chromophoric displays. Since the optical properties[2] and probably the conductivity[3] of phthalocyanines are highly anisotropic, thin films of highly oriented phthalocyanines are of particular interest for electronic device applications.   Cook *et al.* have made monolayer films of phthalocyanines by reacting various surfaces with phthalocyanines that contain pendant functional groups such as thiols and trichlorosilanes.[4] However, the large phthalocyanine molecules are attached to the surface by one tether at the periphery of the phthalocyanine ring, and consequently the rings are tilted at various unpredictable angles to the substrate surface.

In order to force the phthalocyanine ring to lie parallel to the surface, a short tether could be attached to the central metal atom of a phthalocyanine or alternately, several such tethers could be arranged around the periphery of the phthalocyanine ring. These design strategies have been applied to a set of soluble silicon and copper phthalocyanines $XYMPc(OR)_8$, which contain axial alkyl or alkoxide groups and eight alkoxy groups around the periphery of the ring. This report describes the synthesis of several such molecules and gives some early results of studies of gold surfaces that are modified with phthalocyanines.

2 EXPERIMENTAL

2.1 General

The following starting materials were commercially available, and were used without purification: AIBN, bromine, 1-bromopentane, 5-bromo-1-pentene, catechol, copper(I) cyanide, methyltrichlorosilane, potassium carbonate, tetrachlorosilane, thio-acetic acid

(Aldrich); anhydrous ammonia (Air Products). Quinoline (Aldrich) contained a yellow impurity and was vacuum distilled from barium oxide; thioacetic acid was vacuum distilled from 3 Å molecular sieves and stored under Ar; methanol, ethanol and 2-propanol were dried by storage over activated 3 Å molecular sieves, and other organic solvents were used directly without further treatment. Dibromocatechol was synthesized by the bromination of catechol.[5]

^{1}H NMR spectra were taken on Varian Unity+ 300 or GN300 instruments. IR spectra were taken as thin films (from evaporation of chloroform or dichloromethane solutions) on KBr plates, using a Perkin Elmer Paragon 1000 FT-IR spectrometer. UV–Vis spectra were obtained with a Perkin Elmer Lambda 11 UV/Vis spectrometer. Elemental analyses were performed by M-H-W Laboratories, Phoenix, AZ. Mass spectra of the phthalocyanines and their precursors were obtained on a JEOL JUS-AX505 HA mass spectrometer using a JEOL 102517 electrospray source (for phthalocyanines) or FAB ionization. X-Ray photoelectron spectra were obtained using a Kratos Analytical ESCA system with monochromatic Mg K$_v$ radiation at 1253.6 eV. Low-angle X-ray diffraction experiments were carried out using Cu K$_v$ radiation in a Scintag X1 Series 22 diffractometer. A Nanoscope II STM (Digital Instruments) was used to image gold substrates. After each sample, the same tip was used to observe highly ordered pyrolytic graphite (HOPG), for which atomic resolution is obtained if the tip is good (not twinned).

2.2 Substrate Preparation

Clear mica sheets (Woodman Assoc. Inc., East Wakefield, NH, (603) 522-8216) were cleaved with a razor blade and the freshly cleaved sheets were clipped to a custom-built substrate holder in a Ladd evaporator (Ladd Research Industries, VT, (800) 451-3406). Gold (99.999%, Aldrich) was evaporated onto the mica at a base pressure of 1 to 5 x 10^{-6} torr, without external substrate heating or cooling. No shutter was used. Deposition of the gold film was monitored by an STM-100 quartz crystal thickness/rate monitor (Sycon, NY, (315) 463-5297) with the probe crystal and the sample holder equidistant from the gold evaporation basket. The first 50 Å was deposited at a rate of 0.5 Å/s, after which the deposition rate was maintained at 1–2 Å/s to about 500 Å. After deposition, the gold surfaces were annealed by passing a 4 cm long hydrogen flame across the gold about 50 times.

2.3 1,2-Dibromo-4,5-bis(pentenyloxy)benzene

The procedure for this synthesis was adapted from that used by Hanack *et al.*[6] Dibromocatechol (26.8 g, 0.10 mol) and potassium carbonate (20 g, 0.14 mol) were dissolved in 50 mL DMF, stirred for 0.5 h, then 5-bromo-1-pentene(24 mL, 0.21 mol) which was dissolved in 20 mL DMF was added dropwise. The resulting brown solution was heated to 100 °C for 12 h, then the dark brown reaction solution was poured into 500 mL 0.2 N HCl and the product was extracted three times with 100 mL ether. The ether was removed by rotary evaporation to yield about 15 mL brown oil, which was purified on a silica gel column (150 g silica gel, 1:1 hexane/CHCl$_3$). The pure product (28.6 g, yield 71%) was a colorless oil. ^{1}H NMR (CDCl$_3$, shifts in ppm relative to TMS) δ 7.06(s, 2H), 5.84(m, 2H), 5.02(m, 4H), 3.96(t, 4H), 2.24(tt, 4H), 1.91(m, 4H); IR (thin film) 3077(m), 1641(m), 1582(m), 1498(vs), 1467(vs), 1386(m), 1350(s), 1250(vs), 1200(vs), 1013(m), 913(s), 846(m), 651(m)cm^{-1}; MS (FAB) calcd. for C$_{16}$H$_{20}$O$_2$Br$_2$ (average): *m/e* 404.14, found *m/e* 404.14.

2.4 1,2-Dicyano-4,5-bis(pentenyloxy)benzene

The synthesis was a modification of Hanack's procedure. CuCN (13.4 g, 150 mmol) was added to a solution of 1,2-dibromo-4,5-bis(pentenyloxy)benzene (20.2 g, 50 mmol) in DMF (60 mL), and the solution was heated to 145 °C for 12 h under Ar. The green mixture was poured into 600 mL NH_4OH after it cooled to room temperature, and air was bubbled through for 6 h. The green solid was filtered off and washed with H_2O until the filtrate was neutral. After the solid dried in the air, it was extracted with 200 mL methanol in a Soxhlet apparatus for 24 h. The product crystallized from the methanol, and was purified by recrystallization in ethanol as white flaky crystals (7.0 g, yield 47%). ^{1}H NMR(CDCl$_3$) δ 7.11(s, 2H), 5.84(m, 2H), 5.07(m, 4H), 4.06(t, 4H), 2.25(tt, 4H), 1.96(m, 4H) ppm; IR(thin film) 3125(w), 3060(m, vinyl), 2228(s, CN), 1590(vs), 1392(m), 1228(vs), 1217(m), 1093(s), 988(m), 910(m), 536(m)cm^{-1}; MS(FAB) calcd. for $C_{18}H_{20}O_2N_2$(base peak) *m/e* 297.16, found *m/e* 297.16.

2.5 2,3,9,10,16,17,23,24-Octakis(pentenyloxy)phthalocyaninato copper(II) (CuPc(OC$_5$H$_9$)$_8$)

About 3 g of the residue left after extraction of 1,2-dicyano-4,5-bis (pentenyloxy) benzene in the Sandmeyer reaction above was extracted with about 200 mL CH_2Cl_2 in the Soxhlet apparatus for 10 h. The solvent was removed from the extract to yield a green solid. The product (about 2.0 g) was further purified by column chromatography (50 g silica gel, CHCl$_3$). The first green band in the column was the product (1.5 g, yield 10% from 1,2-dibromo-4,5-bis(pentenyloxy)benzene). IR(thin film) 3077(m, vinyl), 1641(m), 1606(m), 1461(vs), 1419(m), 1388(vs), 1354(m), 1278(vs), 1201(s),1109(m), 1057(m), 909(m), 745(s) cm^{-1}; MS(electrospray, 1:1 CH$_3$OH/CHCl$_3$) calcd. for $C_{72}H_{80}O_8N_8Cu$ *m/e* 1250.0, found *m/e* 1249.7; ^{1}H NMR(C$_6$D$_6$) δ 6.1(b, 2H), 5.2(b, 4H), 3.4(b, 2H), 2.4(b, 4H), 2.0(b, 4H) ppm.

2.6 1,3-Diimino-5,6-bis(pentenyloxy)isoindoline

This procedure was modified from Kenney *et al.*[7] Ammonia gas was bubbled through a solution of 1,2-dicyano-4,5-bis(pentenyloxy)benzene (3.0 g, 10 mmol) and sodium methoxide (0.10 g) in absolute methanol (60 mL) at room temperature for 45 min, then the greenish solution was heated to reflux for a further 3 h under NH$_3$. The reaction was stopped and allowed to stand overnight. The reaction solution was cooled to –78 °C and the green precipitate was filtered off and washed with cold methanol. The green product (2.7 g, yield 90%) was of sufficient purity to be used directly for the synthesis of SiPc. A pale yellow form of this product (believed to be a tautomer[8]) may be isolated; it can be used interchangeably with the green form. ^{1}H NMR (DMSO-d$_6$) δ 8.48(b, 2H, NH), 5.83(m, 2H), 5.00(m, 4H), 4.02(t, 4H), 2.21(tt, 4H), 1.83(m, 4H)ppm. MS (FAB) calcd. for $C_{18}H_{23}O_2N_3$ *m/e* 313, found *m/e* 314(M$^+$ + 1), 336(M$^+$ + Na). IR (thin film) 3414(b, NH), 3072(w, vinyl), 1641(m), 1601(m), 1550(s), 1525(s), 1445(s), 1422(s), 1287(s), 1205(vs), 1142(s), 1038(s), 989(m), 909(s), 720(m)cm^{-1}.

2.7 General Procedure for Template Synthesis with SiCl$_4$. Dimethoxy-2,3,9,10,16,-17,23,24-octakis(pentenyloxy)phthalocyaninatosilicon ((OCH$_3$)$_2$Si(OC$_5$H$_9$)$_8$)

The procedure was adapted from that used by Nolte *et al.*[9] All procedures are carried out in the dark as much as possible (flasks and NMR tubes were wrapped in aluminum foil).

1,3-Diimino-5,6-bis(pentenyloxy)isoindoline (0.10 g, 0.32 mmol) was stirred with 1 mL dry quinoline under Ar, then 0.1 mL tetrachlorosilane was added by syringe and the mixture was brought quickly to 180 °C and kept at this temperature for 1 h. After the reaction cooled, methanol was added to give a dark green solid, which was filtered on a glass frit. The crude product was washed in a Soxhlet apparatus with methanol for 24 h. Although completely insoluble in methanol, the product was soluble in $CHCl_3$, CH_2Cl_2 and benzene. The purity by NMR (as estimated from the phthalocyanine ring proton singlet which occurs at 9.20 ppm and the vinyl signals between 5 and 6 ppm) was > 90%. Further purification by recrystallization or chromatography using silica gel or alumina was unsuccessful. ^1H NMR (C_6D_6) δ 9.20(s, 8H), 5.84(m,8H), 5.08(m, 16H), 4.09(t, 16H), 2.28(tt, 16H), 1.90(tt, 16H), –1.20(s, 6H) ppm; IR (thin film) 3077(w, vinyl), 1604(m), 1464(vs), 1427(s), 1392(s), 1280(s), 1208(s), 1099(vs), 892(m), 759(m)cm^{-1}; UV–VIS (CH_2Cl_2) λ_{max} 679.6nm; MS (electrospray) calcd. for $C_{74}H_{86}O_{10}N_8Si$ m/e 1275.6, found m/e 1245(M$^+$ - OCH3).

Other silicon phthalocyanines discussed in this report were synthesized by variations on this method. The yields (based on diiminoisoindoline) range from 40–50% for most diiminoisoindolines we have tried. Solid products can be stored under Ar for months, but solutions must be protected from light and water.

2.8 $(OCH_3)_2Si(OCH_2$-CH_2-CH_2-CH_2-CH_2-$SCOCH_3)_8$ (1)

$(OCH_3)_2Si(OC_5H_9)_8$ (250 mg) and AIBN (100 mg) were dissolved in 10 mL dry toluene under Ar, then dry thioacetic acid (2 mL) was added and the mixture was heated at 70 °C for 2 days. Excess thioacetic acid and solvent were removed under vacuum, and the dark green solid was washed with methanol and hexane and dried under vacuum overnight. For $(OCH_3)_2SiPc(OCH_2CH_2CH_2CH_2CH_2SCOCH_3)_8$: ^1H NMR (C_6D_6, 50 °C) δ 9.27(s, 8H), 4.03(t,b, 16H), 2.91(t , 16H, SCH$_2$), 1.98(s ?4H, COCH$_3$), 1.76(m, 16H), 1.60(m, 16H), 1.52(m, 16H), –0.86(s, 6H, axial methoxide) ppm; IR(thin film) 1689(vs, C=O), 1607(w), 1466(s), 1430(s), 1394(s), 1363(m), 1283(s) 1208(m), 1104(s), 750(m), 668(m), 626(m)cm^{-1}; MS (electrospray) calcd. for $C_{90}H_{118}O_{18}N_8SiS_8$ m/e 1884.5, found m/e 1883.8(M$^+$), 1869.3(M$^+$-CH$_3$), 1853.0(M$^+$-OCH$_3$), 1776.2(M$^+$-OCH$_3$-CH$_3$COSH); elemental analysis observed (calc) C: 57.12 (57.36) H: 6.07 (6.31) N: 5.65 (5.94) S: 13.73 (13.61).

2.9 $CuPc(OCH_2$-CH_2-CH_2-CH_2-CH_2-$SCOCH_3)_8$ (3)

The addition of thioacetic acid (0.3 mL) to the terminal vinyl groups in the $CuPc(OC_5H_9O)_8$ (0.13 g) was carried out following the same procedure as the previous reaction. The green solid (0.14 g, yield 75%) was analysed by MS (electrospray) calcd. for $CuC_{88}H_{112}O_{16}N_8S_8$ m/e 1857.9, found m/e 1857.1, 1781.0(M$^+$-CH3COSH), 1705.4(M$^+$-2CH$_3$COSH); IR (thin film) 1691(vs, C=O stretch), 1605(m), 1462(s), 1420(m), 1386(s), 1354(m), 1279(s), 1202(m), 1110(s), 1053(m), 953(m), 744(s), 627(s) cm^{-1}; elemental analysis observed (calc) C: 56.83 (56.89) H: 6.15 (6.08) N: 5.81 (6.03) S: 13.88 (13.81).

2.10 $(OCH_3)_2Si(OCH_2$-CH_2-CH_2-CH_2-CH_2-$SH)_8$ (2) and
$\quad\quad CuPc(OCH_2$-CH_2-CH_2-CH_2-CH_2-$SH)_8$ (4)

Both of these compounds were made from the corresponding octathioacetate phthalocyanines (**1** and **3**, respectively). In a typical synthesis, 24 mg of **3** and 10 mL of degassed 50% HCl/MeOH were heated at 70 °C for one week under Ar; the dark green product was isolated by filtration, washed with methanol, and dried under vacuum. After hydrolysis, the compound showed no thioacetate peak at 1690 cm^{-1}. MS (electrospray) calcd. for $CuC_{72}H_{96}O_8N_8S_8$ (**4**) *m/e* 1521.7, found *m/e* 1522.3; for $SiC_{74}H_{102}O_{10}N_8S_8$ (**2**) *m/e* 1548.3, found *m/e* 1516.3(M^+-OCH$_3$).

2.11 CH$_3$(HSCH$_2$CH$_2$O)SiPc(OC$_5$H$_{11}$)$_8$ (5)

CH$_3$(*i*-PrO)SiPc(OC$_5$H$_{11}$)$_8$ was prepared by using isopropanol instead of methanol in the washing step described in Section 2.7. 36 mg of this compound was dissolved in dry toluene. 0.5 mL of dry HSCH$_2$CH$_2$OH was added, and the solution was stirred at RT for 0.5 h. The solvents were removed under vacuum, leaving behind 40 mg of analytically pure dark green solid (100% yield).

For CH$_3$(HSCH$_2$CH$_2$O)SiPc(OC$_5$H$_{11}$)$_8$: ^{1}H-NMR (dry C$_6$D$_6$, 25°C) δ 9.06 (s, 8H), 3.91 (t, b, 16H), 1.83 (t, b, 16H, SCH$_2$), 1.29 (m, 32H), 0.95 (t, 24H, -CH$_3$), −0.469 (m, 2H, axial -CH$_2$SH), −1.31 (t, 1H, axial SH), −1.64 (t, 2H, axial -CH$_2$OH), −5.027 (s, 3H, axial methyl) ppm; elemental analysis observed (calc) C: 67.98 (68.15) H: 7.89 (7.93) N: 8.42 (8.48) S: 2.26 (2.42).

2.12 Derivatization of Gold Substrates

Flame annealed substrates were soaked in approximately 0.5 mM solutions of the desired phthalocyanine in toluene (**1, 3, 5**) or toluene/methylene chloride (**2, 4**) for 1–7 days, rinsed with fresh toluene, and blown dry in a stream of argon.

2.13 Contact Angle Measurements

Contact angles were measured using a home-built goniometer. This consisted of a sample stage with attached 50 :L syringe, and an Edmund #P52,219 zoom microscope head attached to an Edmund #P3683 XY translation stage. The scope was provided with a protractor reticule (Edmund #P30,079). Contact angles were determined using 18 MOhm water. 5 μL water were placed on the sample, with the syringe tip remaining in the drop (captive drop). After addition of 0.5 μL water to the drop, the contact angles at both sides of the drop were measured and averaged to give the advancing contact angle; after withdrawal of 1–2 μL water, the contact angles were re-measured as the receding contact angle. This measurement was repeated in at least three locations on the surface.

2.14 XPS

Samples were mounted on sample stubs with conductive carbon tape. Significant charging effects (up to 10 eV shifts in BE) were observed, so the binding energies for each peak were referenced to the Au 4f $_{7/2}$ peak at 84.0 eV.

In addition to a survey scan, each element of interest was studied in a 200 s high-resolution region scan. The area of each peak was measured by fitting the individual region scans with one or more Gaussian peaks. Atom percentages were calculated using Equation 1, shown below; the corrected results are listed in Table 2.

$$C_i = \frac{A_i / S_i}{\Sigma(A_i / S_i)} \times 100 \tag{1}$$

3 RESULTS

3.1 Substrate Preparation

Gold substrates were prepared by thermal evaporation of gold onto freshly cleaved mica substrates. These substrates were then either used directly or annealed with a hydrogen flame. The directly evaporated gold, as expected, was quite rough. Crystallites could not be detected by SEM, indicating that they were probably smaller than the 60 nm resolution of the instrument. STM images showed a mosaic of small crystallites about 50 nm in diameter. Small-angle X-ray diffraction (XRD) showed a very strong (111) line; the (100), (110), and (120) lines were often completely undetectable. Since the Miller planes parallel to the surface diffract the best in this geometry, the XRD indicates a strong (111) orientation of the gold crystallites.

After annealing, scanning electron microscopy showed a mosaic of what appeared to be crystallites with an average diameter of about 500 nm. The gold surfaces showed atomically flat terraces > 100,000 nm^2 in area in the scanning tunneling microscope. X-Ray diffraction showed very strong (111) lines. A weak signal from the (100) plane was also observed. Combining the STM, XRD, and SEM data, these annealed substrates consist of large, very flat crystallites exposing a mostly (111) gold surface.

3.2 Synthesis of Phthalocyanines

Soluble silicon phthalocyanines with eight peripheral ether groups were synthesized from di-substituted diiminoisoindolines.[10] Catechol was dibrominated, then the phenolic oxygen was coupled to an alkyl halide (pentyl bromide or ω-pentenyl bromide) under basic conditions. The aromatic bromides were displaced by cyanide in a Sandmeyer reaction, and the resulting phthalonitriles were converted into diiminoisoindolines by addition of ammonia in warm methanol. The overall yield of diiminoisoindoline from dibromocatechol was 30%. Copper phthalocyanines were obtained as a side product in the Sandmeyer reaction; the yield of the copper phthalocyanine could be increased by increasing the amount of CuCN used in the reaction.

Silicon phthalocyanines were made by refluxing diiminoisoindoline and excess $SiCl_4$ or $RSiCl_3$ in quinoline for an hour. The green precipitates were washed with methanol or isopropanol in a Soxhlet extractor for 24 h to remove impurities; for phthalocyanines with eight ether substituents, this procedure causes alcoholysis of any axial SiCl bonds. Isolated yields (after chromatography) for all the SiPcs were about 50% from diiminoisoindoline. The products are microcrystalline green solids, totally insoluble in methanol but soluble in chloroform and benzene. The silicon phthalocyanines are sensitive to photolysis in wet solvents and must be stored in the dark.

Thiol-containing copper(II) and silicon(IV) phthalocyanine molecules shown in Figure 1 were synthesized from phthalocyanine precursors containing eight pentene groups attached around the periphery of the macrocycle. The octathioacetyl derivatives **1** and **3** were synthesized *via* radical addition of thioacetic acid to these terminal olefins. NMR spectra during the reaction showed the complete disappearance of olefinic signals and the appearance of a thioacetate singlet at 1.98 ppm, with integrated intensity of 24 H, indicating complete derivatization of the eight vinylic groups with >95% anti-Markovnikov selectivity. The deprotected octathiol derivatives **2** and **4** were prepared by refluxing **1** or **3** respectively in degassed HCl/methanol solution for one day; the phthalocyanine macrocycle is stable to these conditions. IR showed complete loss of the thioacetate carbonyl band at 1689 cm^{-1}.

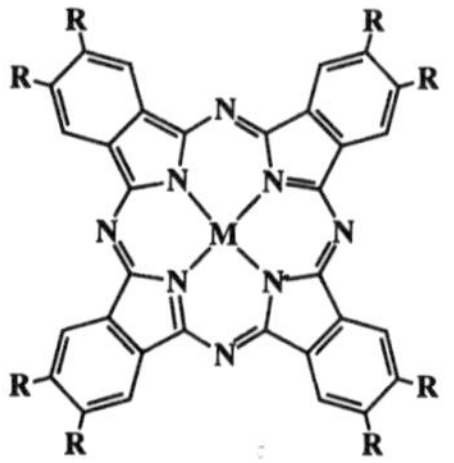

1:   M = Cu, R = $O(CH_2)_5SCOCH_3$

2:   M = Cu, R = $O(CH_2)_5SH$

3:   M = Si(OMe)$_2$, R = $O(CH_2)_5SCOCH_3$

4:   M = Si(OMe)$_2$, R = $O(CH_2)_5SH$

5:   M = Si(OCH$_2$CH$_2$SH)(CH$_3$), R = OC_5H_{11}

Figure 1 *Phthalocyanines 1–5 contain thiol or thioacetate groups for binding to gold surfaces.*

The silicon phthalocyanine **5** with a thiol moiety in the axial position was synthesized by ligand exchange[11] from an isopropoxide silicon phthalocyanine and mercaptoethanol. The isopropoxide was itself prepared by washing crude chlorosilicon phthalocyanine with isopropanol instead of methanol. Both compounds were characterized by NMR and also gave good elemental analyses.

3.3 Derivatization of Gold Surfaces

Gold substrates were derivatized by soaking them for 1–7 days in 0.5 mM solutions of the desired phthalocyanine in toluene or toluene/methylene chloride (for **2** and **4**). Derivatization was followed by contact angle measurement and X-ray photoelectron spectroscopy, and appears to be complete after 36 h.

3.4 Contact Angle Measurements

Contact angles (Table 1) were determined by averaging measurements of at least three drops, the estimated error is ± 2°. The contact angles were observed to change over time, presumably as a result of accretion of contaminants.

Table 1 *Contact Angles for Water on Phthalocyanine-modified Gold Surfaces.*

	Θ_a (deg.)	Θ_r (deg.)
Sample	Freshly made/after 2 days in air	Freshly made/after 2 days in air
Annealed gold	75/80	40/50
1	70/60	50/30
2	55/40	15/10
3	70/52	50/28
4	60/35	18/10
5	70/80	30/42
Methyl-terminated SAM[12]	111°–115°	
HS(CH$_2$)$_{11}$OCH$_3$ SAM[13]	74°	

3.5 X-Ray Photoelectron Spectroscopy

X-Ray photoelectron spectroscopy was used to detect the appearance on the gold surface of elements found in the thiol-modified phthalocyanines, and also to investigate how the thiols in different phthalocyanines interact with the gold surface. Figure 2 shows survey spectra of gold samples soaked in either toluene (a) or a solution of **2**, $CuPc(OCH_2\text{-}CH_2\text{-}CH_2\text{-}CH_2\text{-}CH_2\text{-}SH)_8$ (b). In Figure 2a (bare gold), only gold, carbon, and oxygen peaks are observed. The carbon and oxygen indicate surface contamination ("adventitious carbon") which is present to some extent on all the samples. Figure 2b (**2**) shows additional peaks due to sulfur, nitrogen, and copper. The carbon signals are stronger. On the other hand, the gold peaks are decreased in intensity, indicating that they are covered by some kind of an overlayer.

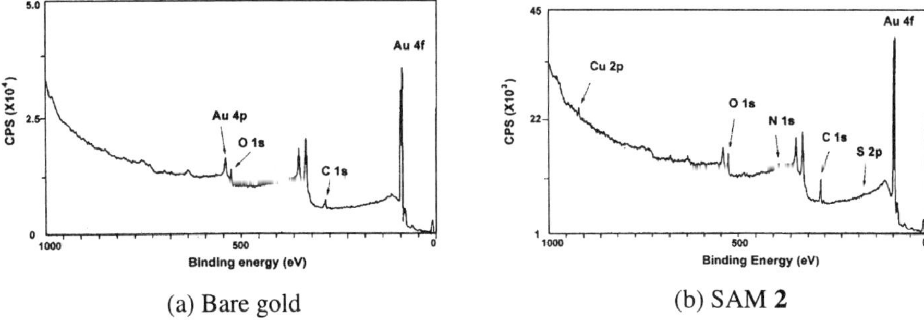

(a) Bare gold   (b) SAM **2**

Figure 2 *XPS spectra of (a) bare gold and (b) gold/CuPc(OCH$_2$-CH$_2$-CH$_2$-CH$_2$-CH$_2$-SH)$_8$ (2). Samples were soaked 36 h in toluene with and without the phthalocyanine, washed with fresh toluene, and dried with a stream of argon. Binding energies are referenced to Au 4f $_{7/2}$ at 84.0 eV.*

Table 2 *Corrected Binding Energies and Intensities of XPS Peaks for Samples on Annealed Gold Substrates.*

Peak positions shown in eV (atom percent)

	Au	S	C	N	O	Cu
Sample	4f$_{7/2}$, 4f$_{5/2}$	2p	1s	1s	1s	2p
toluene	84.0(21.7), 87.7(17.2)	--	284.4(23.3), 286.1(6.7), 288.3 (10.8)	--	531.8(16.4) 536.8(4.0)	
1	84.0(22.0), 87.7(17.2)	162.9(3.7)	284.6(25.9), 286.0(10.1), 287.6(6.0)	398.6(1.8)	531.6(9.2), 532.8(3.0)	945.6 (.2), 934.6 (.3), 952.2 (.4), 931.9 (.4)
2	84.0(13.5), 87.7(10.5)	163.2(1.9), 168.1(.8)	284.5(26.1), 286.0(20.1), 288.2(6.7)	398.8(3.4)	532.0(13.7) 536.1(1.6)	934.4 (.6), 943.2 (.5), 955.4 (.5), 962.8 (.4)
3	84.0(25.3), 87.7(19.9)	162.7 (2.6)	284.5(23.8), 286.1(9.7), 288.3(7.1)	398.8(1.7)	531.6(7.1), 532.6(2.9)	
4	84.0(9.9), 87.7(7.9)	163.6 (3.0)	284.6(38.5), 286.0(15.2), 287.3(4.4)	398.8(4.7), 400.9(1.1)	530.9(3.9), 532.5(11.3)	
5	84.0(22.9), 87.7(18.3)	162.4(.8), 168.5(.4)	284.9(25.7), 286.5(8.0), 288.9(5.6)	398.7(3.4)	530.3(4.4) 532.3 (10.4)	

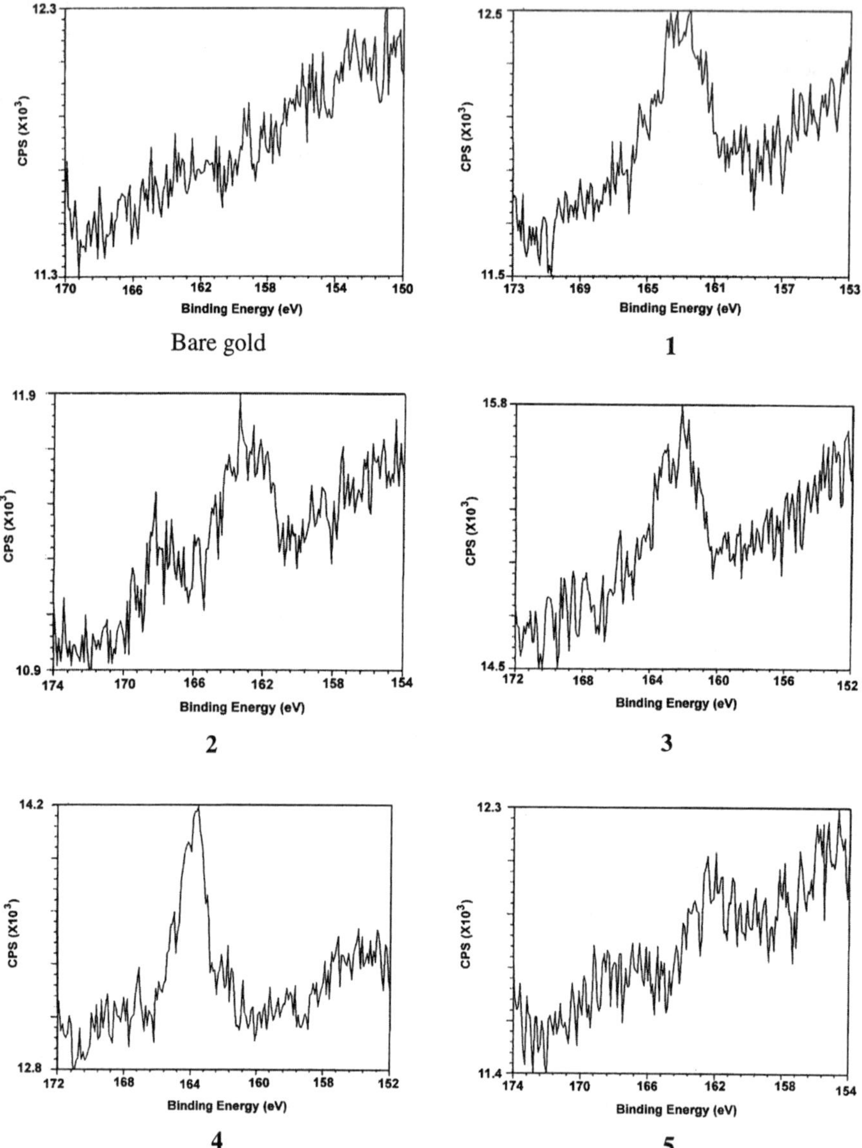

Figure 3 *XPS spectra of bare gold and SAMs of 1–5 in the S_{2p} region.*

High-resolution scans of the sulfur region showed several sets of S_{2p} peaks for each molecule. These scans are reproduced in Figure 3.

For compounds **3–5**, which contain a central silicon atom, the silicon peak was not observed, although its signal is expected to be about as strong as that of a single sulfur atom, which we do observe in the case of compound **5**.

4 DISCUSSION

Two basic approaches to derivatizing gold with large flat molecules were compared (Figure 4): an "octopus" strategy, in which the periphery of the molecule contains eight thiol or thioacetate arms (**1–4**), and an "umbrella" strategy, in which an axial site on the central metal is derivatized with a short thiol (**5**). Ultrathin phthalocyanine films were formed on gold substrates regardless of which approach was used, but the XPS and contact angle data show that the coverage and structure of the resulting films were different.

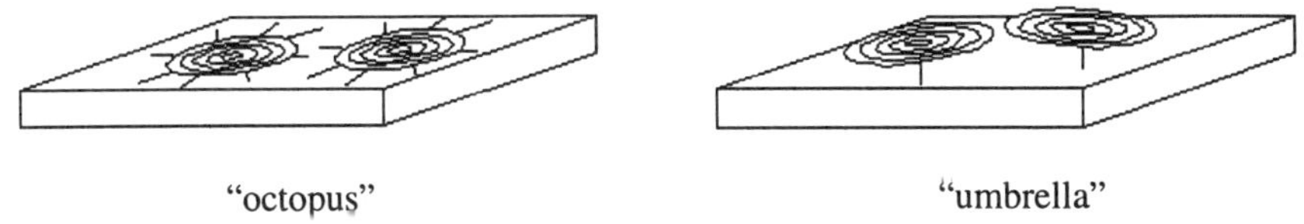

Figure 4 *Two ways to control the orientation of large flat molecules on a surface.*

All of the sulfur-containing phthalocyanines bound to gold surfaces. Plain gold substrates soaked in toluene showed XPS signals corresponding to gold, oxygen, and "adventitious" carbon (Figure 2a). Substrates that were exposed to sulfur-containing phthalocyanines in toluene solution showed enhanced carbon peaks, reduced oxygen peaks, as well as new signals due to sulfur, nitrogen, and copper (in the case of CuPc) (Figure 2b). The XPS signals did not change appreciably when the soaking time was increased from 1.5 to 7 days, implying that the derivatization was substantially complete in 1.5 days.

There were no significant differences between the SAMs derived from copper and silicon phthalocyanines. On the other hand, phthalocyanines that contain eight thioacetate groups gave lower coverages than those that contain eight thiols, as measured by the degree of attenuation of the gold signal and the relative strength of the N and Cu signals. Tour *et al.* found that long-chain thioacetyls were capable of producing good coverages on gold, although higher concentrations were required than for long-chain thiols.[14]

The degree of attenuation of the gold signal is related to the thickness of the monolayer, which in turn depends on coverage and how the molecules are oriented on the surface. Edge-on binding of the octopus phthalocyanines would give the SAM a thickness of about 28 Å, while face-on binding would be about 6 Å. The C_{1s}/Au_{4f} ratios for our SAMs range from about 1.0 (bare gold, **1**, **3**, **5**) to 2.2 (**2**) or 3.3 (**4**). These values are in the ballpark of C_{1s}/Au_{4f} ratios for SAMs of alkanethiols such as octadecanethiol (1.69).[15]

The chemical shifts observed by XPS show that thiol binding to the gold surface is far from ideal. For thiols bound to gold, Castner *et al.*[16] assigned the signal from the $S_{2p\ 3/2}$ electrons to a value of 162 eV; disulfides and unbound thiols were found to come between 163.5 and 164 eV. Oxidized sulfur species occur at higher BEs; for example, the sulfur signal in sulfite salts is seen at a value of 165.5–167.5 eV, and sulfones come around 166.5–170.7.[17] In SAMs derived from compounds **1–5**, from 60–80% of the sulfur groups appear to be bound to the gold as thiols. However, some gold surfaces that were derivatized with phthalocyanines containing a free thiol (**2** or **5**) gave XPS signals between 164.9 and 168.1 eV. From the relative intensities of these signals relative to the total sulfur signal, 20–35% of the sulfur atoms present in these phthalocyanine films were oxidized. The partially oxidized products may have failed to dissolve in the toluene rinse, or they might be attached to the gold through their intact thiols. Oxidation of sulfur was not observed in any of the SAMs derived from thioacetates.

The contact angle measurements of substrates which had been derivatized with the "octopus" thiols **2** or **4** showed advancing contact angles for water of 55–60°, similar to nitrile-terminated SAMs. This hydrophilicity is certainly consistent with the presence of some partially oxidized sulfur. Over two days, the contact angles decreased further; this suggests that these monolayers may contain groups that can undergo further oxidation. The other derivatized substrates, including the "octopus" thioacetates **1** and **4** and the "umbrella" thiol **5**, had advancing contact angles for water of 70–75°, about the same as plain gold, and their contact angles did not decrease over time, which suggests that their thiols are better protected from oxygen.

5 CONCLUSIONS

Self-assembled monolayers can form from phthalocyanines that contain thiol or thioacetate anchoring groups. These SAMs have different structures. The coverages determined by XPS and the surface hydrophilicity as measured by the water contact angle appear to be related to where the anchoring groups are located (ring periphery *vs.* axial) and to whether thiols or thioacetates are used. Thiol anchoring groups are associated with higher coverages, but the thiols on these phthalocyanines seem to be unusually sensitive to oxidation. The next step is to determine the absolute coverages of these SAMs and to elucidate structural details of how the phthalocyanines are anchored on the gold surfaces.

6 ACKNOWLEDGMENTS

Thanks are due to our colleagues who shared their time and instruments: low angle X-ray diffraction spectra were taken by Sharon Yeung and John Fultz on Prof. Paul McGinn's diffractometer; Dr. Yandong Gong took the XPS spectra; Dr. Bill Archer took scanning electron micrographs, and Prof. Eduardo Wolf lent us an STM. This work was funded by the University of Notre Dame and by NSF grant CHE-97-14438.

References

1. A. Kumar, H. A. Biebuyck and G. M. Whitesides, *Langmuir*, 1994, **10**, 1498.
2. M. J. Stillman and T. Nyokong, in 'Phthalocyanines: Properties and Applications', C. C. Leznoff and A. B. P. Lever, eds., VCH, New York, 1989, p. 133.
3. B. N. Diel, T. Inabe, J. W. Lyding, K. F. Schoch, Jr., C. R. Kannewurf and T. J. Marks, *J. Am. Chem. Soc.*, 1983, **105**, 1551.
4. a) T. R. E. Simpson, D. J. Revell, M. J. Cook and D. A. Russell, *Langmuir*, 1997, **13**, 460; b) M. J. Cook, R. Hersans, J. McMurdo and D. A. Russell, *J. Mater. Chem.*, 1996, **6**, 149.
5. M. Kohn, *J. Am. Chem. Soc.*, 1951, **73**, 480.
6. M. Hanack, A. Gul and A. Hirsch, *Mol. Cryst. Liq. Cryst.*, 1990, **187**, 365.

7.   M. K. Lowery, A. J. Starshak, J. N. Esposito, P. C. Krueger and M. E. Kenney, *Inorg. Chem.*, 1965, **4**, 128.
8.   "Dictionary of Organic Compounds," Chapman & Hill, NY, 6th Ed., 1996, p. 3876.
9.   C. F. van N., S. J. Picken, A.-J. Schouten and R. J. M. Nolte, *J. Am. Chem. Soc.*, 1995, **117**, 9957.
10. Z. Li and M. Lieberman, *Supramolec. Science*, 1998, in press.
11. Z. Li and M. Lieberman, Fall 1998 ACS National Meeting, Boston, MA., Inorganic Abstract #256.
12. A. Ulman, 'An Introduction to Ultrathin Organic Films', Harcourt-Brace, NY, 1991.
13. C. D. Bain, J. Evall and G. M. Whitesides, *J. Am. Chem. Soc.*, 1989, **111**, 7155.
14. J. M. Tour, L. Jones II, D. L. Pearson, J. J. S. Lamba, T. P. Burgin, G. M. Whitesides, D. L. Allara, A. N. Parikh and S. V. Atre, *J. Am. Chem. Soc.*, 1995, **117**, 9529.
15. D. A. Hutt and G. J. Leggett, *J. Phys. Chem.*, 1996, **100**, 6657. Our takeoff angle was 45° and Hutt and Leggett's was 70°, so these ratios can only be compared qualitatively.
16. D. G. Castner, K. Hinds and D. W. Grainger, *Langmuir*, 1996, **12**, 5083.
17. J. F. Moulder, W. F. Stickle, P. E. Sobol and K. D. Bomber, 'Handbook of X-ray Photoelectron Spectroscopy', Perkin Elmer Corp, Eden Prairie, MN, 1992.

*   Correspondence: Marya.Lieberman.1@nd.edu

MODIFICATION OF PARTICLE SURFACES BY GRAFTING OF FUNCTIONAL POLYMERS

N. Tsubokawa

Department of Material Science and Technology
Faculty of Engineering
Niigata University
8050, Ikarashi 2-nocho, Niigata 950-2181, Japan

1 INTRODUCTION

Ultrafine particles, such as carbon black and fumed silica, are widely used industrially as fillers and pigments in polymer-based materials. In general, dispersing ultrafine particles uniformly into a polymer matrix or an organic solvent is difficult to achieve. Surface grafting of polymers onto these ultrafine particles has been shown to markedly improve their dispersibility in solvents and their compatibility in polymer matrices.[1-3] In addition, polymer grafting onto ultrafine particles enables us to give various functions to these surfaces, such as photosensitivity, crosslinking ability, and temperature sensitivity.[1-3] In this paper we describe methodologies for the grafting of polymers onto the surface of ultrafine particles and applications of these polymer-grafted particles.

2 METHODOLOGIES FOR THE SURFACE GRAFTING OF POLYMERS

While silica particles have only silanols on the surface, carbon black particles have numerous types of oxygen-containing groups, such as carboxyl, phenolic hydroxyl, and quinonic oxygen groups on the surface.[4] These functional groups on carbon black undergo normal organic reactions, such as amidation, esterification, and neutralization.[4] Therefore, these surface functional groups can readily be converted to various initiating groups for the graft polymerization and used as grafting sites of polymers.[1-3] For example, the conversion of phenolic hydroxyl and carboxyl groups on carbon black to several reactive groups is shown in Scheme 1. In general, one of the following principles may be applied to prepare the polymer-grafted particle, whether carbon black or silica:

(1) Grafting onto particle: This process uses conventional initiators to carry out polymerization in the presence of particles, and is based on the termination of growing polymer chain with functional groups on a particle's surface.

(2) Grafting from particle: This depends on the initiation of graft polymerization from initiating groups introduced onto a particle's surface.

(3) Reaction of functional groups of particle with functional polymers: This method depends on the reaction of surface functional groups on a particle with polymers having terminal functional groups.

Polymer grafting to particles has been achieved by process (1), but the percentage of grafting usually is found to be less than 10% because of preferential formation of ungrafted polymer. Process (2) is the most favorable for the preparation of polymer-grafted particles in that it gives a higher percentage of grafting. However, in this process it is difficult to control the molecular weight of the polymer chains and the number of chains grafted per particle. Two important characteristics of process (3) are that the molecular weight and the number of polymer chains grafted are easily controlled, and also that the grafted polymer is of well-defined structure.

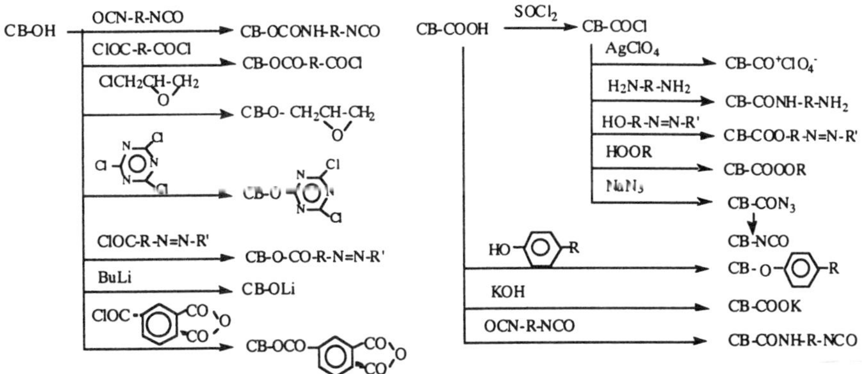

Scheme 1 Conversion of Functional Groups on Carbon Black to Reactive Groups

3 GRAFT POLYMERIZATION FROM SURFACE INITIATING GROUPS

Radical, anionic, and cationic graft polymerizations from ultrafine particle surfaces have been achieved by surface-attached azo or peroxyester groups, potassium carboxylate groups, and acylium perchlorate groups, respectively. We consider each of these three situations in turn below.

3.1 Radical Graft Polymerization

We have achieved the introduction of azo groups onto carbon black surfaces by first treating the carbon black with a diisocyanate to introduce a free, reactive isocyanate group, then allowing 4,4'-azobis(4-cyanopentanoic acid) to react with these isocyanate groups.[5] Radical polymerization of vinyl monomers was successfully initiated by these surface-attached azo groups, so the resulting polymers were tethered to the carbon black surface. Representative results are shown in Table 1. We have already reported similar results using azo and peroxyester groups tethered to ultrafine silica particles.[6,7]

In these radical grafting systems, surface radicals initiate polymerization, but ungrafted polymer was also formed because of formation of fragment radicals by the decomposition of these groups. Therefore, grafting efficiency (proportion of grafted polymer to total polymer formed) was nearly 50% at the initial stage of polymerization, but decreased with time.

An even more effective system for radical graft polymerization consists of $Mo(CO)_6$ and trichloroacetyl groups introduced onto silica and titanium dioxide surfaces by treatment with trichloroacetyl isocyanate (Table 2).[8] Formation of ungrafted polymer was suppressed because no radical fragments were formed. Thus, the initial grafting efficiency was increased to 60% and remained high even at long polymerization times.

Table 1   *Graft Polymerization of Various Vinyl Monomers Initiated by Azo Groups Attached to Carbon Black Surfaces.*

Vinyl monomer	Conversion (%)	Grafting (%)
Styrene	2.0	28.5
Acrylamide	1.7	24.0
Methyl methacrylate	4.7	40.0
Acrylic acid	6.2	64.8
N-Isopropylacrylamide	5.6	35.9
Styrene / Acrylic acid	13.1	76.2

Carbon black-Azo, 0.30 g; Monomer, 10.0 ml; 70°C.

Table 2   *Graft Polymerization of Vinyl Monomers Initiated by Silica-R-COCCl$_3$ / Mo(CO)$_6$, and TiO$_2$-R-COCCl$_3$ / Mo(CO)$_6$ System.*

Inorganic Particle	Monomer	Time (min)	Conversion (%)	Grafting (%)
Silica-R-COCCl$_3$	Styrene	120	9.0	150.5
	Methyl methacrylate	40	12.7	741.4
	N-Vinylcarbazole	120	78.5	35.4
	N-Isopropylacrylamide	40	15.4	99.5
TiO$_2$-R-COCCl$_3$	Styrene	60	4.4	39.2
	Methyl methacrylate	30	10.8	302.7
	N-Isopropylacrylamide	30	3.9	50.1
	N,N-Diethylacrylamide	6	7.6	295.3

Particle-R-COCCl$_3$, 0.10 g; Mo(CO)$_6$, 0.01 g; monomer, 10.0 ml; 70 °C.

3.2 Anionic Graft Polymerization

Ohkita *et al.* reported that anionic graft polymerization of methyl methacrylate and acrylonitrile are initiated by carbon black/*n*-butyllithium complex.[9] We have reported that anionic ring-opening copolymerization of epoxides with cyclic acid anhydrides can be initiated by potassium carboxylate groups introduced onto carbon black and silica surfaces to give polyester-grafted particles.[10,11] Table 3 shows the results of this anionic ring opening graft polymerization. Epoxides used were styrene oxide (SO), glycidyl methacrylate (GMA), epichlorohydrin (ECH), and glycidyl cinnamate (GC). Cyclic acid anhydrides used were phthalic anhydride (PAn), maleic anhydride (MAn), and succinic anhydride (SAn). Table 3 clearly shows that various polyesters can be grafted onto carbon black surfaces. If these grafted polyesters contain a peroxide- or UV-crosslinkable functionality, then the carbon black particles to which they are attached can be incorporated in a cured matrix through covalent bonds rather than just by surface adsorption.

Table 3 *Anionic Ring Opening Copolymerization of Epoxides With Cyclic Anhydrides Initiated by COOK groups on Carbon Black Particles.*

Epoxide	Anhydride	NPNA[a] (g)	Solvent[b] (ml)	Conversion[c] (%)	Grafting (%)
GMA	PAn	0.02	1.0	46.5	52.0
GMA	MAn	0.02	1.0	80.0	105.0
GMA	SAn	0.02	-	46.3	126.2
ECH	MAn	0.02	-	44.7	44.2
ECH	SAn	-	-	10.0	40.1
SO	PAn	-	-	25.2	43.3
GC	PAn			11.3	34.6

[a] N-Phenyl-β-naphthylamine as an inhibitor of radical polymerization.
[b] Nitrobenzene.
[c] Carbon black-COOK, 0.30g; epoxide=anhydride=0.01 mol; 120°C; 1 h.

3.3 Cationic Graft Polymerization

Acylium perchlorate groups were introduced onto carbon black surfaces by the reaction of silver perchlorate with acyl chloride groups.[12] These acylium perchlorate species can initiate cationic polymerization of vinyl monomers, as well as cationic ring-opening polymerization of cyclic ethers (e.g. THF, ECH, and oxepane), cyclic acetals (e.g. trioxane), and lactones (e.g. ε-caprolactone, CL, and β-propiolactone, PL). The corresponding polymers were successfully grafted onto the surface as shown by the data in Table 4. The molecular weight of grafted polymer onto the surface was less than 8.0 x 10^3 in the cationic graft polymerization, because of preferential chain transfer reaction of growing polymer cations.

Table 4 *Cationic Ring-opening Polymerization of Several Monomers Initiated by Acylium Perchlorate Groups on Carbon Black.*

Monomer	Promoter (mol%)	Temperature (°C)	Time (h)	Conversion (%)	Grafting (%)
THF	0.1	40	48	40.7	66.6
ECH	-	40	48	10.0	42.4
Oxepane	0.1	10	120	48.1	98.6
Trioxane	-	50	10	93.6	176.5
CL	-	40	36	52.0	54.3
PL	-	70	60	18.8	28.0

Carbon black-COCl, 0.30g; AgClO₄, 0.20g: nitrobenzene, 5.0 ml; monomer, 0.08 mol. Epichlorohydrin (ECH) was used as a promoter.

Similarly, the cationic graft polymerization of vinyl monomers was also successfully achieved by benzylium perchlorate groups, which are introduced onto ultrafine silica surfaces by the reaction between benzyl chloride groups and silver perchlorate.[13]

3.4 Carboxyl Groups on Carbon Black as an Initiator of Living Cationic Polymerization

Carboxyl groups present on the carbon black surface are able to initiate the cationic polymerization of vinyl monomers, such as isobutyl vinyl ether (IBVE) and *N*-vinylcarbazole.[14] The polymerization is thought to be initiated by proton addition to the monomer; propagation proceeds with carboxylate anion on the surface as the counter anion. However, the molecular weight of polymer formed was less than 5.0×10^3 and the molecular weight distribution was very wide, because of preferential chain transfer reaction. Recently, we pointed out that the living cationic polymerization of IBVE is initiated by the system consisting of the carboxyl group on carbon black and zinc chloride, as shown in Scheme 2.[15,16] Figure 1 below shows the conversion *vs.* molecular weight and molecular weight distribution of polyIBVE. The molecular weight of polyIBVE was found to be directly proportional to monomer conversion. Furthermore, a fresh feed of monomer added to the reaction mixture was smoothly polymerized when the initial supply of the monomer was completely depleted.

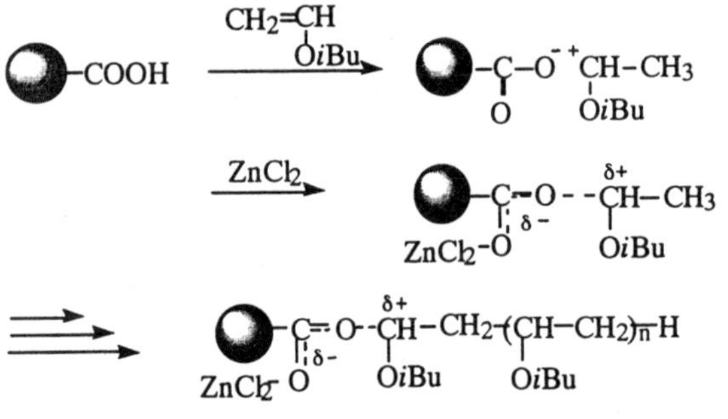

Scheme 2   Living Cationic Polymerization of IBVE Initiated by Surface Grafted Carboxyl Groups and Zn/Cl$_2$

These results clearly show that the system is able to initiate the living cationic polymerization of IBVE.

It is interesting to note that a part of polyIBVE, with controlled molecular weight and narrow molecular weight distribution, was attached to the surface through ester bonds by the termination of the living polymerization with methanol. The percentage of grafting was determined to be 28%.[15,16]

4 GRAFTING REACTION WITH SURFACE FUNCTIONAL GROUPS

Rather than initiate polymerization from the surface, one may attach a pre-formed or living polymer to the surface. The known methods for grafting of polymers onto ultrafine particle surfaces are: (1) reaction of certain functional groups on the surface with commercially available functional polymers, (2) reaction of carboxyl and hydroxyl groups on the surface with polymers having reactive groups, (3) the direct condensation of surface functional groups with functional polymers in the presence of condensing agents, and (4) the termination of living polymer with surface functional groups.

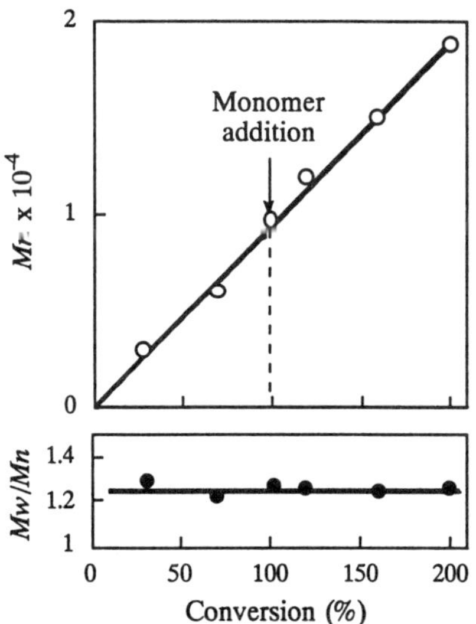

Figure 1 *Conversion vs M_n and M_w/M_n curves in the cationic polymerization of IBVE initiated by the carbon black/ZnCl₂ system.*

As an example of method (4), the grafting of polymers by trapping of polymer radicals by polycondensed aromatic rings of carbon black has proven to be very effective, and is useful for the grafting of polymers onto carbon black surfaces having few functional groups.

4.1 Reaction of Surface Functional Groups with Polymers

We have reported the preparation of reactive carbon blacks and their application for the grafting of commercially available functional polymers having terminal hydroxyl or amino groups.[17,18] For instance, reactive carbon blacks with isocyanate, epoxide, and acid anhydride groups were readily prepared by the reaction of carbon black with diisocyanate, epichlorohydrin, and trimellitic anhydride chloride, respectively.

4.2 Reaction with Polymers Having Terminal Reactive Groups

Polymers having terminal isocyanate, epoxide, or acyl chloride groups readily reacted with carboxyl and/or phenolic hydroxyl groups on carbon black surface to give the corresponding polymer-grafted carbon black.[19] In addition, carbon blacks with phenolic hydroxyl and carboxyl groups on their surface are able to cure epoxy resin[20] and to gel urethane prepolymer.[21] This gives crosslinked materials in which the carbon black is incorporated in the matrix by covalent bonds.

4.3 Direct Condensation

In the presence of *N,N'*-dicyclohexylcarbodiimide (DCC) as a condensing agent, direct condensation of carboxyl groups on carbon black with terminal hydroxyl or amino groups of polymers proceeded to give polymer-grafted carbon black.[22] For example, Figure 2 shows the relationship between the molecular weight of hydroxyl-terminated poly(dimethylsiloxane) (SDO) and the percentage of grafting or the number of grafted polymer chains (*Gn*) by this method. The percentage of grafting and the number of grafted chains decreased with increasing molecular weight of polymer. This may be due to reactive groups being shielded by neighboring grafted chains. This effect on the grafting reaction was considered to be enhanced with increasing molecular weight of the polymer.

The grafting of several polymers onto carbon black by direct condensation using the thionyl chloride/pyridine system as a condensing agent has been also reported.[23]

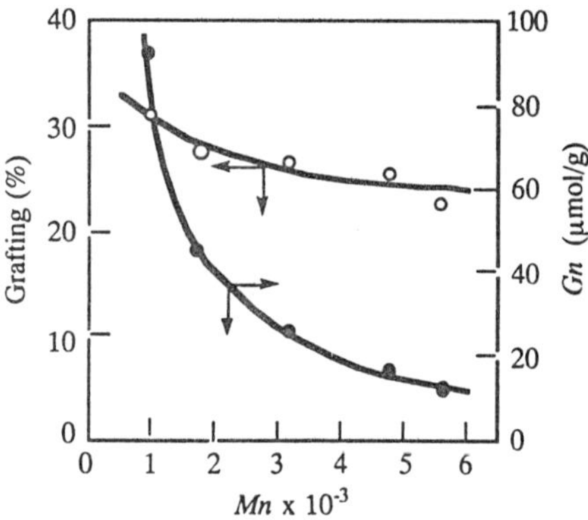

Figure 2   *Effect of molecular weight of SDO on the grafting onto carbon black in the presence of DCC.*

4.4 Reaction with Living Polymers

Donnet and coworkers have reported that living polymer anion is readily reacted with ester groups on carbon black to give polymer-grafted carbon black.[24,25] Recently, we have reported that polymers having controlled molecular weight and narrow molecular weight distribution are grafted by the termination of living polymer cation with nucleophilic groups introduced onto carbon black and silica surfaces.[26,27]

Table 5 shows the result of the grafting of polyIBVE onto carbon black and silica by the termination of living polymer cation with amino groups on the surface. It was found that polyIBVE with controlled molecular weight and narrow molecular weight distribution was successfully grafted onto the surface. In the grafting reaction, the mole number of grafted polyIBVE (*Gn*) decreased with increasing molecular weight (*Mn*) of living polymer. The relationship between *Gn* and *Mn* was found to be given by the following equation:[13]

$$Gn = 0.68Mn^{-1}$$

Table 5   *Grafting Reaction of Living Poly(IBVE) cation with Amino Groups Introduced onto Carbon Black and Silica[a].*

Particle	Poly(IBVE)[b]	Grafting (%)	Polymer grafted (mg/m^2)	R[c] (%)
Untreated	living	trace	trace	-
CB-NH$_2$	quenched	trace	trace	-
CB-NH$_2$	living	16.2	0.18	24.5
Silica-NH$_2$	living	20.0	1.0	10.0
Silica-NHPh	living	17.9	0.90	7.2

a) Silica, 0.01 g; poly(IBVE), 1.2 mmol; toluene, 10.0 mL; 25°C; 1 h.
b) *Mn* = 5.0 x 10^3, *Mw/Mn* = 1.10.
c) Percentage of amino groups used for the grafting reaction.

4.5 Trapping of Polymer Radicals by Carbon Black

4.5.1 Reaction with Azo- and Peroxide Polymers. Polycondensed aromatic rings on the carbon black surface are known to act as a strong radical trapping agents.[28] We have reported that polymer radicals formed by the decomposition of azo polymers[29] and peroxide polymers[30,31] were effectively trapped by carbon black surfaces to give polymer-grafted carbon black.

Figure 3 shows the results of the reaction of carbon black with azo-polymer at several temperatures. When the reaction was carried out below 40 °C, at which temperature the decomposition of azo-polymer was negligible, grafting of polymer was scarcely observed. This may be due to adsorption of polymer onto the surface. On the contrary, when the reaction was carried out at 70 and 100 °C, the corresponding polymer was effectively grafted onto the surface. This suggests that polymer radicals formed by the thermal decomposition of azo-polymer were effectively trapped by carbon black. Furthermore, it was found that the polymer radicals formed by the redox reaction of hydroxyl-terminated polymer, such as poly(ethylene glycol) (PEG), with ceric ion were trapped by carbon black to give the corresponding polymer-grafted carbon black.[32]

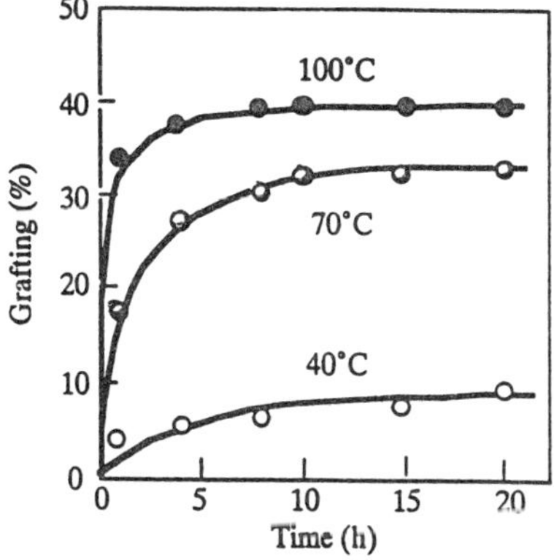

Figure 3   *Grafting reaction of azo-polymer (Mn = 9600) with carbon black. CB, 0.30 g; azo-polymer, 5.0 g; toluene, 20 mL.*

4.5.2 Reaction with Living Polymer Radical.   The grafting of polymers with controlled molecular weight and narrow molecular weight distribution onto the carbon black surface through the trapping of polymer radicals formed by the thermal dissociation of 2,2,6,6-tetramethyl-1-piperidinyloxy (TEMPO)-terminated polystyrene (PSt-TEMPO) by carbon black surface has been reported (Scheme 3).[33]

PSt-TEMPO was prepared by the living radical polymerization of St with benzoyl peroxide as an initiator in the presence of TEMPO.[34]   In the system, PSt radicals formed by the thermal dissociation of the C–ON bond between PSt and TEMPO are trapped by polycondensed aromatic rings of carbon black. The mole number of grafted PSt chain on carbon black surface decreased with increasing molecular weight of the PSt-TEMPO, because the steric hindrance of carbon black surface increases with increasing molecular weight of PSt-TEMPO.

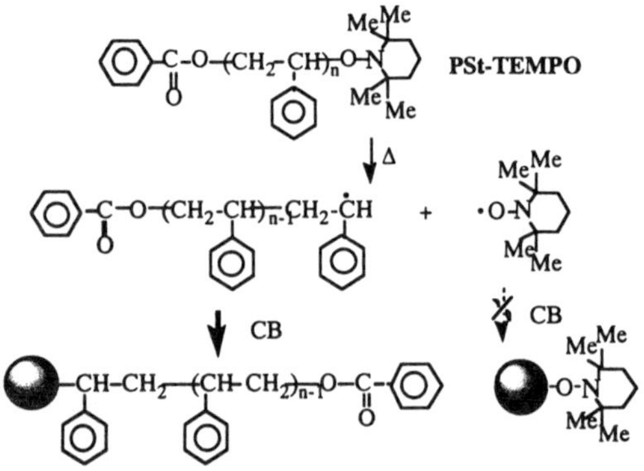

**Scheme 3 Trapping of PSt Radicals Formed by Thermal
Dissociation of PSt-TEMPO by Carbon Black**

5 GRAFTING OF BRANCHED POLYMERS

It is expected that by grafting of branched polymers onto ultrafine particle surfaces, the dispersibility and the wettability of the particle surface can be controlled by the selection of backbone and branched polymer. In this way ultrafine particles with novel functions, such as polymer compatibilizer, may be prepared.

5.1 Grafting of Branched Polymers by Postgrafting

After grafting a polymer with free, pendant carboxyl groups to a carbon black or silica surface, we converted these pendant carboxyl groups azo,[35] potassium carboxylate,[36] and acylium perchlorate groups.[37]   Then, to prepare polymer-grafted carbon black with a higher percentage of grafting, and to graft branched polymer onto these surfaces, the post-polymerization of vinyl monomers initiated by these pendant initiating groups of grafted polymer on the surfaces was examined.

Figure 4 shows the result of the grafting of branched polymer onto a silica surface by radical post-polymerization of MMA initiated by pendant azo groups of previously grafted polymer. The overall grafting of polymer was found to exceed 300%. The pendant azo groups were introduced by the reaction of pendant isocyanate groups with an azo initiator having a carboxyl group.

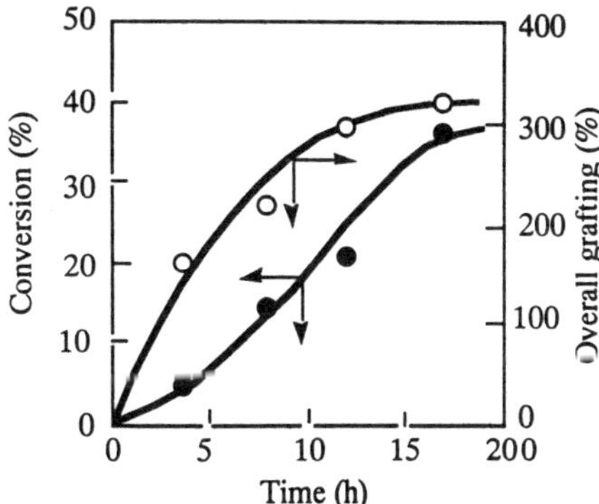

Figure 4　*Post-polymerization of MMA initiated by pendant azo groups of grafted polymer on silica.*

5.2　Grafting of Dendrimers

It was found that poly-amidoamine dendrimer was propagated from carbon black and ultrafine silica surfaces by Michael addition of methyl acrylate to surface amino groups, and amidation of the resulting esters with ethylenediamine (Scheme 4).[38]

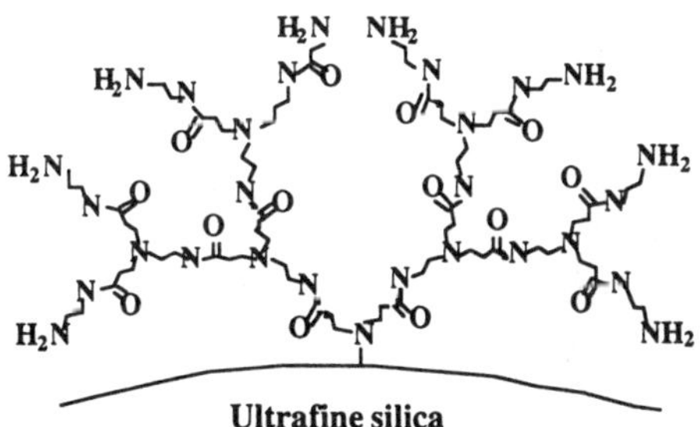

Scheme 4　Polyamidoamine Dendrimer-grafted Silica

Table 6 shows the result of grafting of polyamidoamine dendrimer onto a silica surface. The percentage of grafting increased with each generation, and reached 575.7% after the 10th generation. However, this value was considerably smaller than the theoretical value, suggesting that the propagation of dendrimer grafting from silica surface was not achieved because of steric hindrance of grafted dendrimer.

In this example of grafting of polyamidoamine dendrimer onto silica, a number of amino groups are readily introduced onto the surface. Such dendrimer-grafted silicas with amino groups act as curing agents for epoxy resins.[39]

Table 6 *Grafting of Polyamidoamine Dendrimer onto a Silica Surface.*

Initiator site (mmol/g)	Generation	Grafting (%)		Amino group (mmol/g)	
		Observed	Theoretical	Observed	Theoretical
0	10th	0	0	0	0
0.40	4th	62.3	139.8	1.0	3.2
	6th	126.2	577.5	1.6	12.8
	8th	223.6	2328.6	3.2	51.2
	10th	575.7	9332.7	8.3	204.8

6 PROPERTIES OF POLYMER-GRAFTED PARTICLES

6.1 Dispersibility Control of Polymer-grafted Particle

6.1.1 Effect of Dispersant on the Dispersibility. Figure 5 shows the comparison of dispersibility of polyacrylamide (polyAAm)-grafted carbon black in water with that of polyAAm-adsorbed carbon black and untreated carbon black. Untreated carbon black completely precipitated within 12 h. The stability of the carbon black dispersion was scarcely improved by adsorption of polyAAm. On the other hand, polyAAm-grafted carbon black gave a stable colloidal dispersion in water, which is a good solvent for grafted polyAAm. About 90% of this carbon black remained dispersed in water even after 30 days.

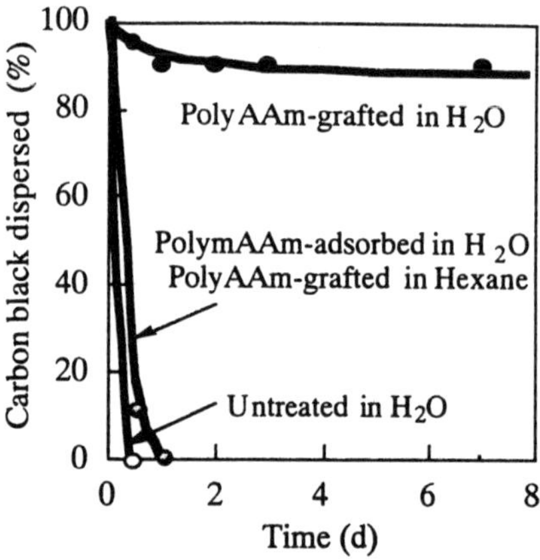

Figure 5 *Dispersibility of polyAAm-grafted carbon black in water and hexane.*

However, even the polyAAm-grafted carbon black precipitated in hexane, which is non-solvent of polyAAm. This suggests that polymer chains grafted onto carbon black lose the ability to interfere with the aggregation of carbon black when the carbon black is dispersed in a poor solvent for the grafted polymer. Accordingly, it seems that grafted polymer chains spread out from the carbon black surface in a good solvent for the grafted polymer, but shrink onto the surface in a poor solvent.

6.1.2 Dispersibility Control by Temperature. It is well known that poly(*N*-isopropylacrylamide) (poly-NIPAM) and poly(*N,N*-diethylacrylamide) (polyDEAM) are soluble in water below the lower critical solution temperature of about 32 °C, but are insoluble in water above this temperature.[40] To test whether this effect could be used to advantage in controlling the dispersibility of ultrafine particles in water, we grafted polyNIPAM to a carbon black surface.

Figure 6(a) shows the dispersibility of polyNIPAM-grafted carbon black in water. As expected, polyNIPAM-grafted carbon black gave a stable dispersion in water below 32 °C, but readily precipitated above about 32 °C.[41] We attribute this to the temperature-dependent solubility of the polymer in water. Yoshinaga *et al.* have reported that polyNIPAM-grafted spherical silica, which was prepared by the reaction of polyNIPAM with poly(styrene-co-maleic anhydride) on the surface, also shows the same tendency toward temperature-controlled dispersibility.[42]

6.1.2 Dispersibility Control by pH. Poly(acrylic acid (AC)-co-styrene(St)) grafted carbon black gave a stable colloidal dispersion in alkaline water, but immediately precipitated in acidic water as shown in Figure 6(b).[36] On the other hand, carbon black modified with amino groups, such as imidazoline groups, uniformly disperses in acidic water, but precipitates in alkaline water.[43]

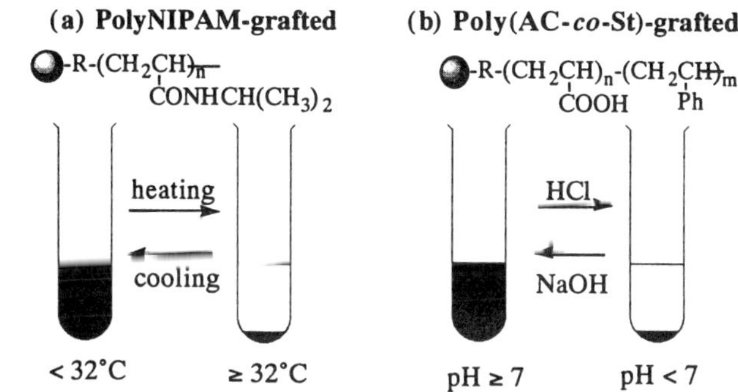

Figure 6 *Dispersibility control of carbon black in water by (a) temperature, and (b) pH.*

6.2 Electric Properties of Polymer-grafted Carbon Black

6.2.1 PTC Characteristic. A polymer-grafted carbon black/polymer composition, which is crosslinked with a variety of crosslinking agents, shows a large positive temperature coefficient (PTC) near the glass transition temperature or melting point of the matrix polymer.[44,45]

We have pointed out that the electric resistance of the composites prepared from PEG-grafted carbon black drastically increases about 10^4–10^5 times initial resistance at the melting point of PEG as shown in Figure 7.[32,46] This may be due to a widening of the gaps between carbon black particles by melting of PEG. A plane heater from the composite produces uniform heat at approximately 50 or 60 °C without any circuit for controlling temperature.

6.2.2 Sensing of Solvent Vapor. The electric resistance of a composite prepared from crystalline polyethyleneimine (PEI)-grafted carbon black drastically increased to 10^3–10^4 times initial resistance in methanol, ethanol, and water vapor, as shown in

Figure 8.[47]   However, the change of electric resistance of the composite was hardly observed in hexane and toluene vapor. In addition, the electric resistance of amorphous PEI-grafted carbon black and the mixture of crystalline PEI and untreated carbon black hardly changed in these vapors.

The resistance of poly(ε-caprolactone) (PCL)-grafted carbon black also has an ability to respond to ester vapor, and that of polyethylene (PE)-grafted carbon black has an ability to respond to butane gas and hexane vapor.

These results suggest the possibility that slight changes of crystalline structure of PEI, PCL, and PE, induced by the absorption of solvent vapor, can be detected by the large increase in electric resistance of the composite.

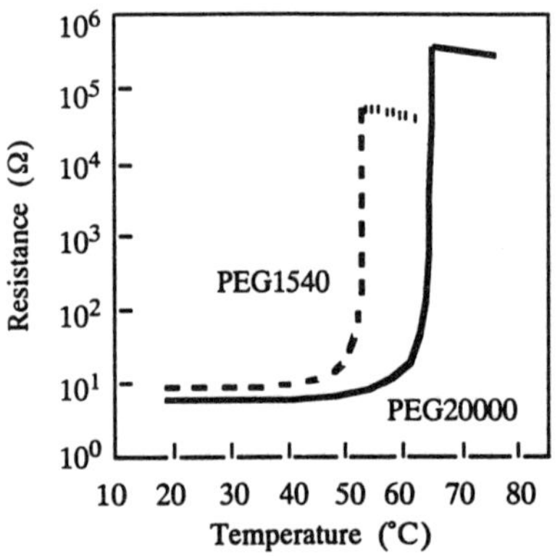

Figure 7 *Relationship between electric resistance of PEG-grafted carbon black and temperature.*

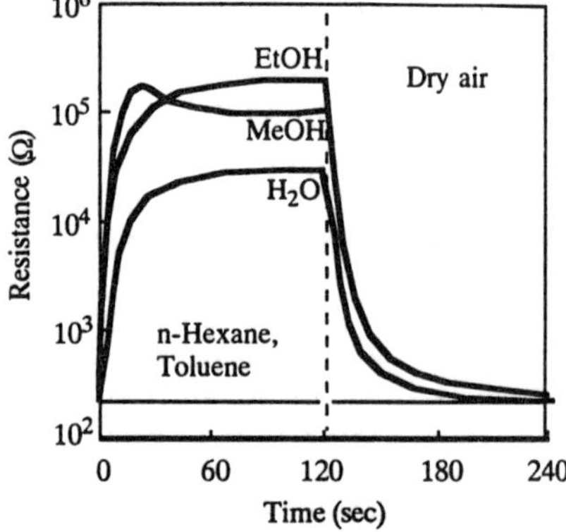

Figure 8 *Effect of various vapors on the electric resistance of crystalline PEI-grafted carbon black.*

6.4 Amphiphilic Property of Branched Polymer-grafted Particle

Pendant glycidyl groups of poly(GMA-co-MMA)-grafted carbon black were found to react with PEI, and the corresponding polymers postgrafted to the grafted copolymer chains on carbon black surface. The percentage of PEI postgrafting was readily controlled by reaction conditions.[48] Figure 9 shows the relationship between percentage of PEI-postgrafting and penetrating rate of water through the column packed with PEI-postgrafted carbon black. It was found that the penetrating rate of water increased with increasing percentage of postgrafting of PEI. The surface of carbon black was hydrophobic when PEI-postgrafting was 0–3.0%, hydrophilic when PEI-postgrafting exceeded 9.5%, and amphiphilic with a tendency to act as an emulsifier when PEI-postgrafting was 3.0–9.5%.[49]

The dispersibility of PEI-postgrafted carbon black in a binary solvent (toluene/water) was examined. Poly-(GMA-co-MMA)-grafted carbon black and PEI-postgrafted carbon black with different percentage of postgrafting were added to toluene/water binary mixture and the mixture was shaken vigorously then allowed to stand at room temperature. The results are shown in Figure 10.

Poly(GMA-co-MMA)-grafted carbon black dispersed in the toluene phase only. No carbon black was observed in the water phase, as shown in Figure 10(a), because of the extremely hydrophobic nature of the poly(GMA-co-MMA)-grafted carbon black.

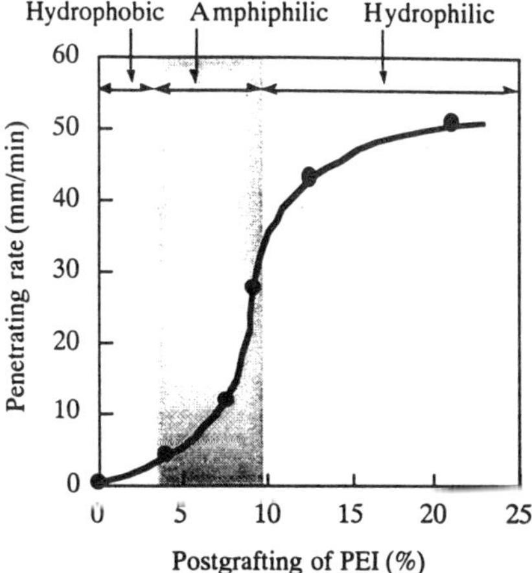

Figure 9 *Relationship between percentage of PEI postgrafting and penetrating rate of water through the column packed with PEI-postgrafted CB.*

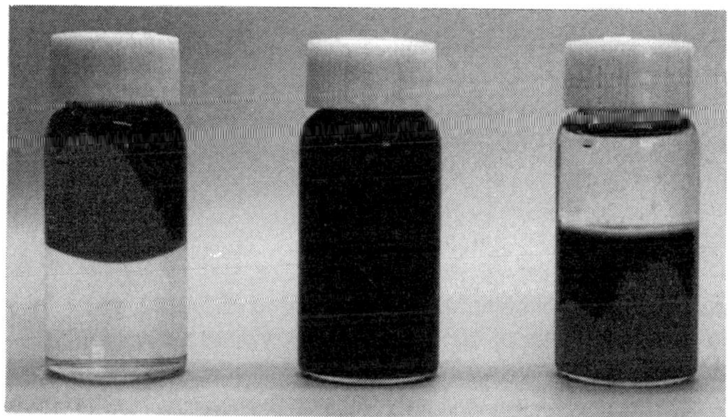

Figure 10 *Dispersibility of (a) poly(GMA-co-MMA)-grafted CB (grafting=24.3%), (b) PEI-postgrafted CB (postgrafting=3.9%), and (c) PEI-postgrafted CB (postgrafting=13.7%) in binary mixture of toluene-water.*

On the other hand, carbon black with 13.7% PEI-postgrafted gave a stable dispersion in water, but not in toluene (Figure 10(c)). It is interesting to note that carbon black with 3.9% PEI-postgrafted was suspended homogeneously in toluene/water mixture for a long time (Figure 10(b)). In this case, the solution was confirmed to be an O/W (oil in water) type-emulsion.

These results suggest that PEI-postgrafted carbon black whose percentage of postgrafting of PEI is 3.9% possesses surface activity. This may be due to the fact that PEI-postgrafted carbon black shows amphiphilic nature and has the most suitable value of the hydrophile-lipophile balance (HLB) to emulsify toluene and water.

7 CONCLUSIONS

In this paper we have given an overview of surface grafting of polymers onto carbon black and ultrafine silica. The methodologies can also be applied to the surface grafting of polymers on other particles, such as organic pigments[50] and polymer powders,[51] as well as fibers, such as carbon fiber[52] and glass fiber.[53]

The molecular design of grafted polymer chains on particle surfaces became possible only in the last ten years. However, there are only a few examples of accurate control of the polymer structure and molecular weight and number of grafted polymer. Furthermore, the characterization of grafted polymer chains on the surface has been investigated very little. These subjects are current focal points of research into the surface grafting of particles.

8 ACKNOWLEDGMENT

This work was partly supported by a Grant-in-Aid for Scientific Research (No.09650745) from the Ministry of Education, Science, Sports and Culture of Japan, which is gratefully acknowledged.

References

1. N. Tsubokawa, *Prog. Polym. Sci.*, 1992, **17**, 417.
2. N. Tsubokawa, in 'Polymeric Material Encyclopedia,' J. C. Salamone, ed., CRC Press, New York, 1996, pp. 941–946.
3. N. Tsubokawa, *Kobunshi*, 1996, **45**, 412.
4. J. B. Donnet, *Carbon*, 1982, **20**, 266.
5. K. Fujiki, N. Tsubokawa and Y. Sone, *Polym. J.*, 1990, **22**, 661.
6. N. Tsubokawa, A. Kogure, K. Maruyama, Y. Sone and M. Shimomura, *Polym. J.*, 1990, **22**, 827.
7. N. Tsubokawa and M. Ishida, *Polym. J.*, 1992, **24**, 809.
8. Y. Shirai and N. Tsubokawa, *React. Funct. Polym.*, 1997, **32**, 153.
9. K. Ohkita, N. Nakayama and T. Ohtaki, *J. Org. Coatings Technol.*, 1983, **55**, 35.
10. N. Tsubokawa, A. Yamada and Y. Sone, *Polym. Bull.*, 1983, **10**, 62.
11. N. Tsubokawa, A. Kogure and Y. Sone, *J. Polym. Sci., Part A: Polym. Chem.*, 1990, **28**, 1923.
12. N. Tsubokawa, *J. Polym. Sci., Polym. Chem. Ed.*, 1983, **21**, 705.
13. N. Tsubokawa, K. Saitoh and Y. Shirai, *Polym. Bull.*, 1995, **35**, 399.
14. N. Tsubokawa, N. Takeda and T. Iwasa, *Polym. J.*, 1981, **13**, 1093.
15. S. Yoshikawa, R. Nishizaka, K. Oyanagi and N. Tsubokawa, *J. Polym. Sci., Part A: Polym. Chem.*, 1995, **33**, 2251.
16. K. Oyanagi, S. Yoshikawa, R. Nishizaka and N. Tsubokawa, *J. Polym. Sci., Part A: Polym. Chem.*, 1997, **35**, 581.
17. N. Tsubokawa, *Nippon Kagaku Kaishi*, 1993, No.9, 1012.
18. N. Tsubokawa, A. Kuroda and Y. Sone, *Polym. J.*, 1988, **20**, 721.
19. N. Tsubokawa, M. Hosoya, K. Yanadori and Y. Sone, *J. Macromol. Sci. - Chem.*, 1990, **A27**, 445.
20. N. Tsubokawa, H. Sakamoto and Y. Sone, *Kobunshi Ronbunshu*, 1984, **41**, 597.
21. N. Tsubokawa, S. Saikawa and Y. Sone, *Kobunshi Ronbunshu*, 1983, **40**, 753.

22.　N. Tsubokawa, M. Hosoya and J. Kurumada, *React. Funct. Polym.*, 1995, **27**, 75.

23.　N. Tsubokawa and J. Kurumada, *Shikizai Kyokaishi*, 1996, **69**, 90.

24.　E. Papirer, N. Tao and J. B. Donnet, *J. Polym. Sci., Polym. Lett. Ed.*, 1971, **9**, 195.

25.　E. Papirer, N. Tao and J. B. Donnet, *Angew. Makromol. Chem.*, 1971, **19**, 65.

26.　N. Tsubokawa and S. Yoshikawa, *J. Polym. Sci., Part A: Polym. Chem.*, 1995, **33**, 581.

27.　S. Yoshikawa and N. Tsubokawa, *Polym. J.*, 1966, **28**, 317.

28.　D. Hey and G. Williams, *Discuss. Faraday Soc.*, 1953, **14**, 216.

29.　N. Tsubokawa and K. Yanadori, *Kobunshi Ronbunshu*, 1992, **49**, 865.

30.　N. Tsubokawa and S. Handa, *Shikizai Kyokaishi*, 1993, **66**, 468.

31.　S. Hayashi, S. Handa, Y. Oshibe, T. Yamamoto and N. Tsubokawa, *Polym. J.*, 1995, **27**, 623.

32.　S. Hayashi and N. Tsubokawa, *Appl. Organometallic Chem.*, 1998, **12**, 743.

33.　S. Yoshikawa, S. Machida and N. Tsubokawa, *J. Polym. Sci., Part A: Polym. Chem.*, 1998, in press.

34.　E. Yoshida and Y, Okada, *Bull. Chem. Soc., Jpn.*, 1997, **70**, 275.

35.　S. Hayashi, T. Iida and N. Tsubokawa, *J. Macromol. Sci., Pure Appl. Chem.*, 1997, **A34**, 1381.

36.　N. Tsubokawa, K. Fujiki and T. Sasaki, *Kobunshi Ronbunshu*, 1993, **50**, 235.

37.　K. Fujiki, N. Motoji, H. Tsuchida and N. Tsubokawa, *Polym. J.*, 1994, **28**, 571.

38.　N. Tsubokawa, H. Ichioka, T.Satoh, S.Hayashi, and K. Fujiki, *Reactive Functional Polym.*, 1998, **17**, 75.

39.　M. Okazaki, T. Takayama, K. Kawagichi, Y. Shirai and N. Tsubokawa, *Handbook of Extended Abstracts in the 7th International Conf. on Composite Interfaces*, 1998, p. 66.

40.　M. Heskins and G. E. Guillet, *J. Macromol. Sci., Chem.*, 1968, **A2**, 1441.

41.　N. Tsubokawa, K. Kawamura, T Kawamura and H. Fukui, *Shikizai Kyokaishi*, 1998, **71**, 156.

42.　Y. Sasao, K. Sueishi, M. Iwasaki, H. Shinkawa and K. Yoshinaga, *Polym. Preprints, Japan*, 1995, **44**, 797.

43.　M. Okazaki, K. Sugimoto and N. Tsubokawa, *Prepr. 74th Spring Annual Meeting Chem. Soc. Jpn.*, 1998, Vol. II, 1386.

44.　K. Ohkita and K. Fukushima, *Japan Plastics*, 1969, **3** (3), 6; 1969, **6** (4), 25.

45.　S. Miyauchi, Y. Sorimachi, M. Mitsui, A. Honma and K. Ohkita, *J. Appl. Polym. Sci.*, 1984, **29**, 251.

46.　N. Tsubokawa, *Nippon Gomu Kyokaishi*, 1997, **70**, 378.

47.　N. Tsubokawa, S. Yoshikawa, K. Maruyama, T. Ogasawara and K. Saitoh, *Polym. Bull.*, 1997, **39**, 217.

48.　S. Hayashi, J. Naitoh, Y. Takeuchi and N. Tsubokawa, *Nippon Gomu Kyokaishi*, 1997, **70**, 403.

49.　Y. Takeuchi, K. Fujiki and N. Tsubokawa, *Polym. Bull.*, 1997, **39**, 217.

50.　S. Yoshikawa, T. Iida and N. Tsubokawa, *Progress Org. Coatings*, 1997, **31**, 127.

51.　N. Tsubokawa and T. Oyanagi, *J. Polym. Sci., Part A: Polym. Chem.*, 1993, **31**, 1633.

52.　N. Tsubokawa, *Carbon*, 1993, **31**, 1257.

53.　N. Tsubokawa, H. Hamada and K. Fujiki, *Polymer*, 1994, **35**, 1084.

GRAFTING PROCESSES STUDIED WITH A NANOTIP: SILANE MOLECULES AND POLYMERS GRAFTED ON SILICA AND SILANIZED SILICA SURFACES

T. Bouhacina and J. P. Aimé*

CPMOH - Université de Bordeaux I
351 cours de la Libération, 33405 Talence Cedex, France

1 INTRODUCTION

In an AFM contact mode experiment, a nanotip is brought into contact with a sample surface. The nanotip is attached to the end of a cantilever and the forces acting between the nanotip and the sample will result in deflection or torsion of the cantilever (Figure 1).

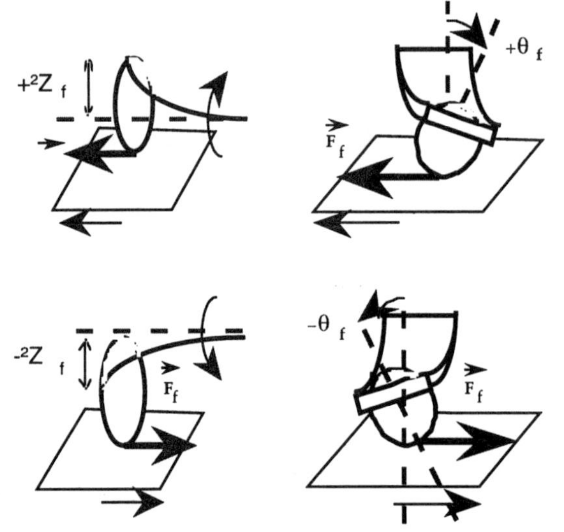

Figure 1

(a) For a scan direction along the symmetry axis of the cantilever beam, the lateral force lying in the contact plane between the tip and the surface and acting at the end of the tip gives an additional deflection either positive or negative.

(b) While a scan along a direction perpendicular to the symetry axis of the beam produces a torsion of the cantilever.

Rubbing a surface with a nanotip provides mechanical information at the nanometer scale. Thus AFM can be a powerful tool to characterize chemical treatments of a surface in a wide range of domain lengths – from the nanometer up to several tens of micrometers. The continued refinement of theory and application of AFM has made it a routinely used tool to optimize a strategy of surface grafting. Tribological investigations provide experimental data in the form of friction loops (Figure 2). Friction loops are obtained by scanning back and forth along a line (deflection or torsion). In both cases, the physical origin of the mechanical response is identical, so the choice of one direction or the other purely technical.[1-3] Note also that lateral forces arise not only from friction but also from surface roughness.[4,5]   For a curved surface, the surface normal force will have a component directed laterally. This complication is avoided by using flat samples.

To get quantitative measurements, or at least reproducible, robust, experimental data, especially when the influence of the tip velocity is investigated, requires experimental constraints described in the Experimental section and in reference 6. In this paper we discuss the observed mechanical responses of polymers grafted on silica or pretreated silica surfaces. Friction loops and force curves are presented and compared to theoretical models whenever possible.

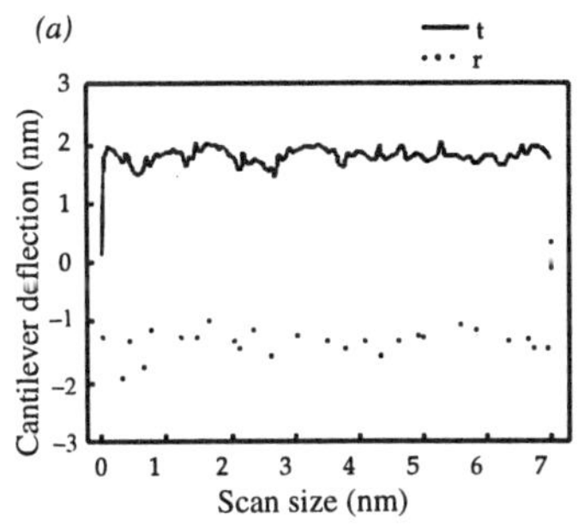

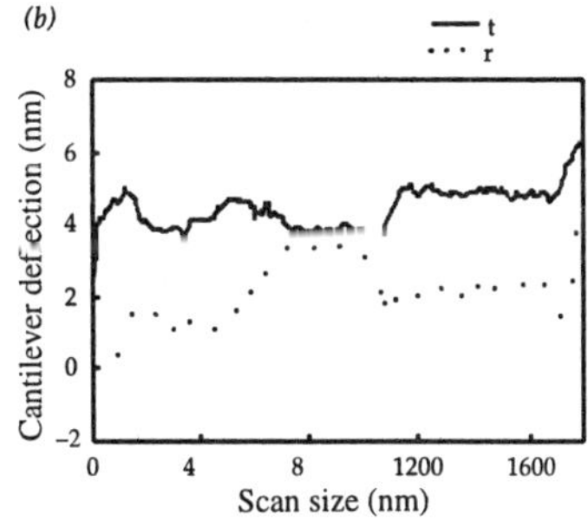

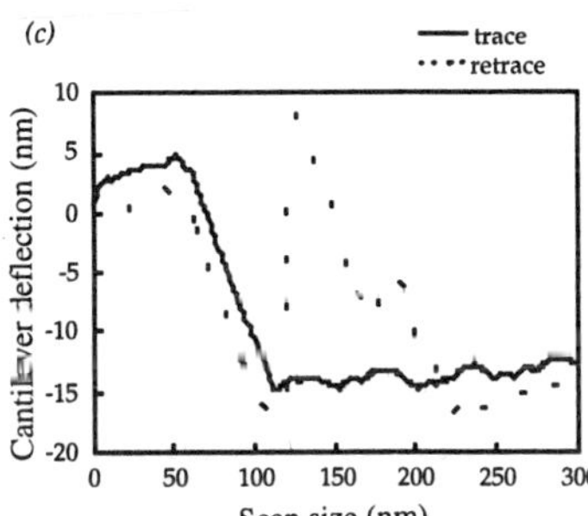

Figure 2 *Examples of friction loops:*

(a) Triethoxysilanes grafted on silica show homogeneous domains.

(b) Triethoxysilanes grafted on silica show inhomogeneous domains.

(c) PDMS (see text) grafted on silica show   unusual   and   irreproducible responses.

2 EXPERIMENTAL

All experiments were conducted in a glove box in which O_2 and H_2O levels were reduced to the PPM range. This condition seems to be necessary since the few experiments we performed in air were not reproducible enough to detect a significant variation of the friction as a function of the tip velocity.[7]

The feed back loop was never used because allowing the piezo actuator to move often resulted in erratic results. The friction loops were recorded using the so-called *"constant height mode"*. Therefore only the deflection of the cantilever was measured and not the corresponding vertical displacement of the piezo actuator. Throughout the tribological experiments, the piezo actuator was kept at rest with a null applied voltage.

2.1 Polymer Grafted on a Pretreated Silica Surface

In this approach, the silica is first treated with an organosilane coupling agent containing a reactive functional group that survives surface attachment, then it is treated with a polymer containing a functional group that is reactive towards the functional group in the organosilane. This is illustrated in cartoon form in Figure 3a. The coupling agent in this case is a triethoxysilane ended with a thiol group, $HS\text{-}CH_2\text{-}Si(OEt)_3$; the polymer is a polyacetylene with substituent side groups terminated with a chlorine atom of generic formula $[HC=C(Cl)(CH_2)_3]_n$. This polymer can be assimilated to a rod-like object with a backbone contour length of 20 nm. The chemical reaction allowing the polymer to be grafted is between the thiol group in the surface-anchored silane and the carbon bearing the chlorine atom in the polymer sidechain, as described by Ono *et al.*[8] The polymer chain can thus be fixed to the silica surface through the coupling agent in several locations along the polymer backbone.

(a)

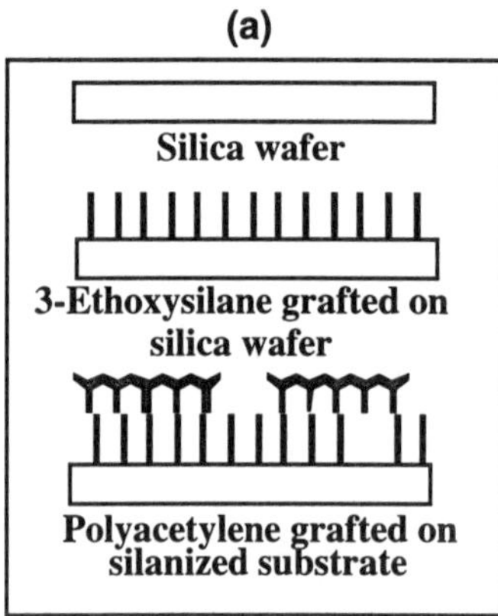

(b)

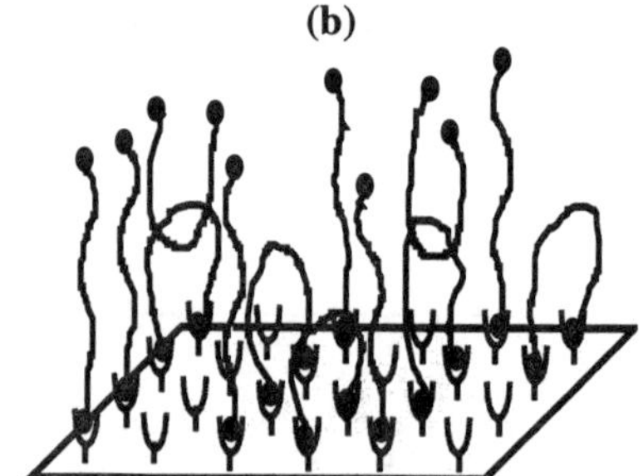

Figure 3 *Sketch of the surfaces investigated : (a) Polymer grafted on a pretreated silica surface, (b) polymer grafted directly on silica surface (PDMS).*

2.2 Polymer Grafted Directly on Silica Surface

The polymers chosen are polydimethysiloxanes (PDMS). The PDMS chains have one functional group at each terminus, and no pendant functional groups. Therefore, the chains can have only one or two points of grafting to the surface. Requiring mono-functional groups at each terminus avoids the possibility of creating a polymer network with chemical cross links. The chemical functionality is either methoxy or silanol. Three experimental situations are possible (Figure 3b): PDMS is grafted at both ends (making a loop onto the silica), the polymer is grafted at one end only, or none of the chemical functions are grafted. For the two latter cases, active ends are available, allowing the polymer to be grafted at the tip apex.

3 RESULTS AND DISCUSSION

3.1 Polymers Grafted on Pretreated Silica Surface

Friction loops were recorded on the three samples, the silica, the organosilanes and the polymer. The lateral force is given by the additional vertical cantilever deflection Δ due to the friction force:

$$F = k_t \frac{\Delta}{2} \qquad\qquad (1)$$

where k_t is the cantilever stiffness that gives the vertical deflection connected to the tangential force in the plane of contact between the tip and the sample. The cantilever stiffness, k_t, is simply related to the announced cantilever stiffness and the geometrical parameters of the lever.[1,8–10] The influence of the tip velocity on the magnitude of the lateral force was systematically investigated. Typical results are reported in Figure 4; each experimental result is an average of ten to twenty measurements.

Before discussing the experimental results it is worth noting that AFM can give a very simple qualitative picture of the efficiency of the coupling agent. For instance, rubbing a silica surface on which the polymer was simply deposited immediately shows that the adsorption is not strong enough to allow the polymer to be firmly fixed. As shown in Figure 5, even after one scan, part of the polymer was moved away. At the opposite extreme, when the polymer is grafted on a monolayer of coupling agents, several hundred scans are unable to remove the polymer and the image remains remarkably stable.

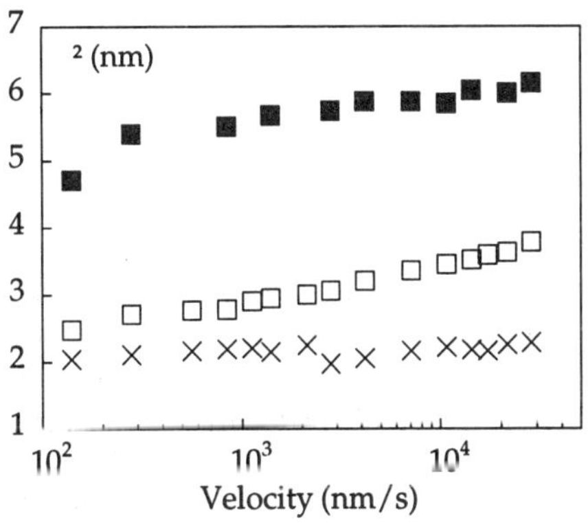

Figure 4 *Friction versus tip velocity, at zero externally applied load on the three samples: silica (crosses), organosilanes grafted on silica (open squares), and polymer grafted on organosilanes (solid squares).*

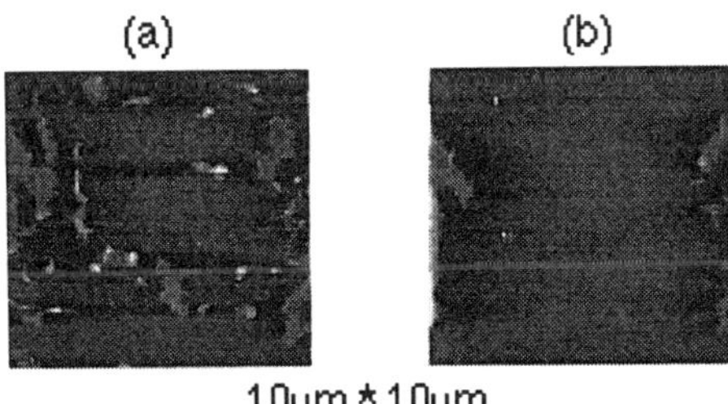

Figure 5 *Polyacetylene directly deposited on silica surface: (a) Image obtained after one scan. (b) Image obtained after ten scans on the same location, polymer is removed.*

The study of the influence of the tip velocity on the tribological behavior provides additional information with which we may obtain a better understanding of the grafting processes.[1,6] Within the velocity range investigated (100 nm/s to 40 µm/s), the friction of the silica surface does not show any variation while the grafted organosilanes and polymers grafted on the organosilane molecules show a sublinear increase. The lateral forces measured on the two grafted surfaces show a linear increase with the logarithm of the sliding velocity. Such a logarithmic dependence suggests a friction behavior described by a viscous flow under the tip.

During the sliding experiment, the interface between the nanotip and the sample experiences a shear force. The relationship between the shear stress τ, the contact area A and the measured force of friction F is given by:

$$F = \tau A \qquad (2)$$

The shear stress may become a function of the temperature, pressure and velocity.[1,9] Logarithmic velocity dependence of the shear stress can be described by an analysis based upon a simple stress-modified thermally activated Eyring model.[10] The relationship between the average velocity of the viscous flow, v, and the shear stress, τ, is given by:

$$v = 2vb \exp\left[-\frac{Q}{kT}\right]\sinh\left(\frac{\tau\phi}{kT}\right) \qquad (3)$$

Combining equations 1, 2 and 3 gives, with the approximation $\tau\phi > k_BT$, the relationship between the observed additional deflection, and the tip velocity becomes:

$$\Delta = \frac{2A}{k_t\phi}kT\ln\left(\frac{V}{V_0}\right)+\frac{2A}{k_t\phi}Q \qquad (4)$$

At room temperature $kT \approx 4.1 \times 10^{-21}$ J and k_t can be estimated. In this equation, we have three unknown parameters, the contact area A, the barrier height Q and the volume ϕ. Equation 4 is useless unless one parameter among the three is fixed. One cannot properly assume the magnitude of the contact area or the barrier height. The stress activation volume ϕ is also difficult to estimate, but there exist experimental measurements of this quantity obtained by Briscoe and Evans.[9] These authors performed their tribological experiments on organic layers with a Scanning Force Apparatus (SFA). Since in that case the area of contact was macroscopic, A was measured allowing the authors to compute the stress activation volume ϕ. A typical value found is $\phi = 5$ nm^3, and is put in equation 4 to fit the experimental data.

Examples of such fits are given in Figure 6. With a stress activation volume $\phi = 5$ nm^3, the AFM results give shear stresses ranging between 8 and 11 MPa, falling within the domain of shear stress measured at the macroscopic scale.[9] In addition, the barrier heights computed, which are independent of the activation volume chosen, are also nearly identical to the ones obtained by Briscoe and Evans. Therefore, as soon as the origin of the force of friction is due to a viscous behavior, we do observe results similar to the ones obtained at a macroscopic scale.

For the polymer grafted on the coupling agent, the calculated shear stress ranges between 17 and 23 MPa – values twice as large as the ones found for the grafted organosilane molecules. These results can be checked by performing another measurement in which the sliding velocity is kept constant and the externally applied load is varied. The externally applied load can be varied by recording friction at different vertical positions of the surface. Letting Δ_p represent the vertical position, the relationship between the friction deflection and the vertical displacement is given by:[1–3]

$$\frac{\Delta}{2} = \frac{3}{2}\frac{h}{L}\mu\Delta_p \qquad (5)$$

where μ is the friction coefficient, h is the tip height and L is the length of the cantilever.

Results are reported graphically in Figure 7, where the zero is set at the position at which the externally applied load is null. This means that the zero corresponds to an effective applied load equal to the adhesive force between the tip and the surface.

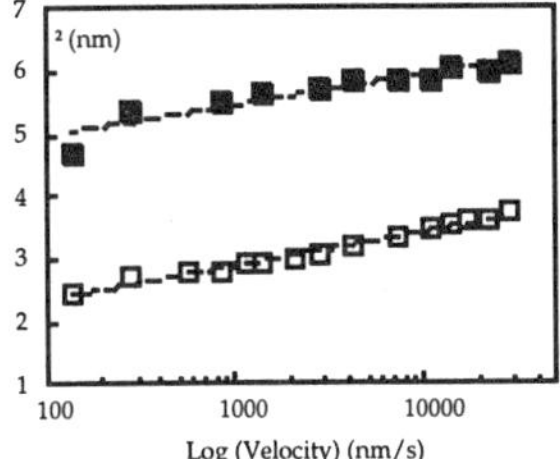

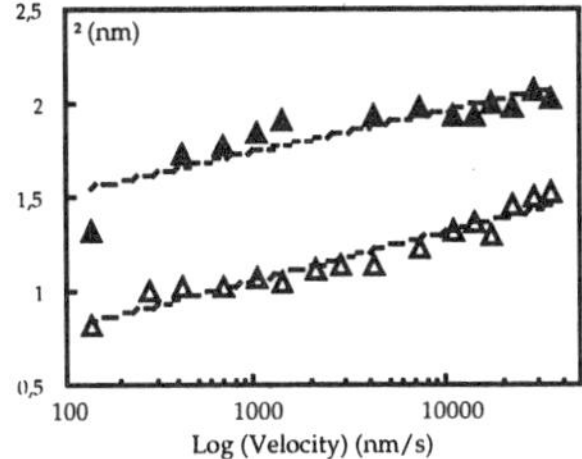

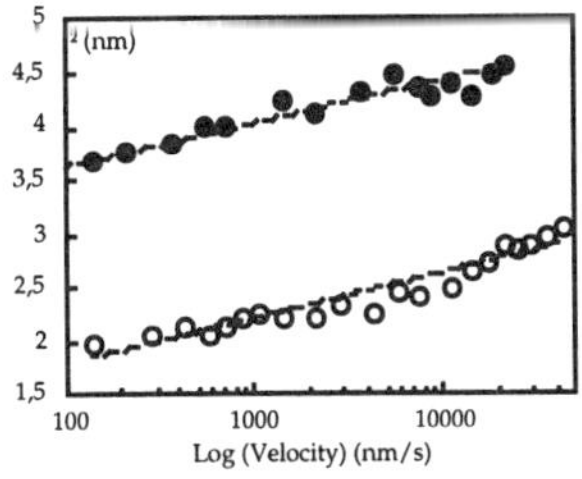

Figure 6 *Semilogarithms of fits obtained with the use of Eyring model and experimental data of friction versus tip velocity for the organosilanes grafted on silica and polymer grafted on the silanized substrate.*

Figure 7 *Frictional force versus vertical piezoelectric displacement. Organosilane (empty symbol) and polymer (filled symbol). At 0.7 μm s⁻¹, the friction coefficient of the grafted polymer surface is 0.13, while the friction coefficient of the silane film is 0.06.*

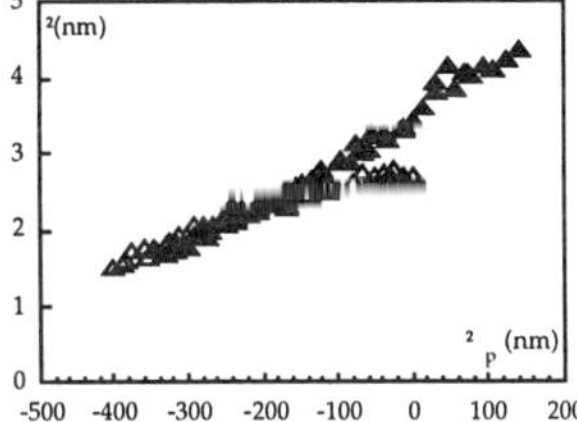

The experimental results are in good agreement with the ones giving the shear strength as a function of the sliding velocity. The slope of each line in Figure 7 gives the friction coefficient μ, which we see is twice as large for the grafted polymer surface (0.13) as for the organosilane film (0.06).

3.2 Polymers Grafted Directly onto the Silica Surface

Tribological investigations were difficult to perform on silica surfaces grafted with PDMS chains. Unusual responses with many accidents were observed, giving friction loops unlike the ones regularly observed in the preceding case (Figure 2c). This is due to the fact that the polymers are not firmly fixed at several locations, thus the elastic response is softer allowing the polymer to have a larger extension. To extract a quantitative, or even qualitative, picture from these measurements is difficult. Therefore, force curve measurements were performed to get a better insight on the effect of grafting.[11]

Force–displacement curves have been described in several papers.[3,11,12] In Figure 8 we show typical force curves obtained with PDMS of various molecular weights. They all show an unusual second elastic response with a second rupture. A sketch of the force–displacement curves is given in Figure 9 in order to explain the way the second elastic response was interpreted.

Several force curves were obtained, among which about 5–10% did show additional features like those reported in Figure 8. The second event occurs when available chemical functions are able to be grafted on the tip.

The elastic response of the neck is simulated with the tube model.[13,14] The tube model introduces a topological constraint without placing it at a specific point. A given chain is forced to have its available configurations within a contorted contour and the structure of the tube gives a parameter equivalent to the average molecular weight M_e between entangled points. The maximum of the uniaxial extension λ_{max} is given by the ratio of the tube diameter, a, and the Kuhn length, b, of the polymer: $\lambda_{max} = a/b$.[10] The entanglements are described as slip links which, if they are strong enough and act as chemical cross links, give a force in uniaxial extension.[14]

The elastic response of a neck of diameter Φ is given by :

$$f_c = N_e\,\Phi^2\,k_BT\,D\left[\frac{1-\alpha^2}{(1-\alpha^2\phi)^2} - \frac{\alpha^2}{1-\alpha^2\phi}\right] \qquad (6)$$

where N_e is the density of entangled points, $\alpha = b/a$, $\phi = \lambda^2 + 2/\lambda$, and $D = \lambda - 1/\lambda^2$. The diameter of the neck can be estimated by considering the strength of the elastic force of the network at a small λ: $f_s \approx 10^{-10}$ N for the $M_w = 27000$ and 110000 polymers, and $f_s \approx 10^{-11}$ N for the $M_w = 310000$ polymer. We have $f_s \approx \Phi 2N_ek_BT$, where k_BT is the thermal energy 4×10^{-21}J, Φ is the diameter of the sample, and N_e is the number of entanglements per unit volume.[16–18] With a polymer density ρ of *ca.* 1 g/cm^3 and a mass M_e of *ca.* 10^4, $f_s \approx 10^{-10}$ N and $f_s \approx 10^{-11}$ N give $\Phi \approx 25$ nm and $\Phi \approx 8$ nm respectively.

A small diameter of 8 nm suggests the following explanation of the particular shape of the force–displacement curve observed for $M_w = 310000$. In many cases, after the first instability a noticeable mechanical response is observed (Figure 8). This force prevents the cantilever from reaching its equilibrium position at rest after the first instability. Because the experiment is performed in air, a very bad solvent for polymer, this force is interpreted as being due to the surface tension. The effect of the surface tension becomes significant when it balances the internal pressure inside the neck and we have:

$$P\,dV = \gamma dS \qquad (7)$$

In this expression, γ is about k_BT/a_u^2, where a_u is the average size of the molecular unit,[3,14] and the pressure P is given by $k_BT/N_ea_u^2$. Taking the geometry of a cylinder, equation 7 becomes:

$$\frac{k_BT}{N_e a_u^3}\pi R^2 dL = \frac{k_BT}{a_u^2}2\pi R dL \qquad (8)$$

The number of chains n_c in the section is given by $n_c\,a_u^2 \sim \Phi^2 \sim R^2$ so that from equation 8 we obtain the order of magnitude of the number of chains/section below which the capillary effect becomes measurable:

$$\sqrt{n_c} \sim N_e \qquad (9)$$

N_e, the number of units between entanglements, is about 135 (corresponding approximately to $M_e = 10^4$ and a molar mass of the unit m = 74 g) which gives $n_c \approx 1.8$ x 10^4. This is a high value compared to the one calculated with the diameter Φ estimated to get a force of 10^{-10} N. The number of chains per section is given by:

$$n_c \sim \left(\frac{\Phi}{a_u}\right)^2 \tag{10}$$

For most of the experiments performed with $M_w = 27000$ and 110000, after the first instability the cantilever reaches a position close to the one at rest. The capillary forces

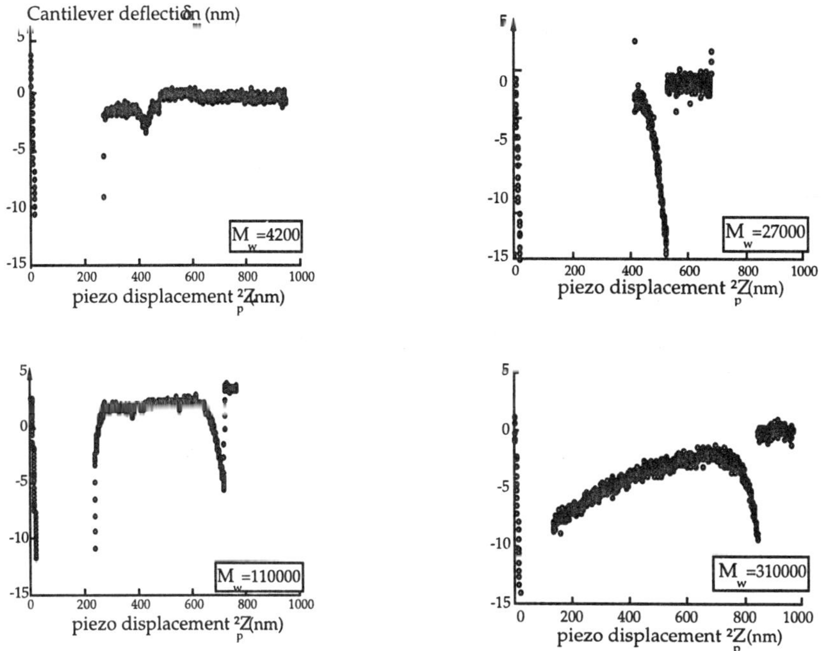

Figure 8　*Example of force–displacement curves obtained with PDMS grafted on silica. The signal is taken at the output voltage of the Nanoscope III head, then is recorded with a numerical oscilloscope. A V-shape cantilever of announced stiffness 0.58 Nm^{-1} is used. The three molecular weights, M$_w$=27000, M$_w$=110000 and M$_w$=310000, show very similar elastic responses just before the second rupture occurs. For M$_w$=4200, the molecular weight is below the average molecular weight M$_e$=17000 between entanglement points.*

can be neglected, suggesting that n_c is larger than the value given by the criterion in equation 9. The use of a unit length $a_u = 0.5$ nm gives a density $\rho \approx 1$ g/cm^3 which in turn leads to a diameter $\Phi \approx 25$ nm to reach the strength $f_s \approx 10^{-10}$ N. Using equation 10, a diameter of 25 nm gives $n_c \approx 2.5$ x 10^3. For $M_w = 310000$, f_s is more likely to be around 10^{-11} N giving $n_c \approx 2.5$ x 10^2. Therefore, the surface force becomes significant.

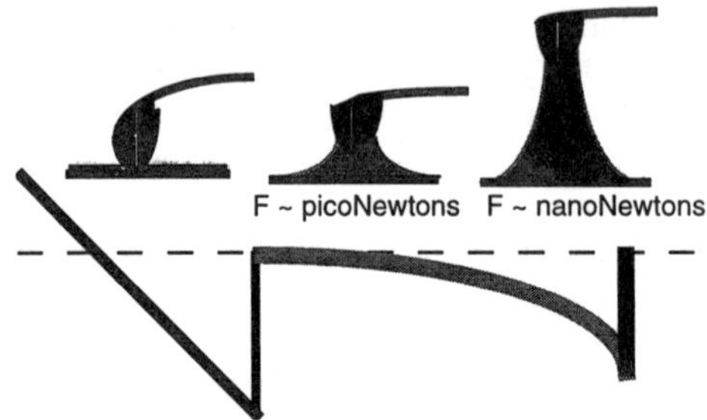

Figure 9 *After the first instability, we assume that the chains make a bridge between the tip and the silica. A network of entangled polymer chains is created behaving as a nano-rubber.*

To evaluate the way the capillary force varies as a function of the uniaxial stretching ratio λ, again we consider the simple geometry of a cylinder and use the incompressibility rule of the rubber elasticity. With $\lambda_R = R/R_0$, the incompressibility of the neck gives $\lambda_R = 1/\sqrt{\lambda}$, so that the capillary force varies as shown in equation 11. The total force of the neck, f_s, is now given by the sum of the contribution of equations 6 and 11, as in equation 12.

$$f_{cap} \sim \frac{\gamma}{\sqrt{\lambda}}\Phi_0 \tag{11}$$

$$f_s \sim f_{cap} + f_c \tag{12}$$

The theoretical variations are shown in Figure 10. Two situations have been simulated, one with a strength of f_{cap} ten times larger than the strength of f_c, and a second with a strength of f_{cap} one hundred times larger than the strength of f_c. The former aims to mimic $n_c = 10^3$, corresponding to $M_w = 110000$, while the second aims to simulate the case $n_c = 10^2$ corresponding to $M_w = 310000$. Using equation 12, good fits can be obtained providing information about the respective influence of the neck entropic elasticity and the capillary forces.[11]

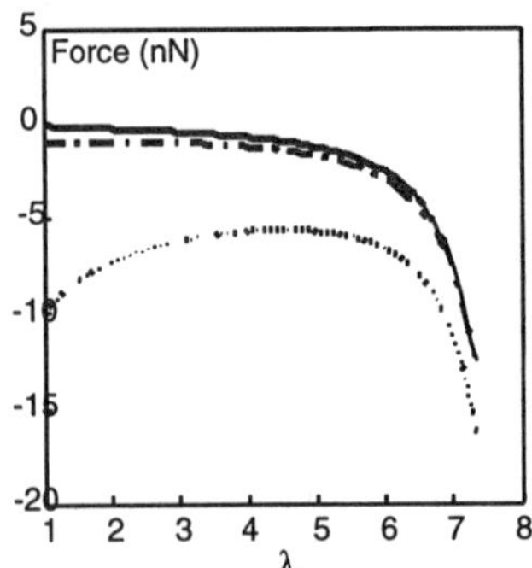

Figure 10 *Theoretical variations of the force–displacement curve taking into account the elastic response of the nanorubber and the surface force. (----) Tube model (equation 6). With a capillary force f_{cap} (equations 11 and 12) : (-- • --) the strength of f_{cap} is ten times larger than the strength of the network f_s corresponding to the case where cross section of the neck is $n_c = 1000$, (••••••) f_{cap} is one hundred times larger than f_s corresponding to the case where $n_c = 100$.*

4 CONCLUSIONS

Steady friction loops on grafted surfaces were easily obtained from which trends in the friction behavior were extracted. Whatever the variation of the lateral force, either as a function of the sliding velocity or as a function of the effective applied load, one gets results in good accordance since they all show the same amount of increase of the lateral force when the grafted polymer is investigated. The shear stress and the friction coefficient with the grafted polymers are twice as large as the ones observed with the coupling agent. In addition the barrier height (equation 4) is found to be larger when the grafted polymer is studied, meaning that the number of units involved in the dissipation process is greater with the polymer. These results allow us to conclude that the available chemical units along the backbone, one per monomer unit, are mostly active grafting the polymer in several points.

The polymer is a rod-like object making the underneath organic layer more rigid as the molecules become chemically connected. Enhanced elastic and viscoelastic response are expected, first because of a larger number of monomer units involved in the shear force, and second because of a more rigid structure. On the contrary, for polymers uniquely grafted on one or two locations, or even simply trapped in the organic layer, the structures of the grafted layer will be ill defined In the present experiments, whatever the tip location and the samples investigated, we obtained the same behavior. Thus, grafting the polymer at several points dominates all the other possibilities and in turn provides the same mechanical response during a sliding experiment.

On the other hand, when the polymer can uniquely be grafted at one or at the two ends, unusual force curves were observed corresponding to an additional elastic response of a polymer network. There exist numerous active chemical functions which are neither grafted on the silica surface nor on the tip, and which through entanglements built a bridge between the tip and the surface. The elastic response of entropic origin occurs because of a loose grafting of the polymer, thus giving another route to check the efficiency of the grafting process on the surface.

References

1.　T. Bouhacina, J. P. Aimé, S. Gauthier, D. Michel and V. Heroguez, *Phys. Rev. B*, 1997, **56**, 7694.
2.　D. Michel, S. Kopp-Marsaudon and J. P. Aimé, *Tribology Letters*, 1998, **4**, 75.
3.　R. W. Carpick, N. Agraït, D. F. Ogletree and M. Salmeron, *J. Vac. Sci. Technol. B*, 1996, **14** (2), 1289; R. W. Carpick and M. Salmeron, *Chem. Rev.*, 1997, **97**, 1163.
4.　R. J. Warmack, X. Y. Zheng, T. Thundat and D. P. Allison, *Rev. Sci. Instrum.*, 1994, **65**, 394.
5.　J. P. Aimé, Z. Elkaakour, S. Gauthier, D. Michel, T. Bouhacina and J. Curély, *Surface Science*, 1995, **329**, 149.
6.　S. Gauthier, J. P. Aimé, T. Bouhacina, A. J. Attias and B. Desbat, *Langmuir*, 1996, **12**, 5126.
7.　T. Bouhacina and J. P. Aimé, unpublished results.
8.　N. Ono, H. Miyake, T. Saito and A. Kaji, *Synth. Comm.*, 1980, **952**; Thèse de l'université Bordeaux I (1996).
9.　B. J. Briscoe and D. C. Evans, *Proc. R. Soc. London A*, 1982, **380**, 389.
10.　H. J. Eyring, *J. Chem. Phys.*, 1937, **3**, 107.
11.　J. P. Aimé and S. Gauthier, 'Scanning Probe Microscopy of Polymers', V. Tuskruk and D. Ratner, eds., American Chemical Society, Chapter 16, 1998.
12.　J. P. Aimé, Z. Elkaakour, C. Odin, T Bouhacina, D. Michel, J. Curély and A. Dautant, *J. Appl. Phys.*, 1994, **76** (2), 754.
13.　P. G. De Gennes, 'Scaling Concepts in Polymer Physics', Cornell University Press, Ithaca, NY, 1979.

14. S. F. Edwards and T. A. Vilgis, *Polymer*, 1986, **27**, 483; S. F. Edwards and T. A. Vilgis, *Rep. Prog. Phys.*, 1988, **51**, 243.

MULTILAYER DENDRIMER–POLY(ANHYDRIDE) NANOCOMPOSITE FILMS

David E. Bergbreiter,* Yuelong Liu, Merlin L. Bruening, Mingqi Zhao and Richard M. Crooks*

Department of Chemistry
Texas A&M University
P.O. Box 300012
College Station, TX 77842-3012 USA

1 INTRODUCTION

The design of functional surfaces is important in many aspects of technology. While such surfaces can be prepared by simple functionalization (*e.g.* oxidation of polyethylene or self assembly of an ω-functional thiol on gold), grafting is a time-honored method to multiply a monolayer of functional groups to produce a higher capacity, more functionalized thin film interface. Grafting processes normally involve the introduction of an initiating functional group on a surface. Subsequent polymerization from this initiator group then leads to a grafted interface (Figure 1).

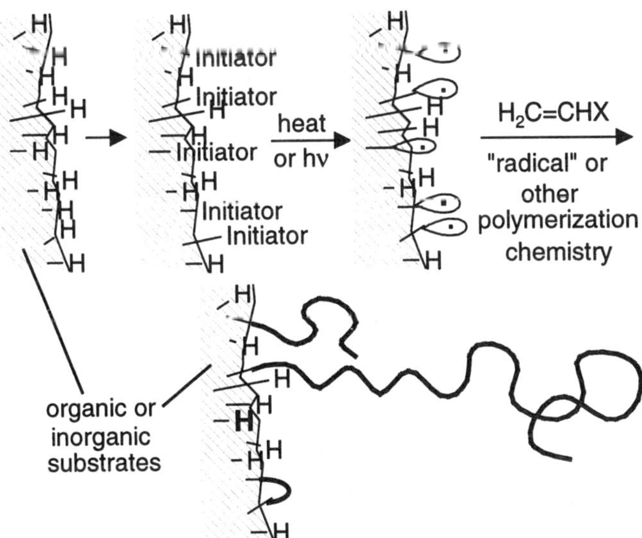

Figure 1 *Conventional linear grafting to a surface.*

While linear grafting, such as the radical grafting shown in Figure 1, is very successful in many applications, this chemistry has some intrinsic deficiencies. For example, the desired long grafts are not necessarily formed from each initiator. Coupling due to a locally high concentration of radicals, chain termination due to chain transfer or termination by other adventitious processes leads to shorter chains. Such shorter chains cannot be removed from the surface. Thus, the final surface contains failure sequences. A fully and completely grafted surface would only be formed if every step of every graft reaction proceeded in 100% yield.

There are several alternatives to conventional grafting. One such alternative is to solution cast a polymer thin film on a support. However, further modification of such films and the stability of such films in varied solvent media can be problematic. An alternative, shown in Figure 2a, is to use a polymer containing groups that can bind to the inorganic support. Thiol and phenolic groups have been used for this purpose with some success. Such 'sticky' polymers bind to the support through multiple binding sites.[1-3] The resulting polymeric thin films are thus firmly attached to the underlying support. However, it is difficult to make films that are very thick using this approach alone.

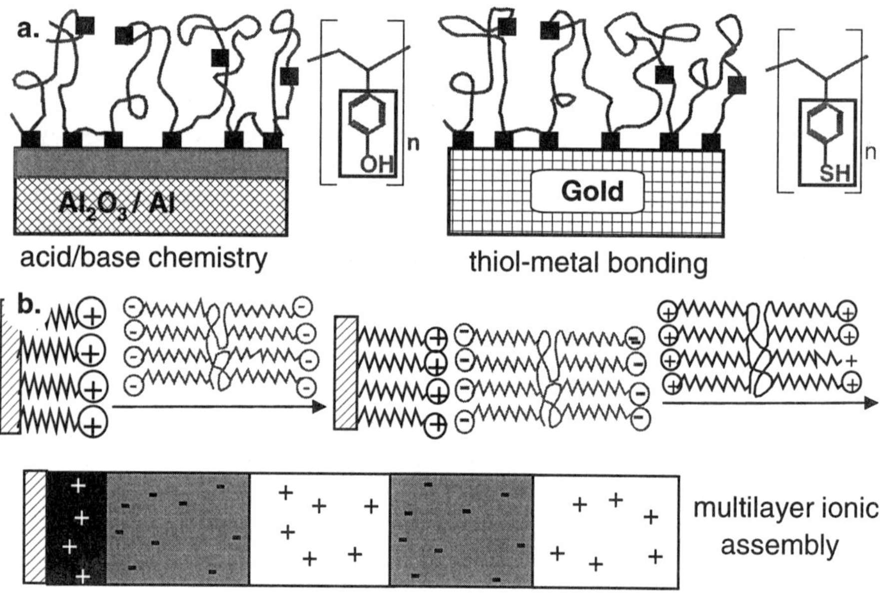

Figure 2 *Alternatives to conventional grafting: a) 'sticky' polymer deposition; b) ionic multilayer assembly.*

A very successful route to thicker films is ionic assembly where alternating layers of a cationic polyelectrolyte and an anionic polyelectrolyte are deposited in successive steps.[4] This chemistry, illustrated in Figure 2b, has been developed extensively by Decher and is now recognized as an efficient and simple route to lamellar arrays on various surfaces. It is successful with organic polymers, organic surfaces, inorganic materials and inorganic surfaces.

The problems of conventional graft chemistry were most apparent to us when we tried to prepare thin films from vinyl-terminated self assembled monolayers using borane initiated radical graft chemistry. This chemistry proceeds via oxygen initiation and the presence of oxygen both promotes and inhibits radical reactions,[5] thus exacerbating the problems of radical grafting chemistry discussed in the context of Figure 1 above.

To address the problems we encountered in grafting onto functional SAMs, we developed a new strategy to prepare functional interfaces relevant to problems in corrosion passivation, adhesion, biocompatibility and sensor synthesis. A key development of this work was the development of grafting chemistry we termed hyperbranched grafting.[6–10] This 'forgiving' synthetic process is illustrated in Figure 3. It has obvious advantages over conventional radical, anionic and cationic graft polymerization for the synthesis of stable covalent thin films on inorganic and organic substrates, and represents a general route to functional interfaces.

By grafting polyfunctional polymers like poly(*tert*-butyl acrylate) to a carboxylic acid-containing surface, a simple linear graft is first prepared. The *tert*-butyl esters are then hydrolysed under mild conditions, the newly formed carboxylic acid groups are reactivated and more poly(*tert*-butyl acrylate) is grafted to this first formed graft. Repetition of this process through several stages leads to highly branched poly(acrylic acid) grafts. The repetitive nature of the grafting and the polyfunctional nature of each graft result in complete surface coverage and films with thickness between 10 and 100 nm. The resultant thin films are easily functionalized using a variety of reagents.

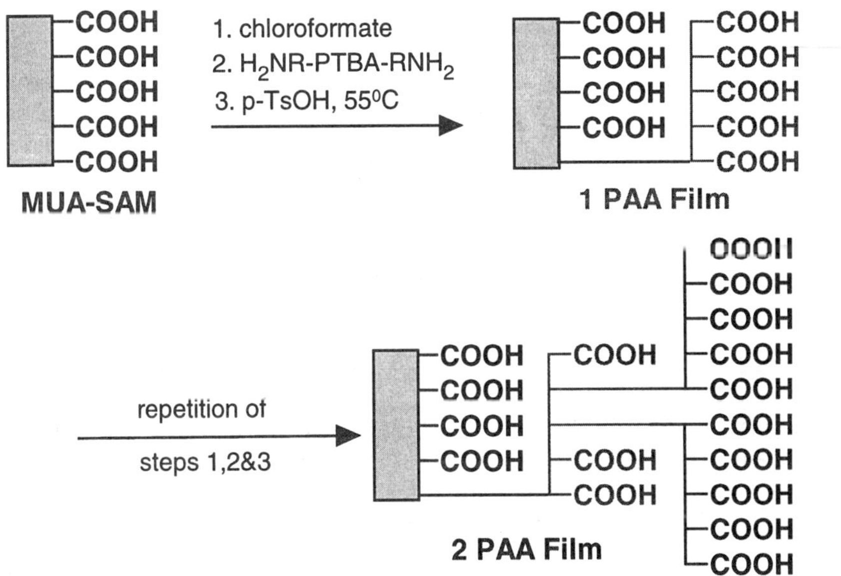

Figure 3 *Hyperbranched grafting of a two-stage poly(acrylic acid) (2 PAA) graft onto a Au-coated Si wafer. The low yields in coupling of the PTBA to the surface are compensated for by the presence of ca. 100 CO₂R groups in each PTBA oligomer.*

2 SYNTHESIS OF POLY(ANHYDRIDE)–DENDRIMER COMPOSITES

As an alternative to our original approach which used poly(*tert*-butyl acrylate), we have also explored approaches that emphasize the use of commercially available polymers. This synthesis begins with a poly(anhydride)-containing polymer (Gantrez, poly(maleic anhydride)-*c*-poly(methyl vinyl ether)). This random coil polymer serves as the glue or mortar that is subsequently covalently crosslinked by polyamine-terminated, fourth-generation PAMAM dendrimers that serve as bricks or structural elements to rapidly build up 10–60 nm thick thin films. Like the poly(*tert*-butyl acrylate) chemistry, this chemistry is forgiving, involving grafts of polyfunctional materials onto other polyfunctional materials. However, in these cases, the macroscopic size of the dendrimer in this case allows us to more rapidly build up dendrimer–Gantrez nanocomposites like that depicted in Figure 4.[11]

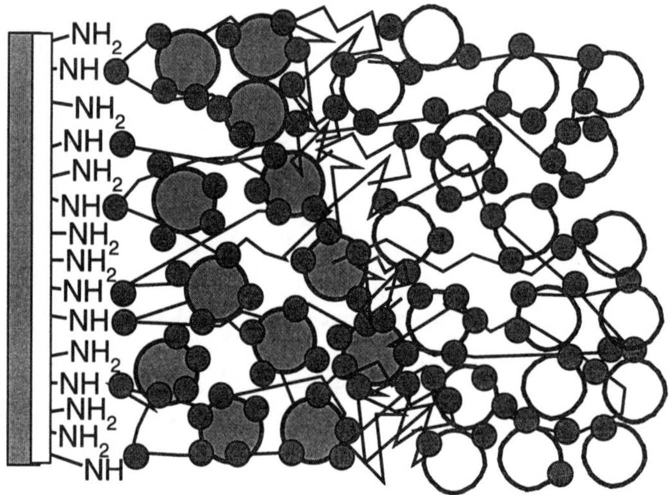

Figure 4   *Dendrimer–Gantrez nanocomposite.*

Several chemical features of the dendrimer–Gantrez composites shown in Figure 4 are noteworthy. First, the assembly is a covalent assembly with the 4.5 nm diameter dendrimer macromolecules coupled to the Gantrez matrix by amic acid groups. Second, we speculate that many and probably most of the amine groups on the dendrimer surface remain unreacted in this synthesis. Third, this nanocomposite matrix also contains many carboxylic acid groups both in the form of amic acids and in the form of succinic acids or half acid-half esters (from water or alcohol hydrolysis of the poly(anhydride) during film formation).

The nanocomposite films formed from these dendrimers and Gantrez grow in a stepwise fashion. Figure 4 illustrates a 2D (2 stages of dendrimer grafting) surface on a silicon wafer. In this case, an amine-containing surface was first prepared by allowing the native oxide layer on a silicon wafer to react with aminopropyltriethoxysilane. Subsequent treatment of this amine-containing surface with a lightly crosslinked Gantrez produces a Gz1 surface that is anhydride-rich (based on contact angles and XPS data). Exposure of this anhydride-rich surface to the dendrimer produces a D1 surface that is amine-rich based on XPS data. Typical N atom% numbers for a dendrimer rich surface are in the range of 16–17% while the N atom% drops to 1–2% in the case of an anhydride-rich surface. There is a corresponding change in the oxygen atom% from 20 to 30%, respectively. Contact angle measurements show that the dendrimer-rich surface is more hydrophilic (Θ_a values of *ca.* 31° *versus* Θ_a values of *ca.* 56° for the anhydride-rich surface). Repetition of this alternating process of anhydride reaction followed by dendrimer reaction produces successively a Gz2 (19.3 nm), a D2 (29.4 nm), a Gz3 (34.9 nm) and finally a D3 (45.9 nm) surface. Use of a Gantrez with a higher M_n (100,000 *vs.* 16,000) does not change film thickness, presumably because the amount of the 'sticky' Gantrez bound to the surface is not a function of the polymer's molecular weight. However, use of a smaller dendrimer (smaller bricks) does slightly change the thickness of the nanocomposites formed. For example, use of a G2 PAMAM dendrimer (2.9 nm in diameter) produced films whose thicknesses were: Gz1, 6.5 nm; D1, 13.8 nm; Gz2, 18.3 nm; D2, 25.3 nm; Gz3, 29.9 nm; D3, 37.8 nm; Gz4, 43.3 nm; D4, 51.8 nm.

In addition to ellipsometric and contact-angle measurements, IR spectroscopy was also used to follow film growth. Figure 5 shows external reflectance IR spectra following the growth of a composite film like that shown in Figure 4. Again, an alternation in functionality is seen with the Gz1–Gz3 films all containing anhydride peaks that are absent in the D1–D3 films. The gradual growth of amide carbonyl intensity reflects both amic acid peaks and the internal amides of the PAMAM dendrimer.

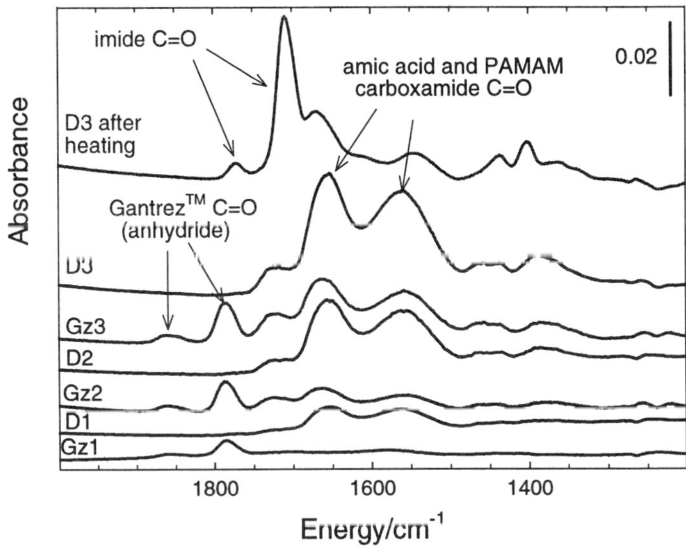

Figure 5 *IR spectra of dendrimer–Gantrez composites at various stages of thin film nanocomposite growth (Gz# = Gantrez graft stage number; D# = dendrimer stage number).*

These dendrimer–Gantrez nanocomposites are intimate covalent mixtures of carboxylic acids and amines. As a result, they are thin films that should have amphoteric characteristics. The carboxylic acid groups should be anionic at pHs greater than about 5 or 6 and the amines should be cationic at pHs less than 10. This means that these thin films could exhibit pH-dependent permeability – a supposition that was confirmed in electrochemical studies using either $Ru(NH_3)_6^{3+}$ as a cationic probe or $Fe(CN)_6^{3-}$ as an anionic electrochemical probe.[12] The results of these studies have been published and are illustrated in Figure 6. As shown here, the cationic probe is largely silent at pH 3 where the thin film has a net cationic charge. At pH 11 where the film has a net negative charge, the cationic probe is reduced and oxidized while the anionic probe is silent. Not shown is an intermediate pH (pH = 7) where both the iron ferricyanide and the ruthenium hexamine are electroactive.

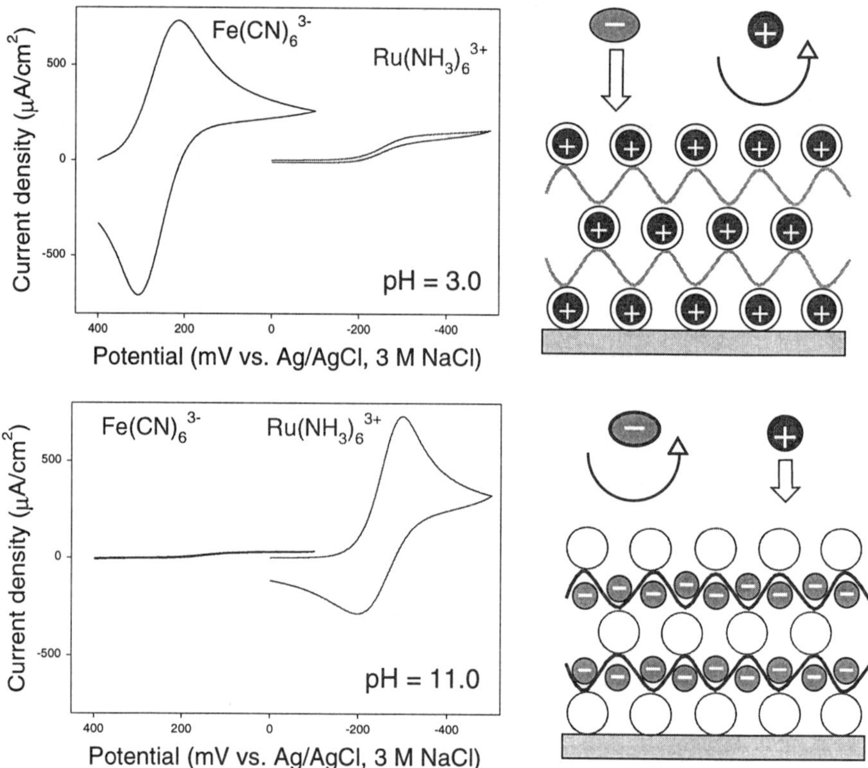

Figure 6   *pH-dependent permselectivity of dendrimer–Gantrez composites with anionic or cationic electroactive probes.*

In addition to synthesis of nanocomposites with PAMAM dendrimers, we have also successfully prepared dendrimer–Gantrez nanocomposites by successive reaction of Gantrez-rich surfaces with hydroxyl-terminated G4 PAMAM dendrimers (64 hydroxyl groups) and fourth-generation Cascade dendrimers. In the case of the hydroxyl-terminated dendrimers, the product films have dendrimers immobilized via ester groups. In the case of the Cascade dendrimers, the dendrimers are bound via amic acid groups but the interiors of the dendrimers contain no amide functionality. When nanocomposites were prepared from a MUA-SAM on gold by initial covalent coupling of the dendrimer to an activated carboxylic acid group[13] followed by repetitive treatments with Gantrez and dendrimer, the Cascade dendrimer–Gantrez and PAMAM (-(NH$_2$)$_{64}$) dendrimer–Gantrez nanocomposites were of comparable thickness (*ca.* 20% thinner than

D3 nanocomposites starting with Si wafers). Nanocomposites formed using the polyhydroxyl PAMAM dendrimer ($-OH_{64}$) were about 30% thinner than nanocomposites of the same stage formed using the amine-containing dendrimers.

All the nanocomposites prepared using the amine-containing dendrimers contain amic acids that on heating become imides. This behavior is illustrated in the IR spectra shown in Figure 5. This imidization along with additional intra- and interdendrimer chemistry changes these thin film nanocomposites into thin film monoliths on heating. The thin film monoliths so formed are somewhat thinner (about 15% shrinkage is noted by ellipsometry) and are much less permeable than the starting thin films. Electrochemical studies show that the pH-dependent permeability of the original film disappears upon heating (Figure 7). This is presumably a result of the increased crosslinking in these films and their lower polarity (θ_a (water) values change from 30 to 97°).

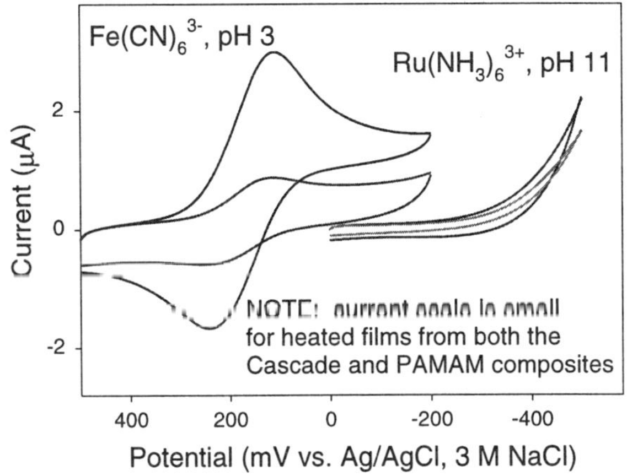

Figure 7 *Electrochemical studies of heated dendrimer–Gantrez composites show permeability is significantly reduced (note scale vs. Figure 6) and that the monolith produced by heating the PAMAM dendrimer–Gantrez thin film nanocomposite is less permeable (lower CV traces) than the thin film monolith derived from a Cascade dendrimer–Gantrez nanocomposite (upper CV traces).*

Numerous electrochemical studies have examined the effects of heating on these films. These studies indicate that dendrimer decomposition in addition to imidization is changing these films structure on heating. Dendrimer decomposition is a result of the structure of PAMAM dendrimers. As is shown in the partial structure of the dendrimer in Figure 8, these dendrimers contain β-amino carbonyl groups. Such groups are known to undergo thermally induced *retro*-Michael reactions. The resulting amines can then either react with half acid-half ester groups derived from the original anhydride (from reaction with the alcohol solvent used in the D1, D2, D*n* synthesis steps) or undergo a new Michael reaction with another acrylamide. The acrylamide itself may also polymerize. Regardless of which chemistry occurs, the result is that the dendrimers can break down into components that can then reassemble. This reassembly can involve species from different dendrimers and this process is what leads to formation of a thin film monolith.

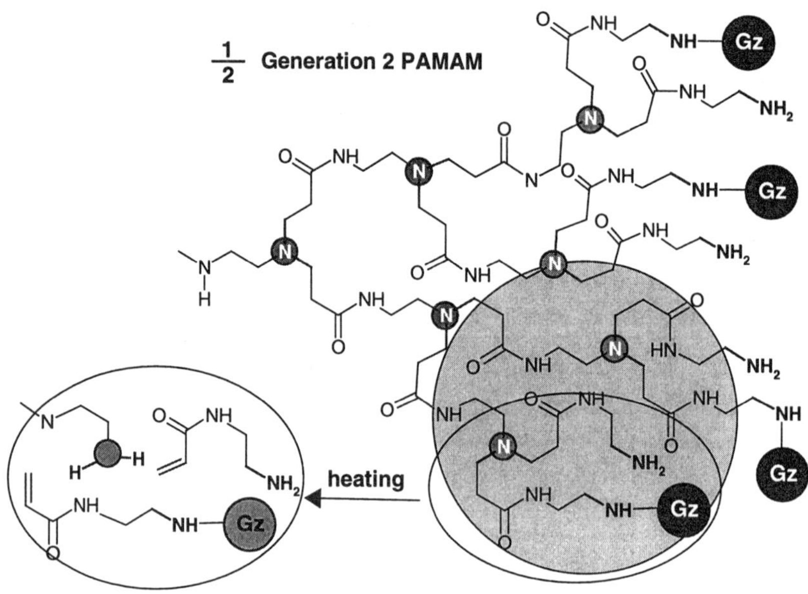

Figure 8   *Retro-Michael chemistry of a PAMAM dendrimers within a PAMAM–dendrimer nanocomposite.*

3 SUMMARY

The reaction of dendrimers with poly(anhydrides) is a convenient way to rapidly build up a functional nanocomposite on various inorganic substrates. Extensions of this work to modify aluminium have also been successful and have shown that addition of a capping hydrophobic layer (*e.g.* octadecylamine) onto the final poly(anhydride) stage in these nanocomposites leads to additional functionality in the nanocomposite. Other work in our group has shown that similar chemistry is effective on organic polymers.

4 ACKNOWLEDGMENTS

We gratefully acknowledge support of this work by the National Science Foundation (RMC, CHE-9313441 and DEB, DMR-9634196), the Robert A. Welch Foundation and the State of Texas (Texas Higher Education Coordinating Board grant number 160307) through the Advanced Technologies Program. We also thank Mr. Mark Kaiser of MMI/Dendritech (Midland, MI) for providing samples of the Starburst PAMAM dendrimers.

References

1. J. Stouffer and T. J. McCarthy, *Macromolecules*, 1988, **21**, 1204.
2. G. Mao, D. G. Castner and D. W. Grainger, *Chem. Mater.*, 1997, **9**, 1741.
3. D. E. Bergbreiter and Y. Zhou, unpublished results describing poly(4-hydroxy-styrene) modification of aluminum.
4. G. Decher, *Science*, 1997, **277**, 1232.
5. D. E. Bergbreiter, G. F. Xu and C. Zapata, *Macromolecules*, 1994, **27**, 1597.
6. Y. Zhou, M. L. Bruening, D. E. Bergbreiter, R. M. Crooks and M. Wells, *J. Am. Chem. Soc.*, 1996, **118**, 3773.
7. M. L. Bruening, Y. Zhou, G. Aguilar, R. Agee, D. E. Bergbreiter and R. M. Crooks, *Langmuir*, 1997, **13**, 770.
8. M. Zhao, Y. Zhou, M. L. Bruening, D. E. Bergbreiter and R. M. Crooks, *Langmuir*, 1997, **13**, 1388.
9. M. Zhao, M. L. Bruening, Y. Zhou, D. E. Bergbreiter and R. M. Crooks, *Isr. J. Chem.*, 1997, **37**, 277.
10. R. F. Peez, D. L. Dermody, J. G. Franchina, S. J. Jones, M. L. Bruening, D. E. Bergbreiter and R. M. Crooks, *Langmuir*, 1998, **14**, in press.
11. Y. Liu, M. L. Bruening, D. E. Bergbreiter and R. M. Crooks, *Angew. Chem. Int. Ed. Engl.*, 1997, **36**, 2114.
12. Y. Liu, M. Zhao, D. E. Bergbreiter and R. M. Crooks, *J. Am. Chem. Soc.*, 1997, **119**, 8720.
13. M. Wells and R. M. Crooks, *J. Am. Chem. Soc.*, 1996, **118**, 3988.

CONDUCTIVITY ENHANCEMENT OF POLYMER COMPOSITES THROUGH ADMICELLAR POLYMERIZATION OF PYRROLE ON PARTICULATE SURFACES

W.B. Genetti, P.M. Hunt, M. Shah, A.M. Lowe, E.A. O'Rear and B.P. Grady*

School of Chemical Engineering and Material Science
The University of Oklahoma
Norman, OK 73019 USA

1 INTRODUCTION

Ultrathin polymer films are often used to change the surface of a substrate in order to enhance properties such as adhesion or corrosion resistance. Other possible applications for materials modified with ultrathin coatings include chromatographic packings, inorganic core ion exchange resins, and substrates for immobilized hydrophobic enzymes.[1] One recently introduced mechanism to add an ultrathin film to a substrate is termed admicellar polymerization.[2] Polypyrrole (PPy) has been shown to form ultrathin conductive films on the surface of both plate and particulate substrates through admicellar polymerization.[3] This paper expands this technology by modifying fillers that are conductive, resistive, and insulating and determining the effect on the conductivity of a thermoplastic composite made with the modified filler.

Admicellar polymerization can be visualized as the two-dimensional surface analogue of emulsion polymerization with the micelle being replaced by a surfactant bilayer. There are four steps in this process: admicelle formation, monomer solubilization, initiation and polymerization, and washing. A schematic representation of the four steps is shown in Figure 1 and described briefly below.

1.1 Admicellar Polymerization

1.1.1 Admicelle Formation. A bilayer of surfactant, called an admicelle, is adsorbed on the surface of the substrate in water. The pH of the solution is adjusted so that the head groups of a surfactant are attracted and held to the substrate surface. An adsorption isotherm is a plot of the surfactant adsorption as a function of concentration in solution at a constant temperature, as shown schematically in Figure 2. In Region I, only random adsorption of surfactant molecules occurs on the surface of the substrate. As concentration of surfactant increases, more structured adsorption occurs as the heads of the surfactant are attracted to the surface and the tails draw towards each other, but uniform coverage of the surface is not attained in Region II. In Region III, a uniform bilayer of surfactant molecules has formed on the surface. In Region IV, no more surfactant adsorbs to the surface of the substrate; in many cases the transition between Regions III and IV corresponds to the critical micelle concentration (cmc) where micelles first begin to form in solution. It is important that micelles do not form in solution so as to prevent emulsion polymerization. To perform an admicellar polymerization, the surfactant concentration must be below the critical micelle concentration, but sufficiently high to favor admicelle formation on the particle surface.

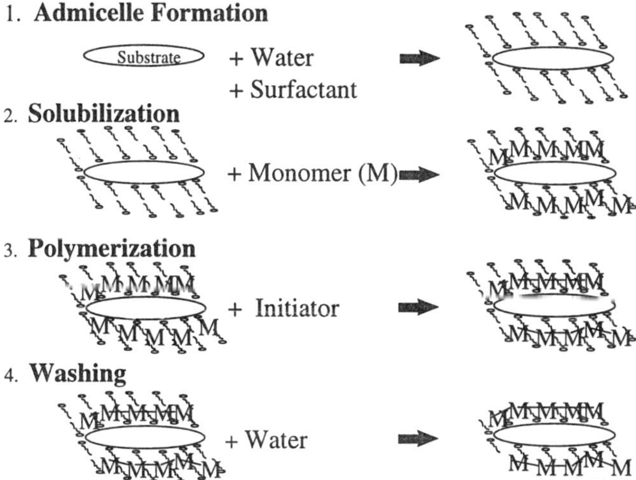

Figure 1 *Schematic diagram of the admicellar polymerization process.*

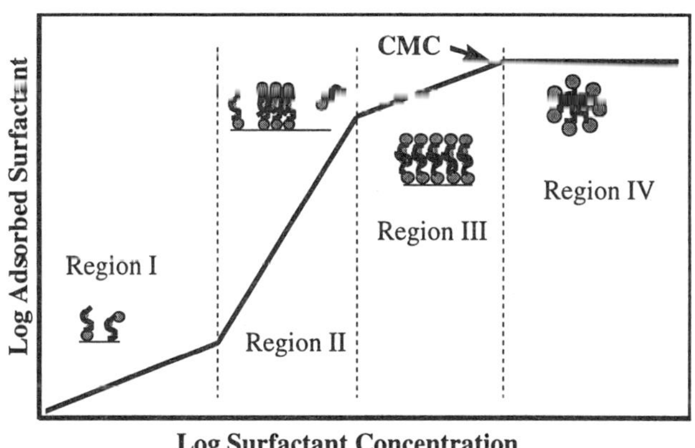

Figure 2 *Schematic diagram of an adsorption isotherm.*

1.1.2 Solubilization. After the surfactant bilayer has formed on the substrate's surface, hydrophobic monomer is added to the solution. The monomer partitions to the adsorbed bilayer.

1.1.3 Polymerization. The polymerization reaction is initiated with a water-soluble oxidizing agent.

1.1.4 Washing. After polymerization, the modified particulates are collected and washed to remove the outer layer of surfactant.

Admicellar polymerization has been performed on silica, glass fibers, nickel, and alumina with monomers such as styrene, ethylene, propylene, tetrafluoroethylene and pyrrole. The kinetics of the admicellar polymerization reaction have been studied and modeled by Wu *et al.*[4] An in-depth study of solubilization and adsolubilization was done by Funkhouser *et al.*[2] More recently, surface modified conductive nickel has been studied as a filler for conductive composites for electronic applications.[3,5]

1.2 Conductive Composites

Electrically conducting thermoplastic composites consist of a conductive filler supported by an insulating polymer matrix. In order to achieve conduction in these systems, conductive pathways of filler particles are required to allow electrons to travel freely through the material. Percolation theory quantitatively relates the electrical conductivity of the composite to the volume fraction of the filler.[6] The critical volume fraction, V_c, also called the percolation threshold, is the lowest concentration of filler that forms continuous conductive pathways throughout the polymer matrix. With increasing filler content, composite conductivity increases slowly until the critical volume fraction is reached.[7] At V_c a very sharp jump in conductivity is obtained over a very small range in concentration, referred to as the critical region. In the critical region the conductivity, σ, and concentration have a power law relationship as given by Equation 1,

$$\sigma \propto (V - V_c)^{t_p} \tag{1}$$

where V is the volume fraction.[8] Power law index, t_p, is a function of the interactions between the polymer and the matrix.[9] The critical region ends when all filler particles are involved in at least one conductive pathway and higher filler concentrations only achieve moderate changes in conductivity.

The purpose of this study was to explore how polymerization of pyrrole on the surface of a particle affects electrical conductivity in thermoplastic composites. Three different types of fillers were tested: an electrically conducting filler (nickel), a moderately conducting filler (alumina) and a non-conducting filler (glass). The supporting matrix in these studies was low-density polyethylene (LDPE).

2 EXPERIMENTAL PROCEDURES

2.1 Adsorption Isotherm

In all cases, an ionic surfactant, sodium dodecylsulfate (Aldrich) was used. The pH of the water solution was adjusted to be below the point of zero-charge for each substrate and is listed in Table 1. Adsorption isotherms for nickel flake (Sigma) and glass fibers (Owens Corning, Fiberglas Reinforcements) were measured; the adsorption isotherm for alumina (Degussa) has already been published.[2] Solutions of sodium dodecylsulfate (SDS) from 1000 to 12000 µM were prepared for the adsorption isotherm measurements and as standards for HPLC calibration. Four grams of nickel flake or two grams of glass fibers were placed in 30 mL of solution for each concentration and allowed to equilibrate for 24 h. A Shimadzu 10-A HPLC with a Waters 486 tunable UV absorption detector was used to measure the final concentration of SDS in solution. The SDS adsorbed per gram of substrate was determined from the difference in the concentration between the standard solution and the solution that contained the sample.

2.2 Admicellar Polymerization

Admicellar polymerization was performed on nickel flake, alumina, and glass fibers at the conditions specified in Table 1. Surfactant concentrations were adjusted so that polymerizations took place in Region III of the isotherm. The pH of the solution was adjusted to the desired value using hydrochloric acid, and SDS was allowed to adsorb on the surface for 24 hours. Pyrrole (Aldrich) was filtered through a packed bed of basic alumina, added to the reaction, and given 24 hours to solubilize into admicelles. Sodium persulfate (Aldrich) was used to initiate the oxidative reaction at a molar ratio of pyrrole to initiator of 1:1. The reaction was carried out for 24 hours at 25 °C to ensure completion. Particulates were collected by vacuum filtration and washed with water, then air dried for 24 hours prior to vacuum oven drying at 70 °C for an additional 24 hours.

Table 1 *Polymerization Conditions.*

Filler	SDS [μM]	Pyrrole [mM]	PH	Filler [g/L]
Nickel Flake	7500	20	4	50
Alumina	11000	76	3	30
Glass Fibers	4250	8	3	40

2.3 Composite Production

2.3.1 Solution Casting. LDPE was dissolved in 1.5 mL xylene (Aldrich) per gram of polymer at 110 °C. Nickel or Ppy-coated nickel was added and the mixture was allowed to reheat to 110 °C. After reheating, the mixture was stirred for 2 min, and then cast onto a 70 °C mercury surface. LDPE quickly formed a gel, which did not allow the filler particles to settle from solution. The composite dried under a chemical hood for 48 h. The composite film was removed from the mercury surface and allowed to dry for another 48 h at 25 °C under the hood. Residual solvent was then removed by placing the sample in a vacuum oven at 70 °C for an additional 48 h. The film thickness was measured using a micrometer and varied from 0.6 to 0.9 mm.

2.3.2 Powder Mixing. Composites of LDPE and alumina were produced by powder mixing the components for 5 min, the sample was compression molded into a 1¼ inch diameter pellet using a mold temperature of 120 °C and mold pressure of 5000 psi. The pellet was then placed in a platen press, heated to 120 °C, then compressed to the desired thickness of approximately 1 mm.

2.3.3 Extrusion. Nickel flake and glass fiber composites were produced in a Killion KL-100 Extruder single-screw extruder with one part of the screw having a back mixing section. The temperatures of three zones were 150 °C, 150 °C, 160 °C back to front and the die head temperature was 120 °C. The extrudate was sent through a pelletizer and the pellets were compressed into sheets on a platen press under the same conditions discussed in Section 2.3.2.

2.4 Conductivity Measurements

Two different geometries were used to measure DC electrical conductivity, the four-point probe and sandwich methods. The four point probe geometry was used to measure conductivities higher than 10^{-9} S/cm, while the sandwich geometry was used for lower conductivities.

2.4.1 Four Point Probe Geometry. Four copper electrodes were attached using a silver conductive epoxy (TRA-CON, BA-2902). An INSTEK Model PS-6010 Power Supply was used to induce current and a Keithley 197A multimeter was used to measure voltage. The resistance of the composite was determined from the linear portion of the current versus voltage curve and the resistivity was determined from sample geometry.

2.4.2 Sandwich Geometry. Thin composite films were placed in a Keithley Model 6105 Resistivity Chamber, which uses one pound of contact pressure and has a guard ring to bleed off surface current. A voltage was induced across the sample using a Keithley 247 High Voltage Power Source and the current was measured using a Keithley 610C Electrometer. The conductivity was calculated as in Section 2.4.1.

3 RESULTS AND DISCUSSION

3.1 Adsorption Isotherms

Figure 3 shows the adsorption isotherm for the nickel flake. Region III adsorption was approximately 17 µmol per gram of nickel flake corresponding to a final SDS concentration in solution of 6600 µM. Data in the plateau region of adsorption were quite scattered and repeated testing did not remove the scatter from the data. The same amount of scatter was not seen in either the glass fibers (Figure 4) or alumina and may be a result of non-uniform particle areas in the randomly cut flakes.

The adsorption isotherm for the glass fibers showed a Region III adsorption of approximately 8 µmol per gram of glass fibers, corresponding to a final concentration of 3900 µM in solution after adsorption. Alumina adsorbed approximately 300 µmol of SDS per gram of alumina, corresponding to a final SDS concentration of 400 µM.

Differences in adsorption are due to the difference in surface areas of the substrates. The surface area of alumina is *ca.* 100 m^2/g as a result of porosity. Nickel flake and glass fibers are non-porous and have a much smaller surface area per gram of material so much less SDS can absorb. Therefore, the polymer coating on the nickel flake and glass fibers will be entirely on the outside of the substrate and much less pyrrole is needed.

Figure 5 shows the percolation diagram for solution cast nickel and Ppy-coated nickel flakes and indicates a percolation threshold for both composites of approximately 5 percent by volume. The conductivity curves have the same shape, but above the critical volume fraction the conductivity of the Ppy-coated nickel-filled composites is 3 orders of magnitude greater (10^{-2} S/cm *vs.* 10^{-5} S/cm) than the composite made from the uncoated nickel. This increased conductivity in the composite films at a given filler loading is a result of either a decrease in the resistivity of the filler or a decrease in the particle–particle contact resistance. The resistivity of both the coated and uncoated nickel flakes was measured as approximately 1.1 ohm-cm,[5] hence the first possibility cannot explain the result.

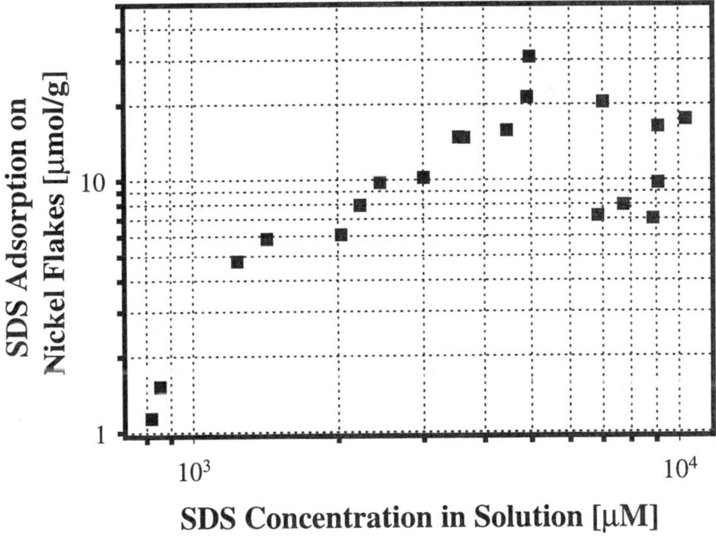

Figure 3 *Adsorption isotherm of sodium dodecyl sulfate on nickel flake.*

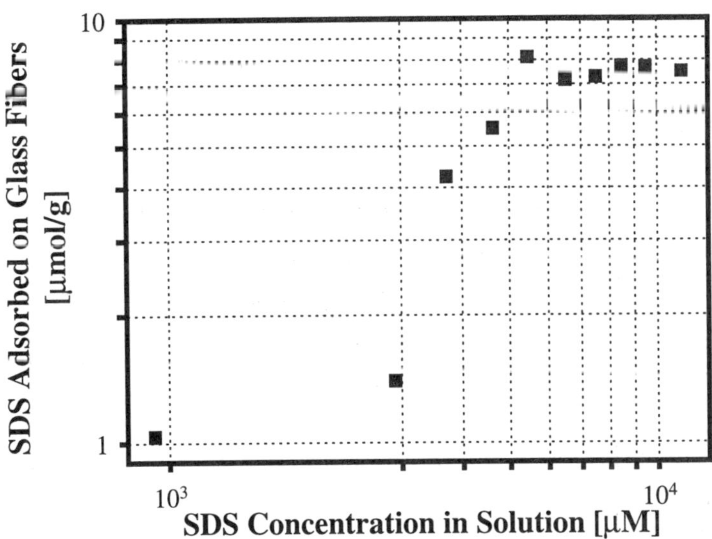

Figure 4 *Adsorption isotherm of SDS on glass fibers at a pH of 3.0.*

3.2 Composite Conductivity

We believe the increase in conductivity is due to the admicellar coating reducing the particle–particle contact resistance by forming PPy entanglements between the particles.

These entanglements increase the contact area between the nickel particles by, in effect, becoming "molecular wires" connecting the filler particles causing an increased contact area due to contacts of PPy particles outside the polymer matrix. Further, the PE matrix shrinks during solvent evaporation and/or crystallization and internal stresses place forces on the filler particles. The elastic nature of the softer Ppy-coated nickel surface may also increase the contact area versus the non-coated material.

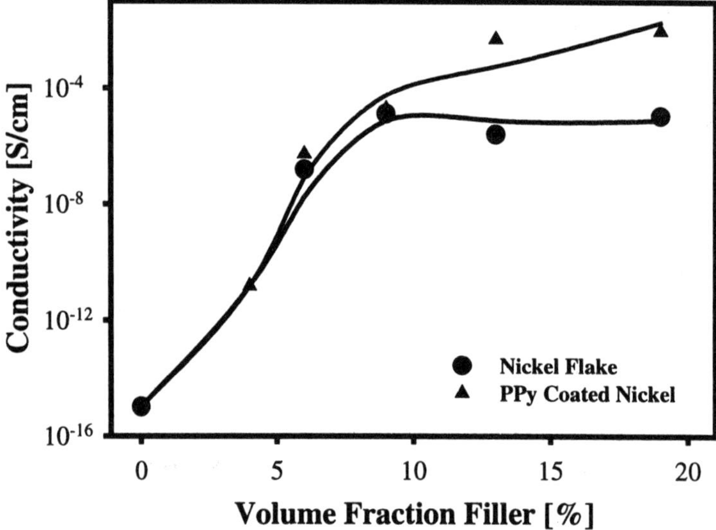

Figure 5 *Percolation diagram for nickel and Ppy-coated nickel via solution casting.*

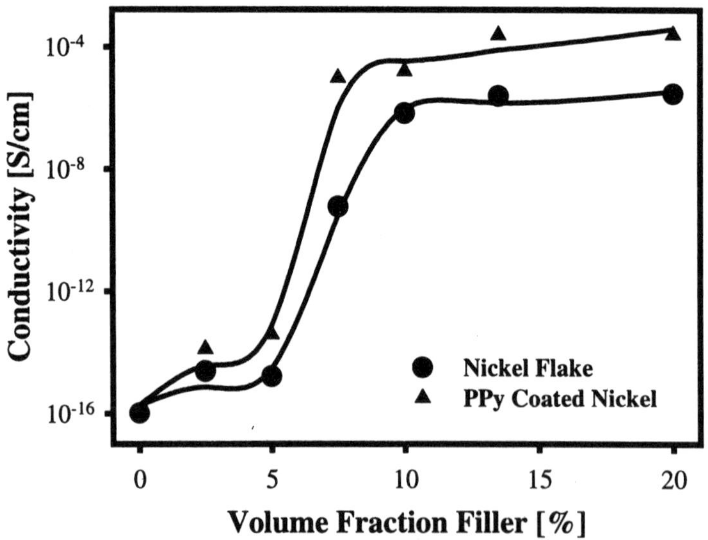

Figure 6 *Percolation diagram for nickel and PPy-coated nickel via extrusion.*

In order to determine how processing effects the conductivity enhancement of the PPy-coated nickel filler, composites of PPy-coated nickel and nickel flake were prepared by melt mixing in an extruder. The 3 order of magnitude incremental increase in composite conductivity due to the PPy coating was decreased to 2 orders of magnitude when the composites were processed by extrusion as shown in Figure 6. The high shear in the extruder may dislodge some of the PPy film or the particles may be broken apart by mechanical forces exposing more non-coated surface area.

Figure 7 shows the percolation diagram for alumina and indicates that the percolation threshold is between 10 and 15 percent by volume and was not affected by the addition of the PPy coating. The significantly higher percolation threshold of alumina compared to the nickel flakes is a result of a much lower aspect ratio of the former. The plateau conductivity for the PPy coated alumina composites was on the order of 10^{-6} S/cm, a 4 order of magnitude increase over the non-coated composites. As discussed above, the powder conductivity of the nickel flakes was not affected by the addition of the PPy coating indicating that the primary electron transfer phase was the nickel particles. The coated alumina had a powder conductivity of approximately 10^{-6} S/cm, compared with 10^{-9} S/cm for the non-coated alumina powder. This result, coupled with the shift in the percolation threshold, indicates that a significant fraction of electron transfer occurs in the PPy phase independent of the alumina.

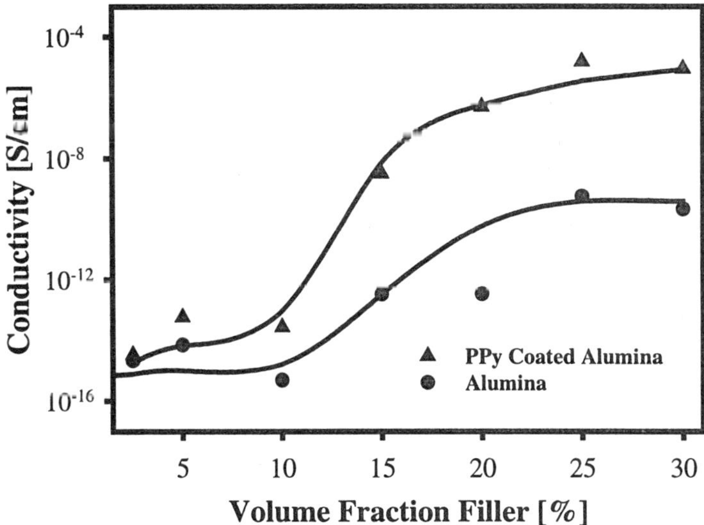

Figure 7 *Percolation diagram for alumina.*

The percolation diagram for the glass fiber composites is not presented here. Above the percolation threshold, composites made from the coated material had a one-order of magnitude higher conductivity than the composite made from the non-coated glass fiber (10^{-13} vs. 10^{-14}). We believe this indicates that even though the glass fiber is above the percolation threshold, no continuous network of PPy molecules exists in this material.

4 CONCLUSIONS

The admicellar polymerization technology has been shown as a viable method for producing ultrathin, conductive films on particulates to be used as fillers for conductive composites. The surface modifications increase the conductivity by at least an order of magnitude over the unmodified films on conductive, resistive, and insulator substrates. The limitation of this technology is the ability to get uniform coverage necessary to obtain greater increases in conductivity on non-conductive substrates.

5 ACKNOWLEDGMENTS

The authors would like to thank the National Science Foundation (cooperative agreements OST-9550478 and CTS-9521985) for providing funding for this project, and PFS Thermoplastics for providing low-density polyethylene.

References

1.   J. Wu, J. H. Harwell, E. A. O'Rear and S. C. Christian, *AICHE J.*, 1988, **34**, 1511.
2.   G. P. Funkhouser, M. P. Arevalo, D. T. Glatzhofer and E. A. O'Rear, *Langmuir*, 1995, **11**, 1443.
3.   W. B. Genetti, B. P. Grady, and E. A. O'Rear, in 'Electronic Packaging Materials Science IX', S. K. Groothuis, P. S. Ho, K. Ishida and T. Wu, eds., MRS, Pittsburgh, PA, Vol. 445, 153.
4.   J. Wu, J. H. Harwell and E. A. O'Rear, *J. Phys. Chem.*, 1987, **91**, 623.
5.   W. B. Genetti, W. L. Yuan, B. P. Grady, E. A. O'Rear, C. L. Lai and D. T. Glatzhofer, *J. Mater. Sci.*, 1998, **33**, 3085.
6.   S. Kirkpatrick, *Rev. Mod. Phys.*, 1973, **45**, 574.
7.   S. M. Aharoni, *J. Appl. Phys.*, 1972, **43**, 2463.
8.   F. Lux, *J. Mater. Sci.*, 1993, **28**, 285.
9.   E. P. Mamunya, V. V. Davidenko and E. V. Lebedev, *Polym. Composite*, 1995, **16**, 319.

FILLER SURFACE CHARACTERISATION AND ITS RELATION TO MECHANICAL PROPERTIES OF POLYMER COMPOSITES

D. Maton,[†] I. Sutherland[†] and D.L. Harrison[‡]

[†]Department of Chemistry, Loughborough University
Leicestershire LE11 3TU, UK

[‡]Zeneca Resins
Northwich, Cheshire, UK

1 INTRODUCTION

The aim of coating a filler surface is to achieve enhanced mechanical properties in the resulting composite material, and a wide variety of fillers and coatings are now in use.[1] Coatings can be reactive and act as coupling agents between filler and polymer. More often they alter the energies of interaction in the interphase region, so as to reduce agglomeration and to improve compatibility and dispersion.

1.1 Material Characterisation

In order to improve our understanding of how the filler coating improves the mechanical properties of the filled polymer composite, it is necessary to characterise the nature and extent of the coating, and relate these directly to some measurable physical property of the material under investigation. This can be done in a number of ways.

1.1.1 Surface Characterisation Techniques. The use of spectroscopic techniques, such as diffuse infra-red fourier transform spectroscopy (DRIFT) and X-ray photoelectron spectroscopy (XPS), to characterise surfaces has been well documented.[2] However, non-spectroscopic (*i.e.* wet) techniques are more commonly used to study the effect of filler coating on the energies of interaction of fillers. Such an approach using a dual BET adsorption isotherm can provide information on the extent of coating.

1.1.2 Mechanical Analysis. A number of mechanical tests exist for the testing of material properties. One of the most sensitive techniques to probe interphase interactions is dynamic mechanical thermal analysis (DMTA).

2 EXPERIMENTAL

2.1 Sample Preparation

Precipitated calcium carbonate was supplied by Zeneca Resins, with a nitrogen BET surface area of approximately 20 m^2g^{-1}. Sodium stearate (BDH purified) was used to coat the sample. The reaction was carried out with constant stirring at 80 °C for 30 min, then the coated filler was filtered and dried at 50 °C at 1 bar. The amount of stearate needed to form a close packed monolayer, in which molecules are adsorbed with alkyl chains approximately perpendicular to the surface, was estimated assuming each molecule occupied 0.2 nm^2 of the surface.[3]

2.2 Sample Analysis

2.2.1 DRIFT. Infra-red spectra were recorded on a Nicolet 20 DXC spectrometer with a Spectratech diffuse reflection attachment. The sample loading was 1% by weight in spectroscopic grade KBr with a consistent particle size of approximately 20 microns.

2.2.2. XPS. Spectra were recorded on a VG ESCALAB spectrometer using Al Kα radiation. Binding energies are referenced to adventitious carbon at 284.6 eV. Quantification was achieved by measurement of peak area after subtraction of a Shirley type background, with appropriate corrections made for photoelectron cross-sections,[4] inelastic mean free paths,[5] energy analyser transmission,[6] and angular asymmetry in photoemission,[7] when required.

2.2.3 Nitrogen Adsorption. Nitrogen adsorption isotherms were measured after outgassing using a Micromeritics ASAP 2010C Analyzer.

2.2.4 Mechanical Analysis. Polymer composite samples were produced using an EVIPOL SH6520, unplasticized poly(vinyl chloride) (uPVC). 'Ideally filled' uPVC polymer composites have been produced by the casting of a solution of the filler and the uPVC in tetrahydrofuran (THF). Two fillers have been tested, an uncoated filler and an ammonium stearate-coated filler, which has been shown to have a similar level of coating to that produced by sodium stearate. The dynamic mechanical properties of the composites were measured using a Polymer Laboratories - DMTA MK 2 analyser in the bending mode at a fixed frequency of 1 Hz.

3 RESULTS

3.1 DRIFT

Diffuse reflection spectroscopy (DRIFT) is the preferred IR technique for studying organic coatings on inorganic powders.[8-10] The experimental values obtained for the IR absorption bands of precipitated calcium carbonate agree well with the literature values for calcite.[11,12] Figure 1 shows the C–H stretching bands increasing in intensity at higher coating levels.

3.2.1 DRIFT Quantification. It is necessary to establish whether the technique can be used quantitatively for each filler coating system studied. In some instances it may be convenient to take ratios of peaks – one peak characteristic of the coating (C–H band, 3000–2800 cm^{-1}) and another peak characteristic of the filler (carbonate band, 2450–2650 cm^{-1}) concentration. In this case a simple ratio of peak areas, in either absorbance or Kubelka–Munk format, increases approximately linearly with extent of coating over the 0–1 monolayer range. Above one monolayer a slight deviation is seen.

3.2.2 DRIFT Summary. DRIFT is both surface specific and very sensitive. On 20 m^{2}g^{-1} particles, detection limits of 0.05 monolayer and less can be achieved, which is more sensitive than XPS. To determine exactly what is happening with the coating a more surface specific technique, such as XPS, is required.

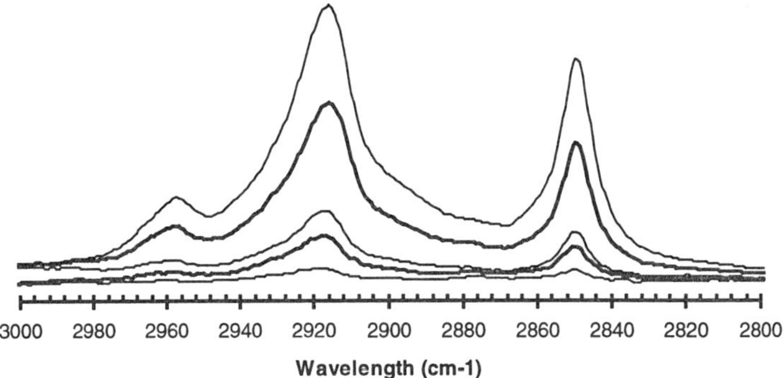

Figure 1   *Quantitative DRIFT for sodium stearate coated calcium carbonate showing the increase in C–H bands with coating level. (A) 2 Monolayers, (B) 1 Monolayer, (C) $^1/_2$ Monolayer, (D) $^1/_4$ Monolayer, (E) $^1/_8$ Monolayer.*

3.2 XPS

XPS is now a well established technique, providing quantitative analysis of the elemental composition of the outer 2–5 nm of solid surfaces. Broad scan analysis of the filler indicates that there is no sodium present at the surface of the filler, or in the coating.

3.2.1 Thickness Measurements. Measurement of the relative intensity of the photoelectron peaks due to coating ($I_{COATING}$) and substrate ($I_{SUBSTRATE}$) may be used to estimate coating thickness. If it is assumed that the coating is uniform over all the particles, the thickness can be estimated by considering the effect of coating on the attenuation of the photoelectrons emitted from the sample.[13] Previous estimates have failed to allow for a variation in attenuation length or sample geometry. Failure to allow for these factors can result in the answer being in error by a factor of two or more. Two models have been investigated: photoemission from a coated flat sample and a coated sphere. These models may be considered to be the most extreme conditions for analysis. A true representation of the sample geometry should lie somewhere between these two extremes. The XPS intensity ratio ($I_{COATING}/I_{CARBONATE}$) has been calculated by curve-fitting the high-energy resolution carbon 1s photoelectron peaks.

3.2.2 Thickness Models. The following equations have been used to estimate the layer thickness, d_{FLAT} for the flat model, assuming normal take off from the flat surface (Equation 1) and $d_{SPHERICAL}$ for the spherical model (Equation 2):

Flat model:

$$\frac{I_{COATING}}{I_{SUBSTRATE}} = \frac{\lambda_c \, n_c}{\lambda_s \, n_s} \frac{\left(1 - \exp\left(-\dfrac{d_{FLAT}}{\lambda_c}\right)\right)}{\exp\left(-\dfrac{d_{FLAT}}{\lambda_c}\right)} \tag{1}$$

Spherical model:

$$\frac{I_{COATING}}{I_{SUBSTRATE}} = \frac{n_C}{n_S} \frac{2\pi r^2}{2\pi r^2 \lambda_S} \frac{\displaystyle\int_0^{\frac{\pi}{2}} \sin\theta\cos\theta \int_0^{d} \exp- \frac{x}{\lambda_C \sin\theta}\, dx\, d\theta}{\displaystyle\int_0^{\frac{\pi}{2}} \sin^2\theta\cos\theta \left[\exp- \frac{d}{\lambda_C \sin\theta}\right] d\theta} \tag{2}$$

In these equations λ_c and λ_s are the attenuation lengths in the organic coating and the inorganic substrate respectively, and n_c and n_s are the number density of carbon atoms present in the coating and the substrate, respectively. We can take $\lambda_c = 45$ Å and $\lambda_s = 26.7$ Å from the work of Schofield.[4]   And if we assume that the carboxylate end groups of the stearate are fixed to the surface, then the orientation of the remaining alkane chains will be such that they will pack to a density similar to that of octadecane, or 777 kg m^{-3}.[14] These models can be used to estimate the coating thickness (Figure 2).

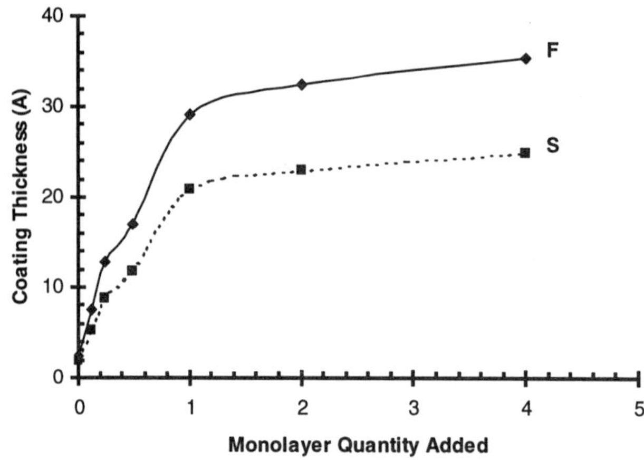

Figure 2   *The effect of sodium stearate added on the coating thickness as calculated by the flat and spherical XPS models. (F – Flat Model, S – Spherical Model).*

3.2.3 XPS Summary.   The coating thickness at one monolayer added is calculated to be 20 Å using the spherical model, and 29 Å using the flat model. These values are in good agreement with the 24.3 Å calculated[15] for the length of an 18 carbon alkyl chain extended perpendicular to the surface, which would be expected if the coating was 100% efficient and all the added coating was present in a close packed monolayer at the surface. Multilayer adsorption does not appear to occur. At higher coating levels it is probable that independent stearate particles are produced that contribute proportionally less to the alkyl carbon 1s peak than stearate molecules spread across the surface.

While the coating would appear to be uniform after sufficient material has been added to form a monolayer, at very low levels of stearate addition, there must be regions of the surface that remain uncoated. The assumptions made in calculating coating thickness from XPS at these levels will then cease to be valid. A technique that is sensitive to surface coverage is therefore required.

3.3 Nitrogen Adsorption

Conventional analysis of nitrogen adsorption isotherms by either Langmuir or BET models rely on an energetically homogeneous surface. Partially coated samples are heterogeneous in nature and as a result require an alternative approach.

3.3.1 Dual BET Adsorption Model. The energy of interaction between nitrogen molecules and an uncoated surface is higher than that for a coated surface. This results in a higher BET c constant for the uncoated surface. By fitting two BET isotherms to the data, an approximation of surface coverage can be obtained:

$$\frac{V}{V_m} = (1-L)\left[\frac{c_H p}{(p_0 - p)\left[1 + (c_H - 1)\left(\frac{p}{p_0}\right)\right]}\right] + L\left[\frac{c_L p}{(p - p_0)\left[1 + (c_L - 1)\left(\frac{p}{p_0}\right)\right]}\right] \quad (3)$$

where L is the fraction of coating and c_H and c_L are the c constants associated with the high energy uncoated filler and the low energy coated filler. Both c_H and c_L can be calculated experimentally when the adsorption isotherms of an uncoated and a fully coated filler are analysed using the BET model.

The validity of this equation has been tested using mixtures of uncoated and fully coated material. Good agreement has been obtained between the proposed model and experimental results (Figure 3).

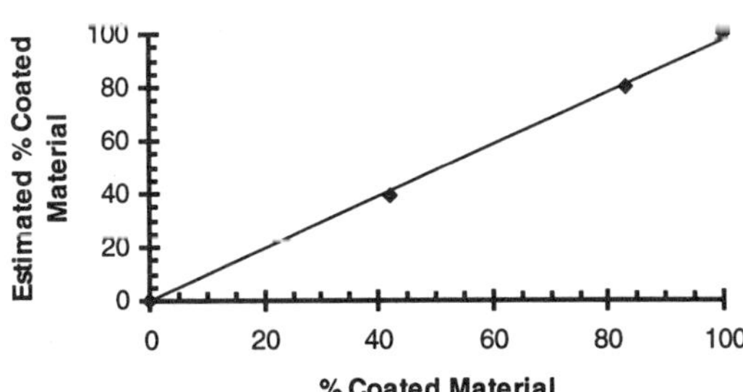

Figure 3 *Validation of the dual BET adsorption isotherm model using mixtures of coated and uncoated calcium carbonate.*

3.2.2 Nitrogen Adsorption Summary. The level of coating L is seen to increase rapidly (Figure 4), reaching maximum coverage well before the proposed close packed monolayer is formed. This would be expected if the stearate were to lay flat rather than perpendicular to the surface. A chain in the horizontal position would occupy approximately 1.1 nm^2 in comparison to 0.2 nm^2 for a vertical orientation.

These results would suggest that the coating is not complete at low levels of stearate addition, and is not completed till approximately 0.4 of a monolayer is added. This means that XPS data for coating thickness will not be valid below about 0.4 monolayer added. This may account for the apparent change in slope in Figure 2 at this level.

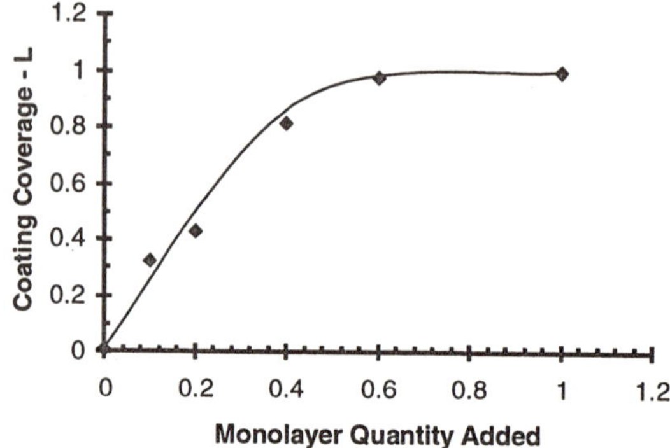

Figure 4   *The effect of sodium stearate added on the fractional coating level L measured using the Dual BET adsorption isotherm model.*

3.4 DMTA

Interactions between the filler surface and the bulk polymer lead to the formation of an interphase region with properties different from those of the bulk polymer matrix. The extent of the interphase is determined by the level of adhesion, which is in turn determined by the morphology and surface energy of the filler, and the type of polymer.

3.4.1 Low Filler Volume Fraction. Initial results for 'ideal' uPVC–PCC composites at low filler volumes, indicated very little variation on mechanical properties on coating of the PCC. From this it may be concluded that the volume of interphase at low filler volumes is so small compared with that of the bulk that it remains undetected.

3.4.2 High Filler Volume Fraction. At higher filler volumes the volume of interphase becomes sufficient to affect the bulk properties, and a change in mechanical properties can be detected (Figure 5).

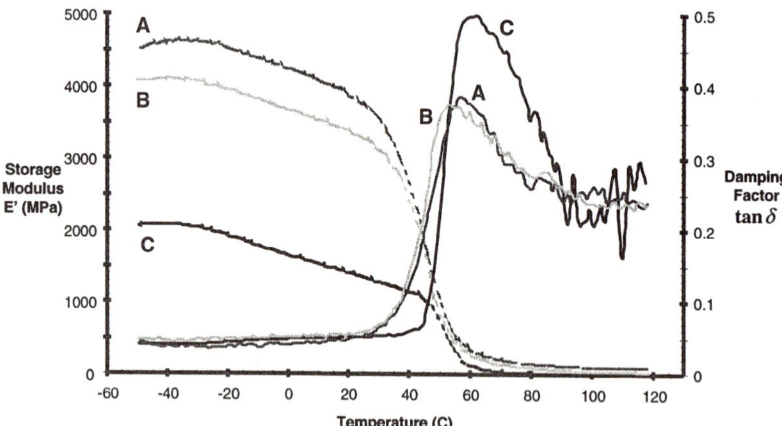

Figure 5   *DMTA results for a 50% loading by weight of ammonium stearate coated an uncoated calcium carbonate in uPVC. (A) Uncoated PCC & uPVC, (B) Coated PCC & uPVC, (C) uPVC.*

The variation in mechanical properties demonstrated in Figure 5 may be interpreted in terms of the constraints imposed by the filler on the molecular motion of the polymer. Tan δ is reduced by the presence of filler, and the decrease in the storage modulus, which has been seen to result from the coating of the filler, could be interpreted as a result of the reduction of the polymer–filler interaction. This is consistent with the concept of a 'high' surface energy uncoated filler undergoing adsorption of polymer molecules, and the formation of a rigid interphase, which in turn leads to further reinforcement. In contrast the coated filler with no 'strong' bonding across the coating–polymer interface will result in a lower level of reinforcement. Confirmation of this effect could only be determined by the use of high surface area fillers, causing the formation of a higher volume fraction of interphase

4 CONCLUSIONS

FTIR DRIFT has been successfully shown to allow the quantification of the extent of system coating. FTIR DRIFT spectra provide detailed information on the surface chemistry of the filler, while not as surface specific as XPS, FTIR has a high signal to noise ratio allowing lower levels of coatings to be detected. The surface specificity of XPS has been shown to allow the thickness of the surface coating to be estimated.

The Dual BET adsorption isotherm model has been shown to allow calculation of the extent of coverage of the coating. For a true estimate of coating thickness at low levels, a combined Dual BET adsorption isotherm model and XPS model is required to allow for the 'patchy' nature of the surface.

The effect of surface coating on the mechanical properties of uPVC filled polymer composites has been demonstrated and interpreted in terms of the reduction in the filler–polymer interaction.

5 ACKNOWLEDGEMENTS

The authors would like to thank Daniel Horner of the Chemical Engineering Department, Loughborough University, and Jim Whitcombe of Zeneca Resins for their help in the collection of the nitrogen adsorption data.

References

1. R. Rothon, 'Particulate-filled Polymer Composites', Longman Scientific and Technical, Harlow, 1995.
2. D. Maton, I. Sutherland and D. Harrison, *Composite Interfaces*, in press, 1998.
3. E. Papirer, J. Schultz and C. Turchi, *Eur. Polym. J.*, 1984, **20**, 12, 115.
4. J. H. Schofield, *J. El. Spec. Rel. Phenom.*, 1976, **8**, 129.
5. M. P. Seah and W. A. Dench, *Surf. Int. Anal.*, 1979, **1**, 2.
6. M. P. Seah, *Surf. Int. Anal.*, 1980, **2**, 222.
7. R. F. Reilman, A. Msezane and S. T. Manson, *J. El. Spec. Rel. Phenom.*, 1976, **8**, 389.
8. H. Ishida, *J. Rubber Chem. Tech.*, 1987, **60**, 497.

9.   J. D. Miller, H. Ishida, *Surf. Sci.*, 1984, **148**, 601.
10. P. Kubelka, *J. Opt. Soc. Am.*, 1948, **38**, 448.
11. W. Forsling *et al.*, *Spec. Acta*, 1994, **50 A11**, 1857.
12. W. Sterzel and E. Chorinsky, *Spec. Acta*, 1967, **24A**, 353.
13. E. Sheng and I. Sutherland, *Surf. Sci.*, 1994, **314**, 325.
14. R. C. Weast, 'Handbook of Chemistry and Physics', CRC Press, 53rd Edition, 1972.
15. J. Israelachvili, 'Intermolecular and Surface Forces', Academic Press, 1985.

GRAFTING OF CRYSTALLINE POLYMERS ONTO CARBON BLACK SURFACES AND ITS APPLICATION FOR GAS SENSORS

N. Tsubokawa,* M.Okazaki, and K.Maruyama

Department of Material Science and Technology
Faculty of Engineering
Niigata University
8050, Ikarashi 2-nocho, Niigata 950-2181, Japan

1 INTRODUCTION

We have reported the grafting of various vinyl polymers onto carbon black surfaces by the polymerization of vinyl monomers initiated by groups such as azo[1] or peroxyester,[2] potassium carboxylate,[3] and acylium perchlorate,[4] introduced onto the surface. In addition, the grafting of polymers onto carbon black surfaces has been achieved by the direct condensation of carboxyl groups on the surface with polymers having terminal hydroxyl and amino groups in the presence of condensing agents.[5,6] The polymer-grafted carbon blacks thus obtained were easily and uniformly dispersed in polymer matrices.[7–9]

It is well known that a polymer composite containing vinyl polymer-grafted carbon black, crosslinked with a variety of crosslinking agents, shows a large positive temperature coefficient of electric resistance (PTC) at or near the glass transition temperature of the matrix polymer.[10,11] That is, the electric resistance of the composite gradually increases with increasing temperature of the composite. Meyer *et al.* reported on the stability of the composite of carbon black with a crystalline polymer, such as high-density polyethylene, for use as PTC resistors.[12,13]

Recently we have achieved the grafting of crystalline polymers, such as poly(ethylene glycol) (PEG) and polyethylene adipate, onto carbon black surfaces by the direct condensation of surface carboxyl groups with terminal hydroxyl groups of these crystalline polymers using N,N'-dicyclohexylcarbodiimide (DCC) as a condensing agent. Furthermore, we pointed out that the electric resistance of the composites prepared from these crystalline polymers and these polymer-grafted carbon blacks increases some 10^4 to 10^5-fold at the melting point of crystalline polymer.[14,15]

This phenomenon at the melting point of the polymer matrix may be due to a widening of the gaps between the carbon black particles in the polymer matrix. Therefore, it is expected that the electric resistance of a composite prepared from crystalline polymer-grafted carbon black may be affected by the absorption of vapor of a good solvent for the matrix polymer, because adsorption of such a vapor may also change of the gaps between carbon black particles.

We have already reported the electric properties of crystalline polyethyleneimine (PEI(C))-grafted carbon black in various vapors.[16] In the present paper we report on the effect of crystallinity on the electric properties of PEI-grafted carbon black. We also report here our investigations into the grafting of crystalline polymers, such as PEG, polyethylene (PE), and poly(ε-caprolactone) (PCL), onto carbon black surfaces and the responsiveness of electric resistance of the composites prepared from these crystalline polymer-grafted carbon black surfaces against solvent vapor and humidity.

Table 1   *Properties of Carbon Black Used.*

Porousblack	Specific surface area (m^2/g)	Particle size (nm)	OH group (mmol/g)	COOH group (mmol/g)
Untreated	447	41	0.03	0
ACPA-treated	447	41	0.03	0.08

2 EXPERIMENTAL

2.1 Materials

The conductive carbon black used was Porousblack, a grade of furnace black obtained from Asahi Carbon Co. Ltd., Japan. The properties of the carbon black are shown in Table 1. The content of carboxyl (COOH) and phenolic hydroxyl (OH) groups was determined by use of sodium bicarbonate[17] and 2,2-diphenyl-1-picrylhydrazyl,[18] respectively. The carbon black was dried in vacuo at 110 °C for 48 h before use.

4,4'-Azobis(4-cyanopentanoic acid) (ACPA), obtained from Wako Pure Chemical Ind. Ltd., Japan was recrystallized from methanol. *N,N'*-Dicyclohexylcarbodiimide (DCC) obtained from Wako Pure Chemical Ind. Ltd., was dried *in vacuo* at room temperature. Tetrahydrofuran (THF) and other solvents were purified by ordinary methods.

2.2 Crystalline Polymers

Crystalline PEI(C) (*Mn* = 5.0 x 10^3) was prepared by the hydrolysis of poly(2-methyl-2-oxazoline).[19] Amorphous PEI (PEI(A)) (*Mn* = 1.8 x 10^3) was obtained from Nippon Shokubai Co. Ltd., Japan. PEG (*Mn* = 1.5 x 10^3), oxidized PE (*Mn* = 1.8 x 10^3), and PCL (*Mn* = 1.2 x 10^3) were obtained from Aldrich Chemical Co. These polymers were used without further purification, but were dried *in vacuo* at 80 °C before use.

2.3 Introduction of Carboxyl Groups onto Carbon Black Surfaces

The introduction of carboxyl groups onto carbon black surface was achieved by the treatment with ACPA. The detailed procedures have been described.[20]

2.4 Grafting of Crystalline Polymer onto Carbon Black Surfaces

The grafting of PEI, PEG, PE, and PCL onto carbon black surfaces was achieved by the direct condensation of carboxyl groups on the surface with terminal functional groups of these polymers, using DCC as a condensing agent. For example, 0.25 g of carbon black treated with ACPA, 1.0 g of polymer, 25 mg of DCC, and 20.0 mL of THF were charged into a flask, and the reaction mixture was stirred with a magnetic stirrer under nitrogen at 60 °C for 48 h. After the reaction, the mixture was centrifuged at 1.5 x 10^4 rpm and the supernatant solution was removed by decantation. The resulting carbon black was dispersed in a good solvent for the polymer and the dispersion was centrifuged again. The dispersion–centrifugation procedure was repeated until no more polymer could be detected in the supernatant solution. The percentage of grafting was determined by the following equation:

$$\text{Grafting } (\%) = (A/B) \times 100$$

where A is the weight of polymer grafted and B is the weight of carbon black charged. The quantity of grafted polymer, A, was estimated from the weight of carbon black after grafting reaction minus that before grafting reaction.

2.5 Preparation of a Composite Resistor from Polymer-grafted Carbon Black

The preparation of a composite resistor from crystalline polymer and crystalline polymer-grafted carbon black was carried out as follows.[5] Polymer-grafted carbon black, 0.25 g, was dispersed in a small amount of THF in a test tube, then 1.00 g of polymer was added. The mixture was stirred with a magnetic stirrer at room temperature to produce a paste of the composite. Mixing of untreated carbon black with PEI in solvent was conducted under ultrasonic irradiation rather than magnetic stirring, because untreated carbon black was much harder to disperse uniformly in the polymer matrix.

2.6 Measurement of Electric Resistance

The composite paste was coated onto a comb-like electrode, which was prepared by screen printing of conductive Ag/Pd paste onto a ceramic plate. The comb-like electrode used in this study is shown in Figure 1. The electric resistance under various vapors was measured by hanging the composite resistor in a glass vessel containing organic solvent or water in the bottom, as shown in Figure 1. The electric resistance was measured at 25 °C by use of a digital multimeter (Advantest Co. Ltd., Japan; type R6871E-DC).

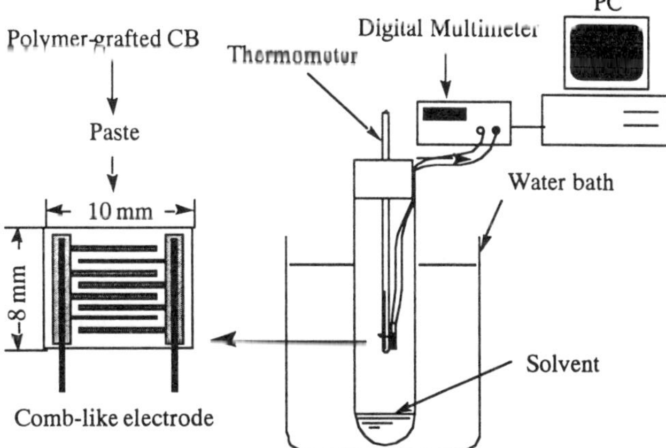

Figure 1   *Comb-like electrode and apparatus for the measurement of electric resistance.*

3 RESULTS AND DISCUSSION

3.1 Grafting of Crystalline Polymer onto Carbon Black Surface

The grafting of PEI(C), PEI(A), PEG, PCL and PE onto carbon black surfaces by direct condensation of surface carboxyl groups with amino or imino groups of these polymers in the presence of DCC was examined (Scheme 1).

$$\bigcirc\text{-COOH} \quad + \quad \text{HO-(Crystalline Polymer)-OH}$$

$$\xrightarrow{\text{DCC}} \quad \bigcirc\text{-COO-(Crystalline Polymer)-OH}$$

Scheme 1 Grafting of crystalline polymer onto carbon black by direct condensation

In general, conductive carbon blacks have no carboxyl groups. Therefore, the introduction of carboxyl groups onto carbon black was necessary, and was achieved through the trapping of 4-cyanopentanoic acid radicals formed by the decomposition of ACPA by the carbon black surface.[20] The carboxyl group content of ACPA-treated carbon black was determined to be 0.08 mmol/g by titration as shown above in Table 1.

In Table 2 we show the results of grafting of PEI, PEG, PE, and PCL onto carbon black surfaces. In the absence of DCC, the grafting of these polymers onto carbon black surfaces scarcely takes place. On the contrary, these polymers were successfully grafted in the presence of DCC as a condensing agent.

Table 2   *Grafting of Crystalline Polymers onto Carbon Black by DCC-induced Condensation with Surface Carboxyl Groups.*

Grafted polymer	$\dfrac{\overline{\text{Mn}}}{10^3}$	Grafting (%)	
		Non-catalyst	DCC
PEI (C)	5.0	trace	7.1
PEG	1.5	trace	11.8
PE	1.3	trace	15.9
PCL	1.2	trace	9.4

Carbon black, 0.25 g; polymer, 1.0 g; DCC, 25 mg; THF, 20 mL; 60°C; 48 h.

Table 3   *Electric Resistance of Polymer-grafted Carbon Black in Dry Air and MeOH Vapor.*

Grafted polymer	$\dfrac{\overline{\text{Mn}}}{10^3}$	Resistance (Ω)	
		dry	MeOH vapor
Ungrafted	5.0	1.5M	700k
PEI (C)	5.0	200	100k
PEI (A)	1.8	4.2M	1.5M

Carbon black, 0.25 g; polymer, 1.0 g; DCC, 25 mg; 60 °C; 48 h.

No gelation was observed during the above grafting reaction. This may be due to the fact that a large excess of polymer is reacted with surface carboxyl groups on carbon black. The same tendency was reported in the reaction of other functional polymers with surface functional groups on carbon black.[5,6,21]

Grafting of PEI and PEG changed the carbon black surface from hydrophilic to extremely hydrophilic, but grafting of PE and PCL produced a hydrophobic surface. In addition, these polymer-grafted carbon blacks readily and uniformly dispersed in solvents, such as methanol, ethanol, and water, and in polymer matrices.

3.2 Electric Resistance of PEI-grafted Carbon Black in Various Vapors

3.2.1 Effect of Crystallinity of PEI. Table 3 shows the effect of crystallinity on the electric resistance of composite resistors prepared from amorphous and crystalline PEI-grafted carbon black under dry air and methanol vapor. It is interesting to note that the electric resistance of composite prepared from PEI(C) and PEI(C)-grafted carbon black was much smaller than that of the amorphous one. On the other hand, even if PEI(C) was used, the electric resistance of the composite containing untreated carbon black was also larger than that from PEI(C)-grafted one. This may be due to the presence of aggregates of carbon black particles in the composite, because untreated carbon black hardly dispersed uniformly in PEI (C).

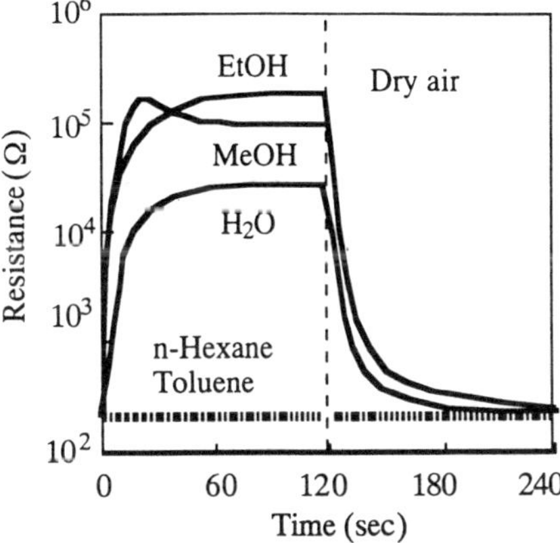

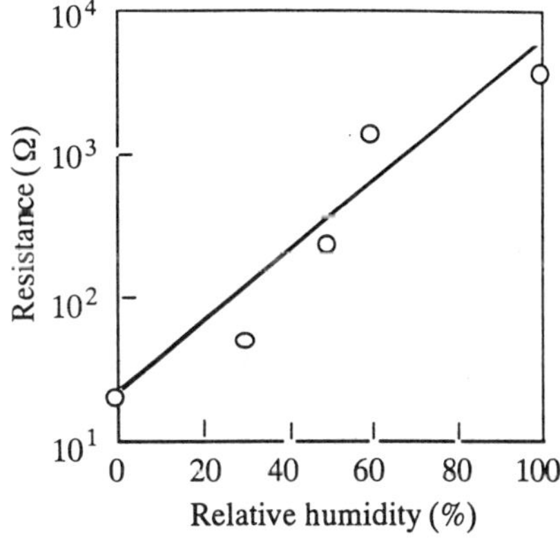

Figure 2 *Effect of various vapors on the electric resistance of crystalline PEI-grafted carbon black.*

Figure 3 *Effect of relative humidity on the electric resistance of PEI(C)-grafted carbon black.*

In addition, the electric resistance of composites from PEI(C) and PEI(C)-grafted carbon black in methanol vapor at 25 °C dramatically increased from 200Ω to 100kΩ. But the electric resistance of composite from PEI(A) and untreated carbon black scarcely changed: the resistance tends to decrease in methanol vapor.

3.2.2 Electric Resistance of PEI(C)-grafted Carbon Black in Various Vapors. Figure 2 shows the effect of various vapors on the electric resistance of composite from PEI(C) and PEI(C)-grafted carbon black at 25 °C. It is interesting to note that the electric resistance of the composite drastically increased in methanol, ethanol, and water vapors, which are good solvents for PEI. On the contrary, the electric resistance hardly changed in toluene and hexane, which are non-solvents of PEI.

3.2.3 Relationship between Electric Resistance and Relative Humidity. Figure 3 shows the relationship between electric resistance of the composite and relative humidity.

It is apparent that the logarithm of electric resistance is linearly proportional to relative humidity. This indicates that the composite can be applied as a humidity sensor.

3.3 Electric Resistance of PEG-grafted Carbon Black in Various Vapors

Figure 4 shows the effect of various vapors on the electric resistance of composite from PEG-grafted carbon black at 25 °C. The electric resistance of the composite also drastically increased in methanol, ethanol, and water vapors, which are good solvents for PEG, and immediately returned to initial resistance when the composite was transferred in dry air. On the contrary, the electric resistance hardly changed in hexane, which is a non-solvent of PEG. In addition, the logarithm of electric resistance is linearly proportional to relative humidity and partial pressure of alcohol.

3.4 Electric Resistance of PE-grafted Carbon Black in Various Vapors

Figure 5 shows the effect of various vapors on the electric resistance of composite from PE-grafted carbon black at 25 °C. The electric resistance of the composite drastically increased in THF, toluene, petroleum ether, and hexane vapors, which have an affinity for PE, and immediately returned to initial resistance when the composite was transferred in dry air. In addition, the resistor of PE-grafted carbon black has an ability to respond to butane gas. On the contrary, the electric resistance hardly changed in polar solvent vapors, such as water and alcohol vapors.

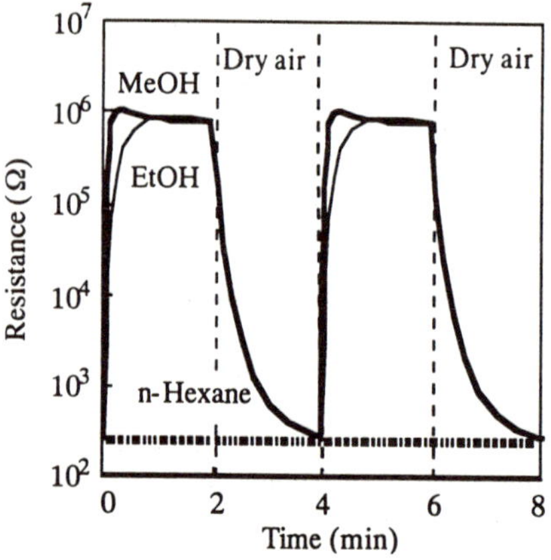

Figure 4 *Effect of various vapors on the electric resistance of crystalline PEG-grafted carbon black.*

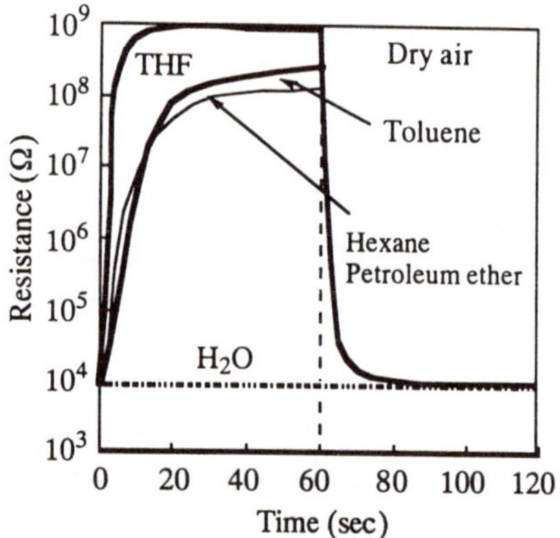

Figure 5 *Effect of various vapors on the electric resistance of PE-grafted carbon black.*

3.5 Electric Resistance of PCL-grafted Carbon Black in Various Vapors

Figure 6 also shows the effect of various vapors on the electric resistance of composite from PCL-grafted carbon black at 25 °C. The resistor has an ability to respond to not only THF and alcohol vapors, but also esters, such as methyl acetate. On the contrary, the electric resistance hardly changed in water and hexane vapors.

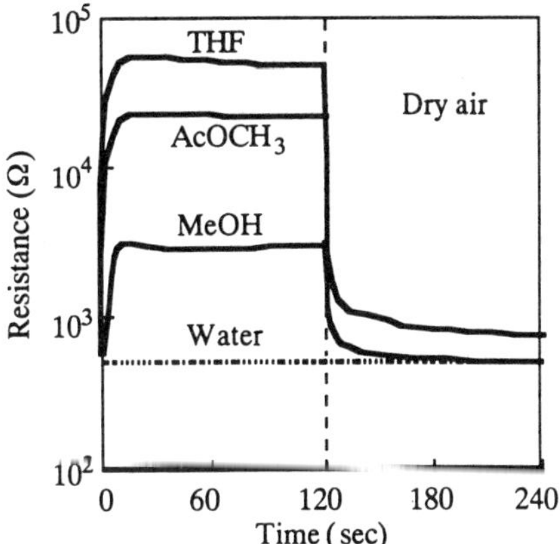

Figure 6   *Effect of various vapors on the electric resistance of PCL-grafted carbon black.*

3.6 Mechanism of Gas Sensing

The response of these crystalline polymer-grafted carbon blacks to good solvent vapor is lost above the melting point of the polymer. As mentioned above, the electric resistance of PEI (A)-grafted carbon black hardly responded to solvent vapor. Based on these results, it appears that the crystalline structure of grafted polymer on the surface plays an important role in the increase of electric resistance in solvent vapor.

These phenomena may be due to a slight change of gaps between carbon black particles based on the absorption of vapor of a good solvent by crystalline polymers. This suggests the possibility of detecting a slight change in the structure of crystalline polymer by the absorption of solvent, as this produces a large change of electric resistance of the composite. This indicates that the composite can be applied as a novel sensor for humidity and alcohol vapors.

4 CONCLUSIONS

1. Crystalline polymers, such as PEI, PEG, PE and PCL were successfully grafted onto carbon black surfaces by direct condensation of surface carboxyl groups with terminal functional groups of these polymers.
2. The electric resistance of crystalline polymer-grafted carbon blacks increased dramatically in the vapor of a good solvent for the grafted chain, and returned immediately to initial resistance when it was transferred in dry air.
3. The logarithm of the electric resistance of PEI-grafted and PEG-grafted carbon black is linearly proportional to relative humidity.

5 ACKNOWLEDGMENT

This study was partly supported by a Grant-in-Aid for Scientific Research (No.08650795) from the Ministry of Education, Science, Sports and Culture of Japan.

References

1. K. Fujiki, N. Tsubokawa and Y. Sone, *Polym. J.*, 1990, **22**, 661.
2. N. Tsubokawa, K. Fujiki and Y. Sone, *Polym. J.*, 1988, **20**, 213.
3. N. Tsubokawa, A. Yamada and Y. Sone, *Polym. Bull.*, 1983 **10**, 62.
4. N. Tsubokawa, *J. Polym. Sci., Polym. Chem. Ed.*, 1983, **21**, 705.
5. N. Tsubokawa, M. Hosoya and J. Kurumada, *React. Funct. Polym.*, 1995, **27**, 75.
6. N. Tsubokawa and J. Kurumada, *Shikizai Kyokaishi*, 1996, **69**, 90.
7. N. Tsubokawa, *Prog. Polym. Sci.*, 1992, **17**, 417.
8. N. Tsubokawa, *Kobunshi,* 1996, **45**, 412.
9. N. Tsubokawa, in 'Polymeric Material Encyclopedia', J. C. Salamone, ed., CRC Press, New York, 1996, pp. 941–946.
10. E. Frydman, UK Patent Specification, 604,695 (1945).
11. K. Ohkita and K. Fukushima, *Japan Plastics,* 1969, **3**, 6; *ibid.*, 1969, **3**, 25.
12. B. Wargotz and W. M. Alvino, *Polym. Eng. Sci.*, 1967, **7**, 63.
13. J. Meyer, *Polym. Eng. Sci.*, 1974, **14**, 706.
14. A. Ueno and S. Sugaya, *Preprints 3rd Symposium on Intelligent Materials*, 1994, p. 107.
15. N. Tsubokawa, K. Maruyama, T. Ogasawara, and M. Koshiba, *Polym. Preprints, Jpn.,* 1996, **45**, 2856.
16. N. Tsubokawa, S. Yoshikawa, K. Maruyama, T. Ogasawara and K. Saitoh, *Polym. Bull.*, 1997, **39**, 217.
17. D. Rivin, *Rubber Chem. Technol.*, 1963, **36**, 729.
18. K. Ohkita and N. Tsubokawa, *Carbon*, 1972, **10**, 631.
19. T. Saegusa, H. Ikeda and H. Fujii, *Macromolecules*, 1972, **5**, 108.
20. N. Tsubokawa, K. Magara and Y. Sone, *Nippon Gomu Kyokaishi*, 1989, **62**, 668.
21. N. Tsubokawa, *Nippon Kagaku Kaishi*, 1993, 1012.

SURFACE MODIFICATIONS TO SUPPORT MATERIALS FOR HPLC, HPCE AND ELECTROCHROMATOGRAPHY

Joseph J. Pesek* and Maria T. Matyska

Department of Chemistry
San Jose State University
One Washington Square
San Jose, CA 95192 USA

1 INTRODUCTION

In some of the most widely consulted texts on the modification of silica surfaces,[1-3] the reaction which receives the greatest amount of discussion is organosilanization. This method has been widely used on a variety of surfaces for decades in both basic research and in commercial preparation of materials. In addition to this approach to surface modification of silica and other oxides, simple esterification (reaction between a surface hydroxyl group and an alcohol) as well as chlorination/organometallation (conversion of the hydroxyl to a halide, X, followed by displacement of X with an organic moiety attached to an active metal as in a Grignard reagent) are often discussed briefly as alternate methods. More recently we have developed another procedure for oxide surface modification that utilizes a two step process in which the hydroxyl groups are first converted to hydrides by silanization followed by attachment of an organic moiety via hydrosilation.[4-6] These two reactions are illustrated below:

$$
\begin{array}{ccc}
\text{—Si–OH} & & \text{—Si–O–Si–H} \\
\text{—Si–OH} \;+\; (EtO)_3Si\text{—H} & \longrightarrow & \text{—Si–O–Si–H} \\
\text{—Si–OH} & & \text{—Si–O–Si–H}
\end{array}
$$

SILANIZATION

$$
\text{—Si–H} \;+\; CH_2{=}CH\text{-R} \;\xrightarrow{\text{Catalyst}}\; \text{—Si–}CH_2\text{-CH–R}
$$

HYDROSILATION

In the hydrosilation reaction the catalyst can be a transition metal complex, the one most often used being hexachloroplatinic acid (Speier's catalyst), or a free radical initiator such as *t*-butyl peroxide.[7]

While the silanization/hydrosilation process theoretically might seem more difficult than organosilanization, it has some advantages which outweigh the presence of a second reaction step. For example, even though the hydrosilation reaction shown above utilizes a terminal olefin, many other functional groups can be used as will be illustrated later. This gives the overall approach a great deal of versatility for the attachment of a wide variety of organic groups. The first step does not require dry conditions and in fact requires some water to be present to achieve the optimum results, *i.e.* formation of a hydride monolayer on the surface. Therefore, the experimental conditions are not particularly rigorous so no extraordinary precautions need to be taken for the reaction to be successful. Another result of the silanization reaction is that most of the surface hydroxyl groups are replaced by hydrides. Silanols are known to be the most important cause of poor chromatographic and electrophoretic performance in the separation of mixtures containing basic compounds. So replacement of Si–OH groups by Si–H will potentially have an important impact on the analysis of biomolecules and pharmaceuticals. Finally, hydrosilation produces a monomeric bonded material (one attached organic group per bonding site, *i.e.* hydroxyl or hydride) which is more reproducible than the polymeric products which result when a trifunctional organosilane reagent reacts with an oxide surface.

In order to illustrate the usefulness of silanization/hydrosilation for the modification of oxide surfaces, a variety of successful reactions will be discussed in the next section. Then some applications of these modified surfaces in both chromatographic and electrophoretic separations will be presented in subsequent sections.

2 HYDROSILATION REACTIONS ON HYDRIDE SURFACES

Porous silica is a highly polar adsorptive material which is often used as a desiccant and functions as a stationary phase in normal phase chromatography. While these properties can often be useful, they are usually detrimental in the vast majority of separation processes. Therefore, the surface is chemically modified according to one of the reaction protocols described above to produce a separation material tailored to the needs of the analytes. In this section, some specific types of surfaces are described as well as how they are produced using the silanization/hydrosilation method.

The most commonly used type of separation material today involves converting the polar adsorptive surface of silica to a hydrophobic medium. This can be accomplished by attaching a long chain alkyl group such as C_8 or C_{18} to the surface. In principle, any alkyl chain length of C_2 or greater can be bonded to a silica hydride surface by hydrosilation of the appropriate terminal olefin. The overall hydrophobicity of the final bonded product depends on both the chain length and the density of attached organic moiety. The bonded phase density (α), usually expressed in terms of $\mu mol/m^2$, can be determined from elemental analysis for percent carbon of the final product if the surface area of the silica is known.[8] Many types of olefins have been bonded to silica hydride via the hydrosilation reaction. A summary of the hydrophobic products synthesized using the silanization/hydrosilation method is shown in Table 1. While terminal olefins comprise the majority of organic moieties used to date by this procedure there are a number of notable exceptions.

Table 1 *Compounds bonded by hydrosilation - hydrophobic surface*

Compound Names and Structures		Catalyst	Reference
1-octene	$CH_2\!\!=\!\!CH\!-\!(CH_2)_5\!-\!CH_3$	Pt, FR	4,11
1-octadecene	$CH_2\!\!=\!\!CH\!-\!(CH_2)_{15}\!-\!CH_3$	Pt	12,13
4-phenyl-1-butene	$CH_2\!\!=\!\!CH\!-\!CH_2\!-\!CH_2\!-\!C_6H_5$	Pt	13
1-docosene	$CH_2\!\!=\!\!CH\!-\!(CH_2)_{19}\!-\!CH_3$	Pt	14
1-triacontene	$CH_2\!\!=\!\!CH\!-\!(CH_2)_{27}\!-\!CH_3$	Pt, FR	11,14
1H,1H,2H-Perfluoro-1-octene	$CH_2\!\!=\!\!CH\!-\!(CF_2)_5\!-\!CF_3$	Pt	15
Squalene		Pt	9
1-octyne	$CH\!\equiv\!C\!-\!(CH_2)_5\!-\!CH_3$	Pt	10
1,9-decadiyne	$CH\!\equiv\!C\!-\!(CH_2)_6\!-\!C\!\equiv\!CH$	Pt	MP
3,9-decadiyne	$CH_3\!-\!CH_2\!-\!C\!\equiv\!C\!-\!(CH_2)_4\!-\!C\!\equiv\!CH$	Pt	MP
Benzonitrile	$N\!\equiv\!C\!-\!C_6H_5$	FR	MP

Pt - Hexachloroplatinic acid
FR - Free radical
MP - Manuscript in preparation

One interesting example is squalene, which has a number of double bonds none of which is in the terminal position.[9] This bonded material certainly has similarities to one of the more important phases used in GC, squalane, and it was shown that the squalene attached to silica hydride via hydrosilation was particularly effective for separating mixtures of light hydrocarbons. Another unique possibility involves the bonding of alkynes to silica hydride.[10] Spectroscopic studies showed that multiple bonding of the organic moiety to the surface is likely, and could occur with the final product having one or more of the following Structures I–III:

Structure I   Structure II   Structure III

If one or more of these structures constitute the majority of the bonded organic moiety, then increased hydrolytic stability should result due to the double attachment to the surface. As can be seen from Table 1, a wide variety of organic moieties bonded to silica hydride via hydrosilation results in hydrophobic surfaces which leads to a range of separation capabilities in reversed-phase HPLC.[4,9–15] Bonding has been accomplished by both metal complex catalysis and free radical initiation.

In order to obtain separation materials with hydrophilic surfaces more reproducible and less adsorptive than bare silica, chemical modification with appropriate polar organic groups is necessary. A summary of the organic moieties which can be attached to silica hydride via hydrosilation for use in normal phase chromatography is shown in Table 2[10,11,15–17] The most common of these phases contain hydroxyl groups but examples of epoxy- and cyano-bonded moieties are also present. The perfluorinated material is shown in Table 1 and Table 2 since it can function in both normal and reversed-phase modes depending on the solutes to be separated.

A rapidly growing area of separation science involves the analysis of chiral compounds. The need for optically pure drugs is becoming increasingly more important in the pharmaceutical industry so a wide array of stationary phases have been developed to assist in meeting the new regulatory guidelines. A variety of different chiral separation materials is necessary because each individual stationary phase is able to resolve a limited number of enantiomeric compounds. This is due to the highly specific nature of the interaction between the chiral stationary phase and the analyte. Therefore, until a new approach is developed in the separation mechanism for optical isomers, there will be a continuing need to prepare more bonded materials for chiral analysis. Table 3 shows that the two most important types of chiral phases have been synthesized via the hydrosilation method on a silica hydride intermediate.[18] These are the inclusion type (cyclodextrin) and Pirkle phases (all other materials shown in the table) which utilize a three-point interaction with the solute for chiral discrimination.[19]

Table 2 *Compounds bonded by hydrosilation - hydrophilic surface*

Compound Names and Structures	Catalyst	Reference
CH_2=CH— CH_2–O—CH_2—CH—CH_2 (with epoxide O bridging CH—CH$_2$) Allyl Glycidyl Ether	Pt	16
CH_2=CH— $(CH_2)_4$–CH—CH_2—OH (with OH on CH) 7-octene-1,2-diol	Pt, FR	11,16
CH_2=CH— $(CH_2)_4$–CH_2—CH_2—OH 1-octene-8-ol	Pt	17
CH_2=CH— CH_2 — CH_2—OH 3-butene-1-ol	FR	11
CH_2=CH—CH—C≡N (with CH$_3$ on CH) 2-methyl-3-butenitrile	FR	11
CH_2=CH— $(CF_2)_5$–CF_3 1H,1H,2H-Perfluoro-1-octene	Pt	15,17
HC≡C—C(OH)(H)—C$_6$H$_5$ 1-phenyl-2-propyn-1-ol	Pt	10

Pt - Hexachloroplatinic acid
FR - Free radical

The final and most unusual class of compounds bonded by the silanization/hydrosilation method consists of liquid crystals. The structures of the organic moieties bonded to date are shown in Table 4.[13,20,21] The first two are attached in a manner similar to most of the other compounds listed in the previous tables, *i.e.* through a terminal olefin. However, the last two utilize a cyano group for bonding to the hydride surface through hydrosilation. This approach was tested for the simple compound, benzonitrile, as shown in Table 1. The advantage of utilizing such a bonding scheme is that many commercially available liquid crystals have cyano groups as well as biphenyl or terphenyl groups similar to the compounds shown in Table 4. It has already been demonstrated that bonded liquid crystals have the ability to separate compounds based on molecular shape.[22,23] This shape discrimination is also a function of temperature and mobile phase composition so that one bonded phase can be adapted to

Table 3 *Compounds bonded by hydrosilation - chiral phases*

Compound Names and Structures	Catalyst	Reference
4-(Allyloxy)benzoyl Chloride + (R)-(+)-(1-Naphthyl)ethylamine	Pt	18
β-cyclodextrin	Pt,FR	MP
(R)-(+)-Acryloxy-β,β,dimethyl-γ-butyrolactone	Pt, FR	MP

Pt - Hexachloroplatinic acid
FR - Free radical
MP - Manusctipt in preparation

different mixtures in order to optimize the separation capabilities. The results obtained to date involve only a few bonded liquid crystals, and polycryclic aromatic hydrocarbons (PAHs) have been the predominant type of solutes evaluated. However, it appears that further testing on these materials is warranted based on the preliminary data.

3 APPLICATIONS OF HYDRIDE-BASED SEPARATION MATERIALS IN HPLC

The wide variety of syntheses described above indicate that hydride-based bonded phases should be applicable to a wide range of applications in HPLC. It would be impossible to give examples of separations for each of the materials listed in Tables 1–4, so some particularly noteworthy separations in each of the four categories will be presented.

Table 3 (cont.) *Compounds bonded by hydrosilation - chiral phases*

Compound Names and Structures	Catalyst	Reference
(8S,9R)-(-)N-Benzylcinchonidinium Chloride	Pt	MP
(R)-(+)-Limonene	Pt	MP

Pt - Hexachloroplatinic acid
FR - Free radical
MP - Manusctipt in preparation

While the vast majority of development of the silanization/hydrosilation process has been done on silica matcrials, it is applicable to modification of other oxide surfaces. In particular, titania has proved to be a suitable porous support material for HPLC bonded phases. In order to make comparisons to silica-based stationary phases, most of the preliminary investigations involved the bonding of C_8 or C_{18} moieties.[24,25] Examples of two separations performed on a C_8 titania-based stationary phase are shown in Figure 1. Figure 1A shows the separation of a reversed-phase text mixture involving a series of aromatic compounds. The results obtained are comparable to those obtained on C_8 silica-based stationary phases. The separation of three anilines is shown in Figure 1B. These highly basic compounds are very sensitive to the presence of residual hydroxyl groups on the surface. The symmetric peak shape indicates that there are very few of these OH groups accessible to solutes on this particular bonded material.

The mono-ol column is a unique example of a hydrophilic material that has only been made by the silanization/hydrosilation process. A variety of mono-ol materials has been synthesized with varying chain lengths and substituents using both alkenes and alkynes. The more common phase which is available commercially is a diol. The

Table 4 *Compounds bonded by hydrosilation - liquid crystals*

Compound Names and Structures	Catalyst	Reference
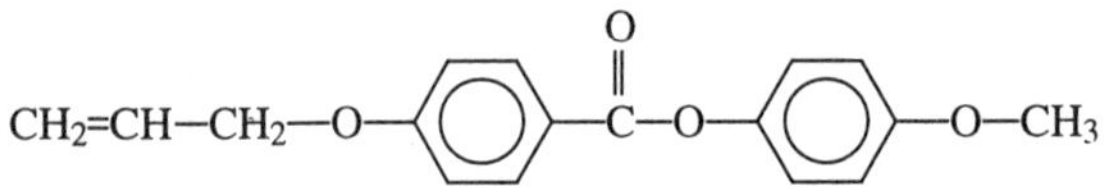 $CH_2{=}CH{-}CH_2{-}O{-}$ (benzene ring) $-C({=}O){-}O{-}$ (benzene ring) $-O{-}CH_3$ 4-Methoxyphenyl-4-allyloxy benzoate (MPAB)	Pt	13,20,21
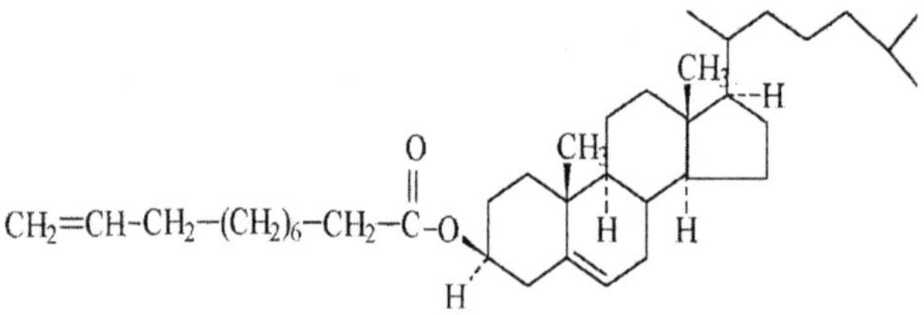 $CH_2{=}CH{-}CH_2{-}(CH_2)_6{-}CH_2{-}C({=}O){-}O{-}$ (cholesteryl structure) Cholesteryl 10-Undecenoate	Pt	13,20,21
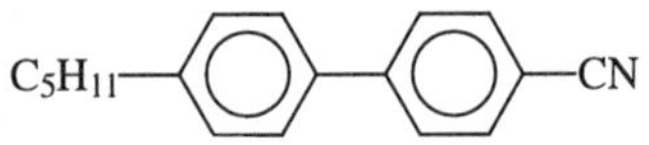 $C_5H_{11}{-}$ (biphenyl) $-CN$ 4-cyano-4'-n-pentylbiphenyl	Pt	MP
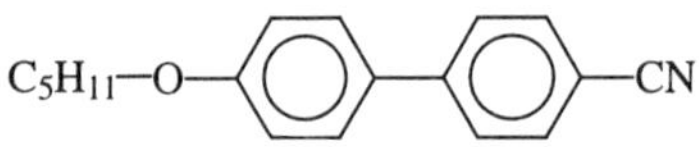 $C_5H_{11}{-}O{-}$ (biphenyl) $-CN$ 4-cyano-4'-n-penthoxybiphenyl	Pt	MP

Pt - Hexachloroplatinic acid
MP - Manuscript in preparation

diol is more polar so the mono-ol phase represents a separation material of moderate polarity that might offer unique properties for certain groups of analytes. An example of an application of the mono-ol phase is shown in Figure 2 for the separation of a group of steroids. These compounds represent a wide range of polarity so that a gradient procedure is still required. However, a fast reproducible separation of this mixture is attained in less than 5 minutes with good peak shape and reasonable efficiency.

A limited number of chiral materials have been synthesized to date and only one has been extensively tested. The most investigated phase is the first one shown in Table 3 and involves the use of the chiral selector (R)-(+)-(1-naphthyl)ethylamine. It was demonstrated that this chiral stationary phase was particularly effective for the separation of the optical isomers of 3,5-dinitrobenzoyl amino acid esters. An example of a chiral separation for the isomers of isoleucine is shown in Figure 3. In order to identify the individual isomers, an optically pure 3,5-DNB-isoleucine methyl ester is analysed under identical conditions. The results of this test are also shown in Figure 3.

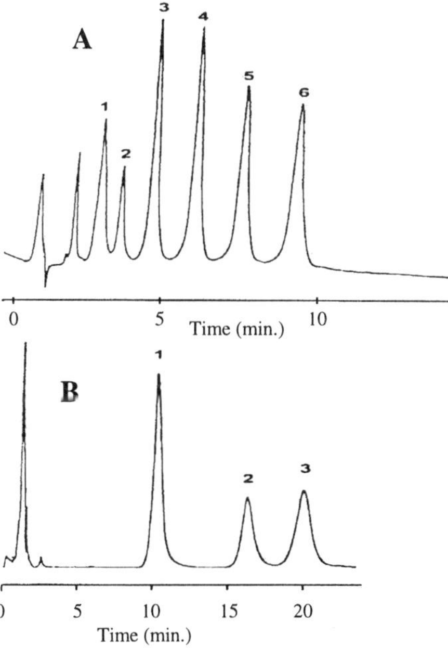

Figure 1 *Separations obtained on octadecyl titania. **A** Aromatic compounds: 1 = benzene, 2 = toluene, 3 = ethyl benzene, 4 = isopropyl benzene, 5 = t-butyl benzene and 6 – anthracene. **B** Aniline mixture. 1 = N-methyl aniline, 2 = N-ethyl aniline and 3 = N,N-dimethyl aniline.*

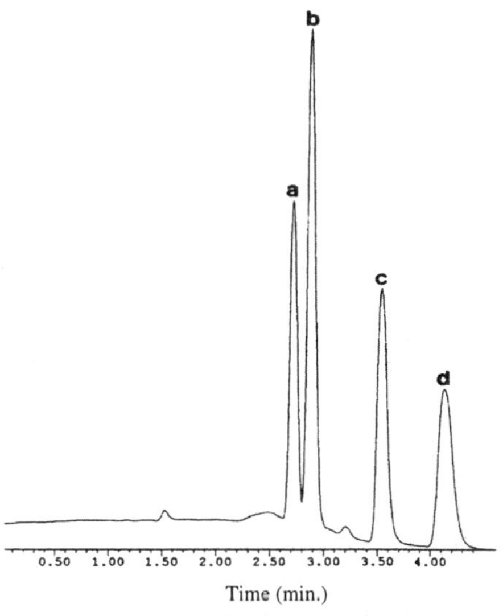

Figure 2 *Gradient separation of steroid mixture on mono-ol column. Solutes: a = 4-androstene-3,17-dione, b = andrenosterone, c = corticosterone and d = prednisone.*

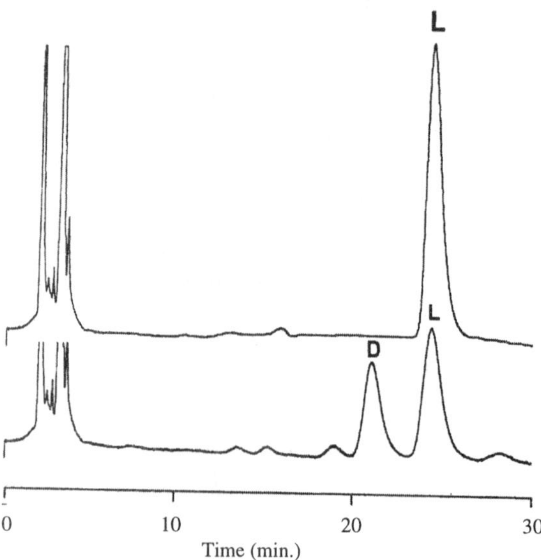

Figure 3   *Separation of the optical isomers of N-3,5-DNB-isoleucine methyl ester on R(+)-naphthylethylamine column.*

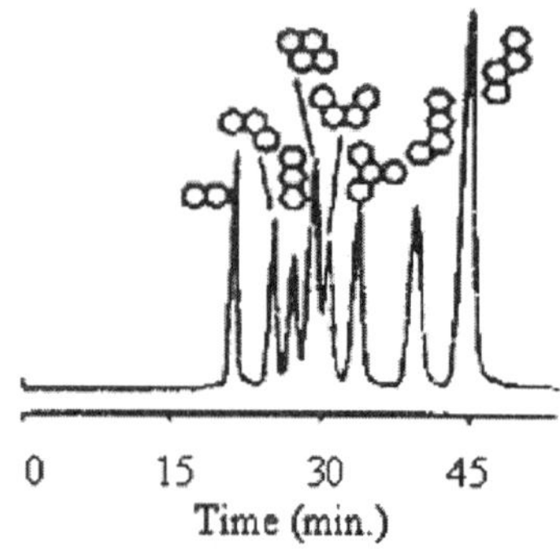

Figure 4   *Separation of a mixture of PAHs on liquid crystal (cholesterol) phase.*

The bonding of liquid crystals to silica surfaces via the silanization/hydrosilation method represents a type of stationary phase that has been almost unique to this approach. Liquid crystals represent an unusual type of material in that they possess a highly ordered structure that under the appropriate conditions of solvent and temperature is intermediate in phase properties between a liquid and a solid. The degree of ordering is subject to change by these variables and presumably could be done in a continuous or discontinuous manner to control retention properties in an HPLC column. It has already been shown that liquid crystal-bonded phases possess a high degree of shape selectivity in the retention of various solutes. This selectivity can be used to advantage in the separation of mixtures of certain types of analytes. An example of such an analysis is shown in Figure 4 for a polycyclic aromatic hydrocarbon (PAH) mixture. The liquid crystal phases retain solutes preferentially which are planar and/or have a large length to breadth ratio. The separation shown in Figure 4 cannot be duplicated by most commercially available reversed-phase materials. In order to determine the overall capabilities that liquid crystal stationary phases may offer in HPLC, a greater variety of bonded materials and potential applications must be tested.

4 APPLICATIONS OF HYDRIDE-BASED SEPARATION MATERIALS IN HPCE AND CAPILLARY ELECTROCHROMATOGRAPHY

The inner wall of a fused silica capillary also possesses silanols which can be modified in a manner similar to porous silica.[26] Therefore, the silanization/hydrosilation reaction sequence should be applicable to the modification of silica tubes for use in high performance capillary electrophoresis (HPCE). The modified surface can be used to decrease adsorption of basic compounds on the inner wall as well as to limit electro-osmotic flow in order to improve separation. In addition to HPCE, a novel approach to capillary electrochromatography (CEC) has been developed which involves etching the inner surface of the fused silica tube before modification of the surface.[27–29] The increased surface area facilitates interactions between solutes and the bonded organic moiety attached to the etched inner wall of the capillary. When an electric field is applied, separation is based on solute retention by the stationary phase if the analyte is neutral or a combination of solute/bonded phase interactions and differences in electrophoretic mobility if the analyte is charged. Applications using hydride-based separation materials formed on the inner wall of a fused silica capillary in both HPCE and CEC are shown below.

In order to demonstrate the advantages of surface modification in HPCE, separations of compounds that are very basic provide the most demanding tests. An excellent separation of two compounds which are often present in physiological fluids, tryptamine and serotonin, is obtained on a capillary modified by the silanization/hydrosilation method (Figure 5). The organic moiety added to the hydride surface is a polymerized sugar derivative of acrylamide.[30] It has been shown that these materials are resistant to hydrolysis and that the linkage to the surface is stable at a pH > 8.0. Small basic compounds like the ones presented in Figure 5 as well as basic peptides and proteins often cannot be analysed by HPCE due to the strong tendency to adsorb on the capillary wall. Therefore, surface derivatization is necessary in order to take advantage of the high efficiency of HPCE for the analysis of these types of compounds. The silanization/hydrosilation method provides both excellent synthetic versatility for bonding a wide range of organic moieties to the inner wall of a fused silica capillary as well as significantly higher stability than modifications done by organosilanization.

The desirable properties of synthetic versatility and high stability can also be exploited in CEC. The most common approach to CEC involves taking a typical HPLC stationary phase such as an octadecyl moiety attached to a porous silica support by organosilanization and packing it into a fused silica capillary.[31] In order to retain the stationary phase, frits must be placed before and after the packed bed. The presence of the particulate silica and the frits can both serve as sites for bubble formation which will disrupt the current flow in the column. In addition, in order to create electro-osmotic flow for the movement of neutral solutes, the bonded phase cannot be endcapped, *i.e.* there must be some residual silanols. The presence of the Si–OH groups precludes the use of these columns for highly basic compounds such as the ones described above. However,

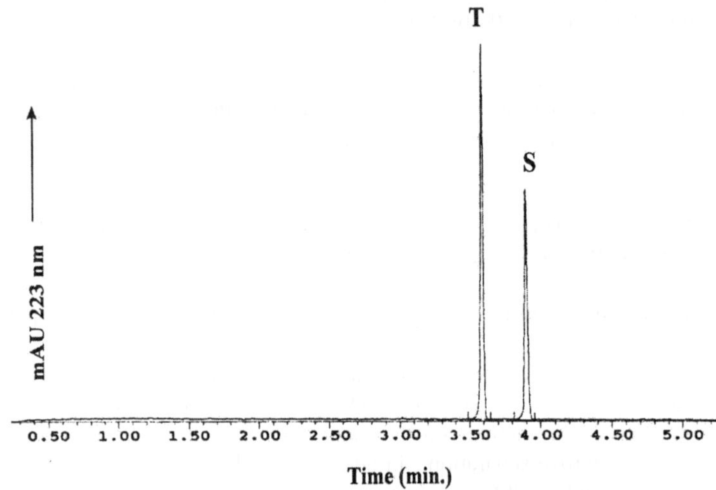

Figure 5   *Separation of tryptamine (T) and serotonin (S) by CE in modified capillary.*

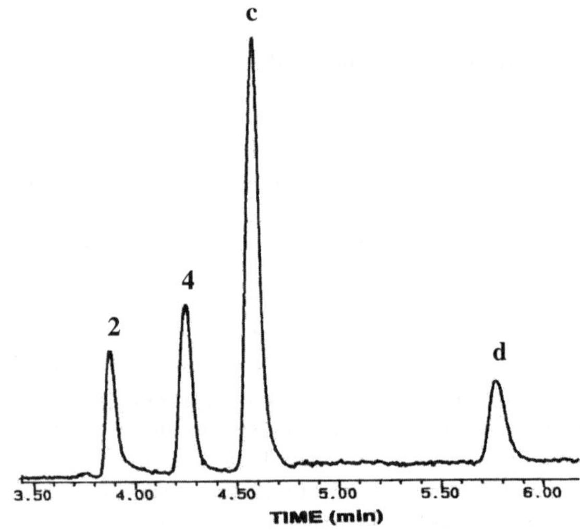

Figure 6   *Separation of a tetracycline mixture using a C_{18} modified etched capillary. Solutes: 2 = tetracyline, 4 = oxytetracyline, c = epioxytetracyline, d = apooxytetracyline.*

the use of etched chemically modified capillaries does not require the use of frits or packing materials which greatly reduces the possibility of bubble formation. In addition, it appears that residual silanols, while apparently present to a limited extent, are not readily accessible to typical solutes. An example of a separation of a mixture of basic compounds, tetracyclines, is shown in Figure 6. If this mixture (typical components found in a commercial oxytetracycline sample) were run in the usual CEC mode, the peaks would be broad and very asymmetric (tailing).

5 CONCLUSIONS

Since 1989 when the first publication appeared discussing the silanization/hydrosilation method,[32] the past ten years have resulted in numerous publications exploiting the basic concepts and expanding the field of application. Originally conceived as a method for making more stable reversed-phase materials, a wide range of bonded organic moieties has been attached to silica and other oxide surfaces which has resulted in stationary phases for most areas of application in HPLC. In addition, the method has been shown to be amenable to the modification of the inner walls of capillaries for HPCE as well as etched surfaces for use as a separation medium in CEC. In all cases, synthetic versatility and high stability of the bonded organic moiety are features which have made this approach increasingly more attractive for the production of separation materials.

6 ACKNOWLEDGEMENTS

The authors wish to thank the National Institutes of Health (Grant R15 GM 49452-01) and the National Science Foundation (CHE 9625628) for partial support of the research reported here.

References

1.   K. K. Unger, 'Porous Silica', Elsevier, Amsterdam, 1979.
2.   R. K. Iler, 'The Chemistry of Silica', Wiley and Sons, New York, 1979.
3.   E. F. Vansant, P. Van Der Voort and K. C. Vrancken, 'Characterization and Chemical Modification of Silica', Elsevier, Amsterdam, 1995.
4.   J. E. Sandoval and J. J. Pesck, *Anal. Chem.*, 1991, **63**, 2634.
5.   C.-H. Chu, E. Jonsson, M. Auvinen, J. J. Pesek and J. E. Sandoval, *Anal. Chem.*, 1993, **65**, 808.
6.   J. J. Pesek, M. T. Matyska, J. E. Sandoval and E. J. Williamsen, *J. Liq. Chromatogr. & Rel. Technol.*, 1996, **19**, 2843.
7.   B. Marciniec, 'Comprehensive Handbook on Hydrosilylation', Pergamon Press, Oxford, 1992.
8.   G. E. Berendsen and L. De Galaan, *J. Liq. Chromatogr.*, 1978, **1**, 561.
9.   S. O. Akapo, M. T. Matyska and J. J. Pesek, *J. Chromatogr. A*, 1997, **773**, 53.
10. J. J. Pesek, M. T. Matyska, M. Oliva and M. Evanchic, *J. Chromatogr. A*, in press.
11. J. J. Pesek, M. T. Matyska, E. J. Williamsen, M. Evanchic, W. Hazari, K. Konjuh, S. Takhar and R. Tranchina, *J. Chromatogr. A*, 1997, **786**, 219.
12. M. C. Montes, C. van Amen, J. J. Pesek and J. E. Sandoval, *J. Chromatogr.*, 1994, **688**, 31.
13. J. J. Pesek, M. T. Matyska, E. J. Williamsen and R. Tam, *Chromatographia*, 1995, **41**, 301.
14. J. J. Pesek, M. T. Matyska and S. Takhar, *Chromatographia*, in press.
15. J. J. Pesek, M. T. Matyska and H. Hemphala, *Chromatographia*, 1996, **43**, 10.
16. J. J. Pesek and M. T. Matyska, *J. Chromatogr.*, 1994, **687**, 33.

17. J. J. Pesek, M. T. Matyska, E. Soczewinski and P. Christensen, *Chromatographia*, 1994, **39**, 520.
18. J. J. Pesek, M. T. Matyska and S. Kamath, *Analusis*, 1997, **25**, 253.
19. C. E. Dagliesh, *J. Chem. Soc.*, 1952, 3940.
20. Y. Saito, K. Jinno, J. J. Pesek, Y. L. Chen, G. Luehr, J. Archer, J.C. Fetzer and W. R. Biggs, *Chromatographia*, 1994, **38**, 295.
21. J. J. Pesek, M. T. Matyska, E. J. Williamsen, R. Tam and Z. Wang, *J. Liq. Chromatogr. & Rel. Technol.*, in press.
22. A. Catabay, Y. Saito, C. Okumura, K. Jinno, J. J. Pesek and E. Williamsen, *J. Microcol Sep.*, 1997, **9**, 81.
23. A. Catabay, C. Okumura, K. Jinno, J. J. Pesek, E. Williamsen, J. C. Fetzer and W. R. Biggs, *Chromatographia*, 1998, **47**, 13.
24. J. J. Pesek and V. H. Tang, *Chromatographia*, 1994, **39**, 649.
25. J. J. Pesek, M. T. Matyska and J. Ramakrishnan, *Chromatographia*, 1997, **44**, 538.
26. P. Camilleri, 'Capillary Electrophoresis. Theory and Practice', CRC Press, Boca Raton, 1998, p.33.
27. J. J. Pesek and M. T. Matyska, *J. Chromatogr. A*, 1996, **736**, 255.
28. J. J. Pesek and M. T. Matyska, *J. Chromatogr. A*, 1996, **736**, 313.
29. J. J. Pesek, M. T. Matyska and L. Mauskar, *J. Chromatogr. A*, 1997, **763**, 307.
30. M. Chiari, N. Dell'Orto and A. Gelain, *Anal. Chem.*, 1996, **68**, 2731.
31. M. M. Dittmann and G. P. Rozing, *J. Chromatogr. A*, 1996, **744**, 63.
32. J. E. Sandoval and J. J. Pesek, *Anal. Chem.*, 1989, **61**, 2067.

SYNTHESIS, CHARACTERIZATION AND APPLICATION OF NEW BONDED PHASES FOR HPLC

K. Albert,* A. Ellwanger, M. Dachtler, T. Lacker, S. Strohschein, J. Wegmann, M. Pursch and M. Raitza

Institut für Organische Chemie, Universität Tübingen
Auf der Morgenstelle 18
D-72076 Tübingen, Germany

1 INTRODUCTION

In reversed-phase liquid chromatography, C_8 and C_{18} bonded phases have proven to be well suited for the solution of a myriad of separation problems,[1,2] but there is still a need for tailored stationary phases to solve demanding separation problems. We use different approaches to synthesize new silica bonded phases. By applying different reaction schemes in the chemical modification of the silica surface, coverage density and ordering of the tethered ligands can be varied systematically. Two basic types of stationary phases have been synthesized which exhibit enhanced capabilities for the separation of carotenoid isomers and the separation of polycyclic aromatic compounds (PAHs). An increase in shape recognition in the separation of carotenoid isomers is obtained with n-alkyl bonded phases with longer ligands such as C_{22} and C_{30} chains.[3,4] In addition, n-alkyl fluorenyl bonded phases show excellent separation characteristics for the separation of PAHs due to a mixed mode separation mechanism of hydrophobic and π–π-interactions.[5,6]

The bonded phases are characterized by solid-state NMR spectroscopy employing Magic Angle Spinning (MAS) together with the Cross Polarization (CP) technique. ^{29}Si CP/MAS NMR spectroscopy[7–9] yields information on the variety and quantity of surface species formed in the modification reaction. ^{13}C CP/MAS NMR,[10] together with ^{13}C suspended-state NMR spectroscopy,[3] provides information about structure and mobility of the immobilized organic ligands in the dry state and in the presence of a mobile phase. The conformational and dynamic behavior of n-alkyl chains tethered to the surface of silica gel can be visualized by computer graphics obtained from molecular modeling calculations.[11] Overall, the combined application of chromatographic and spectroscopic investigation methods provides a deeper and more thorough understanding of the separation mechanism in reversed-phase silica.

2 EXPERIMENTAL

2.1 Synthesis and Materials

Solution polymerization: This synthesis of polymeric C_{30} bonded phases follows a slight variation of the procedure described by Sander *et al.*[12] Triacontyltrichlorosilane (5 g,

  Fundamental and Applied Aspects of Chemically Modified Surfaces

ABCR, Karlsruhe, Germany) was dissolved in 50 mL of xylene at 343 K. Insoluble impurities were removed by decanting. This solution was added to 4 g of activated silica gel (Prontosil, 3 μm particle size, 200 Å pore size, Bischoff, Leonberg, Germany) suspended in 50 mL of xylene. The resulting slurry was kept at 343 K for 15 min, then treated with 1.0 mL H_2O to initiate the polymerization reaction. The slurry was heated at reflux for 10 h, filtered hot through a glass frit (P4), and then washed successively with xylene, acetone, ethanol, H_2O, ethanol, acetone and *n*-hexane. Finally, the interphase was dried for one hour under reduced pressure at 333 K.

Surface polymerization: In this analogous synthetic method water was not directly added to the slurry. Instead, the polymerization reaction was initiated by the addition of the silane to humidified silica gel, with physically adsorbed water on the surface.

The synthesis of the *n*-hexylfluorene phases has been described previously.[5]

2.2 Solid-state NMR Spectroscopy

^{29}Si Solid-state NMR spectra were obtained on a Bruker MSL 200 NMR spectrometer at 4.7 T. The samples were packed into 7 mm rotors of ZrO_2 which were spun at 3000 Hz by a dry-air gas drive. The proton 90 degree pulse length was 5.5 μs, the recycle delay was 1 s and the contact time 5 ms. ^{13}C and ^{1}H Solid-state NMR spectra were obtained on a Bruker ASX 300 NMR spectrometer at 7 T. For ^{1}H NMR measurements 4 mm rotors were spun at 14 kHz, the recycle delay was 4 s and the 90 degree pulse length was 3.8 μs. ^{13}C NMR measurements were also carried out with 4 mm rotors, but the spinning rate was 10 kHz with a proton 90 degree pulse length of 3.8 μs, a contact time of 1 ms, and a recycle delay of 1 s. All chemical shifts were externally referenced to liquid TMS. The Hartmann–Hahn condition for CP was calibrated with glycine (^{13}C) or Q_8M_8 (^{29}Si).

For the 2D-WISE experiment magic angle spinning was performed at *ca.* 4000 Hz, the contact time was 500 μs, delay time 1 s and 64 increments (TD1) at 3 μs were recorded.

2.3 NMR Spectroscopy in the Suspended-state

A Bruker ASX 300 spectrometer was used for the ^{13}C MAS NMR measurements. Magic angle spinning was performed at *ca.* 3000 Hz. The spectra were acquired using 90° proton pulse widths of 3.6 μs, and repetition times of 5 s for the temperature dependent suspended-state NMR experiments and 3 s for the suspended-state NMR experiments in various solvents.

The ^{1}H MAS NMR experiments were performed on a Bruker ARX 400 spectrometer. The sample was measured in a 4 mm rotor with a spinning rate of *ca.* 4000 Hz, a repetition time of 2 s and a 90° pulse length of 10 μs. The phase was suspended in $CDCl_3$.

2.4 Chromatography

HPLC separations were performed with a Merck / Hitachi L-6200 A pump in combination with a variable wavelength detector Merck / Hitachi L-4000 A (Merck-Hitachi, Darmstadt, Germany) and with a Hewlett-Packard HP Series 1100 system (Hewlett-Packard, Waldbronn, Germany). The stationary phases were packed into

250 x 4 mm stainless steel tubes (Bischoff, Leonberg, Germany) by a high pressure slurry packing procedure on a Knauer Pneumatic HPLC pump (Knauer, Berlin, Germany).

The mobile phase composition for the Sander and Wise test/SRM 869 was acetonitrile/water (85/15, v/v) and a 10 µL volume of the test sample was injected. A gradient elution program was implemented for the SRM 1647 test / 16 PAHs, 10 minute hold at methanol–water (70:30, v/v), a 30 minute linear gradient to methanol–water (90:10, v/v), and hold until the end of the separation. A 20 µL volume of the test sample was injected.

The mobile phase composition for the nitro explosives was methanol/water (50/50, v/v), and a 20 µL volume of the sample was injected containing about 5 µg/mL of each compound. All separations were carried out with UV detection at 254 nm, 1 mL/min flow rate and 25 °C temperature.

The separations of β-carotene isomers were performed with acetone/H_2O (82/18, v/v; 92/8, v/v, respectively) as eluent. The flow rate was 1 mL/min and the UV detection was performed at 450 nm. As stationary phase a C_{30} phase or a C_{18} phase was used, respectively. For the separation 10 µL of a 0.1% solution of isomerized β-carotene in $CDCl_3$ was injected.

2.5 Molecular Modeling

Molecular dynamics simulations were performed on a Silicon Graphics workstation (Indy) using Tripos software (Sybyl 6.2, Tripos, Inc., St. Louis, Missouri, USA). Bond angle, torsion angle bond length, and the non-bonded Van-der-Waals and electrostatic interaction terms were included in the energy calculations. The Tripos force field was parameterized for the problem with data determined from the MMP2 parameters of Allinger. The Gasteiger–Hückel method was used to calculate the charges. For interatomic distances > 8 Å, interactions between two non-bonded atoms were disregarded. At temperatures of 295 K and 355 K molecular dynamics calculations were carried out for 20 ps using calculation steps of 1 fs.

3 RESULTS AND DISCUSSION

3.1 Synthesis of New Stationary Phases

Reversed phase materials based on silica gel are not limited to the very well known types of C_{18} phases. Recently developed alternatives[4,5,13–15] to the usual C_{18} phases, some of which are shown in Figure 1, can be divided into two main groups. The first group consists of *n*-alkyl phases with variable chain length of $n = 22$, 30, and 34. These phases are closely related to C_{18} phases, but the extended chain length leads to different geometric configurations of the alkyl chains. Depending on the length and the density of the tethered ligand, either the trans configuration or the more mobile gauche configuration is energetically favored. This results in different chromatographic properties of the respective stationary phases.

The choice of a suitable surface modification process plays an important role in the adjustment of the order and density of the attached alkyl chains to the requirements of the respective separation problem (Figure 2). One approach is the method of 'solution polymerization'. In this method, the silane and the silica gel are suspended in the solvent,

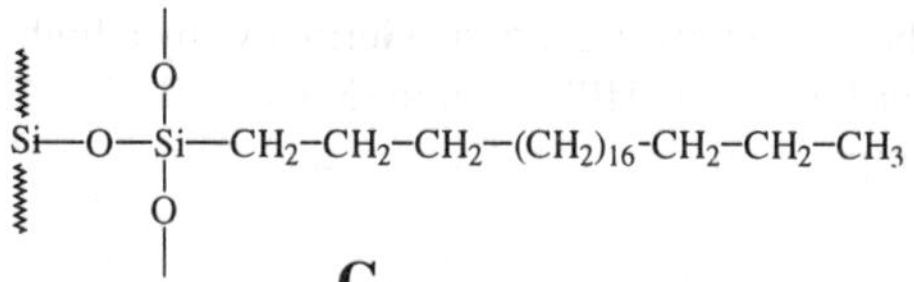

$$Si-O-Si-CH_2-CH_2-CH_2-(CH_2)_{16}-CH_2-CH_2-CH_3$$

C$_{22}$

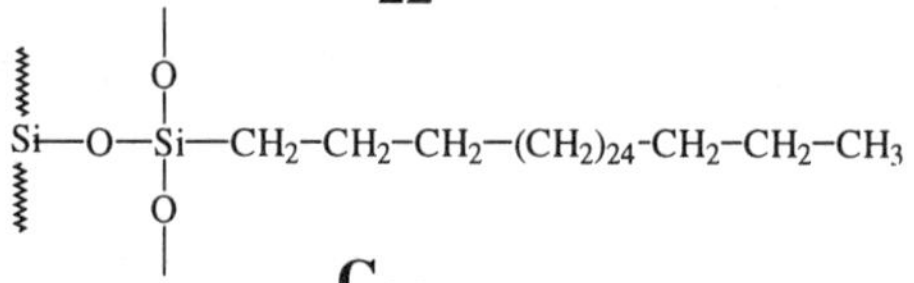

$$Si-O-Si-CH_2-CH_2-CH_2-(CH_2)_{24}-CH_2-CH_2-CH_3$$

C$_{30}$

$$Si-O-Si-CH_2-CH_2-CH_2-(CH_2)_{28}-CH_2-CH_2-CH_3$$

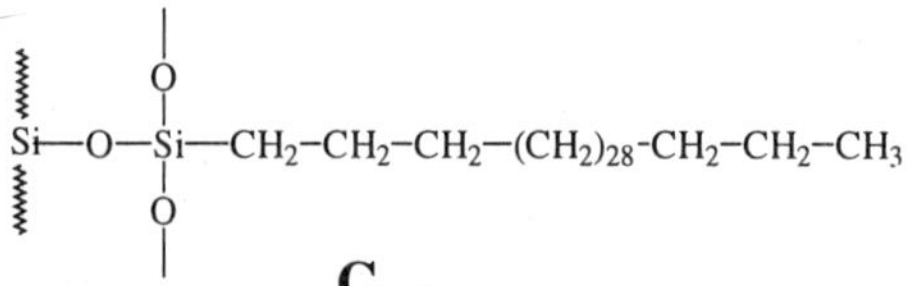

C$_{34}$

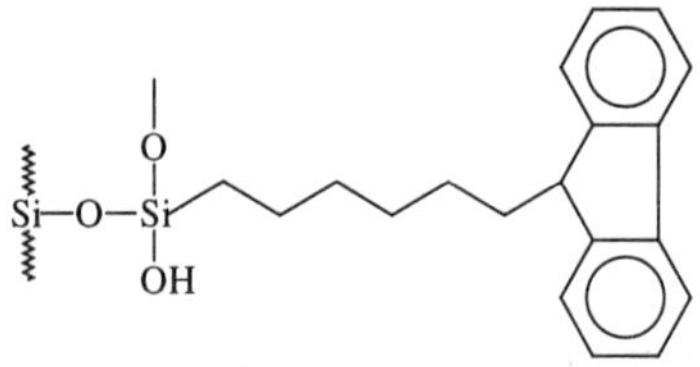

n-Hexyl-Fluorene

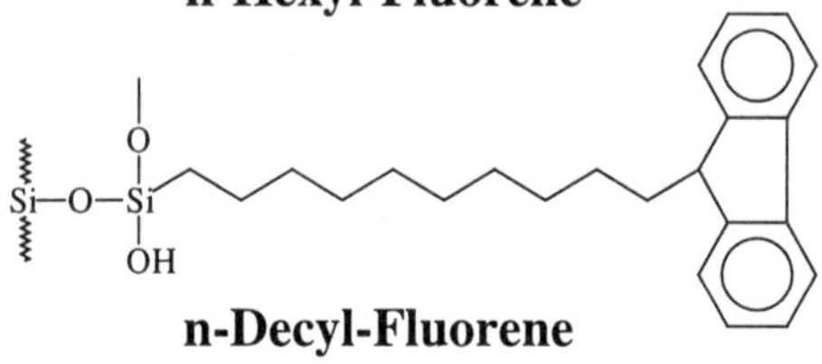

n-Decyl-Fluorene

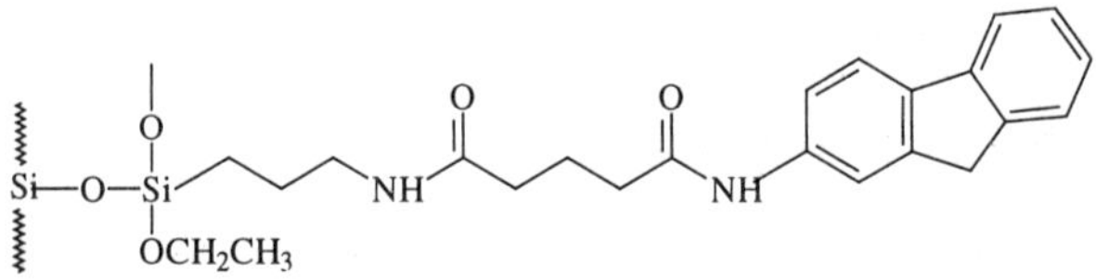

Fluorene (amido coupling)

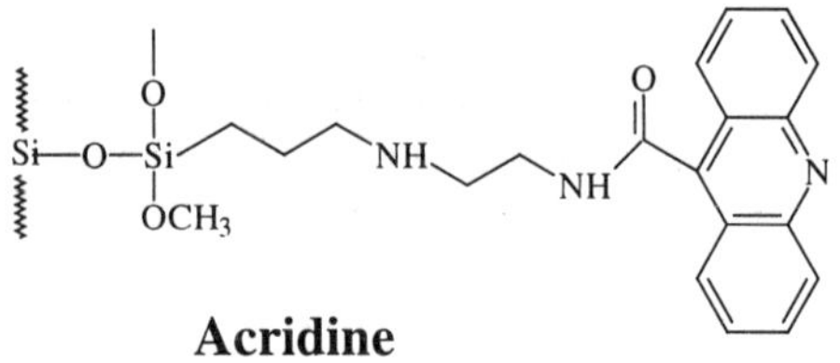

Acridine

Figure 1 *New silica gel-based stationary phases as alternatives to conventional C$_{18}$ phases.*

and water is added to the reaction slurry to initiate polymerization. Silane polymer is formed in the solvent and the resulting silane clusters are irregularly attached on the silica gel surface.

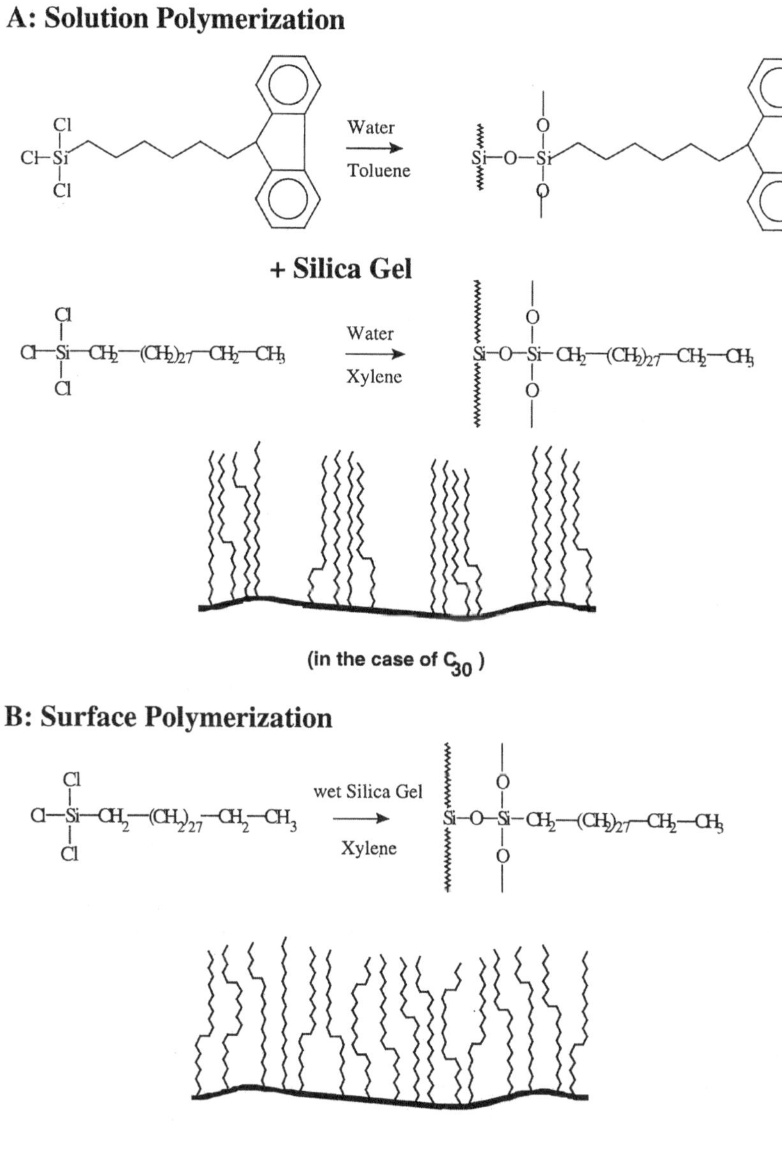

A: Solution Polymerization

+ Silica Gel

(in the case of C_{30})

B: Surface Polymerization

Figure 2 *Comparison of the two different polymerization methods for the synthesis of reversed phase. Whereas a higher ligand density is obtained by surface polymerization (B) in the case of the n-alkyl phases, solution polymerization (A) is preferred for the synthesis of n-alkyl fluorene phases.*

In a different approach, a monolayer of water is first placed on the silica gel and then the silane and solvent are added to the humidified silica gel to form the polymer. This method is referred to as 'surface polymerization' or the 'self-assembled monolayer technique'.[16–18] With this approach the polymerization of the silane must occur on the surface of the humidified silica gel. It is possible with this technique to achieve a higher ligand density and therefore a higher proportion of alkyl chains in the trans conformation compared to the method of solution polymerization using the same amounts of silica and silane.

Stationary phases containing aromatic ligand fragments such as acridine[13,19] or fluorene[5,14] belong to the second group of phases (Figure 1). Whereas solute retention with *n*-alkyl phases depends on the hydrophobic interactions of the alkyl chains, the attached aromatic moieties of fluorene and acridine phases promote a different chromatographic behavior due to the participation of the extended π-electron system into the separation process.

In contrast to *n*-alkyl phases, the organic ligand of stationary phases containing π-electron systems must be prepared in several synthetic steps. The step-by-step synthesis of a π-electron phase from the silica gel surface to its terminal end is one possibility. This was readily utilized for the acridine and fluorene phases, both of which were connected to the silica gel by amido couplings. The immobilization of the aromatic moieties on the aminopropyl intermediate is never be achieved stoichiometrically. This is an obvious disadvantage of the step-by-step method, because the separation behavior of the respective π-electron phase is influenced by the unconverted amino groups. An alternate approach involves immobilization of the complete organic ligand on the silica gel surface, which often requires more complex synthetic efforts. The *n*-alkyl fluorene phases were prepared with this method. Because aromatic ligand fragments occupy more space than alkyl ligands, solution polymerization results in higher bonding densities than surface polymerization, as is observed in the case of *n*-alkyl fluorene phases. An ordered polymerization of *n*-alkyl fluorenyl silanes on the surface of the silica gel is sterically hindered in comparison to the much more flexible *n*-alkyl silanes.

3.2 Characterization by Solid-state NMR Spectroscopy

Solid-state NMR spectroscopy is a very powerful tool for providing information about structure and dynamics of stationary phases.[9,10,13,15,18,20,21] Information about the surface of covalently modified silica gel is provided by ^{29}Si CP/MAS NMR spectroscopy.[7,8] Figure 3A depicts the characteristic silica gel surface signals of the *n*-hexylfluorene phase. The immobilization of the *n*-hexylfluorene ligand becomes clearly visible by the appearance of T-group signals at –58 ppm (T^2) and –66 ppm (T^3) due to the reaction of the *n*-hexylfluorenyl silane with the silica gel surface. The absence of T^1 group signals at –48 ppm is an indication of a high degree of crosslinking of trichlorosilane on the silica gel. Approximately equal crosspolarization characteristics of the respective T group signals allow the comparison of their signal intensity by peak deconvolution. By this method the exact degree of crosslinking (Q = 85.2%) was determined for the *n*-hexylfluorene phase. The resonances at higher field (Figure 3A) represent the signals of the unconverted surface species of the bare silica gel after the modification process, *i.e.* siloxane groups (Q^4) at –110 ppm and silanol groups (Q^3) at –101 ppm. Silanediol groups (Q^2) are detectable as a shoulder at –92 ppm, indicating the high reactivity of these species during the modification process.

Information about the structure of the immobilized *n*-hexylfluorenyl fragment is provided by ^{13}C CP/MAS NMR spectroscopy.[10,18,22] The different fragments of the organic ligand can be unambiguously assigned to signals in the ^{13}C CP/MAS NMR spectrum of the *n*-hexylfluorene phase in Figure 3B.

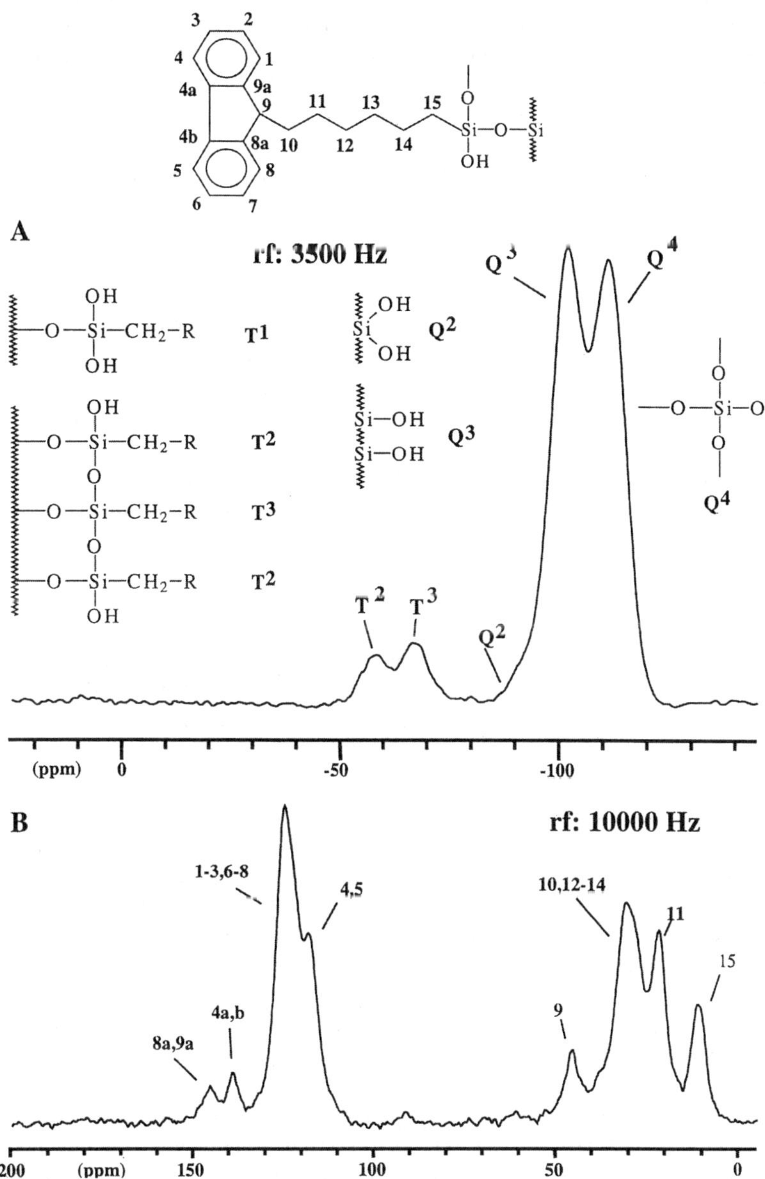

Figure 3 caption:

Figure 3   [29]*Si CP/MAS NMR spectroscopy provides information about the surface of a stationary phase, while* [13]*C CP/MAS NMR spectroscopy provides information about the structures of the organic ligand fragments.*

^{1}H MAS NMR spectroscopy provides additional information about the structure of organic ligand fragments of stationary phases. Very strong homonuclear dipole–dipole interactions are the main problem for the detection of ^{1}H solid-state NMR spectra. However, because the organic ligand fragments are diluted on the surface of the silica gel,[23,24] the homonuclear dipole–dipole interactions are significantly reduced. By employing high rotational frequencies up to 14 kHz, well resolved ^{1}H MAS NMR spectra can be recorded as illustrated in Figure 4A. Further reduction of the homonuclear dipole–dipole interactions can be received by ^{1}H MAS NMR measurements in the suspended-state.[25]

Although the rotation frequency is much smaller compared to conventional ^{1}H MAS NMR spectroscopy, significantly better resolution of the respective signals can be obtained by ^{1}H MAS NMR spectroscopy in the suspended-state (Figure 4B). Unfortunately, this method requires a much greater effort for the measurement than conventional ^{1}H MAS NMR spectroscopy.

The chromatographic separation process occurs in the presence of a mobile phase. Therefore a polymeric C_{30} stationary phase was investigated by suspended-state MAS NMR spectroscopy,[3] with the presence of the mobile phase methanol/MTBE (75/25, v/v) at 295 K, 305 K and 315 K (Figure 5). The assignments are as follows: C-1, 14.2 ppm; C-30, 14.1 ppm; C-29/C-2, 23 ppm; C-3, 33.6 ppm; C-28, 33.8 ppm; and for the main carbon chain, C-4 to C-27, 30.0 ppm and 32.6 ppm.

There should be only one signal for the methylene groups C-4 to C-27, because substituent effects are generally negligible for > 4 bonds. However, there are two signals at 30.0 ppm and 32.6 ppm. At lower temperature (295 K) the signal at 32.6 ppm is more intense than that at 30.0 ppm, whereas at 315 K the signal at 30.0 ppm is the more intense. As a function of temperature, interconversion of the intensities of the two signals is visible (Figure 5).

To obtain more information about the mobility of the main alkyl chain parts assigned to the signals in ^{13}C NMR spectra, experiments using 2D solid-state NMR methods such as ^{13}C-WISE (wideline separation) were performed.[26] Results are shown in Figure 6. The signal line width of the methylene proton ensemble in the second dimension in the ^{13}C-WISE NMR experiment is inversely proportional to the spin-spin relaxation time T_2 and therefore to the averaged overall mobility. The different signal line widths (48 kHz, 15 kHz, respectively) support the assignment of both signals to the two major types of alkyl chain arrangements arising from different methylene group conformations.[27]

Finally, the influence of the solvent used as mobile phase was investigated. Figure 8 shows ^{13}C MAS NMR spectra of a C_{30} phase suspended in various solvents. The relative amount of gauche and trans conformations varies as a function of the nature of the mobile phase.

3.3 HPLC Application

The selectivity of the *n*-hexylfluorene phase was investigated first by the application of the Sander and Wise Test (SRM 869)[28,29] Whereas hydrophobic interactions predominate the separation process using *n*-alkyl phases, π–π-interactions may also contribute to retention with *n*-hexylfluorene phases.[30] Owing to these π–π-interactions, the *n*-hexyl-fluorene phase is well suited to the separation of the three test compounds: BaP (benzo[a]-pyrene), PhPh (phenanthro[3,4-c]phenanthrene), and TBN (tetrabenzonaphthalene). The Sander and Wise Test was designed for a chromatographic characterization of shape selectivity of *n*-alkyl phases as indicated by the relative elution order of its compounds. In terms of the Sander and Wise nomenclature, *n*-alkyl phases are classified either as monomeric ($\alpha_{TBN/BaP} > 1.7$), intermediate ($1.0 < \alpha_{TBN/BaP} < 1.7$), or polymeric ($\alpha_{TBN/BaP} < 1.0$). The chromatogram in Figure 9 illustrates the separation of the three test molecules with the *n*-hexylfluorene phase. The $\alpha_{TBN/BaP}$ value of 1.81 characterizes the phase as monomeric in terms of *n*-alkyl phase selectivity. By ^{29}Si CP/MAS NMR spectroscopy (Figure 3A) the *n*-hexylfluorene phase was identified as typically polymeric with strong T group signals and a high degree of crosslinking. This finding is not a contradiction; it merely proves the important contribution of π–π-interactions in the interaction mechanism. It is clear, then, that the *n*-hexylfluorene phase cannot be classified by the system used to classify *n*-alkyl phases.[5]

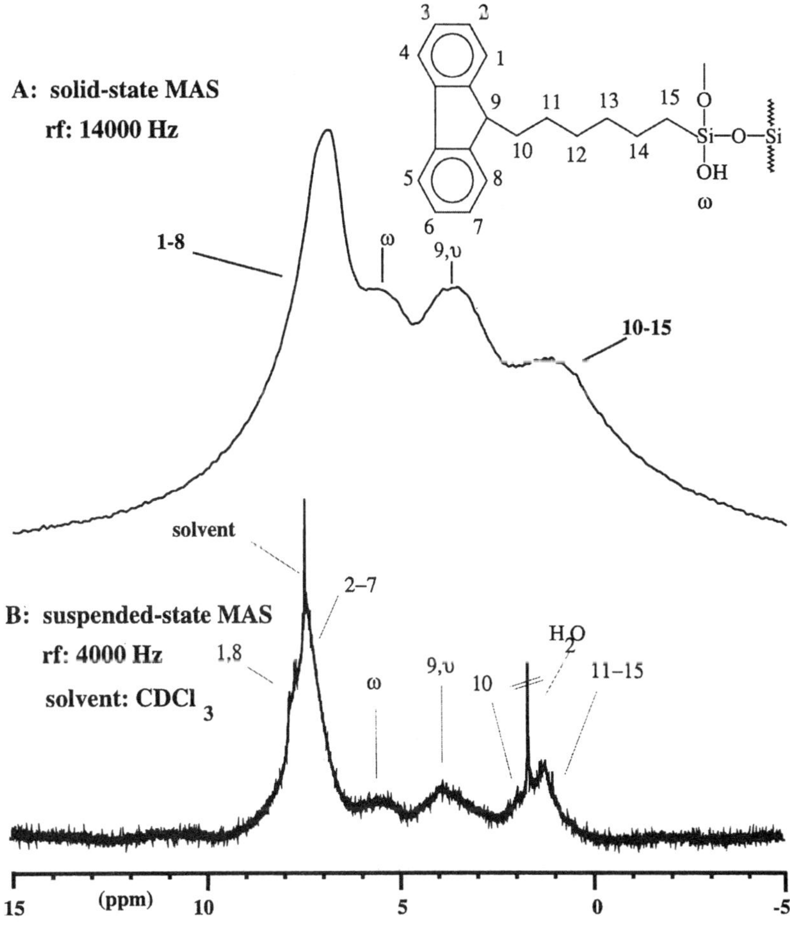

Figure 4 *Comparison of two different ^{1}H MAS NMR measurement techniques applied on the n-hexyl fluorene phase. A: conventional solid state ^{1}H MAS NMR spectroscopy, rotation frequency (rf) = 14000 Hz. B: ^{1}H MAS NMR spectroscopy in the suspended state, rf = 4000 Hz.*

The selectivity of the *n*-hexylfluorene phase was also investigated by a second, more complex test. This test mixture, containing sixteen PAHs regarded as priority pollutants, is widely used as an indicator of stationary phase selectivity. The best separation using the *n*-hexylfluorene phase[6] was obtained by the implementation of a gradient elution program as illustrated in Figure 9B. After the solvent composition was held for 10 minutes at a methanol–water ratio of 70:30, a linear gradient was used for 30 minutes to a methanol–water ratio of 90:10. The later solvent composition was held until the end of the separation. The *n*-hexylfluorene phase was found to be suitable for the separation of all sample molecules except for PAHs 11 and 12, underlining its usefulness in the field of PAH analysis.

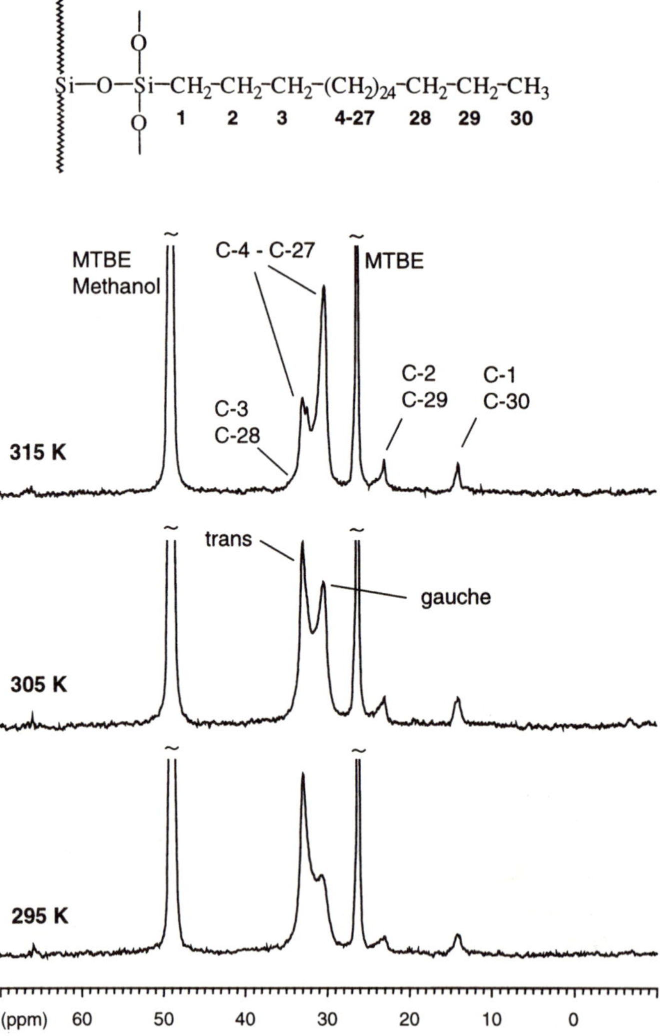

Figure 5 *Temperature dependent* ^{13}*C MAS NMR spectra of a C*$_{30}$ *phase suspended in methanol/MTBE (75/25, v/v).*

The separation of nitro explosives is another important task in environmental analysis. For this separation problem either C_{18}, cyano, or fluorene phases are used as stationary phases. In contrast to C_{18} phases,[31] the elution order of the solute molecules is inverted by the use of fluorene phases which have the aromatic moiety connected to the

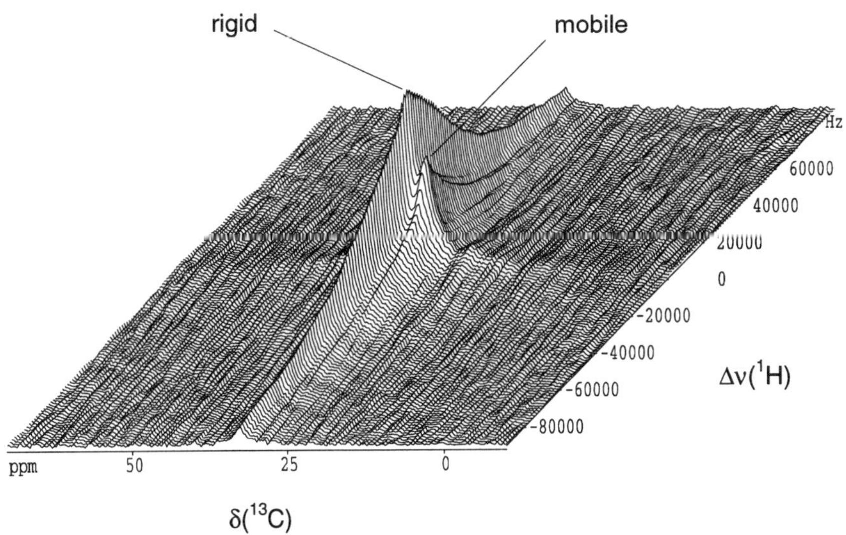

Figure 6 *2D WISE NMR spectrum of a C_{30} phase.*

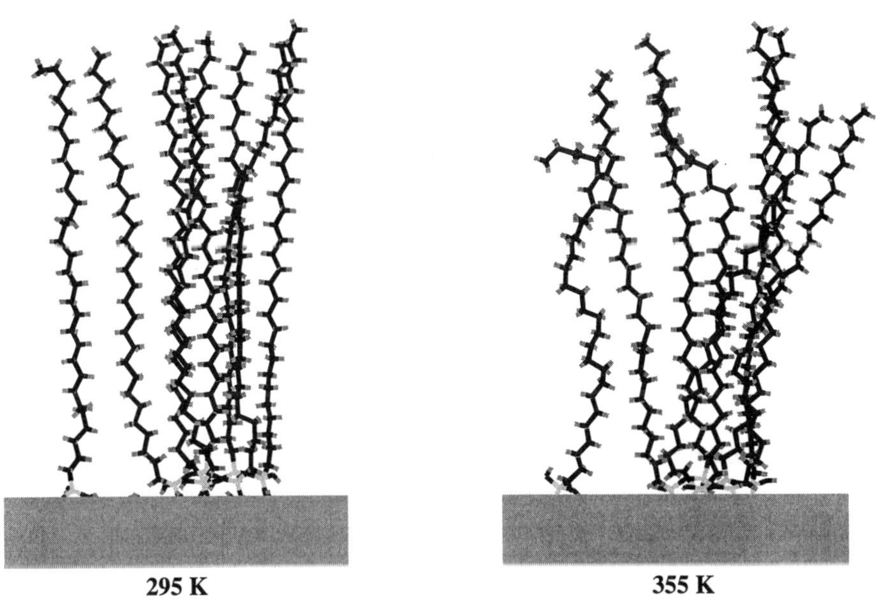

Figure 7 *Molecular modeling of a C_{30} phase at 295 K and 355 K.*

silica gel by amido couplings (Figure 10).[14] Owing to strong π–π-interactions between the electron-rich fluorene ligand fragments of the stationary phase and the electron-poor solutes, the nitro explosives with a single nitro group are eluted first, followed by the molecules with two and, finally, three nitro substituents. An example of such a separation for the *n*-hexylfluorene phase is shown in Figure 10. This separation is a further proof of the predominance of π–π-interactions in the separation process of aromatic solutes.

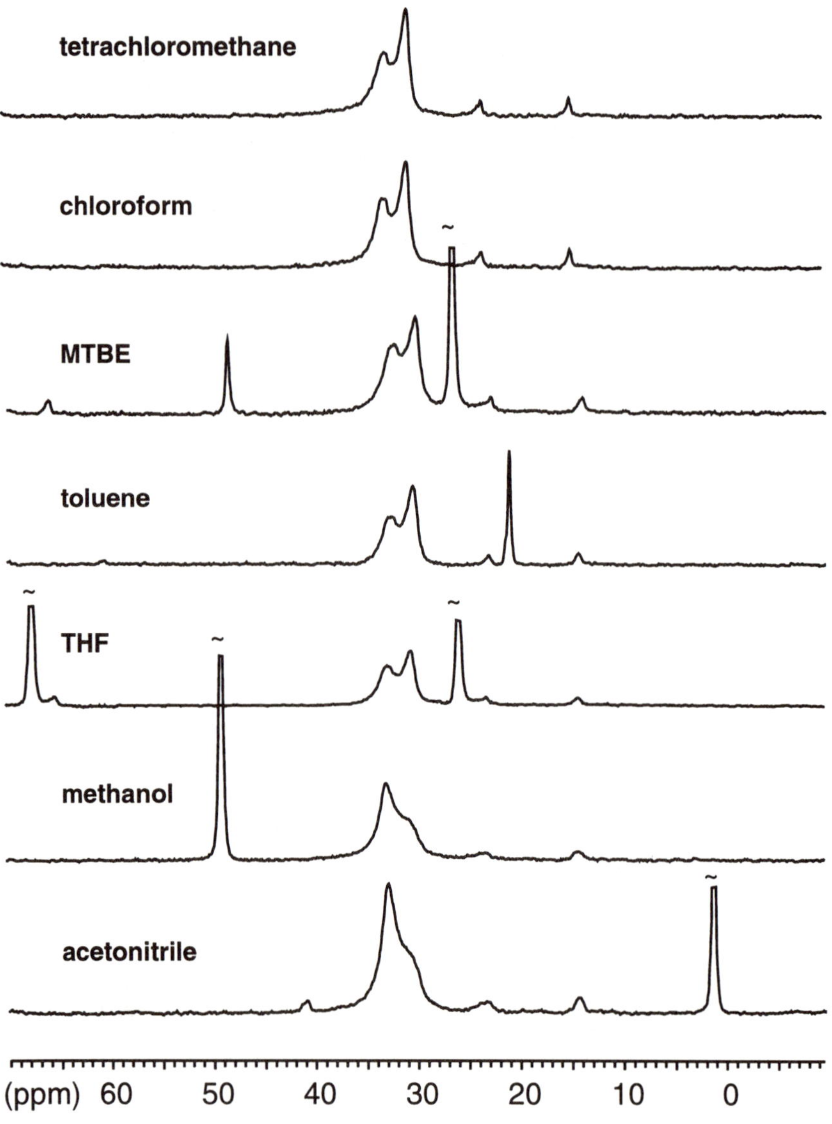

Figure 8   *^{13}C MAS NMR spectra of a C_{30} phase suspended in various solvents.*

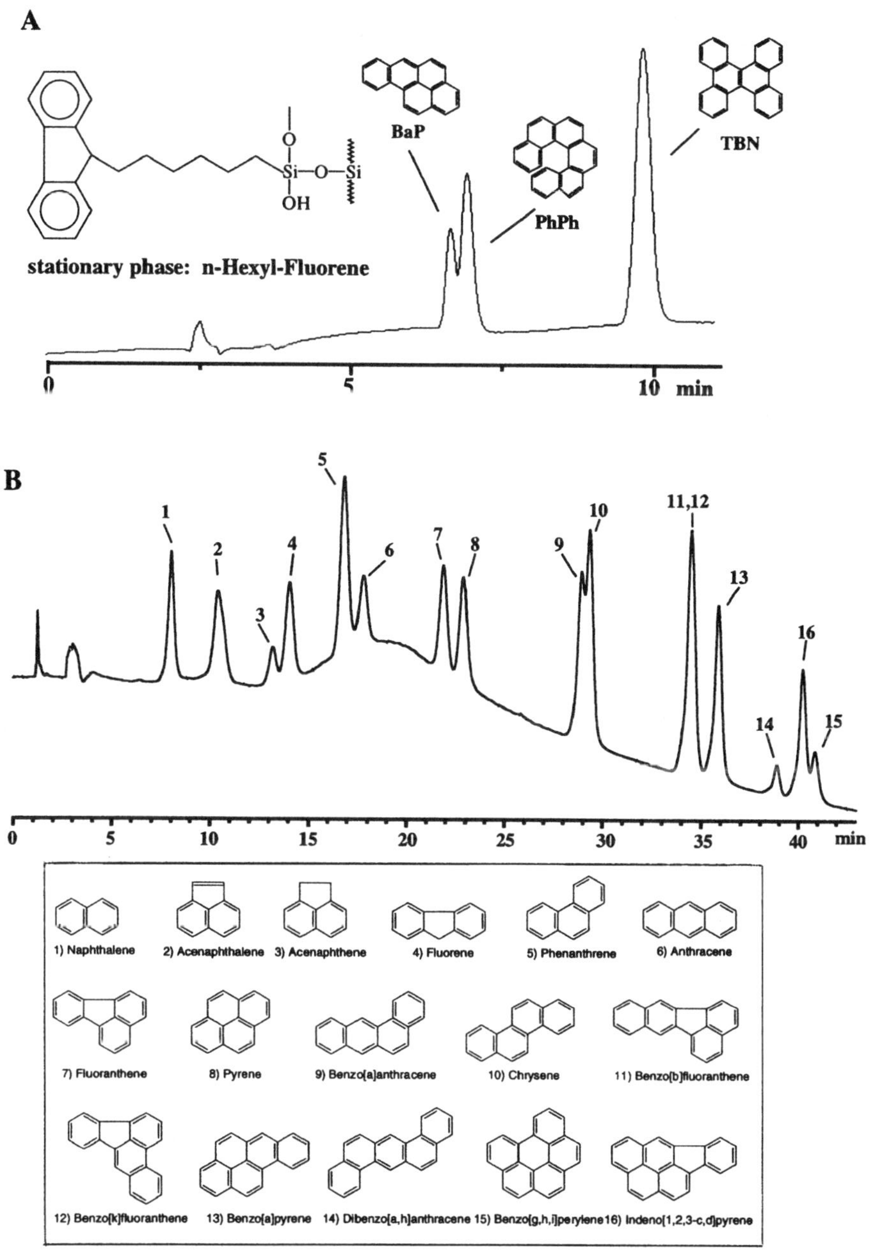

Figure 9   *Separation of PAH samples with the n-hexylfluorene phase. A: test mixture SRM 869, acetonitrile–water (85:15, v/v). B: test mixture SRM 1647, 10 min hold at methanol–water (70:30, v/v), 30 min linear gradient to methanol–water (90:10, v/v), hold at 90:10 to end.*

For the HPLC investigations with C_{30} phases,[32] isomerized β-carotene was used. Figure 11 shows a comparison of the separation of *cis/trans* isomers of β-carotene with C_{18} and C_{30} phases. The resolution of the isomers is clearly dependent on the stationary phase. Better separation of 13-*cis*, all-*trans* and 9-*cis* isomers was achieved with the C_{30} phase. The assignment of chromatographic peaks to the corresponding isomers was performed by LC–NMR. It can be concluded that the interaction between the stationary phase and the solutes is better with C_{30} alkyl chains than those with C_{18} alkyl chains.

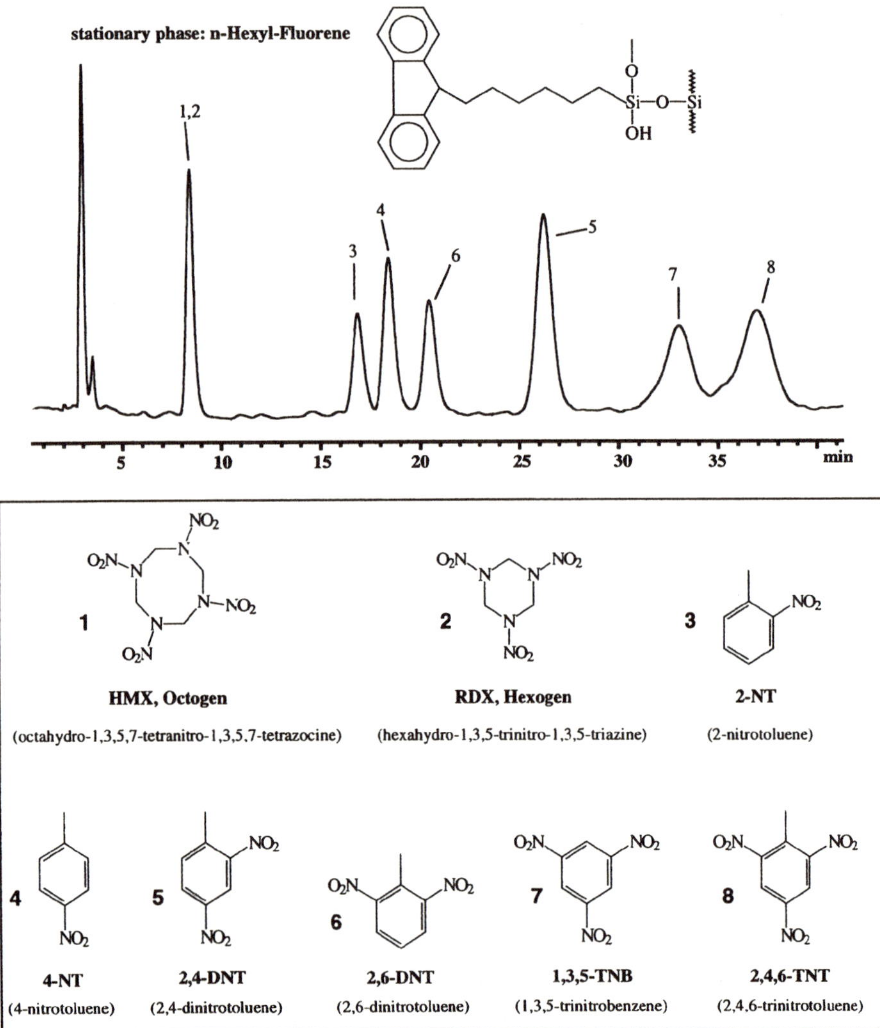

Figure 10 *Separation of nitro explosives with the n-hexylfluorene phase. Mobile phase: methanol–water (50/50, v/v). The elution order is inverted compared to conventional n-alkyl phases.*

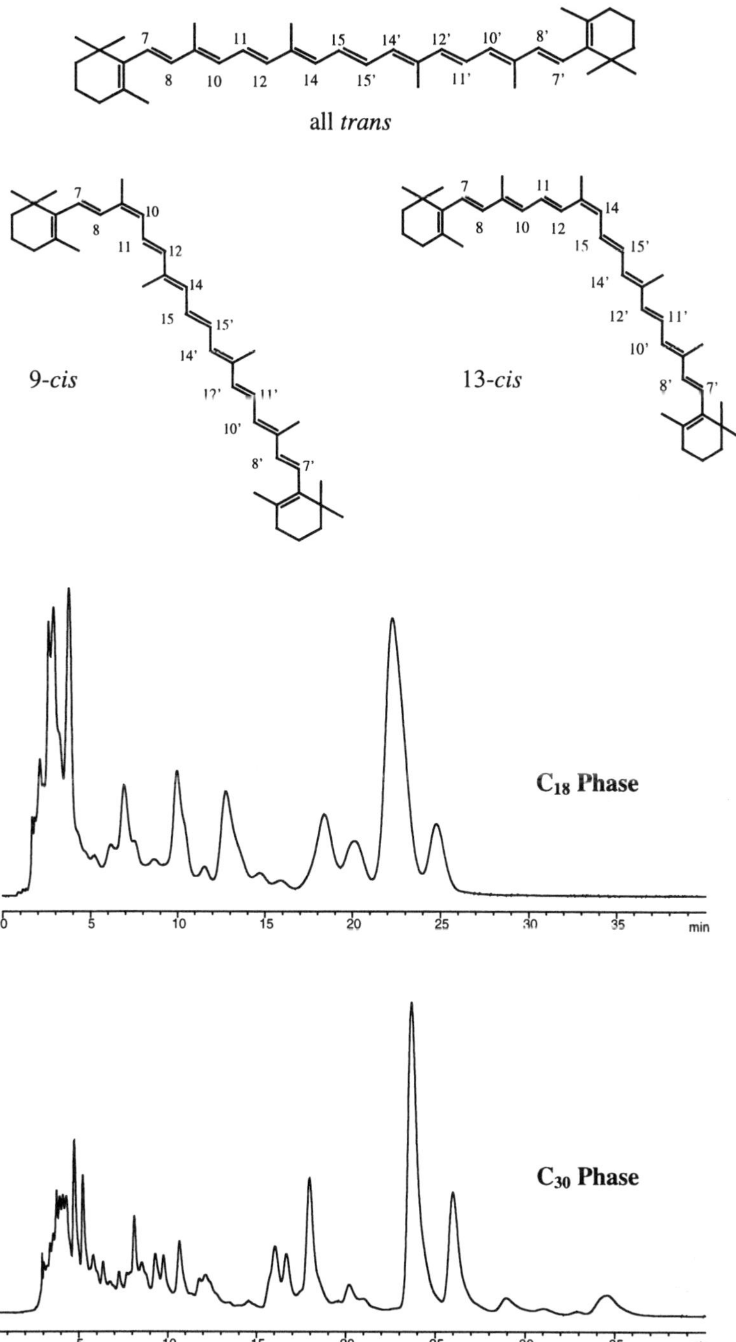

Figure 11 *Comparison of the separation of cis/trans isomers of β-carotene on C18 phase (top) and C30 phase (bottom).*

3.4 Molecular Modeling

For a better understanding of the behavior of the stationary phases, molecular modeling calculations were performed to simulate bonded C_{30} alkyl chains. Recently, Yarovsky *et al.* performed molecular modeling of C_4, C_8 and C_{18} bonded alkyl chains.[11]

Although the influence of solvent was ignored and a zero dielectric constant was assumed in these force field calculations, these qualitative simulations show trends in agreement with the results obtained from NMR and HPLC experiments. Figure 7 shows the C_{30} bonded phase at temperatures of 295 and 355 K. There is an increase of gauche (mobile) *versus* trans (rigid) conformations at higher temperature due to higher kinetic energy. The alkyl chains exhibit gauche conformations much closer to the surface at 355 K than those at 295 K, because small changes in torsional angles close to the silica surface result in larger changes in overall conformation of the C_{30} chains. Therefore, gauche conformations can only occur in the part of the chains close to the silica surface when the whole alkyl chain exhibits large amplitude motions, which is less likely at 295 K than at 355 K.

Temperature can also be seen to influence alkyl chain thickness. Triacontyl chains at 355 K are shorter compared with those at 295 K due to a higher amount of gauche conformations. The average interphase thickness appears to be larger at lower temperature than at higher temperature, which is evidence for a more rigid environment with low amplitude motions. This suggests that surface accessibility of the interphase for solutes varies with temperature. At higher temperature, where the chains possess more gauche conformations, penetration of the β-carotene isomers into the interphase is restricted. With decreasing temperature the mobility of the bonded chains also decreases and the alkyl chains predominantly form trans conformations. Under these conditions the interphase possesses more rigid chains, which favors the insertion of rigid solutes into the interphase.[30] The molecular modeling simulations thus support the results obtained from the NMR and HPLC experiments.

4 SUMMARY

A variety of approaches were utilized to investigate the behavior of stationary interphases. Solid-state NMR spectroscopy, HPLC and molecular modeling investigations provide a coherent model. The utility of the *n*-hexylfluorene phase for HPLC applications was demonstrated by the separation of both PAH and nitro explosive solutes. In addition to the hydrophobic interactions of *n*-alkyl phases, π–π-interactions play a significant role in the interaction mechanism of HPLC separations for *n*-hexylfluorene stationary phases.

The temperature dependent investigations of C_{30} phases show an interconversion of a liquid-like state with a higher amount of gauche conformations, to a solid-like state with a larger fraction of trans conformations. Molecular modeling calculations were performed and show a trend consistent with the results obtained from NMR and HPLC experiments.

5 ACKNOWLEDGMENTS

We gratefully acknowledge the support of B. Schindler and L.-H. Tseng for their help in performing the solid-state NMR measurements. We also wish to thank W. Wieland for his help in synthesis. Additionally we thank Dr. L. C. Sander for the SRM 869 and SRM 1647 standards and Dr. M. Kaiser (BICT) for the nitro explosive standards. We thank Dr. L. C. Sander for helpful discussions and for reading the manuscript. Further, we are indebted to Bischoff Chromatography (Leonberg, Germany) for the silica samples. Financial support by the Deutsche Forschungsgemeinschaft (Forschergruppe, Grant No. Li 154/41-3 and Li 154/41-4) and the Fonds der Chemischen Industrie is gratefully acknowledged.

References

1.   'Chromatogr. Sci.', K. K. Unger, ed., Vol. 47 (Packings and Stationary Phases in Chromatographic Techniques), Marcel Dekker Inc., New York, Basel, 1990.

2.   L. C. Sander and S. A. Wise, in 'Retention and Selectivity in Liquid Chromatography (Journal of Chromatography Library, Vol. 57)', R. M. Smith, ed., Elsevier, Amsterdam, 1995, p. 337.

3.   K. Albert, T. Lacker, M. Raitza, M. Pursch, H.-J. Egelhaaf and D. Oelkrug, *Angew. Chem., Int. Ed. Engl.*, 1998, **37**, 778.

4.   M. Pursch, L. C. Sander, H.-J. Egelhaaf, M. Raitza, S. A. Wise, D. Oelkrug and K. Albert, *J. Am. Chem. Soc.*, submitted for publication (1998).

5.   W. Wielandt, A. Ellwanger, K. Albert and E. Lindner, *J. Chromatogr. A*, 1998, **805**, 71.

6.   A. Ellwanger, R. Brindle, M. Kaiser, W. Wielandt, E. Lindner and K. Albert, *J. Chromatogr. A*, submitted for publication (1998).

7.   D. W. Sindorf and G. E. Maciel, *J. Am. Chem. Soc.*, 1983, **105**, 1848.

8.   G. E. Maciel and D. W. Sindorf, *J. Am. Chem. Soc.*, 1980, **102**, 7606.

9.   E. Bayer, K. Albert, J. Reiners, M. Nieder and D. Müller, *J. Chromatogr.*, 1983, **264**, 197.

10. K. Albert and E. Bayer, *J. Chromatogr.*, 1991, **544**, 345.

11. I. Yarovsky, M.-I. Aquilar and M. T. W. Hearn, *Anal. Chem.*, 1995, **67**, 2145.

12. L. C. Sander, K. Epler Sharpless, N. E. Craft and S. A. Wise, *Anal. Chem.*, 1994, **66**, 1667.

13. A. Ellwanger, R. Brindle and K. Albert, *J. High Resol. Chromatogr.*, 1997, **20**, 39.

14. R. Brindle and K. Albert, *J. Chromatogr. A*, 1997, **757**, 3.

15. M. Pursch, S. Strohschein, H. Händel and K. Albert, *Anal. Chem.*, 1996, **68**, 386.

16. H. O. Fatunmbi, M. D. Bruch and M. J. Wirth, *Anal. Chem.*, 1993, **65**, 2048.

17. L. C. Sander and S. A. Wise, *Anal. Chem.*, 1995, **67**, 3284.

18. M. Pursch, L. Sander and K. Albert, *Anal. Chem.*, 1996, **68**, 4107.

19. H.-J. Egelhaaf, D. Oelkrug, A. Ellwanger and K. Albert, *J. High Resol. Chromatogr.*, 1998, **21**, 11.

20. K. Albert, H. Händel, M. Pursch and S. Strohschein, in 'Chemically Modified Surfaces, Vol. 6', J. J. Pesek, M. T. Matyska and R. R. Abuelafiya, eds., The Royal Society of Chemistry, Cambridge, 1996, p. 30.

21. J. J. Pesek, M. T. Matyska and J. Ramakrishnan, *Chromatographia*, 1997, **44**, 538.

22. M. Pursch, R. Brindle, A. Ellwanger, L. C. Sander, C. M. Bell, H. Händel and K. Albert, *Solid State NMR*, 1997, **9**, 191.

23. S. F. Dec, C. E. Bronnimann, R. A. Wind and G. E. Maciel, *J. Magn. Reson.*, 1989, **82**, 454.

24. R. Brindle, M. Pursch and K. Albert, *Solid-State NMR*, 1996, **6**, 251.

25. M. Pursch, G. Schlotterbeck, L.-H. Tseng, K. Albert and W. Rapp, *Angew. Chem., Int. Ed. Engl.*, 1996, **35**, 2869.

26. K. Schmidt-Rohr, J. Clauss and H.-W. Spiess, *Macromolecules*, 1992, **25**, 3273.

27. W. L. Earl and D. L. Van der Hart, *Macromolecules*, 1979, **12**, 762.

28. L. C. Sander, *J. Chromatogr. Sci.*, 1988, **26**, 380.

29. L. C. Sander and S. A. Wise, *LC GC Int.*, 1990, **3**, 24.

30. A. Tchapla, S. Heron, E. Lesellier and H. Colin, *J. Chromatogr. A*, 1993, **656**, 81.

31. E. S. P. Bouvier and S. A. Oehrle, *LC GC Int.*, 1995, **8**, 338.

32. S. Strohschein, M. Pursch, D. Lubda and K. Albert, *Anal. Chem.*, 1998, **70**, 13.

MICROSCALE SYNTHESIS AND SCREENING OF COMBINATORIAL LIBRARIES OF NEW CHROMATOGRAPHIC STATIONARY PHASES

Christopher J. Welch,* Marina N. Protopopova and Ganapati A. Bhat

Regis Technologies, Inc.
8210 Austin Avenue
Morton Grove, IL USA

1 INTRODUCTION

The current selection of commercial chiral stationary phases (CSPs) for large scale chromatographic separations is rather limited, and most have been developed as general purpose CSPs rather than the best CSP for a particular separation. While new CSPs can be designed for specific separations, the development time is often too long to merit serious consideration by process engineers. As a means of rapidly finding the best preparative CSP for a given task, we have recently developed a new method for the rapid solid phase synthesis and screening of libraries containing milligram quantities of diverse CSPs.[1,2]

1.1 The Emerging Importance of Industrial Scale Preparative Chromatography and Implications for the Design of Chromatographic Adsorbents

Within the past decade the technique of chromatographic enantioseparation has become the method of choice for analytical determinations of enantiopurity.[3] The method is widely used, particularly in the pharmaceutical industry, where most new chiral drugs are manufactured in enantiomerically pure form. In recent years the use of preparative chromatographic enantioseparation has become increasingly popular. While generally more expensive than manufacturing routes employing enantioselective synthesis or classical resolution, chiral HPLC offers the tremendous advantage of speed. Consequently, many pharmaceutical companies use preparative chiral HPLC in the early stages of drug discovery to rapidly produce enantiomerically pure drug candidates for animal testing, metabolism and toxicology studies, *etc.* Once a drug candidate has been selected for larger scale development, alternative manufacturing methods are typically used, although in a few cases chiral HPLC is used to produce enantiopure drugs on the ton/year scale.

Most commercial CSPs have been developed using trial and error methodology, and have been commercialized because they demonstrate some general ability to separate enantiomers. Of these many commercial CSPs, only a small fraction are available in bulk or can be produced in an economical fashion for large scale preparative chromatography.[4] Furthermore, rather than a CSP which has a general ability to separate the enantiomers of a large number of racemates, the process engineer considering a potential manufacturing route for an enantiopure drug is interested in a CSP which can separate the enantiomers of one particular compound.

Practical large scale chromatographic enantioseparation requires highly enantioselective CSPs.[3,4] For example, chromatographic resolution of the enantiomers of a racemate using a CSP with an enantioselectivity of 1.3 can be rather tedious. A comparable CSP having an enantioselectivity of 2 can sometimes afford 5–10 fold greater productivity, allowing the use of smaller and less expensive equipment and requiring the handling of much less solvent.[5] An example of a good preparative chromatographic enantioseparation is shown in Figure 1.

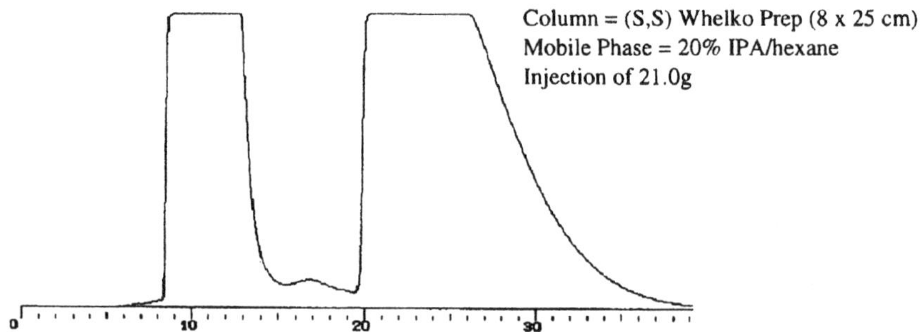

Figure 1   *An example showing an easy preparative chiral HPLC separation. Factors which are important for convenient preparative enantioseparation include high enantioselectivity, good solubility of the analyte in the mobile phase, elution of the desired component before the undesired, and the ability of the product enantiopurity to be upgraded via crystallization.*

The technique of simulated moving bed chromatography (SMB) has recently received attention as a useful tool for large scale chromatographic separations using CSPs with modest enantioselectivity.[6,7] While it is true that SMB can sometimes lead to gains in productivity and can generally reduce solvent consumption by 50%, this technique is an engineering solution for what is essentially a chemical problem. The economically preferable solution is to use a CSP with greater enantioselectivity. The greater the enantioselectivity, the more economical the separation process. There has been considerable recent progress in the design of highly enantioselective CSPs, with several reports of enantioselectivities in excess of 100. With enantioselectivity this great, separation becomes a trivial task. For instance, Pirkle and coworkers have demonstrated that CSPs capable of separating the enantiomers of model compounds with enantioselectivities in excess of 25 can be used for low cost single plate separations to afford nearly enantiopure material by 'batch filtration'.[8] The advantages of such low cost manufacturing approaches utilizing highly enantioselective CSPs are obvious. However, there are a number of difficulties in developing new highly enantioselective CSPs for given separation problems.

2 NEW STATIONARY PHASE DISCOVERY

The pace of discovery and development of new chromatographic adsorbents is dependent upon the time required for synthesis and evaluation of new candidate materials.

Traditional approaches to the discovery of new chromatographic adsorbents have utilized a labor intensive approach in which a candidate adsorbent is designed, prepared on the multigram scale, and the resulting material is packed into a column and evaluated chromatographically.[9] Such an approach can take weeks, months, or even years, and is clearly too slow to develop rapid solutions to actual separation problems in the fast-paced pharmaceutical industry.

2.1 Approaches to Rapid Stationary Phase Discovery

To date, several approaches have been reported to increase the pace of discovery of new chromatographic stationary phases:

2.1.1 Immobilized Guest Method.   Pirkle and Welch have reported a process termed the immobilized guest method.[9,10] In this method, illustrated in Figure 2, a single enantiomer of the compound of interest is immobilized on a solid phase, which is then packed into a chromatography column. Small amounts of candidate selectors are then screened for enantioselective binding using conventional chromatography. The most promising candidates are then prepared in enantiopure form and used to prepare a new CSP which should be useful for separating the enantiomers of the compound of interest. This technique relies on the so-called 'principle of reciprocity', which has long been emphasized by the Pirkle group. The immobilized guest method represents an improvement on the traditional methods used to develop new stationary phases and selectors.

The chief advantage of the immobilized guest method is that submilligram amounts of candidate racemic selectors, which need not be absolutely pure, can be screened using automated HPLC. Disadvantages of the approach include the requirement of gram scale amounts of the enantiopure drug, and the need to immobilize this material on a solid phase. The most serious disadvantage of this method is that the screening results cannot be directly translated to chromatographic performance. For instance, screening of extensive racemate libraries can lead to the discovery of an optimal selector, which when immobilized on a solid phase may afford a CSP with dismal performance. There are at least two ways in which this can happen, the first being that the "optimal selector" is actually the best for enantioselective binding to the tethered guest, and may display little enantioselectivity for the free guest. The second situation occurs when selector immobilization destroys some key interactions required for molecular recognition.

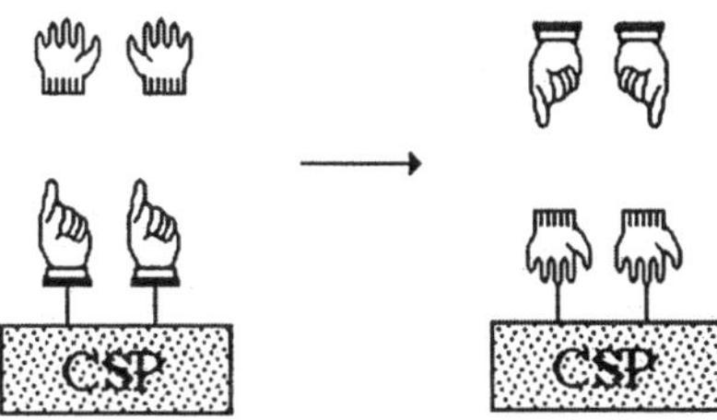

Figure 2   *"Immobilized Guest" approach for developing new CSPs. The compound of interest is immobilized on a solid phase and packed into a chromatography column.*

2.1.2 Screening Encoded Combinatorial CSP Libraries with Dye Conjugates. Still *et al.* have developed a method in which candidate selectors from combinatorial libraries are evaluated using analogs of the target enantiomers to which dye molecules have been conjugated.[11] For example, libraries of candidate CSPs on individual polystyrene beads are screened with a set of analytes derived from the enantiomers of the target compound: one enantiomer contains a pendant blue dye and the other enantiomer contains a pendant red dye. Microscopic examination reveals some beads which become red in color when exposed to a mixture of these two dye conjugates, indicating some preferential binding of one enantiomer, some beads which become blue, indicating preferential binding of the opposite enantiomer, and some beads which become purple, indicating binding of both enantiomers. Deconvolution of an encoded tag on a given bead reveals the structure of the CSP contained on it. This elegant technique allows rapid screening of vast numbers of beads bearing unique selectors in a single assay. However, it suffers from some fundamental drawbacks stemming primarily from the fact that the screening results cannot be directly extrapolated to chromatographic performance. For example, the presence of the pendant dye molecule may influence the binding of the conjugate to the stationary phase; in other words, we may be selecting the best selector for a tethered, rather than a free analyte. Furthermore, enantiopure samples of the two enantiomers of the target molecule must be available for synthesis of the dye conjugates, a situation that rarely exists in the real world. And finally, some compounds are not amenable to the formation of dye conjugates owing to a lack of appropriate functionality.

2.1.3 Phage Display Technology. Phage display technology, or phage panning is an interesting molecular biology technique which is being developed for discovery of new chromatographic selectors.[12] The technique utilizes a bacteriophage which presents a section of protein "knob" outside of the viral capsid. Using molecular biology techniques, the DNA sequence coding for this "knob" is manipulated to afford nearly identical bacteriophages, all presenting slightly different "knob" proteins to the external world. This library of bacteriophages is then screened for their ability to bind to a surface containing an immobilized enantiomer. Phages which show no binding are washed away, then those phages which bind strongly are dislodged from the support and used to infect bacteria, where they proliferate and form millions of copies of themselves. The DNA sequence from these phages is then determined, revealing the sequence of the protein "knob." The technique can be made fairly elaborate. For example, multiple screenings can be done to find phages that bind strongly to a surface consisting of one immobilized enantiomer, but not to a surface consisting of the opposite immobilized enantiomer. The technology is truly impressive, but again suffers from some drawbacks stemming primarily from the inability to extrapolate directly to chromatographic performance. For example, the use of a tethered enantiomer for the capture phase raises the potential for selection of an optimal selector for a tethered, but not a free analyte. Perhaps more important is the fact that once a selector has been identified, it must be prepared on large scale at an affordable cost and immobilized on a chromatographic support without changing its molecular recognition characteristics, a task which could require considerable developmental research.

2.2 A New Approach to CSP Development

We recently reported a strategy for rapid synthesis and screening of libraries of CSPs which offers a tool for the rapid discovery of the CSP that will work best for preparatively resolving the enantiomers of a given compound. The implementation of this strategy consists of two parts: (1) parallel synthesis of libraries of milligram quantities of diverse stationary phases on porous silica particles,[1] and (2) screening the resulting libraries for chromatographic performance.[2] This method for synthesis and screening of CSP libraries has a number of advantages when compared to competing methods used for developing new chromatographic stationary phases. For example, CSP library synthesis on milligram scale results in tremendous savings in materials and reagents. In addition, the parallel nature of the library synthesis means that tens or hundreds of new CSPs can be prepared in the time required for making one full sized CSP. Furthermore, the CSP library screening technique is simple, rapid, inexpensive, and it can be automated – screening of 50 different CSPs can be accomplished in a single day. Another advantage is that after evaluation, the CSP libraries can be washed and reused. Finally, the compound mixture of interest can be directly used in the screening assay without the need for any purification, derivatization, immobilization, or formation of conjugates. Both the CSP and the analyte being evaluated are exactly the same as those that will be used in the full scale chromatographic separation. Therefore a successful screening result can be translated into successful chromatography at the large scale, without the risk of failure due to such things as tether effects or selector immobilization.

2.2.1 Silica-based Solid Phase Synthesis. More than twenty years ago Grushka and coworkers prepared a bonded tripeptide stationary phase using a modified Merrifield solid phase peptide synthesis approach.[13] While it was later shown that a more effective version of this stationary phase could be obtained by first synthesizing and purifying the tripeptide reagent and then immobilizing it on a silica support,[14] the convenience of the solid phase synthesis method for initial discovery of new CSPs is evident. An approach employing modified solid phase peptide synthesis on aminopropyl silica particles was chosen as the most convenient method for preparing combinatorial libraries of CSPs. A necessary prerequisite to begin this work was to learn to make aminopropyl silica with enough free amino groups to be useful, but not so many as to be detrimental. This problem was solved through trial and error.

2.2.2 Silica-based Solid Phase Synthesis of DNB-Leu CSP. Our next goal was to develop coupling, deprotection and acylation conditions that afford viable CSPs free from racemization and with appropriate levels of substitution. For this part of the study, we prepared as a model the well-known 3,5-dinitrobenzoyl Leucine (DNB-Leu) CSP on 5 g scale using the solid phase synthesis protocol outlined in Figure 3. The CSP thus obtained, when packed in a column, separated a group of test analytes nearly as well as the commercial columns.

2.2.3 Silica-based Solid Phase Synthesis of DNB-Peptido CSPs. Thus, confident in our methodology, we set about preparing and evaluating a group of peptido CSPs using a split synthesis type of approach (Figure 4). The group of four CSPs shown below were prepared on 5 g scale, packed into columns and evaluated chromatographically. These phases showed some interesting chromatographic behavior, which is detailed elsewhere.[1]

Two of the most interesting CSPs from this initial group are shown in Figure 5. These CSPs are nearly identical, differing only in one leucine residue. Nevertheless, substantial differences in enantioselectivity are noted for the group of test analytes.

2.2.4 Microscale Silica-based Solid Phase Synthesis of CSPs. The foregoing experiments clearly show the utility of a silica-based solid phase synthesis approach to CSP development. While the cost and time required to make each of these materials on 5 g scale is less than that of conventional CSP development, we required an even more rapid way of sampling the structural diversity of the DNB peptide family. Consequently, we chose to prepare candidate CSPs on 50 mg scale and use our new *ex*-column screening technique to evaluate the enantioselectivity of each CSP.

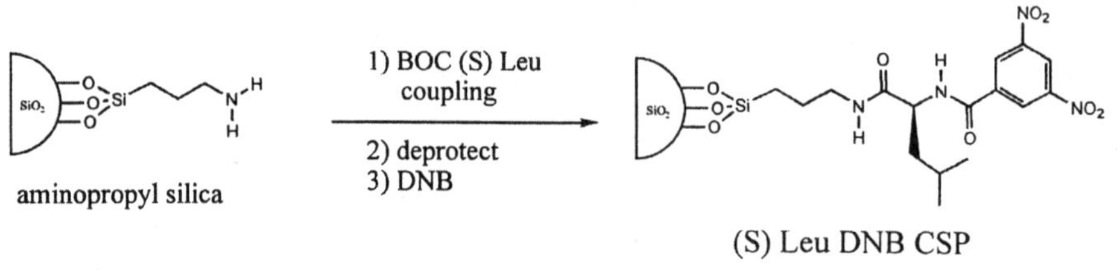

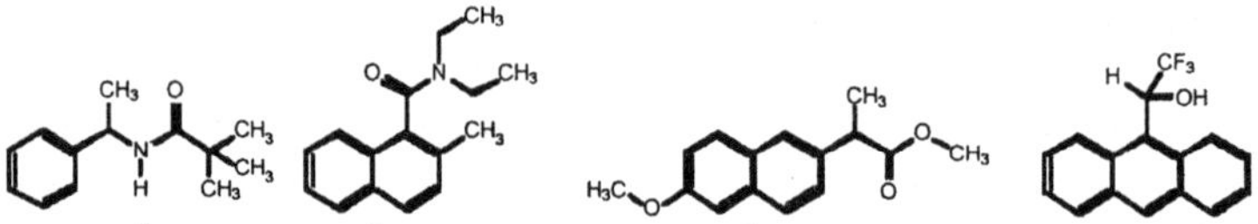

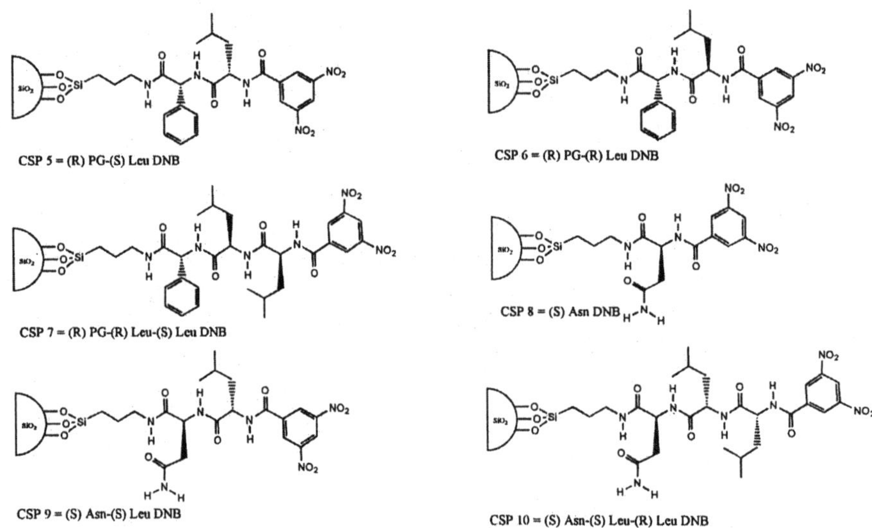

CSP	**1** k'_1	α	**2** k'_1	α	**3** k'_1	α	**4** k'_1	α
Commercial	2.17	1.20	5.67	1.10	3.19	1.33	1.21	1.3
SPS	2.30	1.15	6.48	1.08	3.64	1.33	1.26	1.3

Figure 3   *The "classic" DNB-Leu CSP prepared by solid phase synthesis shows comparable performance to the commercial version which is prepared by a different technique.*

Figure 4   *A variety of DNB-peptido CSP prepared by solid phase synthesis.*

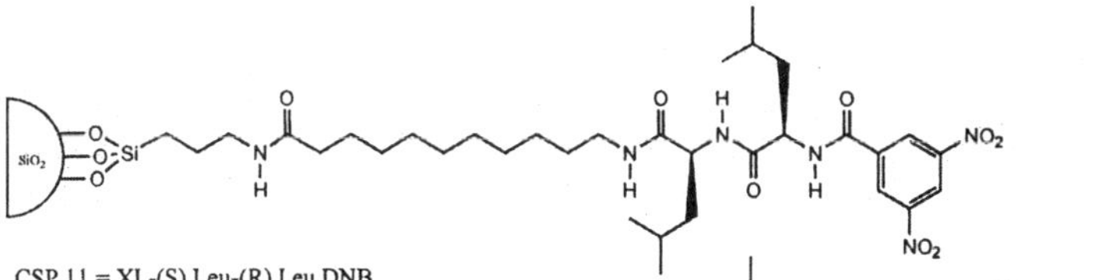

CSP 11 = XL-(S) Leu-(R) Leu DNB

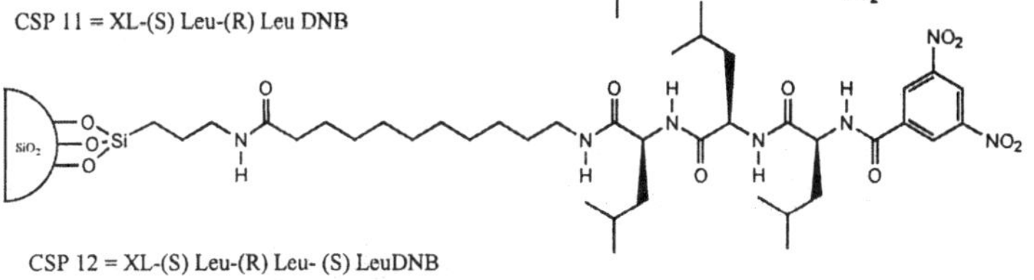

CSP 12 = XL-(S) Leu-(R) Leu- (S) LeuDNB

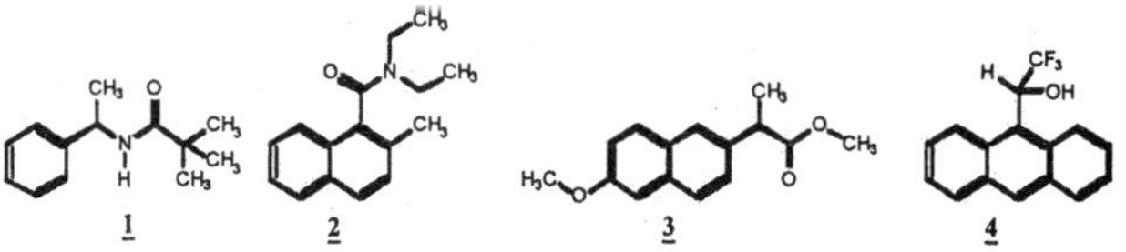

CSP	k'_1	α	k'_1	α	k'_1	α	k'_1	α
	1		**2**		**3**		**4**	
11	2.25	1.16	2.20	1.22	1.32	1.00	2.37	1.2
12	1.70	1.00	3.15	2.22	8.06	1.84	4.83	1.3

Figure 5 *Substantial differences are noted in the performance of structurally similar DNB-dipeptide and tripeptide CSPs.*

We chose to first prepare a library of 50 dipeptide DNB CSPs using all possible combinations of the 5 amino acids: valine, glutamine, phenylalanine, phenylglycine and proline (Figure 6). This set includes sterically bulky, strong hydrogen bonding and aromatic amino acids.

The solid-phase peptide synthesis method which we used in the multigram scale preparation of the CSPs shown in Figures 4 and 5 was scaled down to prepare 50 mg of each of the 50 dipeptide DNB CSPs resulting from all combinations of the 5 amino acids shown in Figure 6. Evaluation of the library was first tested using a model racemate which is known to be well resolved on DNB amino acid types of CSPs.[15-17] The results of the screen are presented in Figure 7.

The vertical axis in Figure 7 represents enantioselectivity, with the tallest bars indicating the best separations. The overall method clearly works, and tells us some useful information. Previous experience with this chiral recognition system had led us to believe that an amide hydrogen on the amino acid closest to the DNB group (aa 2) is essential for good separation.[17] Furthermore, it was suspected that amino acids with a large steric group at this position should work best, with aromatic groups at this position generally being poorer than steric groups. It thus comes as no surprise that the proline in position aa 2 works very poorly, while valine and phenylalanine in this position work best. Some unexpected results are obtained, even though this chiral recognition system has been extensively studied for more than a decade by a variety of techniques in addition to chromatography, including X-ray analysis of co-crystals[18] and nOe NMR analyses of 1:1 complexes.[17] One unexpected result of the screen is the finding that glutamine in position aa 1 seems to have a beneficial effect on enantioselectivity.

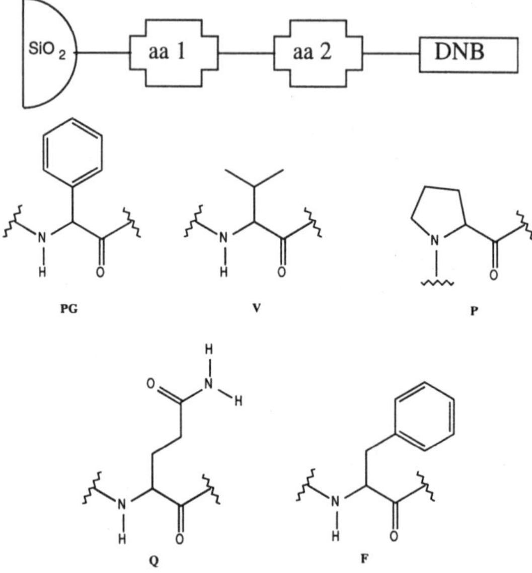

Figure 6   *Five amino acids were used to prepare a library of 50 dipeptide DNB CSPs.*

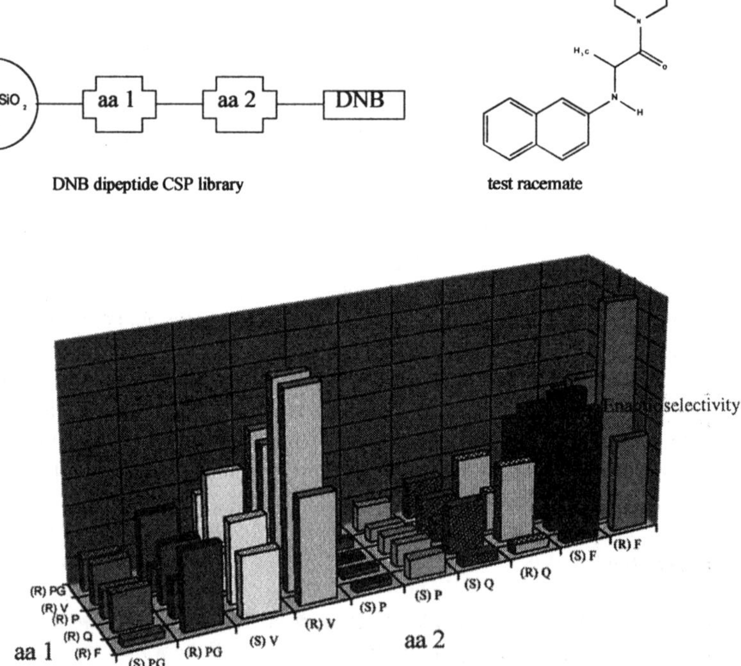

Figure 7   *Results of screening of a library of 50 dipeptide DNB CSPs for the separation of the enantiomers of a test racemate.*

This initial screen clearly points the way to further optimization for this chiral recognition system.   The initial screen indicates that DNB dipeptide CSPs having a strong hydrogen bonding sidechain in the aa 1 position and a sterically bulky sidechain in the aa 2 position work best for the test analyte. A second library based on this motif was prepared and evaluated. As shown in Figure 8, many of the members of this new library show superior enantioselectivity to the DNB Val–Gln CSP, which was the best CSP in the initial library.

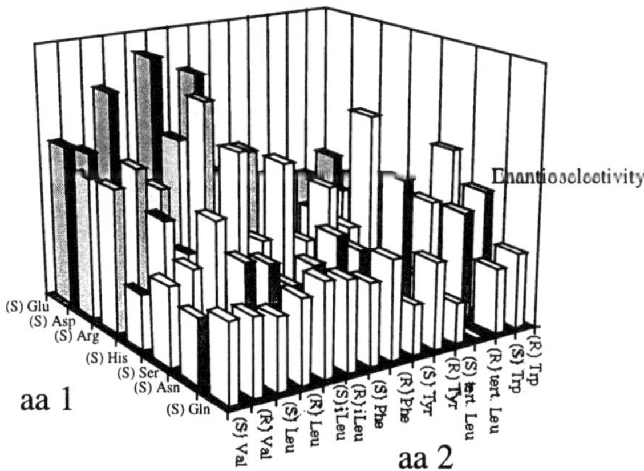

Figure 8   *Results of screening of a new library of dipeptide DNB CSPs containing hydrogen-bonding sidechains in the aa 1 position and sterically bulky sidechains in the aa 2 position. Most of these second-generation CSPs are superior to the best CSPs in the library shown in Figure 7.*

3   SUMMARY AND CONCLUSIONS

The method of combinatorial synthesis and screening of new CSPs represents a turning point in the development of new methods for chiral stationary phase design. It is expected to provide both a wealth of information on the structural requirements for chiral molecular recognition and a method whereby useful chromatographic adsorbents for industrial separation processes can be rapidly discovered. The example presented here clearly illustrates the power of the method. We are currently exploring the use of this approach to provide rapid solutions to real-world process separation problems.

References

1.   C. J. Welch, G. A. Bhat and M. N. Protopopova, *Enantiomer*, in press.
2.   C. J. Welch, M. N. Protopopova and G. A. Bhat, *Enantiomer*, in press.
3.   S. G. Allenmark, 'Chromatographic Enantioseparation: Methods and Applications', Ellis Horwood, New York, 1991.
4.   E. Francotte *J. Chromatogr.*, 1994, **666**, 565.
5.   Personal communication from S. R. Perrin, Regis Preparative Separations Lab.

6. F. Charton and R. M. Nicoud, *J. Chromatogr.*, 1997, **702**, 49.
7. E. Francotte and P. Richert, *J. Chromatogr.*, 1997, **769**, 101.
8. Personal communication from Prof. W. H. Pirkle, University of Illinois at Urbana-Champaign.
9. C. J. Welch, *J. Chromatogr.*, 1994, **666**, 3.
10. W. H. Pirkle, C. J. Welch, J. A. Burke and B. Lamm, *Analytical Proceedings,* 1992, **29**, 225.
11. M. D. Weingarten, K. Sekanina and W. C. Still, *J. Am. Chem. Soc.*, 1998, **120**, 9112.
12. J. Maclennan, *Spec. Chem.*, 1996, **16**, 267.
13. E. J. Kitka and E. Grushka, *J. Chromatogr.*, 1977, **135**, 367.
14. W. A. Howard, T. B. Hsu, L. B. Rogers and D. A. Nelson, *Anal. Chem.*, 1985, **57**, 606.
15. W. H. Pirkle, T. C. Pochapsky, G. S. Mahler and R. E. Field, *J. Chromatogr.*, 1985, **348**, 89.
16. W. H. Pirkle and C. J. Welch, *J. Liq. Chromatogr.*, 1991, **14**, 2027.
17. W. H. Pirkle and T. C. Pochapsky, *J. Amer. Chem. Soc.*, 1987, **109**, 5975.
18. W. H. Pirkle, J. A. Burke and S. R. Wilson, *J. Amer. Chem. Soc.*, 1989, **111**, 9223.

* Correspondence: whelko@aol.com

THE ALKYLATION OF DRUGS AT ION EXCHANGE SITES ON THE SURFACE OF SOLID PHASE EXTRACTION COLUMNS

Thomas August,* Kathy Rymut, Tony Darpino and Michael Telepchak

United Chemical Technologies, Inc.
Bristol, PA 19007 USA

1 INTRODUCTION

Solid phase extraction is routinely used in forensic laboratories to extract a wide variety of drugs from body fluids such as urine and blood. SPE has also gained widespread use in pharmaceutical, clinical and environmental laboratories, particularly since copolymeric extraction columns were introduced in 1980. However, the use of this technology is not without controversy. A recent study indicated that certain commercially available solid phase extraction columns containing ion exchange functional groups may cause alkylation of key metabolites back to the parent drug.[1] The resulting alkyl derivatives may lead to reporting errors, especially at levels at or near the SAMSHA cut-off levels. The objective of the work reported here was to determine whether alkylation of various drugs could be caused by functional groups present on the sorbent.

2 EXPERIMENTAL

2.1 Materials

2.1.1 Reagents.   Sodium hydroxide, glacial acetic acid and concentrated ammonium hydroxide were purchased from J.T. Baker. HPLC grade hexane, ethyl acetate, methanol, methylene chloride, isopropyl alcohol and acetonitrile were obtained from E.M. Science. Dibasic sodium phosphate was obtained from Mallinckrodt, monobasic sodium phosphate and 0.1 N HCl were obtained from Fisher Scientific, and BSTFA in 1% TMCS was obtained from Petrarch.

2.1.2 Standards.   All analytical standards (benzoylecgonine, D3-benzoylecgonine, cocaine, D3-cocaine, norcocaine, codeine, norcodeine, morphine, D3-morphine, diazepam, nordiazepam and prazepam) were acquired from Radian International LLC. Working standards of each drug were prepared at 1 and 10 ng/mL by dilution with methanol, and were used to spike negative urine. Structures of these drugs and the alkylation reaction in question are shown in Scheme 1.

2.1.3 Sorbents.   To determine if the sorbent causes metabolites to convert to their respective alkylated drugs during extraction by solid phase, we used four common copolymeric sorbents, which we will identified only as A, B, C, and D.

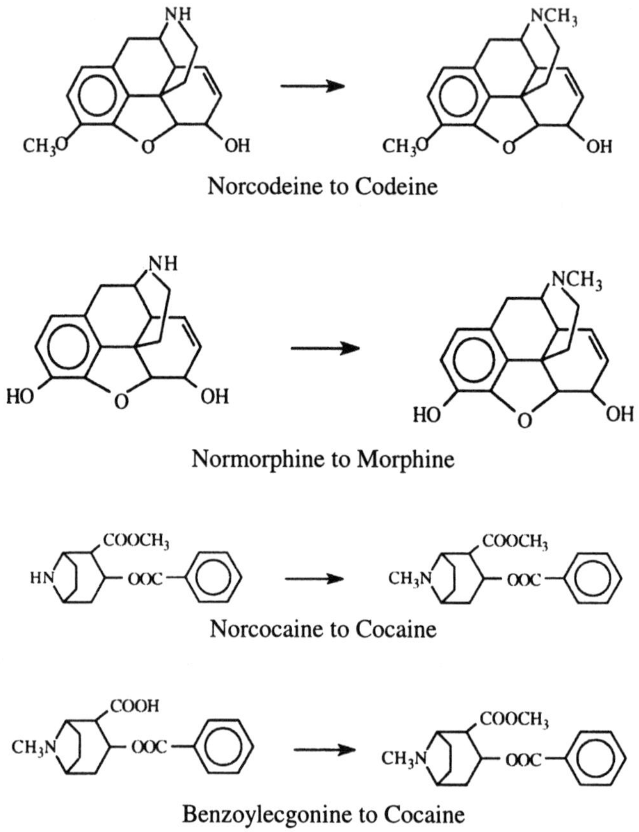

Norcodeine to Codeine

Normorphine to Morphine

Norcocaine to Cocaine

Benzoylecgonine to Cocaine

Scheme 1

2.2 Methods

Exact conditions for sample preparation, column conditioning, *etc.*, may be found in Table 1 for each drug. The following methods are general:

    2.2.1 Sample Preparation. Spiked samples were prepared in 2 mL of negative urine at the following levels: 150 ng/mL norcocaine; 150, 250, 500 and 1000 ng/mL benzoylecgonine; 300 ng/mL norcodeine, normorphine; 200, 400 and 1000 ng/mL nordiazepam. 2 mL of acetate buffer was added to the norcodeine, normorphine and nordiazeparn samples, while 2 mL of phosphate buffer was added to the norcocaine and benzoylecgonine samples.

    2.2.2 Column Conditioning and Sample Application. All SPE columns were activated with 3 mL methanol followed by 3 mL distilled water and 1 mL phosphate or acetate buffer (see Table 1). Samples were loaded at 1 to 2 mL/minute.

    2.2.3 Column Wash. Interferences were removed by washing with 2 mL of distilled water, then 2 mL of either 0.1 N HCl, acetonitrile/phosphate buffer or acetate buffer (see Table 1), and then by 3 mL methanol where appropriate. Maximum vacuum was applied for 5 min, then 2 mL of hexane was added to the nordiazepam tubes before elution.

2.2.4 Analyte Elution. Elution of benzoylecgonine, norcocaine, norcodeine and normorphine occurred by passing and collecting 3 mL of methylene chloride/isopropanol/ammonium hydroxide (78/20/2) through the columns. 3 mL of 2% ammonium hydroxide in ethyl acetate was used to elute nordiazepam.

Table 1 *SPE Procedures.*[2]

DRUG	SAMPLE PREP	CONDITION	WASH	ELUTION
Benzoylecgonine, Norcocaine	2 mL urine 2 mL PB	3 mL MeOH 3 mL DI H_2O 1 mL PB	2 mL DI H_2O 2 mL 0.1 N HCl 3 mL MeOH Dry 5 min.	3 mL MC/IPA/AH
Nordiazepam	2 mL urine 3 mL AB	3 mL MeOH 3 mL DI H_2O 1 mL PB	2 mL DI H_2O 2 mL ACN/PB Dry 5 min 2 mL Hexane	3 mL NH₄OH/EA
Norcodeine, Normorphine	2 mL urine 2 mL AB	3 mL MeOH 3 mL Di H_2O 1 mL AB	2 mL DI H_2O 2 mL AB 3 mL MeOH Dry 5 min	3 mL MC/IPA/AH

Notes: PB = 0.1 M phosphate buffer, pH 6; AB = 0.1 M acetate buffer, pH 4.5; ACN/PB = 20% acetonitrile in 0.1 M phosphate buffer, pH 6; NH₄OH/EA = 2% ammonium hydroxide in ethyl acetate; MC/IPA/AH = methylene chloride/isopropyl alcohol/ammonium hydroxide (78/20/2).

2.2.5 Derivatization. Each eluate was spiked with 200 ng/mL of the appropriate external standard and vortexed before evaporating to dryness at 40 °C. Derivatization was accomplished by adding 50 µL ethyl acetate and 50 µL BSTFA (with 1% TMCS), vortexing and heating at 70 °C for 20 minutes.

2.3 Analysis and Instrumentation

2.3.1 GC/MS. 2 µL of derivatized sample was injected splitless onto a HP 5890/-5971A GC/MS using an Rtx-1® fused silica capillary column (15 M, 0.25 mm ID, 0.25 µm film thickness). Helium flow was adjusted to 0.55 mL/min. The injector and transfer line were kept at 250 °C. Temperature programs were as follows: 150–300 °C at 25 °C/min (benzoylecgonine); 150–300 °C at 20 °C/min (norcodeine, nonnorphine); 150–230 °C at 30 °C/min, 230–250 °C at 10 °C/min, then 250–300 °C at 25 °C/min (norcocaine); 140–250 °C at 25 °C/min, hold 2 min, then 250–300 °C at 25 °C/min (nor-diazepam). The following ions were monitored using SIM analysis (quantitation ion indicated by *):

Drug		Alkylated Product	
norcodeine:	164*, 357, 329	codeine:	371*, 234, 343
normorphine:	415*, 400, 416	morphine:	429*, 324, 430
nordiazepam:	341*, 342, 343	diazepam:	256*, 221, 285
norcocaine:	168*, 136, 289	cocaine:	182*, 303, 198
benzoylecgonine:	240*, 361, 256	cocaine:	182*, 303, 198

2.3.2 Absolute Recovery. 2 mL aliquots of negative urine were spiked at various levels and extracted as previously described. Aliquots of blank urine were run through SPE columns (unextracted drug calibrator) and appropriate concentrations of drug, alkylated derivatives and external standard were then spiked in the eluate before drydown and derivatization. Absolute recoveries were calculated using Equation 1, where 'ratio' is defined as area of drug / area of external standard:

$$\% \text{ Recovery} = \frac{\text{ratio of extracted sample}}{\text{ratio of unextracted calibrator}} \times 100 \tag{1}$$

2.3.3 Determination of Alkylation Conversion. Equation 2 was used to calculate the absolute amount (ng/mL) of alkylated drug produced and Equation 3 was used to determine the percent (%) of metabolite that converted to the parent drug (*e.g.* norcodeine to codeine) through alkylation by the sorbent. In Equation 2, 'ratio' is used as defined above (area of drug / area of external standard), and [AD] = known concentration (ng/mL) of unextracted alkylated calibrator (*e.g.* codeine). In Equation 3 [NAD] = known concentration (ng/mL) of unextracted nonalkylated calibrator (*e.g.* norcodeine), and [AP] = concentration of ng/mL of alkylated product produced.

$$\text{Alkylated product} = \frac{\text{ratio of converted alkylated drug}}{\text{ratio of unextracted alkylated calibrator}} \times [\text{AD}] \tag{2}$$

$$\% \text{ Conversion} = \frac{[\text{AP}]}{[\text{NAD}]} \times 100 \tag{3}$$

3 RESULTS

The results for the conversion of nor-metabolite to the parent drug are presented for each sorbent A–D in Tables 2–5, respectively. Results for each pair of nor-metabolite/parent drug pair are presented in Figures 1–4, and in the text to follow:

3.1 Norcodeine to Codeine (Figure 1)

Sorbent A caused conversion of norcodeine to codeine (52 ng/mL, 17%), while Sorbents C and D liberated 28 ng/mL (9%) and 23 ng/ mL (8%) of codeine, respectively. Sorbent B did not cause conversion of norcodeine to codeine.

3.2 Normorphine to Morphine (Figure 2)

Sorbent A caused conversion of normorphine to morphine (154 ng/ mL, 51 %), while Sorbents C and D liberated 35 ng/mL (1.2%) and 33 ng/mL (1.1%) of morphine, respectively. Sorbent B did not cause conversion of normorphine to morphine.

3.3 Norcocaine to Cocaine (Figure 3)

Sorbent A caused conversion of norcocaine to cocaine (64 ng/mL, 42%), while

Sorbent C liberated 18 ng/mL of cocaine (12%). Sorbents B and D did not cause conversion of norcocaine to cocaine.

3.4 Benzoylecgonine to Cocaine (Figure 4)

None of the sorbents caused conversion of benzoylecgonine to cocaine at levels lower than 250 ng/mL. Conversion occurred at higher concentrations (500 and 1000 ng/mL) using Sorbent A (26 and 29% respectively). Sorbents B and C did not cause conversion of benzoylecgonine to cocaine.

3.5 Nordiazepam to Diazepam

None of the sorbents caused conversion of nordiazepam to diazepam up to concentrations of 1000 ng/mL.

Table 2 *Results of Drug Conversion Studies on Sorbent A.*

Drug	[Drug] (ng/mL)	% Drug Recovered	[Alkyl Derivative] (ng/mL)	% Alkyl Derivative Recovered
Norcodeine	300	46.3	17.3	52.0
Norcocaine	150	54.0	42.0	63.5
Normorphine	300	55.0	51.2	54.0
Nordiazedpam	200	82.0	0	0
	400	104	0	0
	1,000	113	0	0
Benzoylecgonine	150	88.0	0	0
	250	86.0	0	0
	500	76.0	26	26
	1,000	79.0	29	29

Table 3 *Results of Drug Conversion Studies on Sorbent B.*

Drug	[Drug] (ng/mL)	% Drug Recovered	[Alkyl Derivative] (ng/mL)	% Alkyl Derivative Recovered
Norcodeine	300	96.0	0	0
Norcocaine	150	99.0	0	0
Normorphine	300	85	0	0
Nordiazedpam	200	83	0	0
	400	104	0	0
	1,000	104	0	0
Benzoylecgonine	150	86	0	0
	250	92	0	0
	500	93	0	0
	1,000	89	0	0

Table 4 *Results of Drug Conversion Studies on Sorbent C.*

Drug	[Drug] (ng/mL)	% Drug Recovered	[Alkyl Derivative] (ng/mL)	% Alkyl Derivative Recovered
Norcodeine	300	62.3	9.4	28
Norcocaine	150	80.0	12.0	18
Normorphine	300	53	12.0	35
Nordiazedpam	200	78.0	0	0
	400	95.0	0	0
	1,000	90.0	0	0
Benzoylecgonine	150	89.0	0	0
	250	86.0	0	0
	500	86.0	0	0
	1,000	71.0	0	0

Table 5 *Results of Drug Conversion Studies on Sorbent D.*

Drug	[Drug] (ng/mL)	% Drug Recovered	[Alkyl Derivative] (ng/mL)	% Alkyl Derivative Recovered
Norcodeine	300	91.0	8.0	28
Norcocaine	150	98.0	0	0
Normorphine	300	92.0	11	33
Nordiazedpam	200	83.0	0	0
	400	88.0	0	0
	1,000	85.0	0	0
Benzoylecgonine	150	92.0	0	0
	250	95.0	0	0
	500	99.0	0	0
	1,000	89.0	0	0

4 DISCUSSION

A recent study suggested that certain commercially available SPE columns may cause alkylation (methylation, ethylation, and/or propylation) of key metabolites to their parent drug, and thereby lead to reporting errors, especially at levels at or near the SAMSHA cut-off levels.[1] We formulated a hypothesis that this alkylation of drugs may result from certain techniques employed during the sorbent manufacturing processes.

The SPE columns used in this study are copolymeric, and contain both ion exchange and hydrophobic functional groups. The ion exchange portion is typically $R\text{-}SO_2\text{-}OH$ and could convert during specific manufacturing processes to an alkyl sulfonate agent. If conversion were taking place during the extraction procedure, the binding site(s) of the drug would associate with the converted alkyl sulfonate groups of the sorbent. In the process, an alkyl group may be transferred from the sorbent to an accessible nucleophile

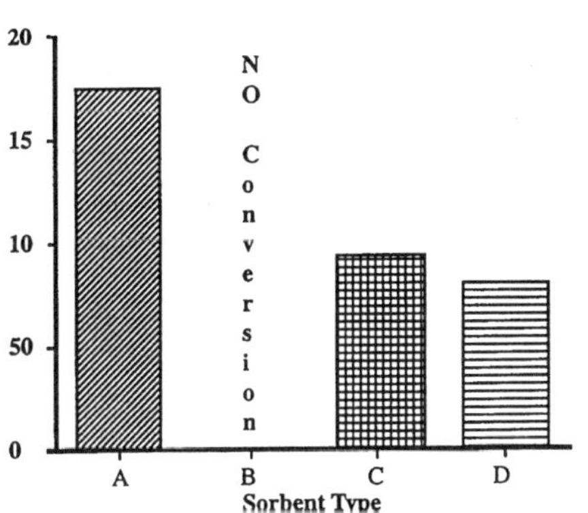

Figure 1 *Conversion of norcodeine to codeine.*

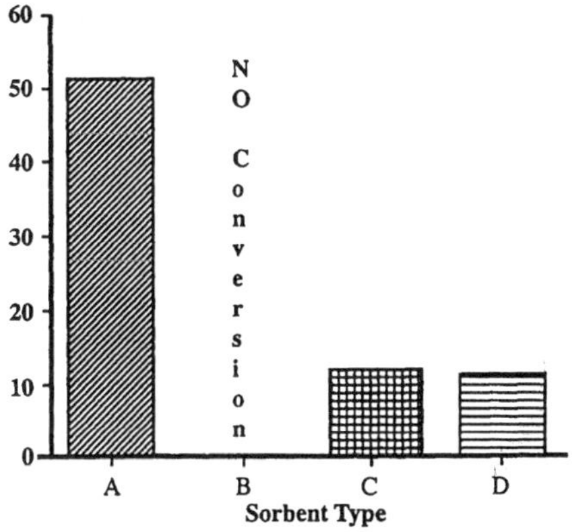

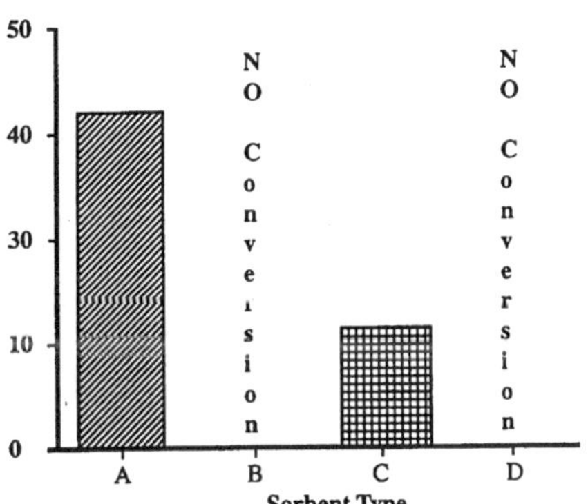

Figure 2 *Conversion of normorphine to Morphine.*

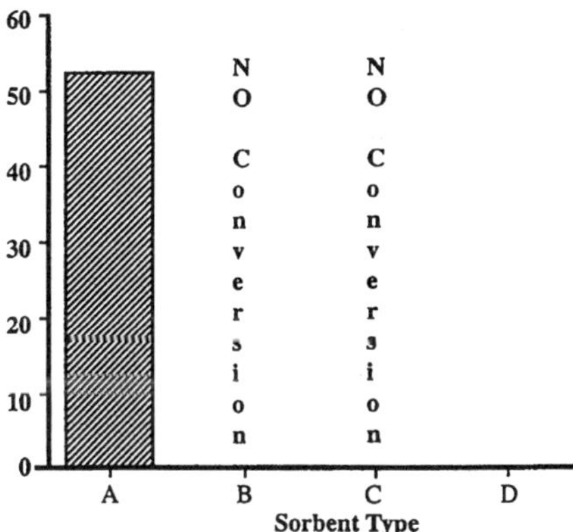

Figure 3 *Conversion of norcocaine to cocaine.*

Figure 4 *Conversion of benzoylecgonine to cocaine.*

of the drug, thus effectively derivatizing the drug to its parent. For example, when passing through the SPE column, norcodeine may accept a methyl group from the sorbent and derivatize to codeine.[3]

To determine if drugs convert to their alkyl derivatives, common drugs (norcodeine, norcocaine, nonnorphine, nordiazepam and benzoylecgonine) were selected because they contain various functional groups (hydroxyl, amide, acid and/or amine). The conversions that could occur upon alkyl in each case are: norcodeine to codeine; norcocaine to cocaine; normorphine to morphine; nordiazepam to diazepam; and benzoylecgonine to cocaine. Unextracted calibrators contained known amounts of nonalkylated and alkylated drug as well as deuterated drug. Deuterated external standards were checked for background noise, alkylated ions, and any other isotopic or adverse contribution(s). None were detected.

Recovery, amount of conversion, and percent conversion of nonalkylated drugs and alkyl derivatives were determined using Equations 1–3. Sorbent A produced the greatest percent conversion of norcodeine, normorphine, norcocaine and benzoylecgonine to their alkylated forms. Sorbent C demonstrated the next highest overall conversion, followed by Sorbent D. Sorbent B showed no ability to convert any drug tested.

Nordiazepam was not converted to diazepam on any sorbent tested, suggesting that the amide functional group of nordiazepam is not readily susceptible to alkylation during solid phase extraction. At levels below 500 ng/mL, benzoylecgonine was not converted on any sorbent, suggesting that the acid functional group is not as susceptible to alkylation as the amine groups of norcocaine, norcodeine and normorphine. The amine group in norcocaine was methylated to form cocaine on Sorbents A and C, indicating that amines may be easily methylated during the use of certain SPE columns. Of all the drugs tested, normorphine and norcodeine were the most susceptible to alkylation on Sorbents A, C, and D. No conversion to morphine or codeine took place using Sorbent B. Again, data suggest that the amine groups are more susceptible to alkylation on three of the four sorbents tested when compared to acid (benzoylecgonine) and amide (nordiazepam) groups.

5 CONCLUSION

During routine use of solid phase extraction, alkylation of certain drugs can occur which may lead to inaccurate quantitative results. The available data indicate that this alkylation is due to reactive functional groups introduced on the sorbent by specific manufacturing practices. These practices are proprietary in nature, but the derivatization problems associated with them may be avoided by careful selection of solid phase extraction columns.

References

1.	L. Marinetti-Scheiff *et al.*, 'Simultaneous Solid Phase Extraction of Whole Blood And Urine For Cocaine, Benzoylecgonine and Morphine Using PFPA/HFIP Derivatization with GC–MS Confirmation', Society of Forensic Toxicologists (SOFT) Meeting, Baltimore, MD, USA, 1995.
2.	Worldwide Monitoring Applications Manual, United Chemical Technologies, Inc., 1997.
3.	J. Warren, J. Macmillan and S. Washburn, *J. Org. Chem.*, 1975, **40**, 743.

CONNECTION BETWEEN SURFACE MODIFICATION OF FUMED SILICA, ITS PARTICLE SIZE DISTRIBUTION, AND ELECTROPHORETIC MOBILITY IN AQUEOUS SUSPENSIONS

V.I. Zarko, V.M. Gun'ko,* E.F. Voronin and E.M. Pakhlov

Institute of Surface Chemistry
31 Prospect Nauki
Kiev, Ukraine 252022

1 INTRODUCTION

Fumed silica is a form of particulate silica that is produced by injection of suitable volatile chlorosilanes into an oxygen/hydrogen flame. It is made up of non-porous, spherical primary particles ($\approx$10 nm) that collide and stick together in the flame to form highly structured aggregates (0.1–0.2 µm). As the aggregates move to cooler regions of the flame and downstream from the burner, they begin to form loosely-bound agglomerates (> 5 µm) via hydrogen-bonding and electrostatic interactions.

The degree to which fumed silica agglomerates can be dispersed back to aggregates bears directly on the performance of the silica in many applications. For example, thickening of liquids by fumed silica is thought to be due to dispersion of agglomerates into a network of hydrogen-bonded aggregates. But freshly-prepared agglomerates are weakly bound and therefore easily over-dispersed, while aged agglomerates are more strongly bound and therefore often underdispersed. Both situations lead to sub-optimal, if not erratic, performance.

It is an object of this work, then, to develop ways to improve the properties of fumed silica, principally through controlling its dispersability. Two approaches to oxide treatment for this goal can be noted.[1,2] The first involves techniques which do not change the chemical structure of the particles, *e.g.* different mechanical treatments of oxides in the gas or liquid phase. The second involves chemical modification of the oxide surface, which typically causes many properties of the oxide to change. In the work we report here, we employed both types of oxide treatment, and studied the characteristics of aqueous suspensions of these samples.

The hierarchy in the fumed silica structure can cause many characteristics of aqueous suspensions to be strongly dependent on the sample preparation technique. For example, large agglomerates (> 1 µm) are observed in freshly prepared aqueous suspension of fumed silica even after ultrasonic treatment for 1 h. Large agglomerates are also observed at pH close to the isoelectric point of silica ($IEP_{SiO2} \approx$ 2.2–2.5), where electrostatic repulsive interaction is absent, and at pH > 10, where larger particles with a high charge can attract smaller particles and dissolution of the surface changes its structure and particle–particle interaction. However, treatment of such suspensions for longer times (3–4 h) decomposes a major portion of these agglomerates to aggregates.

Mechanochemical activation (MCA) of aqueous suspensions also leads to a significant decrease in the particle size, especially in comparison with suspensions made from dry-ball-milled silica. One advantage of MCA is that its suspensions are more stable at high pH than sonicated suspensions. Prolonged MCA treatment can lead to particles at 20–30 nm, larger than primary particles but smaller than primary aggregates, and due presumably to fracture of primary aggregates.

The second approach we took to changing the properties of the oxide surface and controlling particle size distribution was chemical surface modification with certain di- and trifunctional organosilanes, specifically, methacryloyloxymethylenemethyl diethoxysilane (MADES), 3-methacryloyloxypropyl trimethoxysilane (MTES), and 3-aminopropyl triethoxysilane (APTES).

$$(H_5C_2O)_2Si(CH_3)CH_2OC(O)C(CH_3)CH_2$$

MADES

$$(H_3CO)_3Si(CH_2)_3OC(O)C(CH_3)CH_2 \qquad (H_5C_2O)_3Si(CH_2)_3NH_2$$

MTES   **APTES**

Properties of these surface-modified samples were carefully measured and compared to those of the unmodified oxide. Surface grafted MADES and MTES (possessing hydrophilic and hydrophobic groups) would be expected to decompose silica agglomerates by inhibiting the processes that hold them together in air and water: strong hydrogen-bonding and electrostatic interactions, and condensation of surface $\equiv$SiOH groups between aggregates.

2 EXPERIMENTAL

Unmodified fumed silica (Aerosil A-175 and A-380, 99.9%, "Chlorovinyl", Kalush, Ukraine, specific surface area $S \approx 175$ and 380 m^2/g, respectively) and modified silica (A-175) samples with 0.06, 0.11, 0.18, 0.2, 0.24, and 0.3 mmol/g of bound MADES (spectrophotometric grade) were used. The modification of silica samples (weight 220 g) by MADES was performed in a glass reactor vessel. A sample was heated in dry air to 373 K, then the required amount of MADES was charged into the vessel. The temperature was kept at 373 K and the contents allowed to react for 3 h, then the sample was swept with dry air at the reaction temperature for 0.5 h to remove volatile byproducts and any residual silane. This same technique was used to modify silica with MTES (concentration 0.1–0.4 mmol/g) and a combination of MADES and APTES (0.15 mmol/g each).

Electrophoretic and multimodal particle size distribution measurements were performed using a zeta potential apparatus "Zeta Plus" (Brookhaven Instr. Corp., USA). Deionized, distilled water (pH = 6.95) and 0.2 g of the solids per 1 L of the water were used for the suspension preparation in an ultrasonic bath for 2 h. The pH values were adjusted by addition of 0.1 M HCl or NaOH solutions. The pH values of the suspensions of original silica and silica modified by MADES were measured by a Precision Digital pH meter OP-208/1 (Hungary). The effective average diameter (D_{ef}) and polydispersity of the particles in the suspensions were obtained by photon correlation spectroscopy using a Brookhaven MAS OPTION particle sizer (accuracy of $\pm$ 1–2% with monodisperse samples; repeatability of $\pm$ 1–2% with dust-free samples).

3 RESULTS AND DISCUSSION

The available IR data[1] support the notion that MADES reacts with hydroxyl groups on silica as follows:

$$\equiv SiOH + (H_5C_2O)_2Si(CH_3)CH_2OC(O)C(CH_3)CH_2 \rightarrow$$

$$\equiv SiO\!-\!Si(OC_2H_5)(CH_3)CH_2OC(O)C(CH_3)CH_2 + C_2H_5OH \qquad (1)$$

$$\equiv SiO\!-\!Si(OC_2H_5)(CH_3)CH_2OC(O)C(CH_3)CH_2 + H_2O \rightarrow$$

$$\equiv SiOSi(OH)(CH_3)CH_2OC(O)C(CH_3)CH_2 + C_2H_5OH \qquad (2)$$

Bidentate deposition of MADES may also occur through reaction with both ethoxy groups, as in Scheme 1. In the case of MTES three methoxy groups are present to interact with surface $\equiv SiOH$ groups, but formation of three $\equiv Si\!-\!O\!-\!Si\equiv$ bonds between the

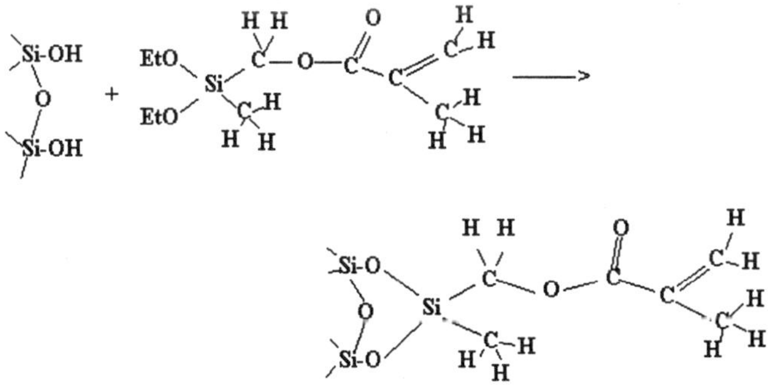

Scheme 1

surface and an MTES molecule is impossible due to the relatively small concentration of $\equiv SiOH$ groups at the surface of fumed silica ($\leq$ 2 OH per nm^2) and the great bond strain involved in forming such a tridentate structure. Therefore, MTES and other trifunctional silanes (such as APTES) are thought to form bridging structures with at most two points of attachment to the surface per silane molecule, just as we see for the difunctional MADES in Scheme 1.

The modification of silica by MADES or MTES has a strong influence on the particle size distribution, as illustrated in Figure 1. Even at very low partial modification of silica (C_{MADES} = 0.06 mmol/g) the particle dispersity in aqueous suspension increases by a factor of 100. We also see in Figure 1 that the effective particle diameter of MADES-modified silica decreases much faster than for MTES-modified silica. This is likely due not only to the difference in the size of the modifiers themselves but also to their linkage to the surface or to neighboring bound groups *via* two or three bonds, respectively. The number of relatively small particles increases and their size decreases with increasing C_{MADES} up to 0.2 mmol/g, and with increasing C_{MTES} up to 0.28 mmol/g (Figure 1). Consequently, even at low levels of bound MADES or MTES, agglomerates can be decomposed easily.

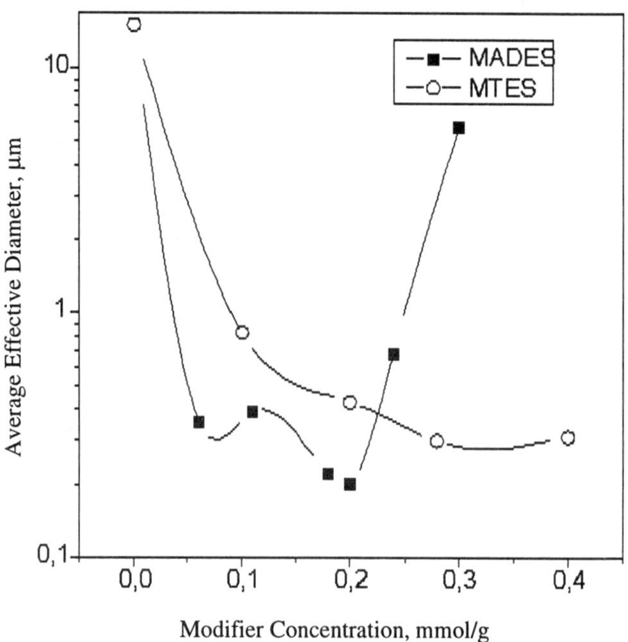

Figure 1   *Average effective diameter (μm) of particles of modified silica in aqueous suspensions as a function of the concentration of bound MADES or MTES. Measurements were performed without addition of electrolytes to the suspension.*

For all studied C_{MADES}, two peaks in the particle size distribution are observed. The first peak (at 0.15–0.25 μm) corresponds to aggregates, and the other peak (at 0.5–1 μm) to agglomerates. The agglomerates are relatively small when C_{MADES} = 0.06–0.24 mmol/g, but D_{ef} increases with increasing C_{MADES}, reaching *ca.* 5 μm at C_{MADES} = 30 mmol/g.

A maximum value of the particle dispersity (D_{ef} = 0.2 μm) is exhibited for C_{MADES} = 0.2 mmol/g (this C_{MADES} is close to half the initial concentration of ≡SiOH groups on heated silica). If MADES adsorbs *via* reaction of two $SiOCH_2CH_3$ groups with surface ≡SiOH groups, then at C_{MADES} = 0.2 mmol/g the amount of residual ≡SiOH at the surface can be minimal and free ≡SiOH groups in bound MADES can be not observed. D_{ef} in this case is close to the aggregate size for unmodified silica (0.1–0.2 μm). However, further increasing C_{MADES} ≥ 0.24 mmol/g leads to a decrease of the particle dispersity that can be caused by the aggregate linkage to agglomerates *via* the interaction of bound

MADES from neighboring aggregates. At the same time modification of the surface by MTES does not significantly increase D_{ef} at $C_{MTES} > 0.2$ mmol/g; however, MTES gives a larger D_{ef} at $C_M < 0.25$ mmol/g (Figure 1).

The D_{ef} value as a function of pH depends on the nature of the modifier (Figure 2). In the cases of MTES or APTES, each of which has three reactive groups $\equiv$SiOR, the decrease in D_{ef} is more pronounced than in the case of MADES. This result can be explained by the formation of bridging structures between the primary particles from different aggregates by the modifier molecules. It should be noted that the time of ultrasonic treatment of the suspension (t) was 3.5 h for unmodified A-175 and for MADES+APTES-modified silicas, but 2 h for MADES- or MTES-modified silicas.

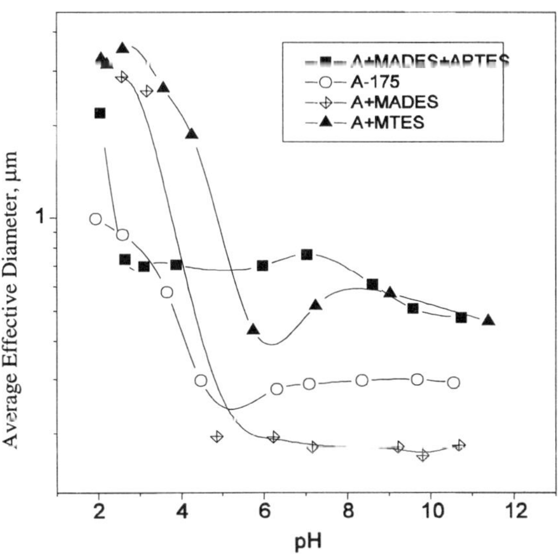

Figure 2 *Average effective diameter of particles of silica (A-380) modified by a blend of MADES+APTES (0.15 mmol/g each), A-175 unmodified and modified by MADES (0.2 mmol/g) or MTES (0.2 mmol/g) as a function of pH; suspension was sonicated for 2–3.5 h.*

The maximum difference in electrophoretic mobility or zeta potential (ζ) for unmodified and modified silica is observed at pH 2–4 (Figure 3). For this pH region, the D_{ef} value is maximal for a sample with $C_{MADES} = 0.2$ mmol/g; this can lead to a reduction of ζ as the diffusion coefficient depends on D_{ef}^{-1}. In addition, an increase of C_{MADES} changes $\zeta(pH)$ due to a reduction in the $\equiv$SiOH concentration (surface charges on the particles are formed due to dissociation of the $\equiv$SiO–H bonds) leading to a decrease in surface acidity and an increase in the isoelectric point value. This dependence is non-linear, however. Minima of $\zeta(pH)$ at pH = 3–4 for the original silica and some modified samples could be caused by the influence of the electric double layer (EDL) states at different pH. For instance, a thick EDL for a pH near the IEP (we did not add neutral electrolyte, *e.g.* NaCl, to the suspensions) gives a $\kappa R < 1$, where κ is the Debye–Huckel parameter and R is the particle radius, but an increase of pH gives $\kappa R > 1$ due to the EDL compression. Additionally, the specific fractal structure of fumed silica can have some influence on the $\zeta(pH)$ shape, especially at $\kappa R < 1$. Consequently, for a large EDL and a pH near the IEP, a weak screening of the particle charges is possible due to the large EDL thickness, but increasing pH gives the EDL compression (*i.e.* particle screening grows) and the value of zeta potential can decrease. Modification of fumed alumina/silica (with 1.5 wt% of Al_2O_3 = SA1) gives exponential grows of ζ with increasing pH (Figure 3) due to the high basic properties of the $Si(CH_2)_3NH_2$ group.

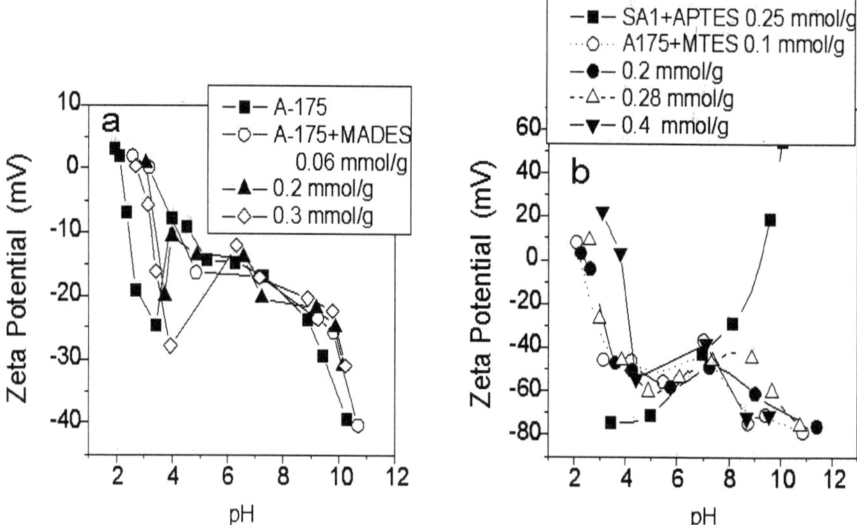

Figure 3 *Zeta potential of fumed silica (A-175) modified by (a) MADES and (b) MTES at different concentrations, and (b) fumed alumina/silica with 1.5 wt% of Al_2O_3 (SA1) modified by APTES (0.25 mmol/g) as a function of pH.*

Modifiers such as MADES bound to the surface possess low solvation energy. AM1-SM1 quantum chemical calculations taking solvation into account[3,4] show weak changes in the solvation energy (E_s) for silica modified by MADES. These effects and the absence of hydroxyl groups in MADES bound to the surface or neighboring bound groups *via* two bonds lead to the effective decomposition of the hydrogen-bonded agglomerates and reduce electrostatic interaction between them. However, the hydrophobic properties of MADES or MTES bound to the surface are too weak to inhibit the wettability of the modified silica particles.

4 SUMMARY

The modification of fumed silica by a relatively small amount of a modifier such as MADES or MTES can strongly change the particle size distribution in aqueous suspensions and the dependence of zeta potential on pH of colloidal particles. Surface modification of fumed silica with MADES is more effective at decreasing the particle (agglomerate) size than is treatment with a MADES/APTES blend or MTES by itself over the concentration range of 0.05–0.2 mmol/g.

References

1. V. M. Gun'ko, V. I. Zarko, R. Leboda, E. F. Voronin, E. Chibowski and E. M. Pakhlov, *Colloid. Surf. A*, 1998, **132**, 241.
2. V. M. Gun'ko, E. F. Voronin, V. I. Zarko, E. M. Pakhlov and A. A. Chuiko, *J. Adhesion Sci. Technol.*, 1997, **11**, 627.
3. M. J. S. Dewar, E. G. Zoebisch, E. E. Healy and J. J. P. Stewart, *J. Am. Chem. Soc.*, 1985, **107**, 3902.
4. C. J. Cramer and D. G. Truhlar, *J. Am. Chem. Soc.*, 1991, **113**, 8305.

HIGHLY EFFICIENT ACID GAS-REMOVING SHAPED FIBER FILTERS

Lixin Xue,[†]* Ronald Rohrbach,[†] Daniel Bause,[†] Peter Unger,[†] Russell Dondero[†] and Gordon Jones[‡]

AlliedSignal Inc.
[†]Research and Technology, Morristown, NJ USA
[‡]Filters and Spark Plugs, Perrysburg, OH USA

1 INTRODUCTION

Acidic pollution of the atmosphere is currently an important environmental issue.[1–3] Acidic gases such as SO_2 and NO_x released into the atmosphere from coal and oil burning power stations, mineral ore processing and chemical manufacture pose lasting threats to human beings and animals. These acidic gases attack the throat and lungs, leading people with respiratory problems to suffer severe illness. They also affect plants, turning leaves yellow, drying, bleaching and even killing foliage. The Acid Rain, caused by the atmospheric pollution of acidic gases, spreads, and the damage involves weather, chemistry, soil, and the life cycles of plants and animals both terrestrial and aquatic.

One strategy to control atmospheric acidic pollution is to limit industrial emission sources. Title IV of the Clean Air Act sets as its primary goal the reduction of annual SO_2 emissions by 10 million tons from the 1980 levels.[4] To achieve this reduction, the law requires a two-phase tightening on the emission levels of fossil fuel-fired power plants. Many high capacity commercial methods using chemisorptive reagents in dry or wet forms are currently utilized in Phase I of the program begun in 1995.[5] The intrinsic demands of the processes on maximum gas reaction interface and minimum reagent carryover call for development of compact high surface area supports for chemisorptive reagents. This demand on compact support will be even higher when the program moves into and beyond Phase II in the year 2000, which will further tighten the emission levels on not only those large higher emitting plants but also smaller, cleaner plants fired by coal, oil, and gas.

Another strategy to alleviate damage from atmospheric acidic pollution is to improve indoor air quality (IAQ) for human beings, animals and plants via removing acidic pollutants from designated areas such as cabins of automobiles, airplanes, and submarines. The concern for cleaner cabin air in automobiles has driven the development of filtration media that can not only remove particulates but also control odors, including acidic pollutants of such as SO_2 and H_2S. European auto-manufacturers have taken the lead in introducing both particulate and combined particulate/odor control filters into their designs. US auto manufacturers are now following a similar approach to passenger compartment air quality improvement. The technologies of granular activated carbon or chemically impregnated dry media are currently employed with limitations in capacity, manufacturability, and flow restriction.[6] Although chemisorptive systems have the advantage in capacity and manufacturability, they need compact structural supports to allow minimum flow restriction, maximal gas–reagent contact and reaction capacity in those applications.

Responding to the growing demand on compact, high surface area support media, we have studied a variety of porous low density materials, including spherical supports and porous sheets, and finally arrived at shaped fibers with multi-lobal cross sections such as the tri-lobal wicking fiber, termed triad, shown in Figure 1. This patented shaped fiber can not only tenaciously hold liquid or solid active reagents in its channels but also present a large surface area to the air flow for pollutant removal.[7] Based on the unique properties of the triad fiber, we have developed new air quality technologies targeting a variety of pollutants.[8-13] In this paper, we will discuss our progress in the removals of SO_2 and H_2S from air.

2 EXPERIMENTAL METHODS

2.1 Filter Media

Point bonded non-woven fiber filter mats with densities of 1.5 and 3.0 oz/yd^2 were prepared from polypropylene (PP) triad fibers of 3 and 6 denier per filament (fiber weight per 9000 meters of filament length). The triad filter media will be denoted as PP n/m in following sections, where m is the denier per filament of the fibers and n is the packing density (oz/yd^2) of the point bonded media mat.

2.2 Filtration Test Apparatus

Several laboratory scale test stands were used to permit simultaneous analyses of the test gases. The test stands were constructed of stainless and/or PTFE tubing connected by appropriate stainless steel fittings and valves. All tests were done at ambient room temperature with compressed air humidified to 50 ± 5 % relative humidity. Concentrated forms of the test gases were introduced into the humidified air stream via flow meters (AALBORG Instruments, Model GFM37) to yield the desired challenge concentrations. The challenges of H_2S and SO_2 were introduced at a flow rate of 11.5 m/min, and concentrations of 400 ppb and 40 ppm from the sources of compressed gas cylinders (Liquid Carbonics, Inc.) of 100, and 200 ppm concentrations, respectively. The circular filter samples to be tested were secured by an in-house designed holder whose outlet ports were connected to a differential pressure meter (Cole-Parmer Instruments, Model DA-1-09E-o-RRRR-15-007), which measured the pressure drop across the filter in inches of water. The circular flat filter pad for testing had an outside diameter of 5.1 cm (*ca.* 2 inches) and a circular exposed area of 3.80 cm diameter (11.3 cm^2) to the test gas. The filters made from triad media generally have a depth of about 0.15 cm and a residence time of about 8 milliseconds. The benchmark commercial carbon filter sample (BM) was also cut into circles of 5.1 cm in diameter, 1.0 cm in depth and about 4.0 g in weight for testing.

The effluent concentration for H_2S or SO_2 was monitored with a total sulfur analyser (Columbia Scientific, Model SA260) containing a flame photometric detector. The analyser was calibrated by measuring both the compressed air stream and the challenge gas concentration with no filter in place. The progress of the breakthrough was monitored by a chart recorder.

2.3 Air Flow Restriction Measurement

A modified test stand was constructed of stainless and/or PTFE tubing connected by appropriate stainless steel fittings and valves. A sample to be tested was secured in an in-house designed circular holder connected to a Mass Flowmeter (Eldrige Products, Model 8669SSS). The pressure taps on the filter holder were connected to a differential pressure meter (Cole-Parmer Instruments, Model DA-l-09E-o-RRRR-15-007), which measured the pressure drop across the filter in inches of water. The circular flat filter pad for testing had an outside diameter of 5.1 cm (*ca.* 2 inches) and a circular exposed area of 10.5 cm^2 (3.66 cm in diameter).

2.4 Filter Preparation

All the reagents used in these experiments were of the best commercially available. The filters discussed in this research will be denoted as Type A, B, C and D filter according to the following preparation procedures.

2.4.1 Type A Filters. The filters were prepared by dried impregnation of triad filter media with finely ground solid adsorbent particles pre-ground and passed through a 400 mesh sieve. A typical procedure was as follows: A 2-inch diameter circular pad of PP3/3 media (0.2850 g) was mixed and shaken with excess amount of Calgon Centaur carbon power in a plastic bag until a homogeneous dark color was observed. The excess powder was then removed by agitation and blowing with air, using an air gun supplied with 20 psi compressed air, until a loaded weight of 0.3312 g was obtained.

The loading was generally controlled in the range of 15–30%. The filters were equilibrated at 50 ± 5% relative humidity prior to testing. Results of H_2S (400 ppb) uptake tests of some of the Type A filters are shown in Figure 3(a). A typical SO_2 (40 ppm) breakthrough profile is shown in Curve A of Figure 10(b).

2.4.2 Type B Filters. The filters were prepared by wet impregnation of the triad filter media with active chemical solutions. Three typical procedures were developed during the course of exploring the most effective filtration chemistry:

<u>Procedure I</u> involved the impregnation of alkali solutions at very high concentration levels (Examples 1–4 in Table 1). The virgin triad PP3/3 media (wt, W1) was first soaked with an excess of 1 % dodecabenzene sulfonic sodium salt (DBS) solution for 5 min, wrung to remove the excess then dried at 100°C for 20 minutes. The media were then cut into 2.0 inch diameter circles and soaked in an alkali solution. After the excess solution was removed by careful compression, the circles were tested for H_2S removal both fresh (wt. W2) and equilibrated in 50% RH air (wt. W3). The results are shown in Figure 4.

<u>Procedure II</u> involved the impregnation with solutions at relatively lower concentration levels (Example 5–6 in Table 1). A virgin triad PP3/3 media was first cut into 2-inch diameter round testing pads (virgin wt. W1), then soaked and squeezed in 20% KOH or K_2CO_3 solution to obtain full impregnation. After the excess solution was removed by compression, the pad (wt. W2) was dried in an evacuating oven at 90–95 °C for 1 h and equilibrated at 50% RH overnight (wt. W3). H_2S uptake results of these filters are shown in Figure 5.

Table 1   *Data for Type B Filters Prepared from Alkaline Solutions.*

Example	Solutions	Fiber wt. W1 (g)	Fiber&Solution wet wt. W2 (g)	50%RH eq.wt. W3 (g)
1*	50% KOH	0.24	0.97	0.82
2*	50% K_3PO_4	0.21	1.01	0.96
3*	50% K_2CO_3	0.23	1.04	0.90
4*	sat. $KHCO_3$	0.23	1.03	0.44
5	20% KOH	0.20	0.50	0.32
6	20% K_2CO_3	0.21	0.70	0.36

DBS surfactant pretreated.

Table 2   *Data for Zn(Ac)₂ Dual Impregnated Type B Filters*

Example	Fiber wt. W_0 (g)	Zn(Ac)₂ solutions	Treated dry wt. W1 (g)	Alkaline solution	Treated dry wt. W2 (g)	Test Samp. wt. W3 (g)
7	2.69	13%	3.02	20% KOH	4.42	0.60
8	3.58	13%	4.00	20% K_2CO_3	5.25	0.42

<u>Procedure III</u> involved dual-impregnation of $Zn(Ac)_2$ and alkaline compounds (Examples 7–8 in Table 2). A virgin triad PP3/3 media mat (wt. W_0) was first soaked in an excess of 13 wt% solution of $Zn(Ac)_2 \cdot 2H_2O$ for 25 min. After removal of the excess solution, the mat was dried under vacuum (Dynamic pressure 8–12 inches of Hg) at 80 °C for one hour (wt. W1). The treated mat was then soaked in 20% KOH or K_2CO_3 solution for about 10 min. The excess solution was removed, and the mat was dried under vacuum (Dynamic Pressure 8–12 inches of Hg) at 95–100 °C for one hour (wt. W2). The dried mat was then cut into 2 inch diameter circular pads before put in a 50% RH chamber to equilibrate overnight (wt. W3). Filters thus formed were tested for H_2S and SO_2 removals. The results are shown in Figures 5, 7, 8 and 10.

2.4.3 Type C Filters. This type of filter was prepared by dry-impregnation of the triad media with chemically pretreated solid fine powders.

A typical treatment procedure is as follows: 5.0 g of Calgon Centaur carbon powder was soaked in 10.0 g of 13 wt% zinc acetate dihydrate solution for 1 h and separated using a Buchner funnel. The solids were further soaked in 15 g of a 20 wt% potassium carbonate solution and the liquids were removed using a Buchner funnel. After being dried in a vacuum oven (Dynamic Pressure: 5–10 inches of Hg) at 100 °C for 3 hours, the solid (5.60 g) was ground and sieved through a 400 mesh sieve. Similar carbons with chemical treatment were also obtained directly from commercial sources in granular form such as Calgon's IVP, S-sorb 12, or FCA, and Pica's Narca-O or Narca-P. For this application, they were ground in a Zirconia ball mill to pass through a 400 mesh sieve.

The impregnation procedure was similar to that for Type A filters. For example, a two-inch diameter circular triad PP3/3 media (0.220 g) was dry impregnated with above treated carbon powder to a weight of 0.269 g (Example 9). The loading for the filters was generally controlled in the range of 20–30%. The filters were equilibrated at 50 ± 5% relative humidity prior to testing. Their H_2S breakthrough test results are shown in Figure 3(b). A typical SO_2 breakthrough profile for them is curve C of Figure 10(b).

2.4.4 Type D Filters. These filters were prepared by stepwise impregnation. As summarized in Tables 3 and 4 (Examples 10–18), a triad PP3/3 media mat (except Examples 15 and 16) with a weight of W_0 was first dry impregnated with solid fine powders to a weight of Ws, and then treated with $Zn(Ac)_2 \cdot 2H_2O$ solution to a dried weight of W1 followed by treatment with an K_2CO_3 or KOH solution to a dried weight of W2 according to <u>Procedure III</u> above in Section 2.4.2. The filters were equilibrated at 50 ± 5% relative humidity prior to testing. Their SO_2 and H_2S breakthrough results are shown in Figures 7, 8 and 11.

Table 3   *Data for Type D Filters with $Zn(Ac)_2/K_2CO_3$ Reagent.*

Examples	Fiber wt. W_0 (g)	Powder Type	Powder load wt. Ws (g)	Zn-treat. wt. W1	Zn/K treat. wt. W2 (g)	Test Samp. wt. W3 (g)
10	4.01	ZnO	4.57	5.62	8.28	0.74
11	3.44	Al_2O_3	3.83	4.96	6.69	0.64
12	3.40	PTFE*	4.14	4.93	6.68	0.71
13	3.90	$NaHCO_3$[&]	4.80	6.31	9.04	0.71
14	3.80	Centaur Carbon	4.37	6.21	9.90	0.75
15[I]	0.1098	Centaur Carbon	0.1310	0.24	0.40	0.40
16[II]	0.1246	Centaur Carbon	0.1406	0.35	0.44	0.44

Notes: *Unless other stated, PP3/3 media used.*
[I] *PP 1.5/3 media,* [II] *PP 1.5/6 media.*
* *A few drops of 1% $C_8F_{17}SO_3Na$ solution were added in the $Zn(Ac)_2$ solution to facilitate wicking.*
[&] *Flowing baking soda powder containing 5% silicates.*

Table 4   *Data for Type D Filters with $Zn(Ac)_2/KOH$ Reagent.*

Examples	Fiber Virgin wt. W_0 (g)	Powder Type	Powder loaded wt. Ws (g)	Zn-treat. dry wt. W1 (g)	Zn/K treat. dry wt. W2 (g)	Test samp. eq wt. W3 (g)
17	1.12	Centaur	1.26	1.61	2.20	0.60
18	1.24	GS-1	1.39	1.83	2.36	0.49

2.4.5   Type D Filters with Various Divalent Metal Acetates. A no heating procedure was designed to study the effect of different divalent metal cations in Type D filters (Examples 19–22). As summarized in Table 5, a triad PP3/3 filter mat was first dry impregnated with 16 wt% of centaur powder to a weight of Wc and then soaked into excess divalent salt solutions for 25 min. After excess solution was removed, without drying, the mat with a wet weight of W1w was directly soaked in excess 20% K_2CO_3 solution for about 10 min. After removal of excess solution, the mat with a wet weight of W2w was then cut into 2-inch diameter round pads and equilibrated at 50% RH overnight. The equilibrated test sample weight was W3. Their SO_2 breakthrough results are shown in Figure 9. A typical H_2S breakthrough profile for them is as shown in curve D of Figure 10(c).

Table 5   *Data for Type D Filters with Various Divalent Acetates.**

Examples	Fiber&C wt. Wc (g)	Divalent compound	M^2 treat. wet. W1w(g)	M^2/K treat. wet. W2w(g)	Test samp. wt. W3
19	2.15	–	–	6.32	0.40
20	2.24	$Zn(Ac)_2$	11.89	13.24	0.73
21	2.49	$Mg(Ac)_2$	8.00	8.40	0.60
22	2.31	$Mn(Ac)_2$	8.18	11.56	0.63

**Without heating.*

2.5 Optical and Electron Microscopy

The optical microscopy images parallel to fibers in the filter mat were taken with a Zeiss Stemi-IV Stereomicroscope. The fiber cross sectional images were taken with a JEOL 6300F Field Emission SEM or an Axiophot Compound Optical Microscope.

3 RESULTS AND DISCUSSION

3.1 Surface Structure of the Virgin Triad Media

After screening a large number of fibers with various cross sections, we found that the cross section in Figure 1, termed the triad, demonstrated the greatest capacity in retaining both liquids and solids. Clearly the length and depth of the channels have an effect on the loading capacity. The addition of caps on the exterior lobes had a significant effect on the ability to entrap reagents irreversibly.[10] Graphical analysis of the cross section revealed that the triad fibers had roughly 50% void volume within the channels. This internal void volume can retain high capacity of reagents without restricting the free gas flow across the media in filtration applications.

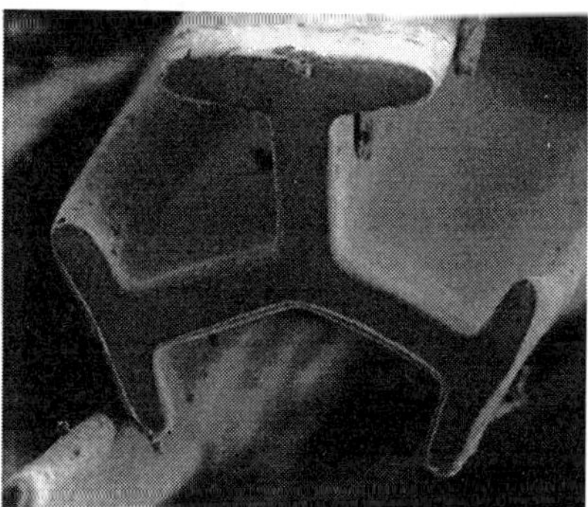

Figure 1   *A scanning electron microscope cross section of a 3 denier per filament tri-lobal polypropylene fiber (32 microns diameter)*

3.2 Development of Filters with Various Surface Structures

Using the unique triad media discussed above, we have developed four types (Type A, B, C, and D) of filters with different surface structures. Their schematic differences in cross sections are illustrated in Figure 2.

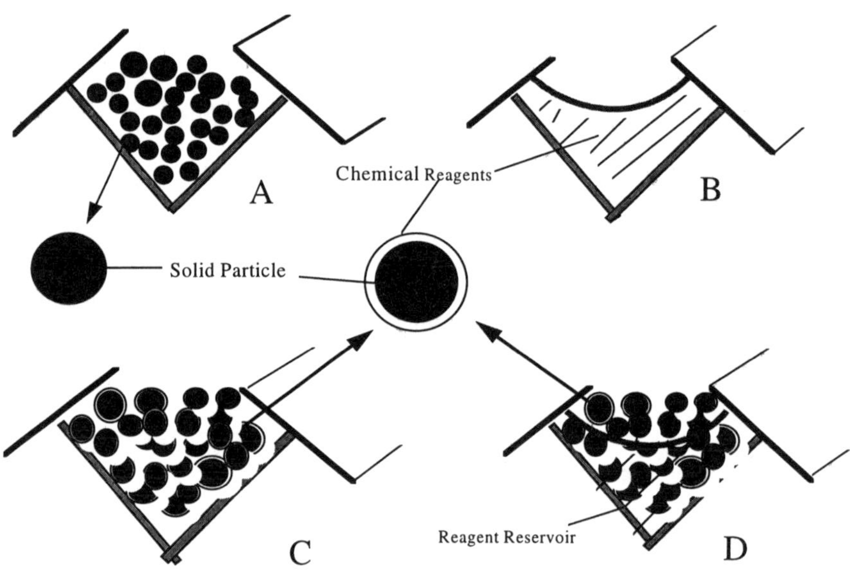

Figure 2	*Schematic Cross Sectional Channel Structures from Various Impregnation.*

3.2.1 Type A Filters. This type of filter was originally developed to serve as a low cost alternative to conventional carbon fiber filters. As demonstrated in Figure 2A, they were prepared by impregnation of finely ground virgin activated carbon powders.[10] The fine solid particles provided very large surfaces for gas adsorption. Type A containing 16% fine carbon particles (average surface area: 1000 m^2/g) may have a surface area of 138 m^2/g. A competitive evaluation using cost, butane capacity, and pressure drop as criteria has revealed its highest potential as replacement for the high cost carbon fibers in indoor air quality (IAQ) applications targeting various organic odors.[14]

When tested with 400 ppb of H_2S, as shown in Figure 3, most of them show relatively low capacity comparing to Type C filters because of the intrinsic low reactivity of virgin carbon powder surface to acidic gases. Calgon Centaur powder gave the best performance with about 5 minutes of zero breakthrough, which was probably a result of its designed high catalytic functionality.[15] The lack of capacity of this type filters to low molecular weight inorganic acidic gases was more obvious in the tests with 40 ppm SO_2, where immediate breakthrough to above 90% was observed for all carbon-only filters including centaur carbon as shown in Curve A of Figure 10(b).

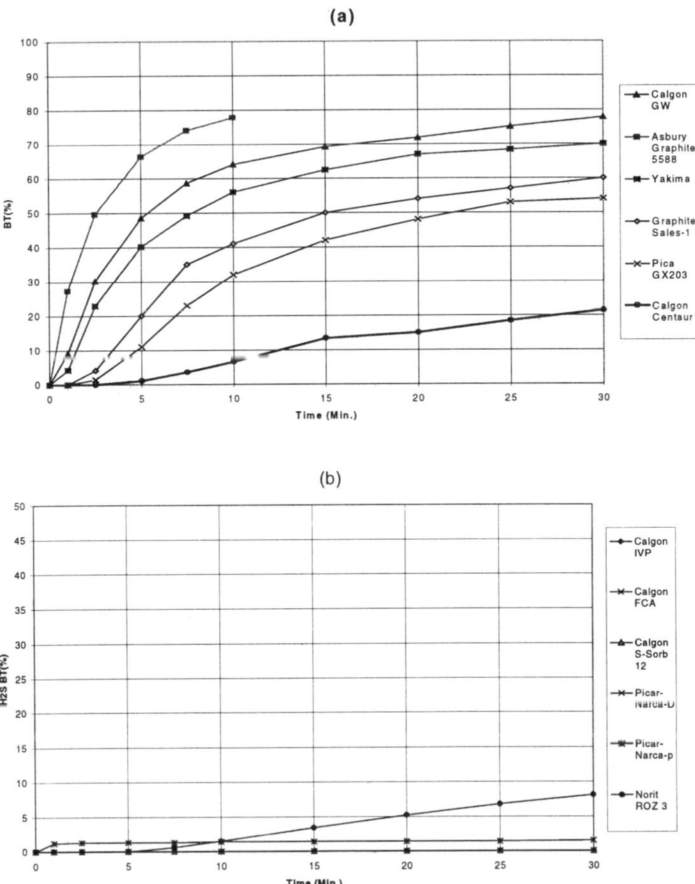

Figure 3 *H_2S breakthrough results of filters prepared from carbon powder impregnated triad PP3/3 media: (a) Type A; (b) Type C.*

3.2.2 Type B Filters. To obtain high capacity filters for acidic pollutant removal, we have impregnated close-to-saturated solutions of various alkalis into the triad media as demonstrated in Figure 2B. Generally, the triad PP3/3 media can retain 3–4 times its own weight of alkaline solutions with the help of DBS surfactant (Table 1). Clearly, from Figure 4(a), 50% KOH solution gave excellent activity to 400 ppb of H_2S (100% removal in 30 minutes) when supported on the triad media. This provides a groundwork for the construction of high efficient and high capacity cross flow continuous acidic gas scrubbing systems for industrial emission control.

To explore the IAQ application of above Type B filters, we also tested the filters for H_2S removal after they were equilibrated in 50% RH air. As shown in Figure 4(b), the superiority of KOH in H_2S (400 ppb) removal to $KHCO_3$, K_2CO_3 and K_3PO_4 was greatly reduced. All of them gave poor performance (flat breakthrough around 70%), which may be the results of CO_2 saturation.

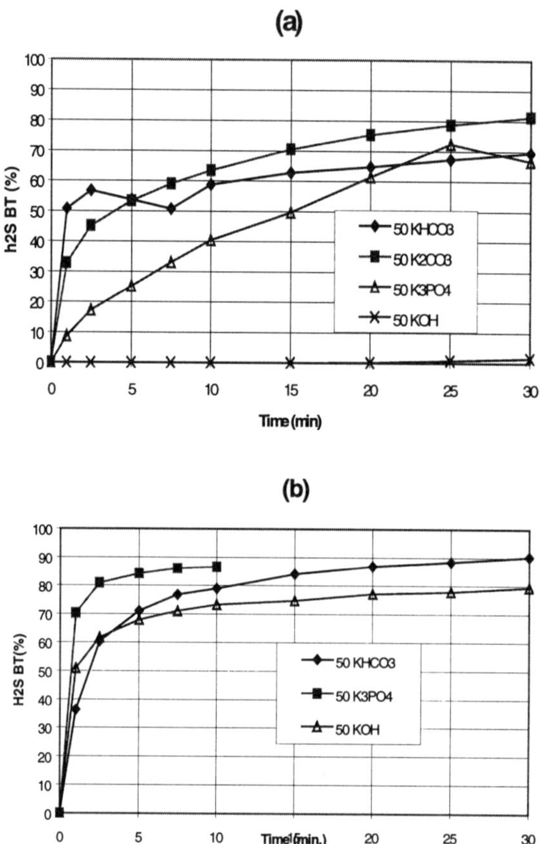

Figure 4   *H_2S breakthrough results of Type B filters prepared from impregnation of PP3/3 media with 50 % solution of KOH, K_3PO_4, K_2CO_3, and $KHCO_3$ according to Procedure I: (a) Freshly prepared: (b) Equilibrated in 50% RH air.*

To improve the performance of Type B filters in air, $Zn(Ac)_2$ was co-impregnated with the K_2CO_3 and KOH solutions. As shown in Figure 5, much improved H_2S performance (flat 15 to 25% breakthrough) was obtained for both systems. This improvement may be a result of the improved reagent dispersion from the good mediation of acetate anions to PP surface, and the improved chemical reaction mechanism from the high affinity of S^{2-} anions to divalent Zn^{2+} cations.

The flatness of the breakthrough curves of Type B filters implies high capacity of gas removal and good communication of reactants and reaction products between the surface and reagent body. If external reagent reservoirs are added to the system, the reaction capacity may be infinitely increased until continuous cross flow gas scrubbing is achieved.

As demonstrated in Figure 2B, Type B impregnation possesses limited direct reagent–gas contact area compared to Type A impregnation. The gas–reagent direct contacting area may maximally be the surface areas of the virgin media (0.307 m^2/g) assuming minimum loading and maximum dispersion, which is only 450th of that of Type A filters. This limitation on direct reagent–gas contact results in immediate by-pass breakthrough in the filtration tests carried out in this research where the resident time was very short (about 8 milliseconds). The initial zero H$_2$S (400 ppb) breakthrough period shown for some of the Type A filters disappeared for all Type B filters. The lack of direct gas–reagent contacting surface area is also expressed in 40 ppm SO$_2$ uptake test. As shown in Figure 10(b), the breakthrough showed obvious increase long before the regent capacity was used up.

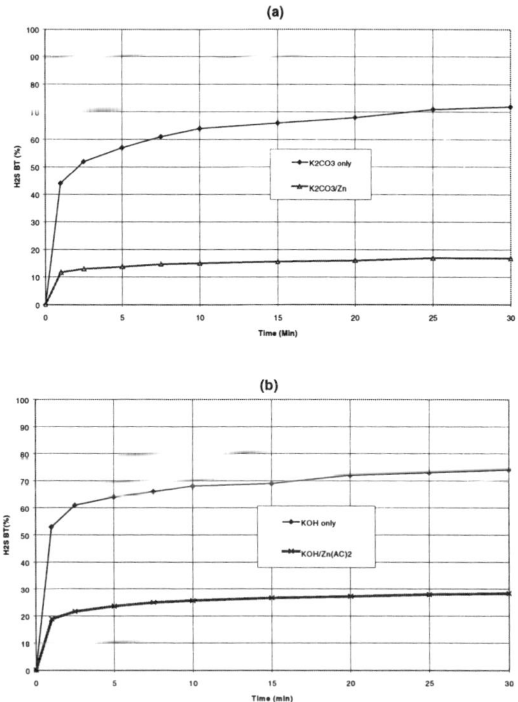

Figure 5   *Effect of Zn(Ac)$_2$ pre-treatment on H$_2$S breakthrough of Type B filters prepared from PP3/3 media with Procedure II and III. (a) K$_2$CO$_3$; (b) KOH.*

3.2.3 Type C Filters. As described in Figure 2C, the surface of the impregnated carbon powders was pretreated with chemical reagents to increase their reactivity to H$_2$S and SO$_2$. They showed much improved H$_2$S removing results (Figure 3) compared to Type A filters. The filters have not only high reaction surface area, but also enough reagent capacity to remove almost 100% of 400 ppb H$_2$S in 30 minutes. However, the capacity was too short under the challenge of 40 ppm SO$_2$ where 100 % removal only lasted for about 1 minute as shown in Curve B of Figure 10(b). This shortage in capacity agreed well with their low reagent loading from impregnation. If a Type C filter (Example 9) was loaded with 22% pretreated carbon powder carrying 12% chemical reagent loading, its chemical reagent content was less than 2.0%, while similar Type B filter (Example 8) had 16 times greater reagent content (about 32%).

3.2.4 Type D Filters. As described in Figure 2D, excess chemical impregnation was conducted after the impregnation of solid fine powders, so that chemical reagent reservoirs formed within the channels of the fibers. The chemical reagents provided high reaction capacity while the co-existing solid fine particles supplied large reagent supporting surface area for direct gas contact. It combined the high capacity of Type B filters and the high surface area of Type C filters. In real situations, the chemical reagent reservoirs are not necessarily found on the bottom as depicted. From the optical microscopic images shown in Figure 6(a) and 6(b), we can see that the channels of the triad fibers are actually heterogeneously filled with white spots of chemical reagent reservoir and black spots of thinly coated carbon particles. This can be further confirmed in the cross sectional views shown in Figure 6(c) and 6(d).

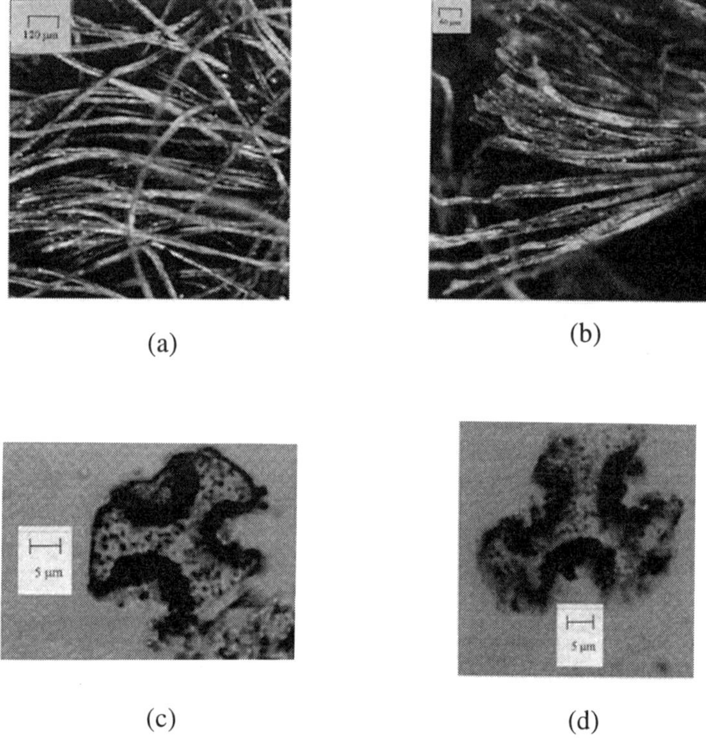

(a)   (b)

(c)   (d)

Figure 6   *Optical microscopic enlarged images of Type D filters prepared from PP3/3 media and Centaur/Zn(Ac)2/$_{K2CO3}$ reagents (a) Part of impregnated filter mat; (b) Strands of impregnated fiber; (c) Fiber section with excess chemical reagent reservoir; (d) fiber cross section without excess chemical reagent reservoir.*

Similar to Type A and Type C filters, the existence of fine ground particles in Type D filters dramatically increased the direct reagent–gas contact area and greatly improved the reaction kinetics. The great advantage realized by the co-impregnation of various solid fine particles is clearly demonstrated in Figures 7 and 8, where both H_2S and SO_2 breakthroughs showed obvious reduction due to the inclusions of ZnO, Al_2O_3, PTFE,

NaHCO$_3$ and carbon powders. When co-impregnated with K$_2$CO$_3$/Zn(Ac)$_2$ reagent system, carbon powders gave the best performance, almost 100% removal to both 400 ppb H$_2$S and 40 ppm SO$_2$ in the testing period of 30 minutes. This excellent performance was most likely the result of the high surface area and high stability of carbon powders in alkaline environment. Type D filters outperformed not only previous types of filters we discussed (Figure 10), but also the benchmark commercial carbon filters (BM) with 4 times greater weight (Figure 11).

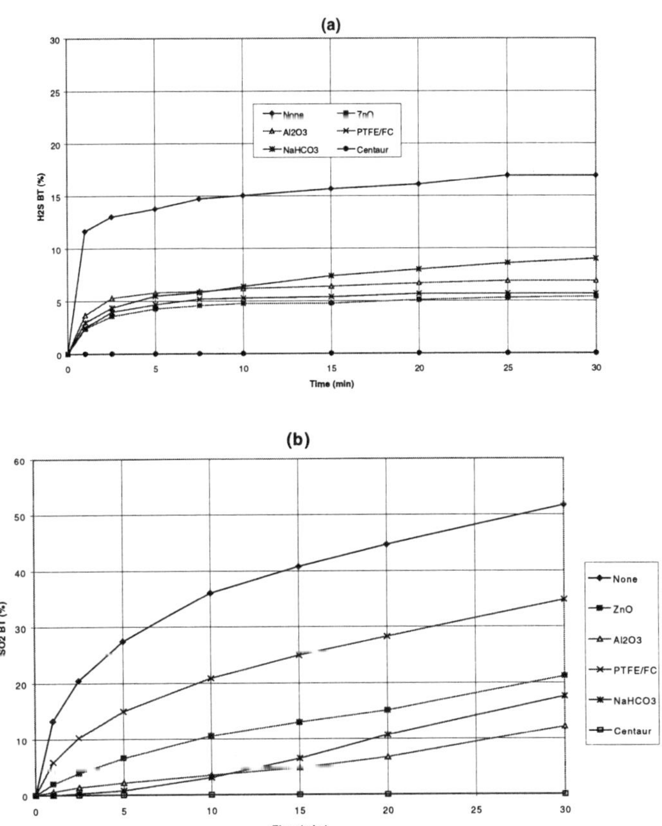

Figure 7 *Effects of solid particles on breakthrough results of Type D filters prepared from PP3/3 media and K$_2$CO$_3$/Zn(Ac)$_2$ reagents: (a) H$_2$S, (b) SO$_2$.*

The effect of divalent acetate salts on the activity of alkaline reagents filters to acidic gasses was studied in Type D. Similar to Type B filters, obvious performance enhancement was observed when various divalent acetate salts were co-impregnated with K$_2$CO$_3$ (Figure 9). It may involve the formation of reactive intermediate complex resulting from the interaction between M(Ac)$_2$ and K$_2$CO$_3$. The detailed mechanism is still under investigation.

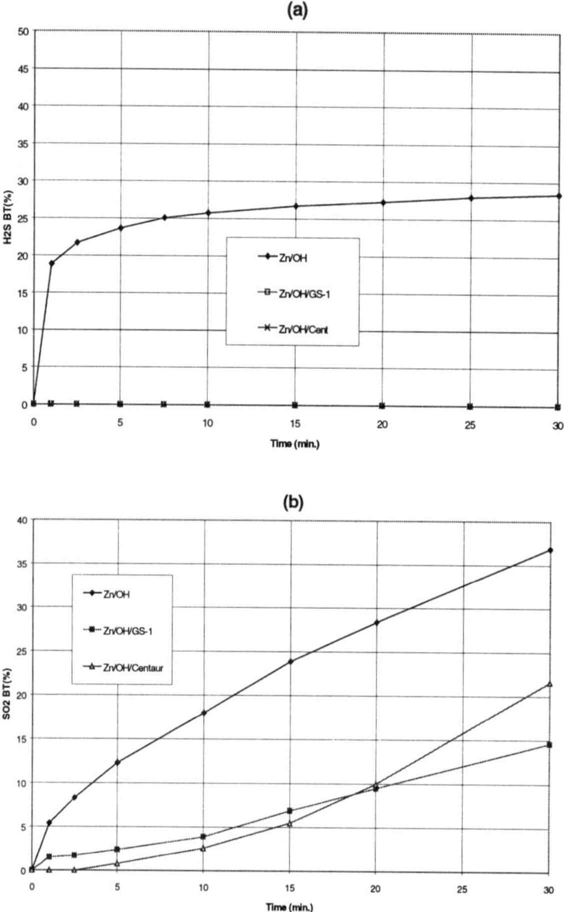

Figure 8   *Effects of solid particles on breakthrough results of Type D filters prepared from PP3/3 media and KOH/Zn(Ac)$_2$ reagents: (a) H$_2$S, (b) SO$_2$ (GS-1 is a carbon sample sold by Graphite Sales, Inc).*

Interestingly, the impregnated solid particles also increased the reagent retainability of the triad fiber media. The reagent loading of Type D filter in Example 14 (Table 3) reached 1.45 g/g virgin fiber. This value is 3 times greater than that of the Type B filters in Example 8 that did not have the pre-impregnated carbon powders. This higher reagent retention may come from the much-increased capillary force generated by the intra-channel impregnation of fine solid particles. Compared to the reagent loading of Type C filters in Example 9, above value of reagent loading for Type D filter in Example 14 is much greater (about 56 times). The super high reagent retention power of Type D filters accounts for their super high capacity in gas removal, which was clearly reflected in the breakthrough results in Figure 10.

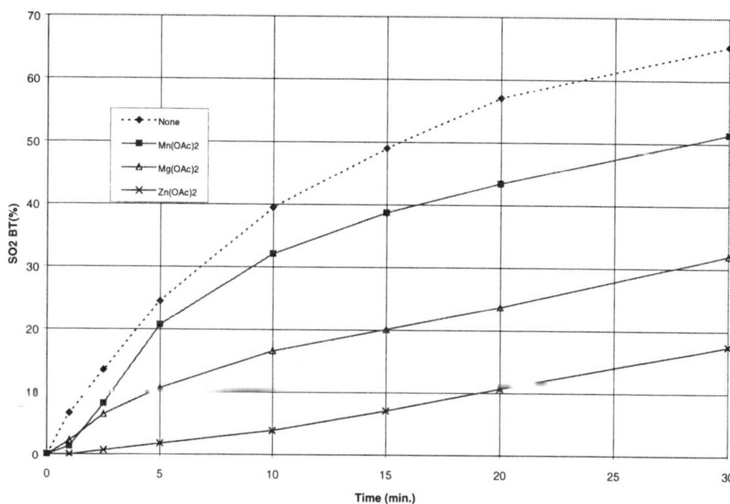

Figure 9   *Effects on SO$_2$ removing performance of divalent cations in Type D filters prepared from PP3/3 media and K$_2$CO$_3$/Centaur carbon.*

3.3 Flow Restriction Analyses

The flow restriction, as well as typical H$_2$S and SO$_2$ breakthrough curves of filters made from triad PP3/3 media, are compared in Figure 10. Compared to virgin media, Type A, B and C filters showed no increase in flow restriction because intra-fiber impregnation posed no blockage to the inter-fiber spaces. The Type D filter showed slightly increased flow restriction, close to that of a round cross section media, because of its heavy solid loading (usually 160% or more of the virgin fiber weight). This result agrees with the cross sectional image for the Type D filter in Figure 6(c), in which the surface of the heavily impregnated chemical reagent approached the openings of the triad channels. Thus, heavily loaded triad fibers almost possessed round cross sections when most of their internal void space was filled.

As shown in Figure 11, the flow restriction of Type D filters developed from triad PP3/3 was higher than the benchmark commercial filters (BM) despite their excellent H$_2$S and SO$_2$ uptake performance. In an attempt to reduce flow restriction, lower density triad filter media of PP 1.5/3 and PP 1.5/6 were produced and used to construct low density Type D filters. To our satisfaction, these low density Type D filters (Examples 15 and 16, Table 3) showed not only lower flow restriction than the benchmark carbon filter (BM), but also superior H$_2$S and SO$_2$ removing performance. Especially as shown in Figure 11(c), the H$_2$S uptake performance of the low-density Type D filters is remarkable. Throughout the test period of 30 minutes, > 99 % of the 400 ppb H$_2$S was removed, which means only <1% of the H$_2$S had passed through the substantially increased inter-fiber spaces in a very short resident time of 8 milliseconds. Further optimizations on fiber denier, packing density and packing pattern of the fiber media are ongoing. We believe the flow restriction of current filters can be further decreased, maintaining the excellent gas removing performance.

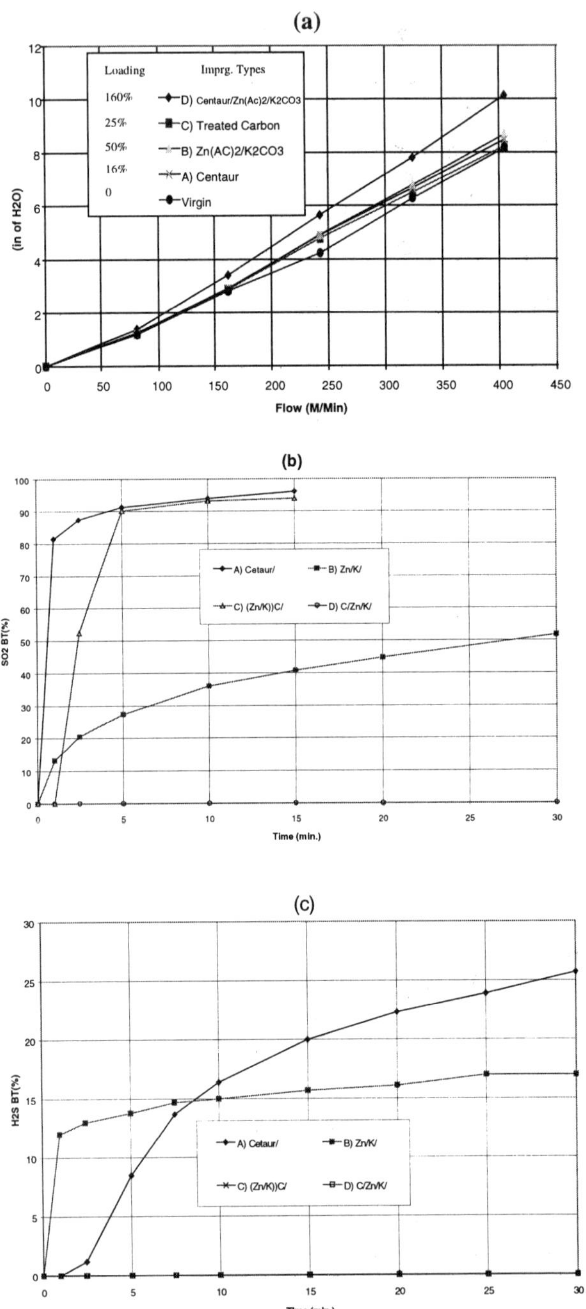

Figure 10 *Comparison among filters with various types of impregnation of PP3/3 media, centaur carbon powder, $Zn(Ac)_2$ and K_2CO_3. (a) Flow restriction; (b) SO_2 removal; (c) H_2S removal.*

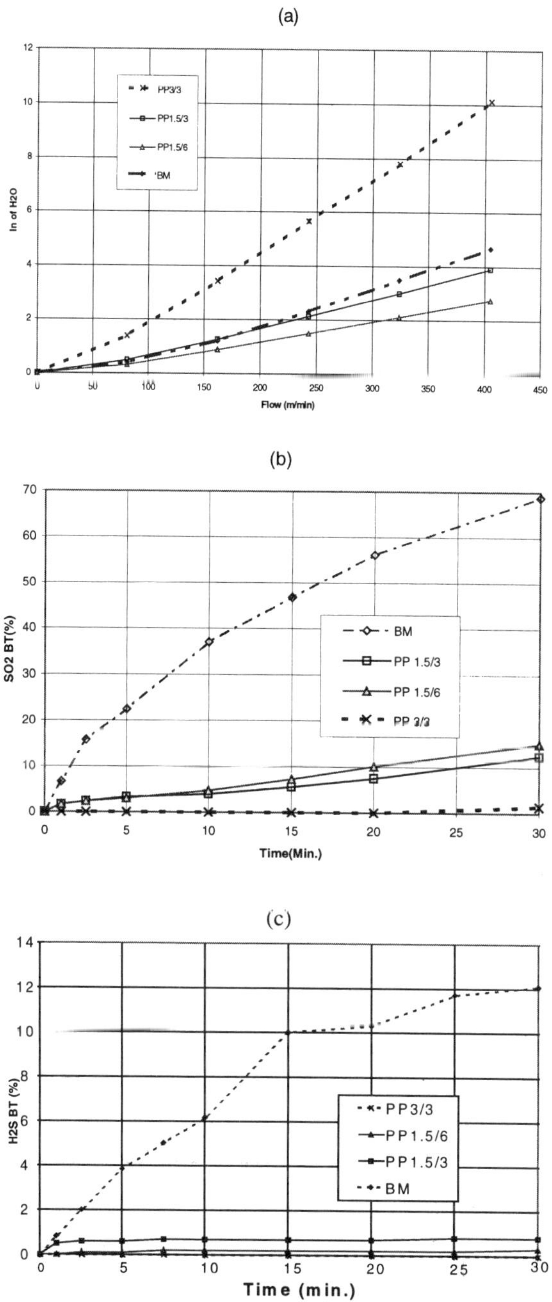

Figure 11 *Comparison between Type D filter with various densities and benchmark commercial carbon filter (BM): (a) Flow restriction; (b) SO_2 removal; (c) H_2S removal.*

3.4 Application of the Filters

The surface structure and chemical treatment level of the filters determined their reaction kinetics, capacity, and application potential. As summarized in Table 6, four types of impregnation procedures generated differences in surface area, surface reactivity and reagent capacity of the resulted filters. Type A filters have very low surface reactivity to SO_2 and H_2S, but their virgin carbon surfaces possessed great affinity to organic odors such as butane and toluene. They have shown high potential in removing organic odors for Indoor Air Quality (IAQ) applications.[11,13,14] Type B filters are suitable for industrial source acidic emission control because of their high reaction capacity. As discussed earlier, if the systems were coupled with additional liquid reservoirs, continuous cross flow gas scrubbing systems may be constructed. Research in these areas is currently being carried out in our laboratory. Type C filters are excellent in achieving zero breakthrough at low challenge level, so they may be good choices for acidic pollutant control in IAQ applications where the challenge concentrations are usually at the ppb levels. Type D filters have the advantage of both Type B and Type C filters and are excellent choices for both industrial and IAQ applications.

Table 6 *Comparison of Filters with Various Types of Impregnation.*

Type	Surface Area	Surface Reactivity	Reagent Capacity	Examples	Possible Applications
A	VH	VL*	VL	Carbon powders	IAQ for organic odors
B	L	M	H	KOH or K_2CO_3	Industrial Emission
	L–M	H	H	w/ $Zn(Ac)_2$	
C	VH	VH	L	Centaur/$Zn(Ac)_2$/ K_2CO_3	IAQ for acidic odors
D	M	H	H	K_2CO_3 or KOH on Carbon powders	Industrial Emission or IAQ
	H	VH	H	w/ $M(Ac)_2$	

V = very; H = high; M = medium; L = low

4 SUMMARY

Highly efficient acid gas-removing filters for industrial emission control and indoor air quality improvement were developed by supporting chemisorptive reagents on shaped triad fiber media. The filters were characterized into four types (Type A, B, C and D) according to their surface structure and chemical treatment level.

The reactivity of the impregnated alkaline chemical reagents (*i.e.* KOH and K_2CO_3) to H_2S and SO_2 could be dramatically improved by co-impregnation of divalent acetate salts such as $Zn(Ac)_2$ and solid fine powders. We have achieved excellent H_2S and SO_2 removing performances by co-impregnating carbon powders and $Zn(Ac)_2$ with alkaline reagents such as K_2CO_3 and KOH in the triad fiber media.

We did not observe obvious gas flow restriction increase from our reagent impregnation that mainly occupied internal void space of the triad fibers. The flow restrictions of the formed filters come mainly from the virgin media and may be reduced by optimizing the denier and packing density of the fibers. We have developed low density

Type D filters that possessed both better gas removing performance and lower flow restriction than a benchmarking commercial carbon filter.

5 ACKNOWLEDGMENTS

We are grateful to Ms. Caroline Bednarczyk for conducting the Optical and Electronic Microscopy of the filters.

References

1. G. T. Wolff, in 'Kirth-Othmer's Encyclopedia of Chemical Technology', Fourth Edition, John Wiley & Sons, Inc., 1991, Vol. I, p. 731.
2. G. D. Thurston, K. Ito, C. G. Hayes, D. V. Bates and M. Lippmann, *Environ. Res.*, 1994, **65**, 271.
3. R. T. Burnett, R. E. Dales, M. E. Raizenne, D. Kreski, P. W. Summers, G. R. Roberts, M. Raad-Young, T. Dann and J. Brook, *Environ. Res.*, 1994, **65**, 172.
4. 'Acid Rain Program', US EPA at http.//www.epa.gov/acidrain/overview.html, September 1997.
5. B. B. Crocker, in 'Kirth-Othmer's Encyclopedia of Chemical Technology', Fourth Edition, John Wiley & Sons, Inc., 1991, Vol. I, p. 763.
6. M. J. Abler, *US Patent 5344626*, 1994.
7. T. Largman, F. Gefri and F. Mares, *US Patent 5,057,368*, 1991.
8. G. Jones, R. Rohrbach, P. Unger, D. Bause, Lixin Xue, R. Williams, R. Dondero and L. Leyrer, 'Wicking Fiber Chemisorption for Air Quality Improvement', *Topics in Automotive Filtration Design*, Society of Automotive Engineering, Feb. 1997, p. 49.
9. G. Jones, R. Rohrbach, P. Unger, D. Bause, Lixin Xue , R. Williams, R. Dondero and H. Sherrel, 'The Development of Active Chemistries for Wicking Fiber Odor Control', *Advances in Filtration and Separation Technology*, 1997, Vol. 11, E. Bauman and L Weisert, eds., in 'Advancing Filtration Solutions' by the American Filtration and Separation Society, p. 234.
10. R. Rohrbach, P. Unger, A. Lobovsky, Lixin Xue, D. Bause and R. Dondero, 'Properties and Applications of Externally Impregnated Shaped Fibers', *Composites*, Material Research Society, in press, 1998.
11. D. Bause, R. Rohrbach, P. Unger, Lixin Xue, G. Jones, and R. Dondero, 'Use of A Chemically Impregnated Shaped Fiber to Improve Air Quality', Environmental Chemistry Division, *215th ACS National Meeting*, Dallas, Texas, March 1998, Poster #47.
12. P. Unger, R. Rohrbach, A. Lobovsky, Lixin Xue, D. Bause and R. Dondero, 'Formation of Novel Hybrid Materials Using Complex Cross Section Fibers', *Material Research Society*, Spring Meeting, San Francisco, CA, April 13–17, O10.8.
13. S. L. Penrod, P. Unger, R. Rohrbach, D. Bause, Lixin Xue and R. Dondero, 'Filtration of Aircraft Cabin Air Odors in 100% Relative Humidity, Low Temperature Conditions', to be published.

14. D. Bause, R. Dondero, R. Rohrbach, P. Unger, Lixin Xue and G. Jones, 'Competitive Evaluation of Carbon Impregnated Wicking Fibers with Activated Carbon Fiber', <u>Research Report 97–12</u>, AlliedSignal, Inc., October 31, 1997.
15. D. R. McAlister, J. J. Tully, M. D. Howes, A. F. Mazzoni, R. T. Lynch and N. E. Magonnell, 'Centaur^{TM} SO$_2$ Removal Process, a New Approach to an Old Problem', at <u>http://www.enviro-chem.com/plants/centaur</u>, August 13, 1997.

MAXIMIZING THE EXTENT OF HEXAMETHYLDISILAZANE REACTION WITH SILICA: AN EXPERIMENTAL DESIGN STUDY

F.R. Jones, M.S. Vedamuthu and J.P. Blitz*

Eastern Illinois University
Department of Chemistry
Charleston, IL 61920 USA

1 INTRODUCTION

There are many uses for materials comprised of silica surfaces chemically modified with organofunctional silanes. Previous symposia[1] of this type have described many of these applications, which include their use as chromatographic stationary phases, catalyst supports and composite materials.

Silica surfaces are naturally polar and hydrophilic.[2] Many applications require a non-polar hydrophobic silica surface to achieve desirable results. One common means of imparting hydrophobic properties to the silica surface is to silylate the silica using hexamethyldisilazane, as in Equation 1.

$$(CH_3)_3\text{-}Si\text{-}NH\text{-}Si(CH_3)_3 + 2Si(s)\text{-}OH \;\rightarrow\; 2Si(s)\text{-}O\text{-}Si\text{-}(CH_3)_3 + NH_3 \qquad (1)$$

Hexamethyldisilazane is a fairly reactive silylating reagent because the ammonia product generated during reaction is a catalyst for further reaction.[3] Even though hexamethyldisilazane is widely viewed as a reactive silylating reagent, many surface silanols are not readily reacted by this or other organosilicon compounds. Hertl and Hair[4] were the first to show that hexamethyldisilazane reacts isolated (non-hydrogen bonded) silanols, but does not silylate vicinal (hydrogen bonded) silanols. This work has been generally confirmed and extended by subsequent work,[5,6] and has been shown to be true for a variety of silylating reagents.[7] This is important because unreacted silica surface silanols often impart undesirable properties to the material. These silanols can act as water adsorption sites on a material designed to be hydrophobic. Much has also been written concerning the role of residual surface silanols and poor chromatographic performance of basic compounds.[8] Improved synthetic conditions to enhance the extent of surface silylation reaction are of considerable interest.

There are many variables that can influence the extent of silylation reaction with the silica surface. Studies have been reported describing the role of solvent,[9] amine catalyst,[9–11] reaction temperature,[8a,12] and silica pretreatment temperature.[13] To our knowledge no studies have been reported which attempt to optimize all of the variables to find the optimum reaction conditions for maximizing the extent of silylation within realistic time constraints.

A total of seven potential variables that may affect the extent of hexamethyldisilazane reaction with silica have been identified. These include reaction temperature, presence of an external amine catalyst, presence or absence of surface adsorbed water, reaction time, reactant concentration, solvent type and dry or wet solvent. To optimize the reaction conditions given these seven potential variables, one would traditionally hold six of the variables constant while optimizing the seventh. This scenario would be repeated for all seven variables to find the overall optimum reaction conditions. There are two reasons why this is not an optimal approach to this problem: (1) it would require a tremendous number of experiments, (2) one could not predict how the variables affect one another. We have thus chosen a factorial design approach to study this problem.[14]

2 EXPERIMENTAL

2.1 Materials

All solvents, hexamethyldisilazane and butylamine were distilled in nitrogen prior to use. These distilled solvents were stored under nitrogen and assumed to be free of water. Water saturated solutions of hydrocarbon solvents were prepared after distillation. The acetonitrile/water mix was prepared by adding 400 μL of distilled water to 40.0 mL of distilled solvent.

Cab-O-Sil HS-5 fumed silica (325 m^2/g, Cabot Corporation) was heated in vacuum at 200 °C for 2 h. One portion of this dry silica was stored in a dessicator, the other portion was exposed to 50% relative humidity atmosphere for at least two weeks prior to use.

2.2 Silylation Reactions

Typically, 0.500 g of silica was slurried with 40.0 mL of solvent. When catalyst was used, 50 μL of *n*-butylamine was then added to the silica/solvent slurry. The desired concentration of hexamethyldisilazane was then added and the reaction allowed to proceed under the prescribed conditions for a set time. After the allotted reaction time the silica was filtered and washed thrice with 12 mL of solvent each time. The sample was then dried in vacuum at 200 °C for 1 h prior to analysis.

2.3 Analysis

Transmission infrared spectra of modified silicas were obtained using a Perkin-Elmer Spectrum 2000 FT-IR spectrometer purged with dry air and equipped with a liquid nitrogen cooled mercury–cadmium–telluride detector. Typically, 64 scans were co-added at a nominal resolution of 4 cm^{-1}. The sample was suspended on a wire mesh for analysis.

Quantitative information concerning the extent of silylation was obtained using infrared spectroscopy. The asymmetric CH_3 stretching band was integrated (2992–2927 cm^{-1}) and ratioed to a Si-O-Si combination band (1927–1750 cm^{-1}) used as an internal standard. At least five aliquots of the same sample were analyzed and averaged. The infrared results were calibrated using combustion analytical data from Galbraith Analytical Laboratory (Knoxville, TN).

3 RESULTS AND DISCUSSION

Typical infrared spectra of silica before and after hexamethyldisilazane reaction are shown in Figure 1. Before reaction a sharp absorbance at *ca.* 3745 cm^{-1} due to isolated silanols is detected. The broad absorbance adjacent to the isolated silanol peak is due to both hydrogen-bonded silanols and physically adsorbed water. Two small peaks, seen in both spectra, are due to gaseous CO_2 in the spectrometer that is of no importance to this study. The peak at *ca.* 1860 cm^{-1} is the Si-O-Si combination band referred to earlier used as an internal standard. The absorbance at *ca.* 1620 cm^{-1} contains contributions from silica (Si-O-Si symmetric stretching overtone) and water (H-O-H bending mode). After reaction with hexamethyldisilazane the isolated silanol peak disappears, the broad O-H absorbance is attenuated, and trimethylsilyl groups can be detected via absorbances attributable to CH stretching (*ca.* 2920 cm^{-1}) and bending (*ca.* 1500 cm^{-1}) modes. The sloping baseline, which could have been digitally corrected, is due to the scattering of infrared radiation from the method used to obtain transmission spectra.

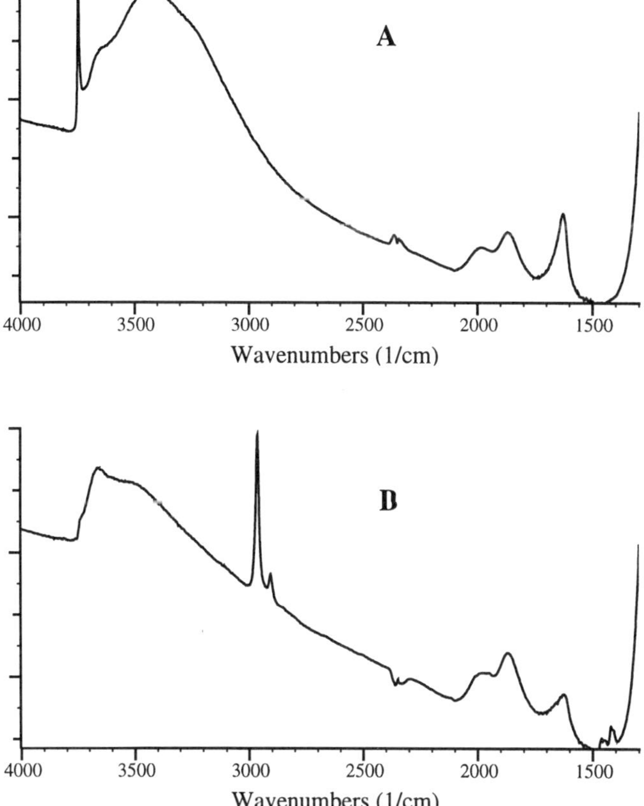

Figure 1   *Transmission infrared spectra of (A) unmodified silica, and (B) hexamethyldisilazane-modified silica.*

The calibration line for the infrared analysis using combustion data obtained by Galbraith Analytical Laboratory is shown in Figure 2. The extent of hexamethyldisilazane reaction is plotted as μmoles/m^2 of trimethylsilyl (TMS) groups versus the infrared band area ratio. Highly reproducible data are obtained by this method. All TMS loading values reported in this work were determined by infrared spectroscopy using the calibration line shown in Figure 2.

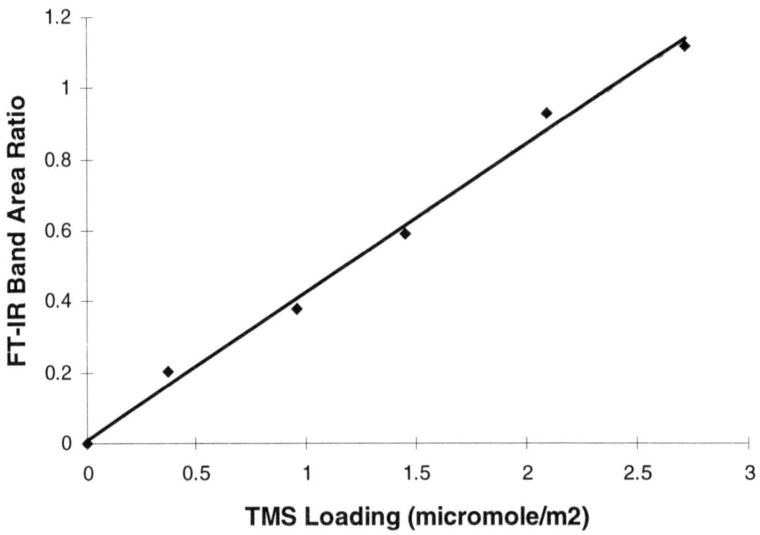

Figure 2 *Calibration line for extent of hexamethyldisilazane reaction by FT-IR.*

Since seven different variables or factors have been identified as possibly effecting the extent of hexamethyldisilazane reaction with silica, a two-level full factorial design would require 2^7 or 128 experiments. Since this full factorial design would most likely contain redundant information, a 2^{7-3} resolution IV design was devised. This matrix, requiring 16 experiments, can distinguish main effects from two-factor interactions.[14] Although information on two-factor interactions is obtained, exactly *which* two-factor interactions are important may require additional experiments. In Table 1 we detail the specifics of this initial design and the results of these first 16 experiments.

Analysis of the data in Table 1 using ANOVA and a normal probability plot (not shown) indicate the following factors are significant: (1) the only main effect not involved in a two-factor interaction is solvent; (2) two factor interactions include catalyst–time, temperature–time and temperature–concentration. It was previously stated that the resolution IV design matrix does not allow one to predict which two-factor interaction effects are significant. Additional studies have shown that the two-factor interaction effects indicated above bring about changes in the extent of hexamethyldisilazane reaction. The other variables considered, water in solvent and water on silica surface did not have a significant effect.

Figure 3 details the two-factor interaction effects from the fractional factorial design results. These data, plotted as wt %C (μmol/m^2 is readily obtained by multiplying the %C values by 0.85), show that for the two-factor interactions involving temperature, higher

Table 1 *Fractional Factorial Screening Design Matrix and Results.*

Reaction Temp (°C)	Catalyst	Water on Silica	Rxn Time (h)	[Reactant] (mmol/g)	Solvent*	H_2O in solvent	TMS loading ($\mu mol/m^2$)
81	Y	N	5	26	CH	N	2.47±0.08
25	N	Y	1	2.6	B	Y	1.20±0.10
81	N	Y	1	26	B	N	1.94±0.15
81	N	Y	5	26	CH	Y	2.44±0.03
25	N	N	1	26	CH	N	1.55±0.08
81	Y	Y	5	2.6	B	Y	2.08±0.08
81	Y	N	1	26	B	Y	2.65±0.06
81	N	N	5	2.6	B	N	2.58±0.18
25	Y	N	1	2.6	B	N	1.54±0.03
25	Y	N	5	2.6	CH	Y	2.27±0.25
25	N	N	5	26	B	Y	2.25±0.13
25	Y	Y	1	26	CH	Y	2.32±0.13
81	N	N	1	2.6	CH	Y	2.20±0.08
81	Y	Y	1	2.6	CH	N	2.80±0.20
25	Y	Y	5	26	B	N	2.39±0.08
25	N	Y	5	2.6	CH	N	2.08±0.08

* CH = Cyclohexane, B – Benzene

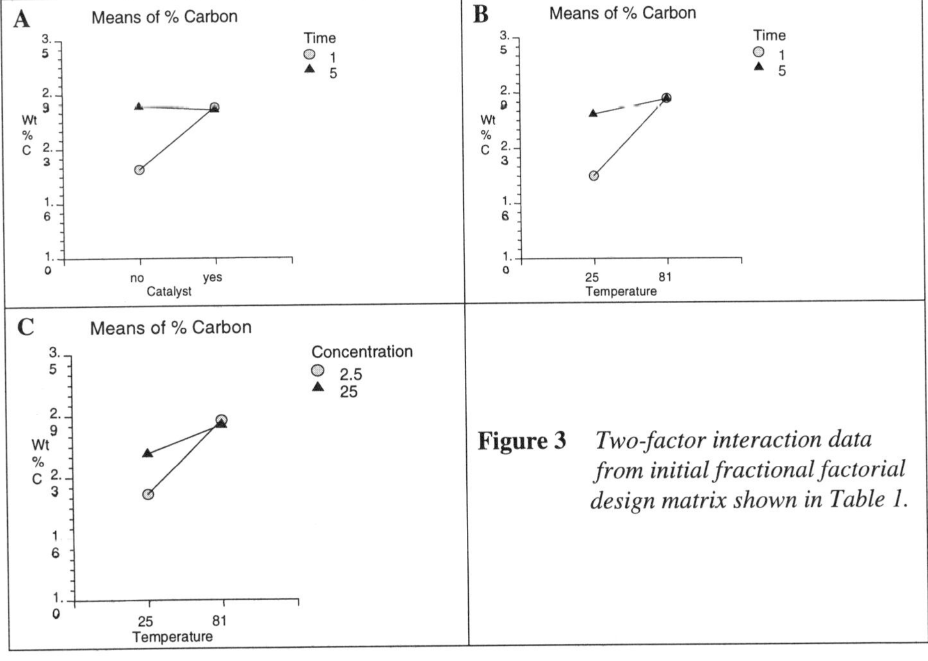

Figure 3   *Two-factor interaction data from initial fractional factorial design matrix shown in Table 1.*

temperatures are always beneficial to enhanced silylation. It was therefore decided that further studies should be carried out at higher temperatures.

Rather than view the temperature–concentration data as a strict two-factor interaction, we believe that the data depicted in Figure 3C is a simple kinetic phenomenon. Both temperature and reactant concentration enhance the rate of reaction. At higher temperature the kinetic effect of reactant concentration is not seen because the enhanced rate due to the increased temperature swamps out the reactant concentration effect. We thus chose to study the effect of hexamethyldisilazane reactant concentration as a main effect at higher temperature. Results of this study were obtained at 138 °C in refluxing *m*-xylene (Figure 4). All other variables were held constant since the initial design results indicated that no other factors would have an effect on this study. The results shown in Figure 4 suggest that an optimum hexamethyldisilazane concentration is approximately 10 mmol/g.

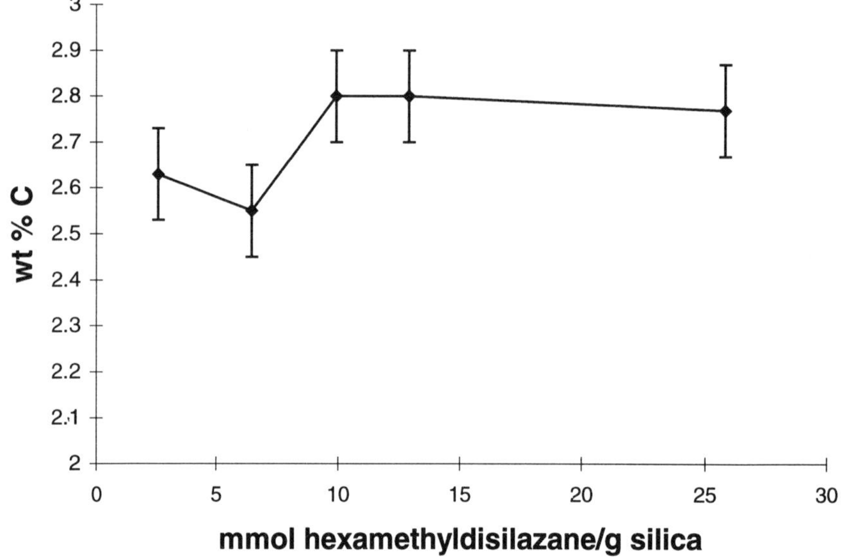

Figure 4   *The effect of hexamethyldisilazane concentration on the extent of reaction with silica.*

The catalyst–time two-factor interaction (Figure 3A) has also been interpreted as a kinetic phenomenon. The data show that the difference in reaction time of one and five hours plays an important role in the absence of *n*-butylamine catalyst, no difference in the extent of reaction is seen when the catalyst is present regardless of reaction time. The *n*-butylamine catalyst acts solely to enhance the rate of reaction. Further studies investigating the catalyst–time interaction were done at higher temperatures as suggested from the results of the initial design, and at higher reactant concentration as suggested by results shown in Figure 4. A simple 2 × 2 catalyst–time design study done in refluxing *m*-xylene with 10 mmol hexamethyldisilazane/g silica is shown in Table 2. The trends are similar to what was found in the initial design. In the absence of an external catalyst reaction time plays an important role. When an external catalyst is added, however, the difference in yield between one and five hour reaction times is within experimental error.

It is interesting to note that the yields of approximately 2.5 μmole/m^2 seen here are near the high end of what was seen in any of the initial design experiments. This suggests that the reaction conditions used here are closer to the optimum conditions for maximum silylation.

Table 2   *Effects of Catalyst and Time on the Yield of Hexamethyldisilazane Reaction with Silica.*

Catalyst	Reaction Time (h)	TMS loading (μmole/m^2)
No	1	2.31 ± 0.13
No	5	2.50 ± 0.03
Yes	1	2.55 ± 0.10
Yes	5	2.50 ± 0.10

Additional investigations of the temperature–time two factor interaction found in the initial design (Figure 3B) at higher temperature were undertaken. These reactions were done between 88 and 138 °C in *m*-xylene. Figure 5 illustrates the results of these studies as a response surface. These results support the interpretation from the screening design. Low temperatures and long reaction times give similar yields to reactions performed at high temperatures regardless of reaction time. The Figure 5 plot indicates that the best yield was found at the lowest temperature and longest reaction time. This was not a consistent trend. Additional studies mapping out the response surface in this direction may be warranted however.

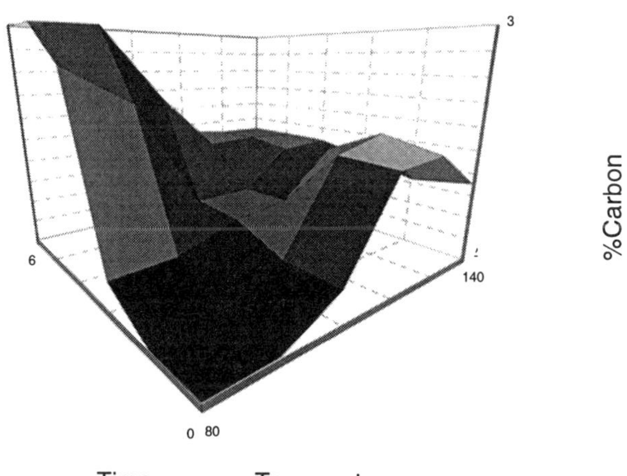

Figure 5   *Response surface plot of the two-factor temperature time interaction.*

It was stated earlier that the only main effect found was that of solvent. It was also found that water in the solvent did not play an important role. Since the initial design looked at two very similar solvents, benzene and cyclohexane, further investigations into the role of solvent were necessary. Given the non-polar nature of both benzene and cyclohexane, and the fact that neither solvent can dissolve an appreciable amount of water, it was decided that the polar solvent acetonitrile should be investigated. Since the data for benzene and cyclohexane had already been evaluated from the initial design, the eight experiments utilizing benzene shown in Table 1 were substituted for acetonitrile. This allows for all the effects to be evaluated for acetonitrile, in particular the role of water in acetonitrile, and provides the ability to compare all three solvents for their effect on reaction yield. The data from these studies are compiled in Table 3.

Table 3 *Repeat of Initial Design Substituting Acetonitrile for Benzene as Solvent.*

Reaction Temp (°C)	Catalyst	Water on Silica	Rxn Time (h)	[Reactant] (mmol/g)	Solvent*	H$_2$O in solvent	TMS Loading (μmol/m^2)
81	Y	N	5	26	CH	N	2.47±.008
25	N	Y	1	2.6	ACN	Y	0.58±.008
81	N	Y	1	26	ACN	N	2.14±.006
81	N	Y	5	26	CH	Y	2.44±.003
25	N	N	1	26	CH	N	1.55±.008
81	Y	Y	5	2.6	ACN	Y	1.98±.020
81	Y	N	1	26	ACN	Y	2.31±.020
81	N	N	5	2.6	ACN	N	2.29±.022
25	Y	N	1	2.6	ACN	N	1.29±.013
25	Y	N	5	2.6	CH	Y	2.27±.025
25	N	N	5	26	ACN	Y	1.67±.015
25	Y	Y	1	26	CH	Y	2.32±.013
81	N	N	1	2.6	CH	Y	2.20±.008
81	Y	Y	1	2.6	CH	N	2.80±.020
25	Y	Y	5	26	ACN	N	2.74±.025
25	N	Y	5	2.6	CH	N	2.08±0.08

** CH = Cyclohexane, ACN = Acetonitrile, B = Benzene*

Interpretation of the data in Tables 1 and 3 leads to two conclusions. First the role of water in the solvent (acetonitrile) has a negative effect on the yield of hexamethyldisilazane reaction with silica. Thus a dry solvent is preferred. Second, on average the yield increases by solvent in the order $CH_3CN < C_6H_6 < C_6H_{12}$. Reactions performed in cyclohexane give on average a 0.2 μmole/m^2 greater yield than reactions performed in benzene. Benzene provides a 0.2 μmole/m^2 improvement in yield compared to reactions performed in acetonitrile.

Results indicate that the yield of the surface silylation reaction is improved by performing reactions in an aliphatic hydrocarbon solvent at high temperatures. It was therefore decided to perform additional reactions using long chain aliphatic hydrocarbon solvents under reflux conditions. It was found that long chain aliphatic hydrocarbon solvents larger than nonane are not readily washed from the surface after reaction. Various studies using nonane under reflux conditions (151 °C) were performed. The results of TMS yield were consistently near the highest obtained ranging from 2.47 μmole TMS/m^2 to 2.62 μmole TMS/m^2.

Recent studies have shown a surface silanol concentration of approximately 2.8 SiOH/nm^2 on Cab-O-Sil HS-5 fumed silica[15] corresponding to a concentration of 4.6 μmol/m^2 (surface area = 325 m^2/g). The observed level of TMS loading is approximately 2.6 μmol/m^2. Thus about 55% of all the silanols are reacted under these conditions.

There are at least two reasons why only slightly more than half the silanols are trimethylsilylated with hexamethyldisilazane. Recent work by Liu and Maciel[15] showed on Cab-O-Sil fumed silica, 32% of the silanols are inaccessible to proton exchange with D_2O. It is reasonable to conclude that at least this high a percentage of silanols are not accessible to hexamethyldisilazane and so are unsusceptible to silylation. From this argument a *maximum* of 65–70% of all silanols are accessible to reagent. Given the relative sizes of hexamethyldisilazane and D_2O, a larger proportion of silanols are probably inaccessible to hexamethyldisilazane than D_2O. A second reason explaining the fact that approximately 55% of the silanols have been trimethylsilylated may be the relative bulkiness of a trimethylsilyl group. Maciel *et al.*[16] have determined that the projected free-rotation area of a trimethylsilyl group is approximately 0.43 nm^2. A trimethylsilylated vicinal silanol thus may sterically hinder silylation of an adjacent silanol. This potential problem may be lessened using the techniques of hydrosilylation described by Pesek[17] in these proceedings. Further work to directly compare the extent of reaction possible via hydrosilylation and conventional silylation may help clarify this steric factor.

4 CONCLUSIONS

The experimental design studies described have provided some insight into the role of how a variety of variables affect the extent of surface silylation reaction. The variables that have an effect, either alone or with other variables, include solvent, reaction time, reaction temperature, external catalyst and reactant concentration. Surprisingly water had little or perhaps a negative effect on the extent of silylation reaction. Many of these conclusions are probably peculiar to hexamethyldisilazane. Systems involving other organosilanes are likely to act very differently in response to these variables.

Although we believe that one cannot get much more surface modification reaction in this system than that attained here, further optimization work could be done. Higher temperatures could result in greater silylation, though solvent removal would be a problem. One possibility would be to perform the reaction in a low boiling alkane solvent under pressure. Another possibility would be to allow the reaction to proceed for much longer reaction times as the results in Figure 5 imply. Since it is desirable to maximize the extent of silylation for so many applications, a direct comparison of organosilylation and hydrosilylation should be undertaken.

References

1.   Previous volumes are: a) 'Silanes, Surfaces and Interfaces,' D. E. Leyden, ed., Gordon and Breach, New York, 1985;   b) 'Chemically Modified Surfaces in Science and Industry,' D. E. Leyden and W. T. Collins, eds., Gordon and Breach, New York, 1987; c) 'Chemically Modified Oxide Surfaces,' D. E. Leyden and W.

T. Collins, eds., Gordon and Breach, New York, 1989; d) 'Chemically Modified Surfaces,' H. A. Mottola and J. R. Steinmetz, eds., Elsevier, Amsterdam, 1992; e) 'Chemically Modified Surfaces,' J. J. Pesek and I. E. Leigh, eds., The Royal Society of Chemistry, 1994; f) 'Chemically Modified Surfaces: Recent Developments,' J. J. Pesek, M. T. Matyska and R. R. Abuelifiya, eds., The Royal Society of Chemistry, 1996.

2. See, for example: a) R. K. Iler, 'The Chemistry of Silica,' John Wiley and Sons, New York, 1979, Chapter 6; b) 'The Colloid Chemistry of Silica,' H. E. Bergner, ed., American Chemical Society, 1994; c) E. F. Vansant, P. Van Der Voort and K. C. Vrancken, 'Characterization and Chemical Modification of the Silica Surface,' Elsevier, Amsterdam, 1995.

3. J. P. Blitz, R. S. S. Murthy and D. E. Leyden, *J. Am. Chem. Soc.*, 1987, **109**, 7141.

4. W. Hertl and M. L. Hair, *J. Phys. Chem.*, 1971, **75**, 2181.

5. J. P. Blitz, *Colloids Surf.*, 1992, **63**, 11.

6. S. Haukka, E. L. Lakomaa and A. Root, *J. Phys. Chem.*, 1993, **97**, 5085.

7. a) W. Hertl, *J. Phys. Chem.*, 1968, **72**, 1248; b) W. Hertl, *J. Phys. Chem.*, 1968, **72**, 3993; c) M. L. Hair and W. Hertl, *J. Phys. Chem.*, 1969, **73**, 2372.

8. A few of the published reports include: a) T. Welsch and H. Frank, *J. Chromatogr.*, 1983, **267**, 39; b) M. L. Miller, R. W. Linton, G. E. Maciel and B. L. Hawkins, *J. Chromatogr.*, 1985, **319**, 9; c) J. Nawrocki and B. Buszweski, *J. Chromatogr.*, 1988, **449**, 1; d) J. Nawrocki, D. Moir and W. Szczepaniak, *J. Chromatogr.*, 1989, **467**, 31; e) T. Welsch, H. Frank and G. Vigh, *J. Chromatogr.*, 1990, **506**, 97; f) J. Nawrocki, *Chromatographia*, 1991, **31**, 193.

9. J. N. Kinkel and K. K. Unger, *J. Chromatogr.*, 1984, **316**, 193.

10. J. P. Blitz, R. S. S. Murthy and D. E. Leyden, *J. Colloid Interface Sci.*, 1988, **126**, 387.

11. W. O. Bascom, *J. Org. Coatings and Plastics Chem.*, 1967, **27**, 27.

12. Y. Sudo, *J. Chromatogr. A*, 1996, **737**, 139.

13. a) M. Mauss and H. Engelhardt, *J. Chromatogr.*, 1986, **371**, 235; b) C. H. Lochmuller and M. T. Kersey, *Langmuir*, 1988, **4**, 572; c) S. P. Landreau and W. T. Cooper, *Anal. Chem.*, 1989, **61**, 41.

14. G .E. P. Box, W. G. Hunter and J. S. Hunter, 'Statistics for Experimenters,' John Wiley and Sons, New York, 1978.

15. C. C. Liu and G. E. Maciel, *J. Am. Chem. Soc.*, 1996, **118**, 5103.

16. D. W. Sindorf and G. E. Maciel, *J. Phys Chem.*, 1982, **86**, 5208.

17. See the article by J.J. Pesek in these proceedings.

SURFACE MODIFICATION OF HIGHLY DISPERSE TITANIA AND TITANIA/SILICA, AND EFFICIENCY OF THEIR APPLICATIONS

V.M. Gun'ko,* V.I. Zarko, E.F. Voronin, E.M. Pakhlov and A.A. Chuiko

Institute of Surface Chemistry
31 Prospect Nauki
Kiev, Ukraine 252022

1 INTRODUCTION

Metal oxides are widely used as pigments, catalysts and fillers. Fumed metal oxides are especially useful because of their high dispersity, high specific surface area, and high adsorptivity. It is common to surface-modify these oxides by organic compounds (OC) or organosilicon compounds (OSC) in order to further improve their performance in many applications. Surface modification by alcohols or other OCs is of commercial interest mostly because these modifying agents in general are less costly than OSCs. But for many applications the hydrolytic stability of these OC-modified surfaces is an issue. Specifically, the $\equiv$M-O-C$\equiv$ linkage formed between oxide surface and alcohol is far more susceptible to hydrolysis than the $\equiv$M-O-M$\equiv$ linkage, formed, for example, by silane or siloxane modification of silica.

OSCs commonly used for modification of oxide surfaces are of the forms $XSiR_3$, X_2SiR_2, and $XSiR_2Y$ (where X, Y = Cl, NCO, NCS, N_3, CN, OR$'$, $etc.$, and R is a hydrocarbon group), but more complex OSCs have been used. The properties of a surface to which an OSC is bound cannot always be predicted from the properties of the pure OSC, because the effect of the bound groups can depend on such variables as the degree of surface coverage (Θ) due to lateral interactions, and on the nature of the substrate surface. For example, unmodified silica and titania, or mixed oxides such as TiO_2/SiO_2, have different OH groups $\equiv$MOH, $\equiv$MO(H)M$\equiv$, $etc.$, and other surface sites with different acidic and basic properties and reactivities; the effects that the OSCs bound to the surfaces of these will differ as well.

An example of a specific problem involving metal oxides in coatings is the tendency of titania, widely used as a pigment, to coagulate at pH values typical in aqueous coatings. Unmodified titania particles, $e.g.$ Degussa P25, can coagulate in aqueous suspensions at pH $\geq$ 5. At pH = 7–10, the average effective diameter, D_{ef}, is 1.2–2.3 μm for this oxide. Surface modification of P25 by inorganics or organics can allow us to stabilize the particle size distribution at D_{ef} = 0.2–0.5 μm – close to the optimal value for pigments – over the pH range of 3 to 11.[1-6]

Another relevant example is the problem of adhesion of paints and coatings to the substrates to which they are applied. Adhesional interactions at the solid/polymer interfaces are first and foremost adsorption interactions between the solid surface and polymer molecules.[3] After polymerization there is a low molecular-weight fraction of coupling agents, which can decrease the cohesion and adhesion of the polymer film. If the molecules from this fraction interact with the filler particles preferentially (which can be reached *via* the filler surface modification) instead of with the material surface covered, then the boundary layer of the film can be free from this fraction and adhesion increases as strengthening the boundary layer of the coating leads to stronger adhesion of the coating to the covered surfaces.

The aim of this work is to explore ways to improve the protective properties of coatings by surface-modifying the highly disperse oxide fillers used in the coatings. In this work we focus on two metal oxides most commonly used in coatings: titania and titania/silica.

2 EXPERIMENTAL

Metal oxides used were (S = surface area of the metal oxides, measured by the BET method using N_2):

- commercial anatase titania, $S = 20 m^2/g$
- fumed titania, $S = 50\ m^2/g$
- fumed titania/silica ("Chlorovinyl", Kalush), labeled as TS^P, containing 9, 14, 20, 29 and 36 wt % of TiO_2, with corresponding S values of 215, 137, 70, 60, and 90 m^2/g
- titania/silica obtained by the chemical vapor deposition (CVD) method using a fumed silica substrate, labeled as TS^H, with 9, 20 and 30 wt % of TiO_2, with corresponding S values of 273, 187, and 154 m^2/g
- commercial rutile titania, labeled as $TiO_2(rt)$, obtained from the Sherwin-Williams Co., USA
- commercial rutile titania covered by a layer of SiO_2 and Al_2O_3, labeled as $TiO_2(rt-M)$, obtained from the Sherwin-Williams Co., USA
- fumed silica (A-300, $S = 300\ m^2/g$, and A-380, $S = 380\ m^2/g$) manufactured by "Chlorovinyl"

Surface modification agents used were:

- 3-aminopropyltriethoxysilane (APTES or AP)
- hexamethyldisilazane (HMDS or DS)
- methacryloyloxymethylene-methyl diethoxysilane (MADES or MA)
- a mixture (MD) consisting of 15 wt% hexamethylcyclotrisiloxane, 80 wt% octamethylcyclotetrasiloxane, with the remaining 5 wt% consisting of unspecified amounts of decamethylcyclopentasiloxane and dodecamethylcyclohexasiloxane
- common alkanols: butanol, spectrophotometric grade methanol, di(ethyleneglycol) (DEG), and glycerol

Some of these materials and the surface treatments applied to them are summarized in Table 1. TiO_2 (or TS) was activated by heating at 373 K under vacuum (10^{-4} torr) in sealed ampoules for several hours to remove the adsorbed water. The ampoules with the heated samples were opened at room temperature and the required amount of OC or OSC was used for their modification. After being dosed with treating agent, the system was evacuated to 10^{-2} torr. Reaction temperature was 373 K and the reaction time was 3–4 h. To remove the gaseous byproducts and excess treating reagents, the samples were held under vacuum for 1 h at the reaction temperature. Silica was modified as described elsewhere.[4]

Table 1 *TiO_2 and TiO_2/SiO_2 (Unmodified and Modified) Used as Fillers in Ethyl Cellulose and Nitrocellulose for Tests of Their Effect on Adhesion, UV Stability, Corrosion Resistance, and Water Absorption of the Filled Films.*

Oxide	S, m^2/g	Wt % TiO_2	Modifier	$T_p = T_r$, K	Treatment, mmol/g
TiO_2	20	100	APTES; MADES; HMDS;	373	0.2
	50	100	MADES + APTES;		0.1 + 0.1
			MADES + HMDS;		0.1 + 0.1
			APTES + HMDS		0.1 + 0.1
SiO_2/TiO_2	215	9	APTES	373	0.4
SiO_2/TiO_2	137	14			
SiO_2/TiO_2	70	20	APTES;	373	0.3
			MADES + APTES		0.15 + 0.15
SiO_2/TiO_2	60	29			
SiO_2/TiO_2	90	36	APTES	373	0.3
$SiO_2/CVD\ TiO_2$	273	9			
$SiO_2/CVD\ TiO_2$	187	20			
$SiO_2/CVD\ TiO_2$	154	30			

Adhesion was determined by first applying the coatings in narrow bands on steel plates, and then removing the cured coating with an adhesive tape. The estimation of adhesion was made from the number of the bands that remained on the steel plate after peeling off the tape. Every test was repeated 2–4 times.

Water absorption experiments were carried out in a box with $p/p_s = 0.95$ at 290 K for 72 h. The films were weighed to ± 0.1 mg before and after absorption of water.

Resistance of films to light was evaluated by accelerated aging under UV light of varying intensity. The simulated times of sample exposure under UV light were equal to 2, 4, 7, 12, and 17 light years with normal moisture. The reflection spectra of the coatings were recorded with an SF-18 (Russia) spectrophotometer. For simplicity, we assumed that the reflection coefficient for the samples without UV light action was equal to 1, *i.e.* 100% of reflection or 0% of absorption. The reflection spectra were recorded in the 400–500 nm range and then converted into absorption spectra according to the Kubelka–Munk formula. An estimation of the light resistance of the coatings was made on the basis of the difference in the peak intensity of the absorption band at 420–460 nm, which is sensitive to the coating degradation under UV light.

Corrosion resistance was measured by coating steel plates with films of filled ethyl cellulose, then placing them in a 3 wt % of NaCl aqueous solution for 72 h. The protection properties of the coatings were estimated in accordance with the appearance and the intensity of corrosion using some scale.

3 RESULTS AND DISCUSSION

3.1 Adhesion of Filled Coatings

In Table 2 we summarize the relative average estimations of the influence of unmodified or modified oxide additives on the adhesion properties of the filled polymer films relative to those for unfilled film.

The addition of 0.5–2.5 wt % of unmodified silica (A-380) reduces the adhesion, an effect that may be caused by weak interaction between the hydrophilic silica particles and the hydrophobic CH groups of ethyl cellulose. This assumption is supported by the increase in adhesion after the addition of modified silica. When TS^P are used as an additive, the adhesion increases for C < 1 wt %, but for C > 1.5 wt % TS gives the same

Table 2   *Influence of Oxide Fillers on the Relative Adhesion to Steel of Filled and Unfilled Ethyl Cellulose Coatings.*

Oxide	0.5 wt %	1 wt %	1.5 wt %	2 wt %	2.5 wt %
TS_9^H	0.74	0.62	0.69	0.80	0.55
TS_{20}^H	0.80	0.95	0.42	0.54	0.55
TS_{30}^H	0.95	1.15	0.78	0.74	0.40
TS_9^P	0.69	0.46	0.46	0.31	0.23
TS_{14}^P	0.88	0.80	0.85	1.38	0.80
TS_{20}^P	1.23	0.92	0.38	0.46	0.38
TS_{29}^P	1.15	0.46	0.23	0.34	0.42
TS_{36}^P	0.77	0.95	0.31	0.19	0.42
TS_9^P-APTES	0.20	0.62	0.15	1.04	1.15
TS_{20}^P-APTES	1.14	1.18	0.95	0.99	1.13
TS_{36}^P-APTES	1.15	1.09	0.98	1.13	1.04
TS_{20}^P-AP/MA*		1.03	0.79	1.01	0.99
A-380	0.78	0.77	0.65	0.58	0.42
A-380-MADES		1.03	0.96	0.90	0.99
A-380-DEG		1.91	1.27	2.17	3.21

* AP = APTES, MA = MADES

effect as unmodified silica. The modification of TS or silica by OSCs or alcohols increases adhesion (Table 2). Analysis of the data obtained shows that good results may be obtained using TS^P (20–30 wt % of titania) modified by OSCs with the active polar groups. Analogous modification of silica improves the efficiency of the additive.

3.2 Water Absorption

Protection properties of the coatings depend not only on adhesion characteristics but also on the absorption of water as a redox agent and a carrier of oxygen, which can promote the coating degradation. This effect depends on the structures of the coupling agents and surface layers at the pigment and filler particles, which can have active polar groups (OH, NH, *etc.*) or hydrophobic non-polar groups (CH). Water absorption by unfilled and filled films (weight 0.3–0.5 g, thickness 0.03–0.04 mm) was studied, and results are summarized in Table 3. For the most part, the use of oxide fillers leads to a decrease in water absorption by the ethyl cellulose coatings which is a positive effect in improving the protection properties of the coatings. However, the addition of hydrophobic oxides sometimes gives an increase in the water absorption by the filled films, *e.g.* samples TiO_2 + HMDS, A_{BUT} (A-300 + butanol), *etc.*

Thus, water absorption by the filled films depends not only on the hydrophilic/hydrophobic properties of the individual components, but also on their interaction in and specific spatial structure of the films. For example, the coupling/filler interaction has an influence on the packing density of molecules and particles in the films and on their absorption properties. The amount of water absorbed by the nitrocellulose films filled by alcohol-modified silica is higher than that for pure nitrocellulose. This effect can be explained by the hydrophilic properties of the oxide fillers such as A_{DEG}, but A_{BUT} is hydrophobic. However, A_{BUT} can adsorb some amount of water. The water absorption by the films filled by the modified oxides has a complicated nature. Such films

Table 3 *Influence of Oxide Additives on the Relative Water Absorption by the Films with Ethyl Cellulose and Nitrocellulose(*) in Comparison with Non-filled Films.*

Oxide	0.5 wt %	1 wt %	1.5 wt %	2 wt %	2.5 wt %
TS_9^H	0.13	0.13	0.06	0.17	0.22
TS_{20}^H	0.30	0.48	0.80	0.53	0.46
TS_{30}^H	0.48	0.78	0.83	0.97	1.12
TS_9^P	0.18	0.23	0.48	0.47	0.58
TS_{14}^P	0.38	0.38	0.43	0.30	0.38
TS_{20}^P	0.35	0.46	0.27	0.36	0.38
TS_{29}^P	0.30	0.29	0.38	0.13	0.26
TS_{36}^P	0.015	0.26	0.29	0.37	0.22
TS_9^P-APTES	0.69	0.12	0.41	1.64	0.87
TS_{20}^P-APTES	1.00	1.04	0.92	0.83	0.77
TS_{36}^P-APTES	1.10	1.10	0.99	0.94	0.91
TS_{20}^P AP/MA		1.75	1.17	1.17	1.08
TiO_2 APTES		1.12	0.87	1.12	1.21
TiO_2-HMDS(0.4)		0.98	1.10	1.13	1.13
TiO_2-HMDS(0.2)		0.70	0.74	0.70	0.91
A-380	0.31	0.14	0.15	0.14	0.24
A-380-MADES		1.60	0.96	0.90	0.99
A_{BUT}^*		2.30	1.60	2.00	1.10
A_{DEG}^*		0.30	1.80	1.70	2.50

have new adsorption sites at the oxide filler surface, which can form the hydrogen bonds with water molecules (energy 30–60 kJ/mol), but the hydrophobic groups such as $SiO(CH_2)_nCH_3$ or $Si(CH_3)_3$ change entropy of adsorbed water (their cluster structure relative to that on pure oxides) due to hydrophobic interaction; in addition, the filler particles alter the polymer film structure. For example, hydrophobic CH groups on the surfaces of modified fillers interact with the hydrophobic fragment of polymers and alter the structure of the films and promote interaction between the water molecules and hydrophilic polar groups of polymers. In the case of hydrophilic groups on filler surfaces, the water molecules can interact with them, *i.e.* the overall effect for differently modified fillers may be the same.

3.3 Light Resistance of the Coatings

Coating stability depends on the degradation rate of the polymer binder, which in turn is connected to occurrence of redox reactions with the participation of water and oxygen.

These processes have an influence on photochemical reactions in the surface layer of the oxide pigments as coating degradation occurs as a result of photostimulated redox reactions with the participation of pigment (titania) surfaces, H_2O, O_2, different radicals and ions such as $HO\bullet$, $HO_2\bullet$, H^+, and HO^-. The light resistance of the titania pigments is usually improved by the thin Al_2O_3, ZnO, and SiO_2 layers or patches, which are deposited on the titania surface from liquids with pH = 6.5–7.1. The mechanism of photochemical stabilization of TiO_2 is complicated. Stabilization may also be improved by hydrophobing the pigment surfaces by organosilicon compounds. This lowers the amount of water on the particle surface, which therefore lowers the content of HO^-, $HO\bullet$ and $HO_2\bullet$. Basic groups in OSCs, *e.g.* amine groups, can enhance this effect, due to neutralization (blocking) of the acidic surface sites, as those interact strongly with adsorbed water. In addition, deposition on surfaces of compounds (*e.g.* substituted phenols such as $C_6H_2(CCH_3)_3OH$, which can be radical or electron traps, leads to inhibition of redox reactions.

We have studied the influence of chemical modification of TiO_2 by some OSCs on the light resistance of the filled coating with nitrocellulose (Table 4). These coatings were examined by accelerated tests for light resistance using climatic equipment with controlled intensity of UV light.

Table 4   *Relative UV Stability of the Coating Filled by Unmodified and Modified Anatase (*) or Rutile (rt) as a Function of Exposure (Years, t).*

Filler	α	t = 0	t = 2	t = 4	t = 7	t = 12	t = 17
TiO_2*		1	0.93	0.80	0.67	0.66	0.64
TiO_2-MA*	0.5	1	0.88	0.70	0.63	0.56	0.55
TiO_2-MA*	1	1	1.00	0.93	0.63	0.63	0.63
TiO_2-AP*	0.5	1	0.90	0.85	0.74	0.66	0.65
TiO_2-AP*	1	1	1.00	0.87	0.80	0.78	0.78
TiO_2-AP/MA*	1	1	1.00	0.83	0.78	0.75	0.75
TiO_2-AP/DS*	1	1	0.93	0.80	0.72	0.65	0.64
TiO_2-MA/DS*	1	1	1.00	0.95	0.75	0.67	0.66
TiO_2-DS*	0.4	1	1.00	0.95	0.84	0.77	0.75
TiO_2-DS*	0.2	1	0.93	0.85	0.70	0.64	0.60
TiO_2(rt)		1	1.00	0.94	0.86	0.83	0.79
TiO_2(rt-M)		1	1.00	1.00	0.88	0.84	0.82
TiO_2(rt-M)-MA	0.5	1	1.00	0.88	0.78	0.73	0.71
TiO_2(rt-M)-AP	0.5	1	1.00	0.86	0.80	0.73	0.72
TiO_2(rt)-MA	0.5	1	1.00	1.00	0.90	0.84	0.80
TiO_2(rt)-AP	0.5	1	0.98	0.93	0.80	0.75	0.74
TiO_2(rt)-MA	1	1	0.95	0.90	0.82	0.78	0.76

The results summarized in Table 4 show that, for the most part, polymer films filled by the modified pigments exhibit improved light-resistance. Differences between TiO_2 modified by HMDS, APTES, and MADES lead to variations of the coating degradation rates under UV light. The data presented in Table 4 show that the modification of anatase surface by OSC (degree of substitution of surface OH by a modifier, α, equals 1) increases the photochemical stability of the coating to close to that for rutile. The modification of the anatase surfaces by MADES ($\alpha = 1$) and HMDS ($\alpha =$

0.4) is more efficient than that by APTES ($\alpha = 1$). More effective modification of TiO_2 surfaces is observed for HMDS at $\alpha = 0.4$. The light-resistance of the samples with APTES is lower but as the α value increases (as in the case of TiO_2-DS) the light-resistance increases. We have studied the influence of the modification of TiO_2 by the OSC mixture on the light resistance of the coatings. Good results in the use of such mixtures for the anatase modification were obtained for MADES and HMDS (TiO_2-MA(DS)). TiO_2 was modified by APTES ($\alpha = 0.5$), then by MADES (TiO_2-AP(MA)) or HMDS (TiO_2-AP(DS)) and the other sample was TiO_2 modified by MADES ($\alpha = 0.5$) and HMDS ($\alpha = 0.5$) – TiO_2-MA(DS). Examination of these samples shows that additional modification of TiO_2-MADES by HMDS improves the light resistance of the coatings, but TiO_2 modified by APTES and then by MADES or HMDS does not give such an effect. It should be noted that the efficiency of the modification of TiO_2 by MADES for the light-resistance of the filled films decreases when the effective exposure of the coatings under UV light is above 7 years, whereas the polymer film containing TiO_2 with bound methylsilyl groups (TiO_2-DS) is more stable. It may be expected that the use of such OSCs in combination with a surface layer of a second inorganic oxide (silica, alumina, *etc.*) on the TiO_2 particles will be more efficient.

3.4 Corrosion Resistance of the Coatings

The study of adhesion and water absorption gives important, but indirect, information about the protection properties of the coating. Direct information of the type we desire may be obtained *via* investigation of the corrosion resistance of the coatings on metal plates (see Experimental). The degree of corrosion (χ) was measured on a scale of 1 to 6. A value of six on this scale corresponds to a coating with no corrosion; a value of five corresponds to a coating with point traces of corrosion; a value of four corresponds to small patches of corrosion; a value of three corresponds to corrosion of less than 5% of the coating; a value of two corresponds to corrosion of up to 10%; and a value of 1 is assigned if the corroded area is above 10%.

Diphenylamine (DPA) as a corrosion inhibitor was adsorbed in addition on oxide filler surfaces. It was found that the most stable coatings were those filled by $A_{MD,DPA}$ ($\chi = 5$–6), A_{DPA} ($\chi = 5$), TS_{20}^{P} ($\chi = 4$–5), TS_{29}^{P} ($\chi = 4$–5), TS_{20}^{H} ($\chi = 4$–5) (1–2 wt%). Lower χ is observed for unmodified TS_{36}^{P} and silica A-380. It should be noted that some correlation between the corrosion resistance of the filled coatings and absorption of water and conductivity of them is observed for unmodified and modified TS or individual oxides. For example, specific conductivity of the coating filled by an effective filler (*e.g.* modified silica with the addition of DPA) did not change after exposure in the electrolyte solution during 80 days, but in the case of a poor filler (*e.g.* unmodified silica) this parameter increased by half.

4 CONCLUSIONS

The addition of small amounts (0.5–2 wt%) of highly disperse (fumed) modified titania, TiO_2/SiO_2 or silica allows us to improve the adhesion properties and light and corrosion resistance of filled coatings. Unmodified and modified fillers can have different influences on different protection properties of the filled polymer coatings. In other words, we can improve certain properties of the coatings by using fillers but other properties may be made worse. The efficiency of the coating depends on the nature of the pigment, filler and modifier and the degree of substitution of the active surface sites of the solid particles. The modification of surfaces gives a stronger effect for pigments with weaker initial light resistance, such as anatase relative to rutile. The effectiveness of chemical modification of pigment surfaces can be improved by the use of radical (or electron)-trap compounds as corrosion inhibitors along with the OCs or OSCs. It was found that highly disperse titania/silica modified by APTES or MADES, or fumed silica modified by butanol or di(ethylene glycol), at certain levels of filler treatment and filler loading, do improve protection properties of filled coatings.

References

1. E. P. Plueddemann, 'Silane Coupling Agents', 2nd ed., Plenum, New York, 1991.
2. E. F. Vansant, P. VanDerVoort and K. C. Vrancken, 'Characterization and Chemical Modification of the Silica Surface', Elsevier, Amsterdam, 1994.
3. Yu. S. Lipatov and L. M. Sergeeva, 'Adsorption of Polymers', Naukova Dumka, Kiev, 1972.
4. V. M. Gun'ko, E. F. Voronin, V. I. Zarko, E. M. Pakhlov and A. A. Chuiko, *J. Adhesion Sci. Technol.*, 1997, **11**, 627.
5. V. M. Gun'ko, V. I. Zarko, E. Chibowski, V. V. Dudnik, R. Leboda and V. A. Zaets, *J. Colloid Interface Sci.*, 1997, **188**, 39.
6. V. M. Gun'ko, V. I. Zarko, R. Leboda, E. F. Voronin and E. Chibowski, *Colloid. Surf. A*, 1998, **132**, 241.

SURFACE MODIFICATION OF MICRON-SIZE POWDERS BY A PLASMA POLYMERIZATION PROCESS

W.J. van Ooij,* Ning Zhang and Sheyu Guo

Department of Materials Science and Engineering
University of Cincinnati
Cincinnati, OH 45221-0012 USA

1 INTRODUCTION

Plasma polymerization by a variety of means has been an active area of research over the past decades.[1,2] This is especially true in industrial laboratories, where the advantages of pore-free, uniform films of superior physical, chemical, electrical, and mechanical properties were appreciated at an early stage. Plasma deposition of polymer films holds potential advantages in the areas of adhesive bonding between substrates and polymers,[3] wettability characteristics,[4] printability of polymers, and so on. Indeed, the ability to selectively modify surface properties whilst retaining desirable bulk properties is unique,[4] and the technique therefore is widely used.

As opposed to the wealth of information published on plasma deposition of polymer films on flat substrates,[5-7] there are very few reports on such deposition on powdered substrates.[2,8-10] Powders are difficult to handle in the plasma treatment because of aggregation and large surface area per unit mass of the powders. Inagaki *et al.*[8] reported surface modification of polyethylene powders using a plasma reactor with a fluidized bed. They discovered that oxygen plasma treatment of polyethylene powder in the fluidized state changed the polyethylene surface from hydrophobic to hydrophilic. Sato *et al.*[9] studied plasma polymerized octamethylcyclotetrasiloxane (OMCTS) films deposited on the surface of mica powders under DC plasma conditions. They estimated the formation of polymerized films using an AFM technique and suggested that plasma polymerization involves two kinds of reactions. First, small particles (clusters) are formed through the agglomeration of active species on the film surface. Second, the particles formed through the first reaction become the nucleus for polymerization. Nutsch *et al.*[10] introduced an induction plasma deposition method to deposit plasma polymer films on powder substrates. They used thermal plasma to heat powders, which were then injected into the flow of plasma, heated and accelerated to form, upon impact with the component surface, a contiguous layer with material properties superior to those of the substrate material. They found that an induction-type plasma generator produced completely different deposition parameters for the powders compared to the DC plasma generator normally used for plasma deposition. Low velocities, long residence times and the large dimensions of the powder particles modify their behavior on impact with the component surface and thus improve coating quality. They also showed that induction plasma deposition could be used to obtain coatings with good adhesion and low porosity. Anders *et al.*[2] used cathodic arc plasma deposition to coat 60 mesh Al_2O_3 powders with platinum. The powder particles were moved during deposition using a mechanical system operating at a resonance frequency of 20 Hz. Analyses showed that all particles are completely coated with a platinum film having a thickness of about 100 nm; the deposition rate was very high because of the short deposition time.

Many micron-size inorganic powders have commercial value as inert fillers, including clay particles (aluminum silicates), talc (magnesium silicate), mica (aluminum potassium silicate) and silica. Surface modification of these powders is carried out usually for the purpose of improving surface wetting and compatibility between the powder fillers and the matrix materials. Success in this endeavor often hinges on the ability to treat the surface of the particles uniformly, and without affecting the bulk properties.

We report here our recent work using an RF plasma reactor to coat three of these important materials – talc, mica and silica powders – with a thin layer of plasma polypyrrole (PPy) film. Pyrrole was chosen as a monomer for plasma polymerization because films of electrically conductive polypyrrole deposited by other processes, such as electropolymerization, have been widely studied in recent years.[11] Our primary objective was to investigate the possibility that uniform surface modification of powdered materials can be achieved by plasma polymer film deposition without affecting the bulk properties of the powders.

2 EXPERIMENTAL

2.1 Materials

The substrate materials were: BENWOOD 2202 talc (95% pure), Suzorite mica (100% pure, from Zemex Industrial Minerals, Boucherville, Quebec), and Hi-Sil 233 silica powder (rubber grade, from PPG, Monroeville, PA). The major difference between these powders is the surface area, which is low for mica and talc powders (3–9 m^2/g) and high for silica (150 m^2/g). Talc (D50 = 7 µm) is a fine powder, while silica (D50 = 16 µm) and mica (D50 = 21 µm) are coarser. Furthermore, talc is rather hydrophobic, while mica and silica are hydrophilic. Thus, these three types of common powders cover a wide range of powder properties.

Pyrrole (98%) and potassium bromide (KBr) disk (99+ %) were obtained from Aldrich Chemical Company (Milwaukee). This substrate was used for FTIR investigation of the PPy films on insulators. Silicon wafers (Cincinnati Milacron, Cincinnati, OH) were used for measurements of the deposition rate and density of the PPy film.

2.2 RF-Plasma Deposition

Plasma polymerization of pyrrole was performed in an RF plasma reactor of the type shown in Figure 1, which is based on tumbling the powders in the RF plasma. In this case the powders are treated remotely, *i.e.* away from the most intense glow. After introduction of the powder substrate, the chamber was evacuated to less than 10 mtorr, the monomer gas was introduced into the reaction chamber at steady flow conditions while maintaining the chamber pressure at a certain value, and radiofrequency (RF) was applied. Once ionized, excited gas species react with the surface of the powders placed in the reaction chamber. These experiments were replicated several times under high-power/low-pressure (HW/LP) or low-power/high-pressure (LW/HP) plasma conditions.

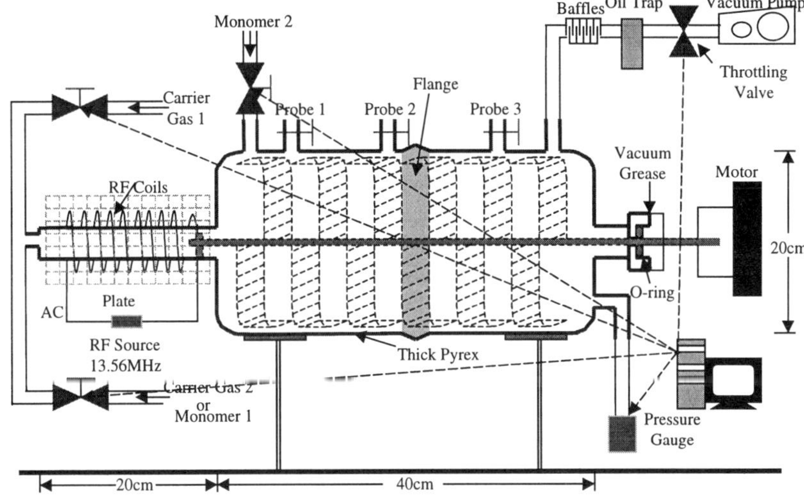

Figure 1　*Schematic of RF plasma reactor for treatment of coarse powders.*

Several parameters such as the monomer pressure and RF excitation power were varied in these experiments. The plasma polymer deposition conditions applied in this work are listed in Table 1.

Table 1　*Plasma Polymer Deposition Conditions.*

Denotation	RF Excitation Power (W)	Monomer Pressure (mtorr)
HW/LP	30	80
LW/HP	2	300

2.3 Contact Angle Measurement

A Lesca dynamic wicking meter (model WET-300) was used to measure the deionized water contact angle of the powders. The principle of the dynamic wicking meter as well as the specific experimental procedures have been described in detail elsewhere.[8] In the determination of the dynamic wicking property, the weight of the penetrating water into the column was measured using an electric balance as a function of time after water contact. The contact angle of water on the surface of the powder was calculated from the following equations:

$$\mathrm{Cos}(\theta) = S_p \frac{\rho_p}{\rho_l^2} \frac{(1-\varepsilon)}{c^3} \frac{2\eta}{\gamma_l} \frac{1}{t}\left(\frac{W}{S}\right)^2 \tag{1}$$

$$\varepsilon = 1 - \frac{W_p}{hS\rho_p} \tag{2}$$

where θ is the contact angle of water, S_p is the surface area of the powder per unit weight, ρ_p is the density of the powder (for talc, ρ_p is 2.73 g/cm^3, for mica, ρ_p is 2.90 g/cm^3), ρ_l is the density of water, ε is the porosity of the powder column, η is the viscosity of water, γ_l is the surface energy of water, t is the time after water contact, W is the weight of the penetrating water, and S is the cross section of the cylindrical tube. The porosity of the powder column, ε, was estimated from equation (2), where W_p is the powder tapped in the column, h is the height of the powder in the column, S is the cross section of the cylindrical column, and ρ_p is the density of the powder.

2.4 TGA and DSC Measurements

An STA 1500H instrument was used to acquire the STA curves for each sample. The samples were heated from 25 to 1,000 °C at 10 °C/min in an oxygen atmosphere. After the experiment, the oven was slowly cooled to room temperature.

2.5 Deposition Rate Evaluation

Thin film interferometry was applied to measure the deposition rate using a silicon wafer simulator. Two reflected beams interfere with each other, depending on path difference. We recorded the changes of detector voltage *vs.* time, made the plot and recorded the maximum and minimum detector voltages in the plot as well as the time change periodicals. The following equations were used to calculate the deposition rate and the refractive index of the film:

$$r = \frac{R_m}{R_M} \tag{3}$$

$$x = \left(\frac{n_2 + 1}{n_2 - 1}\right)\frac{1}{\sqrt{r}} \tag{4}$$

$$n_1 = \sqrt{\frac{n_2(x-1)}{(x+1)}} \tag{5}$$

$$D(\mathring{A}/min) = \frac{\lambda}{2n_1 t} = \frac{6328}{2n_1 t(min)} \tag{6}$$

where R_m is the minimum detector voltage, R_M is the maximum detector voltage, n_1 is the refractive index for the deposited pyrrole film, n_2 is the refractive index for Si, $n_2 = 3.86$, λ is the wavelength of the HeNe red line (632.8 nm), t is the periodical time of the detector voltage changes, and D is the deposition rate of the film.

2.6 PPy Film Density and Thickness Measurements

The density of the PPy films was determined by weighing the silicon wafer substrate of known surface area before and immediately after deposition. Assuming the deposition was homogeneous, the density of PPy film can be obtained through the following equation:

$$\rho_{film} = \frac{W_{After} - W_{Before}}{S \cdot D \cdot t} \tag{7}$$

where ρ is the density of the PPy film, W_{After} and W_{Before} are the weight of the silicon wafers after and before the deposition, respectively, S is the total surface area of the silicon wafers, D is the deposition rate of the PPy film (which can be obtained from the deposition rate measurement data), and t is the total plasma deposition time.

With the density of the PPy film under different deposition conditions, the thickness of the film can be calculated using the following equation:

$$T_{film} = \frac{f}{S_p \cdot \rho_{film}} \tag{8}$$

where T is the thickness of the film, f is the weight percent of the PPy film (which can be obtained from TGA data), S_p is the specific area of the powder, and ρ is the density of the PPy film.

2.7 Infrared Spectroscopy

FTIR spectra were acquired on a BIO-RAD FTS-40 FTIR spectrometer with a BIO-RAD transmittance attachment. The spectra were obtained using a resolution of 8 cm^{-1} and were averaged over 128 scans. A background spectrum obtained with an uncoated KBr pellet was subtracted from the acquired spectra in all cases.

2.8 Scanning Electron Microscopy and Energy Dispersive X-Ray Spectroscopy

SEM/EDX was performed on untreated silica, talc and mica powders and PPy film deposited silica, talc and mica powders. The powders were fixed on the sample holder by double conductive adhesive type, then gold-coated and secondary electron images were recorded in a Hitachi model 90 scanning electron microscope using 5 keV acceleration voltage. Regional average EDX Spectroscopy was obtained at the same time.

2.9 X-Ray Photoelectron Spectroscopy

XPS was performed using an AMICUS model XPS spectrometer equipped with a Mg/Zr dual anode, manufactured by Kratos. The spectra reported here were recorded with the MgK$_\alpha$ anode operated at 300W. The survey spectra were obtained with pass energy of 89.45 eV and the narrow scans were acquired at 36 eV and 0.05 eV step size. The energy resolution of the narrow scans was 0.7 eV.

2.10 Particle Size Distribution and Specific Surface Area Measurements

Particle size distribution was measured on a Malvern 2600 particle size analyser, which can analyse particle sizes in the range of 0.5–1880 microns. This system uses a low power helium neon laser to analyse particle size by the Fraunhofer light scattering method. A collector lens gathers the light and projects it onto 31 detector rings. These concentric annular rings gather the diffracted light and record the light intensity on each ring. Large particles diffract more light, but in smaller angles than small particles. So the method depends only on the particle's volume (*i.e.* not the movement or place in the beam). A proprietary algorithm finds the volume size distribution.

The untreated or PPy film coated talc or mica powders were fed into the Malvern in the dry form through a venturi eductor feeder. Multiple measurements were made for each sample and the equipment was aligned and zeroed before each sample was run. Particle size distribution was based on light concentration on the detector rings. The size distribution gives the median particle size, as well as 90th and 10th percentile size. The particles are treated as homogeneous spheres for this measurement. No shape correction factors were used. The BET surface area for each of the six samples was measured separately, using a Micromeretics Gemini 2375 surface area analyser. This process uses monolayer nitrogen adsorption measurements to determine surface area of samples.

2.11 Performance of Treated Powders

Our usual performance test for experimentally treated powders involves substituting them for standard fillers or pigments in some standard polymer. Mechanical and other properties of the filled polymer film, sheet or particle are then measured and compared to the properties of the standard filler/polymer system using, for instance, ASTM D256-93a, D638-95 and ASTM D648-95 test procedures. This part of the work was carried out in collaboration with industry.

3 RESULTS

3.1 Contact Angle Measurements

The contact angle calculation results, listed in Table 2, rely on the de-ionized water adsorption curves of talc and mica powders shown in Figures 2 and 3. These results confirm that the treated powders are always hydrophobic after PPy film deposition, even though the starting materials display widely differing contact angles – 86 for talc, 15 for mica, and 0 for silica. The LW/HP condition always produces a more hydrophobic PPy film than the HW/LP condition. Moreover, different powders treated by the same technique display the same contact angle, indicating the deposited PPy film is uniform and homogeneous. Another experiment was done for silica and mica powders whose hydrophobicity increases significantly after coating. Before treatment, the particles were very hydrophilic and sank immediately when placed in water. After treatment, the powders floated on water. This effect was stable, demonstrating that the increased hydrophobicity, and therefore the surface treatment, was permanent.

Table 2   *Properties of PPy-treated Powders.*

Material	Film Density g/cm^3	Contact Angle	Film Thickness(Å)
Talc, untreated	–	86°	–
Talc, HW/LP PPy	1.754	65°	55
Talc, LW/HP PPy	1.563	74°	116
Mica, untreated	–	15°	–
Mica, HW/LP PPy	1.754	62°	136
Mica, LW/HP PPy	1.563	76°	194
Silica, untreated	–	0°	–
Silica, PPy	–	72°	–

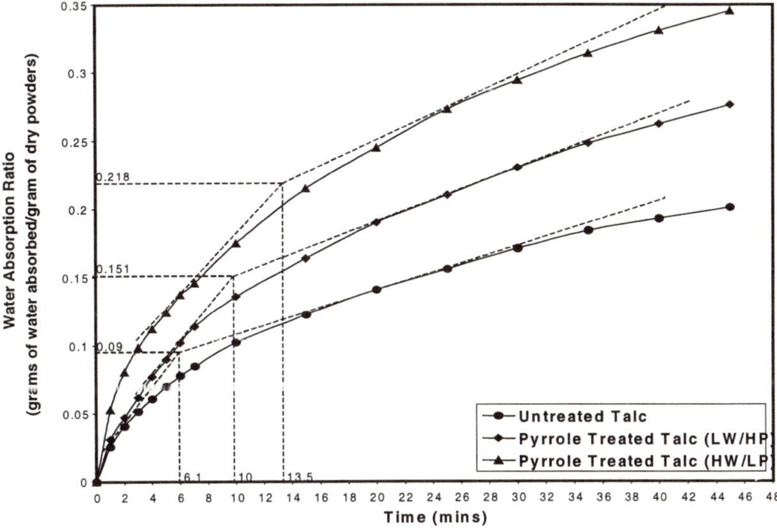

Figure 2   *Water absorption of untreated and PPy-treated talc.*

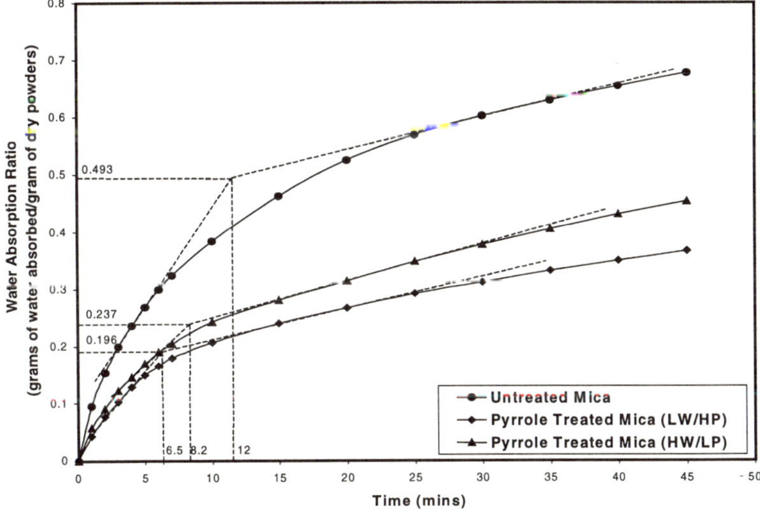

Figure 3   *Water absorption of untreated and PPy-treated mica.*

3.2 TGA and DSC Measurements

The STA curves of silica and mica powders are shown in Figures 4 and 5, respectively. The result for talc is very similar to that of mica and is not shown. We see that the silica contained 16 wt% of deposited PPy film (20% total weight loss, of which 4% is adsorbed water) while mica contained 9 wt% and 12 wt% of deposited PPy film under HW/LP and LW/HP conditions, respectively.

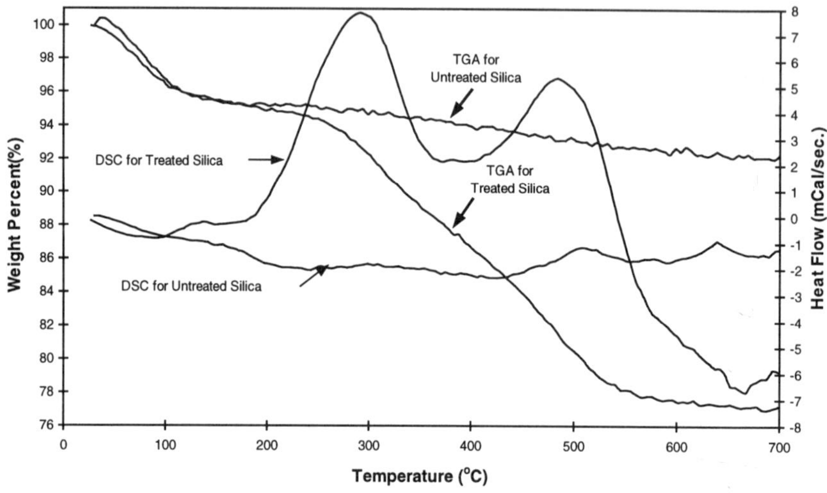

Figure 4   *STA analysis of untreated and PPy-treated silica.*

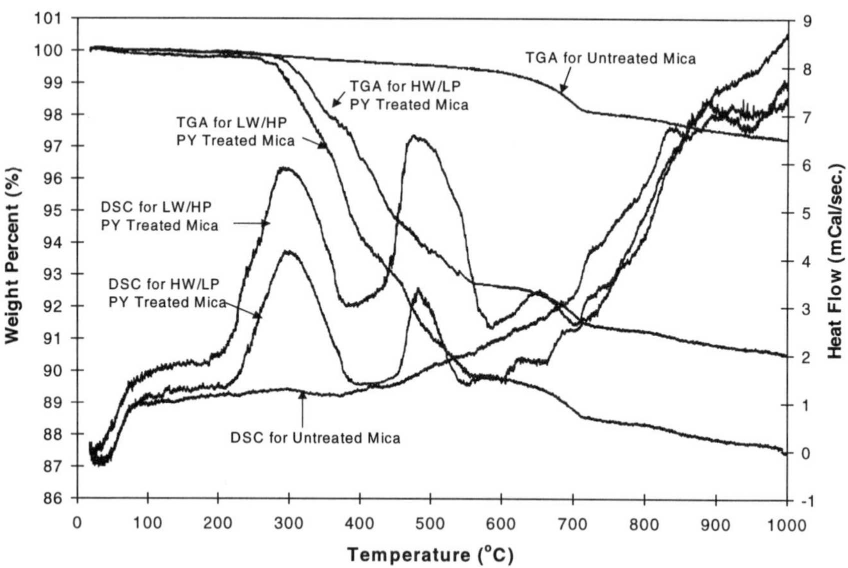

Figure 5   *STA analysis of untreated and PPy-treated mica.*

In these two sets of experiments we see that the PPy film comes off at two well-defined temperature ranges, as is typical of PPy films heated in air. This seems to suggest that there are two forms of PPy polymer present, which could well be an aliphatic and aromatic fraction. The relative amounts of these two fractions were found to be dependent on the film deposition conditions, in good agreement with this hypothesis. Focusing on the results with mica for a moment, we observe from Figure 5 that the second peak is lower in the case of the HW/LP deposition. The same effect was observed for talc. Furthermore, the deposition rate appears to be higher for the LW/HP conditions, under which 12 wt% was deposited, as opposed to 9 wt% for the HW/LP conditions. Again, the effects were very similar for talc. The differences in deposition rate are probably related to stronger concurrent etching effects, which occur under HW/LP conditions. Assuming that the films were completely homogeneous and that the particles can be approximated by spheres, the measured weights can be converted to film thicknesses. The film thickness results are given above in Table 2.

3.3 Deposition Rate of PPy Films

Thin film interferometry was carried out for the purpose of investigating the effects of plasma parameters on the plasma deposition process in an RF reactor, the ultimate goal being to deduce the possible mechanism that dominates this process. Figure 6 shows the recorded detector voltage *vs.* time for the two deposition conditions, and Table 3 contains the derived film refractive index and deposition rate. The HW/LP condition produces a denser film (higher refractive index) while the LW/HP condition produces a greater deposition rate, probably as a result of the increased residence time and/or collision activation. The very periodical changes in the curves shown in Figure 6 indicate that the film deposited homogeneously onto the substrate surface. All these results are in agreement with the results from other characterization methods.

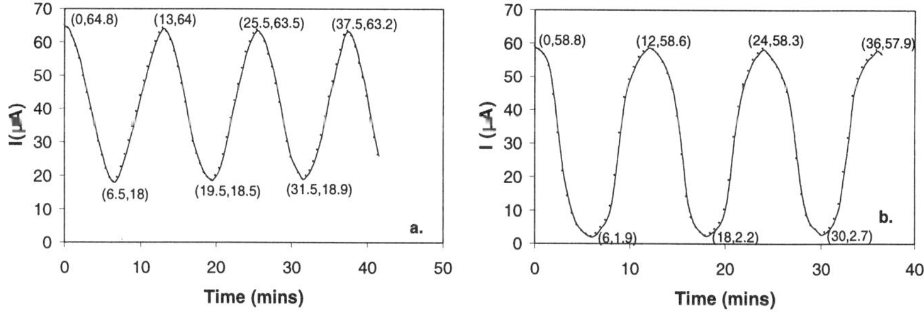

Figure 6   *Deposition rate measurement results: (a) LW/HP, (b) HW/LP.*

Table 3   *Deposition Rate Measurement Results of PPy Films.*

Deposition Condition	R.I.	Deposition Rate, D(Å/min)
HW/LP	1.767	149
LW/HP	1.434	170

3.4 Density and Thickness of PPy Films

The density measurement results of PPy films are shown in Table 2. It can be seen that HW/LP PPy film is denser than LW/HP PPy film, which correlates with the film refractive index calculation results that the HW/LP condition produces a more dense film than the LW/HP condition.

The film thickness results are also shown in Table 2. Since the plasma deposition time is the same in all cases, we infer that the LW/HP condition caused PPy to be deposited faster. Perhaps mica has a greater film thickness than talc in either case because of its smaller specific surface area. It should also be noticed that the thickness of PPy films is much smaller compared with powder particle size, illustrating why surface modification through plasma polymer film deposition is a powerful technique to modify the surface properties of the materials without changing bulk properties. The surface characteristics of the material are determined solely by the composition of the outermost few monolayers.

3.5 Bulk Analysis of PPy Films by FTIR

Typical FTIR spectra of PPy films deposited on KBr pellets are shown in Figure 7. Care was taken to record these spectra immediately after deposition because of reports that exposure of such films to the atmosphere results in the appearance of a new IR band at 1710 cm^{-1} (carbonyl stretch) with diminution of the band at 1630 cm^{-1}, the result presumably of oxidation of the film.[12]

The sample of Figure 7a was obtained under HW/LP deposition conditions. The percentage transmission of the most prominent peak at 1628 cm^{-1} was about 5%. The peak assignment for this spectrum is given in Table 4. The spectrum of Figure 7b was obtained from the sample treated under LW/HP conditions, but otherwise similar conditions as the sample of Figure 7a. The peak assignments for this material are given separately in Table 5. The transmission intensity for the most prominent peak here at 725 cm^{-1} was 9%. Since the films deposited in a comparable time span were thicker under LW/HP conditions, the difference in absorption intensities is understandable.

Table 4   *Peak Assignments for PPy Film of Figure 7a (HW/LP).*

Frequency, cm^{-1}	Assignment	
3377	N–H stretching vibrations of primary and secondary amines and imines	
2960	Asymmetric CH_3 stretch	Symmetric and asymmetric C–H stretching vibration of saturated hydrocarbons
2929	Asymmetric CH_2 vibration	
2879	CH_3, CH_2 symmetric stretch	
2219	C≡N conjugated stretching vibration and C≡C stretching vibration	
1628	C=C conjugated and C=N conjugated stretch and N–H deformation vibration	
1451	Alkane C–H deformation region, $-CH_2-$ scissoring vibration	
1100	CO stretch	
900	C=C–H deformation vibration of a vinylidene group	
745	C–H out of plane bending(cis –CH=CH–) and $-(CH_2-)_n$, $n > 3$ rocking vibration	

Both spectra exhibit a broad band between 3300 cm^{-1} and 3400 cm^{-1}, which is due to the N–H stretching vibration of primary and secondary amines and imines. In the film deposited under HW/LP conditions (Figure 7a), the relative intensity of this band is lower than that in the LW/HP film (Figure 7b). The LW/HP film spectrum also displays a weak band around 3110 cm^{-1}, assigned to the amide N–H stretch, and a fairly strong band at 1030 cm^{-1}, assigned to the C–H or N–H out of plane bending. These bands are absent in the HW/LP film spectrum.

Table 5   *Peak Assignments for PPy Film of Figure 7b (LW/HP).*

Frequency, cm^{-1}	Assignment	
3388	N–H stretching vibrations of primary and secondary amines and imines	
3114	Amide N–H stretch	
2971	Asymmetric CH$_3$ stretch	Symmetric and asymmetric C–H stretching vibration of saturated hydrocarbons
2951	Asymmetric CH$_2$ vibration	
2871	CH$_3$, CH$_2$ symmetric stretch	
2220	C≡N conjugated stretching vibration and C≡C stretching vibration	
1617, 1570	C=C conjugated and C=N conjugated stretch and N–H deformation vibration	
1428	Alkane C–H deformation region, C–H deformation vibration of CH$_3$ group.	
1320	C–H symmetric deformation	
1100	CO stretch	
1030	C–H out of plane bending (vinyl), N–H out of plane bending	
783	C–H out of plane bending	
725	C–H out of plane bending(cis –CH=CH–) and –(CH2–)$_n$, $n > 3$ rocking vibration	

In both spectra, several intense bands appear slightly below 3000 cm^{-1}. They are attributed to the symmetric and asymmetric C–H stretching vibration of saturated hydrocarbons. The PPy film deposited under LW/HP (Figure 7b) shows three clearly distinguished peaks around 2971 cm^{-1}, 2951 cm^{-1}, and 2871 cm^{-1}, with the 2971 cm^{-1} and the 2951 cm^{-1} peaks having almost the same intensity. In the HW/LP film (Figure 7a), however, the peak at 2960 cm^{-1} is much less pronounced. Since the peak around 2930 cm^{-1} can be attributed to the asymmetric CH$_2$ vibration, while the band around 2960 cm^{-1} represents the asymmetric CH$_3$ stretch, we conclude that the HW/LP film contains relatively more methylene groups than the LW/HP film.

This is also corroborated by the alkane C–H deformation region around 1430 cm^{-1}. In the case of the PPy film under HW/LP, there seems to be only one band at 1451 cm^{-1}, which is assigned to the –CH$_2$– scissor vibration. By contrast, the LW/HP spectrum exhibits bands at 1428 cm^{-1} and 1320 cm^{-1}, which are due to the C–H deformation of a -CH$_3$ group. This again shows that the HW/LP film contains relatively more methylene groups than the LW/HP film.

Both spectra exhibit a band around 2220 cm^{-1}, which is assigned to the C≡N and C≡C stretching vibration, it being slightly more intense in case of the HW/LP film. Common to both spectra is a strong band around 1620 cm^{-1}, which appears more intense in the case of the HW/LP films. This peak is a combination of the C=C and C=N conjugated stretching, and the N–H deformation. Its higher intensity in the HW/LP films could be due to conjugational effects.[13] In addition, Figure 7a also shows a fairly broad

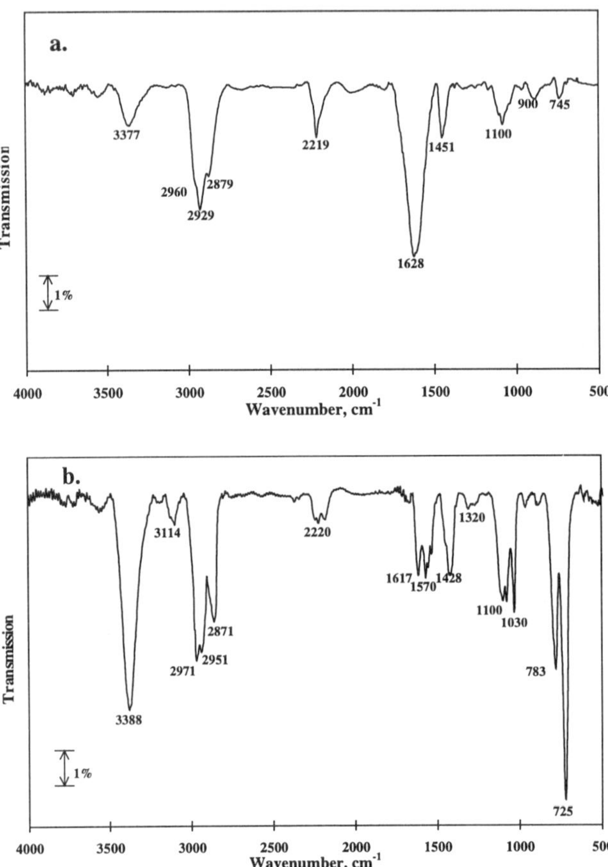

Figure 7 *FTIR spectra of RF PPy film deposited on KBr pellet: (a) HW/LP, (b) LW/HP.*

band at 900 cm^{-1}, which is interpreted as the C=C–H deformation vibration of a vinylidene group. This band is not found in the spectrum of the LW/HP film.

The two spectra exhibit absorption around 730 cm^{-1}, which is much more intense in Figure 7b. This absorption can be atttributed to the –CH$_2$– rocking vibration of long aliphatic chains, just as we see in polyethylene.[14] In the HW/LP case this vibration is weak, while in the LW/HP case it is much more intense. Figure 7b also shows a fairly strong peak at 783 cm^{-1} which represents the C-H out of plane bending. Thus, the films of which Figure 7b is a typical example can be described as consisting of more strongly hydrogenated polymers, whereas the HW/LP films, exemplified in Figure 7a, have a higher content of unsaturation and conjugation.

Other features of these IR spectra suggest substantial differences in the film structures. All singlet peaks in the HW/LP film spectrum were rather broad, indicating a range of closely related structures rather than one well-defined chemical state. This is a sign of a higher degree of crosslinking with more unsaturation and conjugation. Such an effect is to be expected for plasma-polymerized films. Conversely, the LW/HP film spectrum shows a less dense structure that seems to consist to some extent of long hydrocarbon chains. This plasma condition was so mild that, combined with the absence of high energy positive ions, the monomer seems to have linearly polymerized to a certain extent after ring-opening and not to have crosslinked extensively.

Our PPy films produced by plasma polymerization have not been duplicated by electropolymerization methods. Electropolymerized pyrrole films do not show an absorption in the 2850–3000 cm^{-1} range, as our PPy films do, and the main feature in their FTIR spectra is a band at around 800 cm^{-1}, representing the pyrrole ring vibration.[15] This vibration is absent in our PPy films, although nitrogen in other forms still seems to be present, as for example in the bands at around 2220 cm^{-1} assigned to the nitrile group.

Other features of the spectra in Figures 7a and 7b clearly indicate that the pyrrole ring structure is not maintained in either case. The IR spectra of most substituted pyrroles usually have three absorption bands in the 1600–1375 cm^{-1} region, which may be assigned to in-plane ring deformation modes analogous to those observed in the spectrum of pyrrole. These bands are all absent in Figures 7a and 7b.

Power and pressure markedly effects the structure of the deposit, as evidenced by the differences in Figures 7a and 7b. Other workers have subjected films deposited under different conditions of pressure and power to FTIR analysis.[12]   Their results generally are in accord with ours: for the deposition of PPy films from an RF plasma reactor, power and pressure have the greatest effects on the film structure.

3.6 Scanning Electron Microscopy and Energy Dispersive X-Ray Spectroscopy

Figure 8 shows SEM images of each sample before and after deposition of PPy film. Even at very high magnification, the PPy films are not detected: there is no difference between the SEM images of the treated and untreated powders.

Typical EDX spectra of each sample are shown in Figures 9 and 10. The acceleration voltage was 5 keV for all samples. After PPy film coating, the carbon content increased sharply, indicating a significant amount of the PPy film. This confirmed that the film was actually there, although it was too thin to be observed clearly in the SEM images. Nitrogen could not be detected on any sample by EDX, but was detected by XPS and TOF-SIMS. It also can be seen that the relative intensity of carbon is higher for LW/HP condition than that for HW/LP condition. Since the film deposited under LW/HP conditions is much thicker than that deposited under HW/LP conditions, this difference is easily explained and the difference between these two samples is in agreement with the observed weight difference in TGA.

3.7 XPS Analysis of PPy Films

XPS wide scan spectra are shown in Figure 11a, b, and c for uncoated and PPy film coated mica under HW/LP and LW/HP conditions, respectively. For talc and silica the results are so similar that they need not be shown here. A comparison of these Figures indicates that after PPy film coating, signals for all the elements in the bulk material, such as Si, K, Al, Mg, decrease, whereas elements in the PPy film, such as C and N, increase significantly. It should also be noticed that the presence of the PPy film does not suppress the Si, K, Al, Mg signals in XPS.

3.8 Particle Size Distribution and Specific Surface Area Measurements

Measurements of the particle size distribution and the specific surface area were conducted on the talc and mica powders before and after coating with PPy films. From

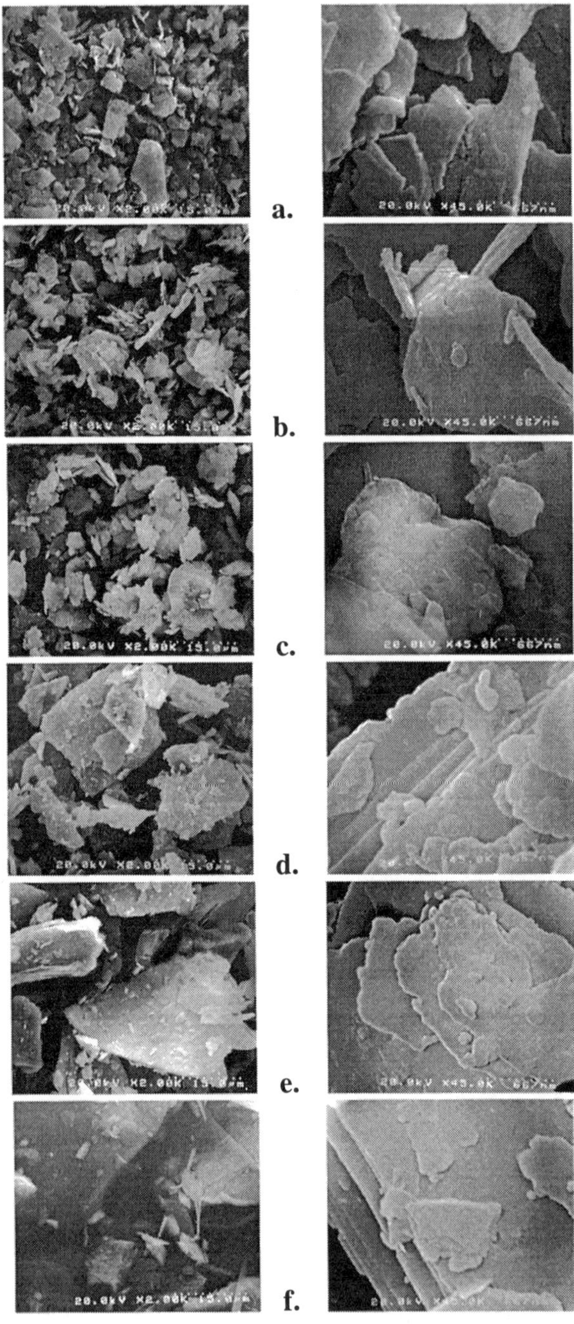

Figure 8   *SEM images of talc and mica: (a) talc, untreated; (b) talc, LW/HP PPy; (c) talc, HW/LP PPy; (d) mica, untreated; (e) mica, LW/HP; (f) mica, HW/LP.*

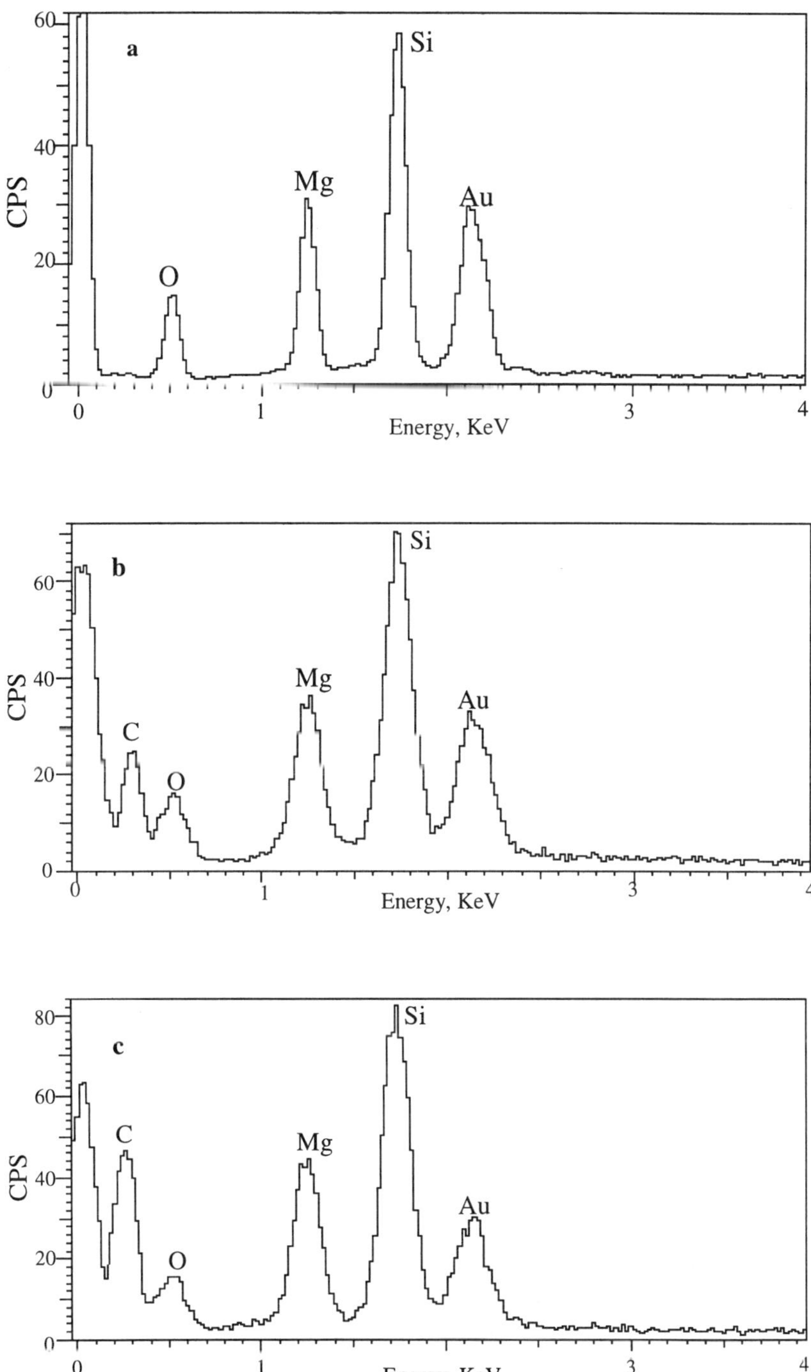

Figure 9　*EDX spectra of talc: (a) untreated, (b) HW/LP, (c) LW/HP.*

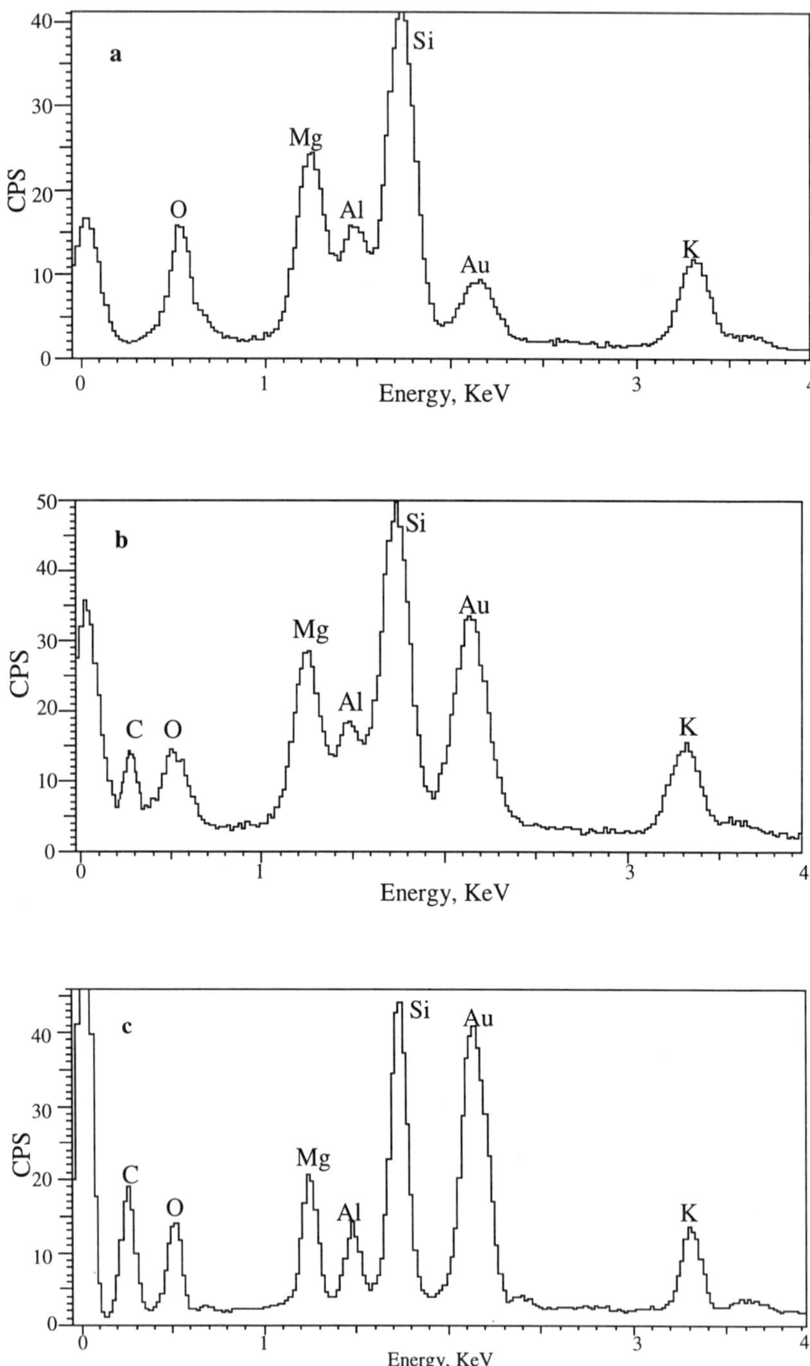

Figure 10 *EDX spectra of mica: (a) untreated, (b) HW/LP, (c) LW/HP.*

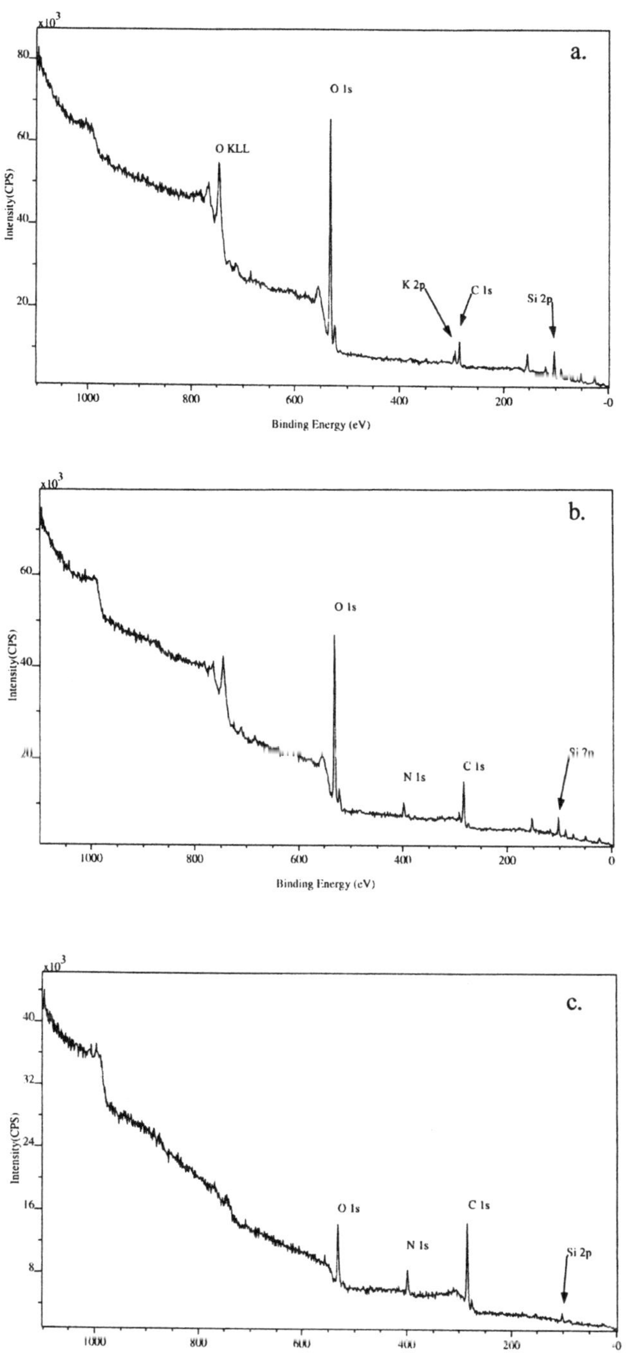

Figure 11 *XPS spectra of mica: (a) untreated, (b) HW/LP PPy, (c) LW/HP PPy.*

the data in Table 6 it is clear that no change in average particle size or size distribution has occurred for either type of particle under either set of deposition conditions. This result demonstrates the unique capability of the plasma deposition technique to delicately alter surface properties, even for high specific surface area particles, without affecting bulk properties.

Table 6   *Particle Size Distribution and Specific Surface Area of Treated Talc and Mica Samples.*

Material	Sp. Area, m²/g	Particle Size, μm		
		D90	**D50**	**D10**
Talc, untreated	9.69	10.0	6.4	2.7
Talc, HW/LP PPy	9.61	14.2	7.3	3.0
Talc, LW/HP PPy	9.67	12.6	6.8	2.8
Mica, untreated	3.17	46	21	6.7
Mica, HW/LP PPy	2.84	50	22	6.5
Mica, LW/HP PPy	3.23	44	21	6.6

3.9 Performance of PPy Film-treated Powders

Only a limited amount of preliminary data can be presented here. Most of this work is in progress, and most of the results are being held in confidence to be considered for patent applications.   However, one key result with treated powders will be presented and discussed.

Table 7 shows rubber property data obtained with the PPy-treated silica. Many properties of a 70/30 NR/SBR rubber blend were determined in which untreated Hi-Sil 233 was compared with silica coated with PPy. One batch was made with DC-plasma-coated, another one with RF-plasma-coated silica. The results were compared with those of the same rubber filled with a commercially available silica that is surface-treated with an organofunctional silane, Si-69. The results were obtained in the laboratories of PPG in Monroeville, PA. Many properties other than those listed in Table 7 were evaluated, but only those that show a significant effect are presented.

We see that the commercial material (compounds 4 and 5) accelerates the cure, but the PPy-treated (mainly compound 6) retards the cure somewhat. Also, the effect on the scorch time is much less than the commercial silica. The elongation is improved and the effect on the 300% modulus is less than the commercial silica. In general, these results, while not yet a spectacular improvement, do indicate that the properties of the rubber can be modified by treating the silica with a hydrophobic film.

4 DISCUSSION

There are many plasma parameters affecting the plasma process, among which the excitation power and monomer pressure were found to have the greatest effect.[12] The excitation power has an effect on the film structure by regulating the energy of the ions that impact the growing film. The increase of the excitation power results in an increased energy of activated plasma species that impinge the growing film and also causes a more intense bombardment of the substrate surface by energetic species.   The monomer

Table 7　　*Effect of PPy Film on Properties of Silica-Filled 70/30 NR/e-SBR Rubber Compounds.*

Compound No.　→	1	2	3	4	5	6
Compound Additive						
Hi-Sil 233, phr	45	27	27	45	45	9
RF-Plasma Hi-Sil 233, phr		18				
DC-Plasma Hi-Sil 233, phr			18			36
Si-69, wt/wt silica, %				4	8	
Property						
ML-1+4 (100 °C)	49	56	58	36	32	56
Scorch (TS2 at 130 °C), min.	16.02	12.77	16.4	14.21	10.83	15.18
Min. Torque, dN/m	3.58	4.76	4.62	2.53	2.2	4.51
T50, min.	23.47	20.27	25.26	21.58	17.32	23.58
T90, min.	34.42	30.67	38.69	31.56	27.93	36.18
Cure Rate, T50-TS2, , min.	7.45	7.5	8.86	7.37	6.4	8.4
Tensile, Mpa	23.8	24.4	25.7	24.9	23.9	25.6
Elongation, Mpa	648	663	683	588	542	679
Energy to Break, J	14.1	16.5	18.2	15.6	13.7	16.8
Modulus @ 100%, Mpa	1.61	1.61	1.64	2.27	2.48	1.61
Modulus @ 300%, Mpa	4.79	5.17	5.11	8.31	9.61	5.14
300/100 Modulus Ratio	2.98	3.21	3.12	3.66	3.88	3.19
MG Tear, kN/m	7.3	7.9	7.4	4.7	3.9	7.1
Pico Abrasion Index	65	69	67	82	90	70
Hardness at 23 °C	59	59	60	62	64	60
Hardness at 100 °C	57	58	59	61	62	59
Tan δ at 60 °C	0.0815	0.0814	0.074	0.0833	0.0723	0.0728

pressure has an effect since it controls the intensity of the bombarding ion flux. Increasing the monomer pressure will increase the residence time of the molecules in the reactor. In addition, the average electron energy will be decreased, leading to less energetic bombardment of the growing film. This is why oily, non-crosslinked films are usually obtained at high monomer pressures in RF plasma.[16]

In our present work we have applied two deposition conditions, HW/LP and LW/HP. HW results in higher ion energy and more ions that bombard the growing film, while LP results in less collisional deactivation of the high-energy ions. Therefore, HW and LP both promote intense bombardment by high energy ions, and the films produced by HW/LP were expected to possess the usual properties of plasma-polymerized organic materials. LW and HP both promote a lower ion intensity and energy, thus the films tend to be softer, more organic, less crosslinked, less unsaturated and more conjugated, of lower density and with lower refractive indices. We had confirmed these effects for PPy films deposited in DC plasma;[17,18] we now confirm these effects again in RF plasma and for powdery substrates.

The oxidation of both PPy films follows a somewhat typical pathway. As we have described previously, two characteristic oxidation peaks of PPy film show up in the DSC curves, the second peak which corresponds to a higher temperature is lower in the case of HW/LP deposition, indicating that the HW/LP PPy film oxidizes more easily. The possible reason correlates with their different electron donating ability. In fact, electron-donating polypyrrole is more susceptible to oxidation than electron withdrawing polypyrrole. Experimental evidence was presented above that HW/LP PPy film can be described as a carbon-like polymer having a higher relative carbon content while LW/HP PPy film consists of more strongly hydrogenated polymers. Therefore, the HW/LP PPy possesses stronger electron donating ability which facilitates its oxidation by lowering the activation energy for oxidation.

Another observation is that the PPy film does not suppress the Si, K, Al, Mg signals in XPS though the surface energy of the powders changed significantly after the PPy film coating. We have seen this effect with silica, too, even in TOF-SIMS analysis. A possible explanation for this is that the film deposition leads to a mixing of the PPy and inorganic substrates. The PPy film intimately mixed with the powder while it dominated the outmost layer of the surface because of its lower surface energy.

The most notable result of this work is that the particle size distribution and specific surface area measurements confirm the unique features of plasma polymer deposition process. The plasma process does not affect the substrates. The surface modification technique through plasma deposition is a very effective technique to selectively modify the surfaces of materials without posing detectable influences on the properties of the bulk materials.

5 SUMMARY

An RF reactor for direct plasma polymerization of organic monomers (*i.e.* in the absence of carrier gases) is a simple but versatile tool in terms of mechanism of polymerization. The films produced in such a device are very homogeneous, and can be very hydrophobic. Perhaps most importantly, the surface modification of particles by this technique does not affect the bulk properties or particle size distribution. Even with one monomer, the properties of the deposited coatings can be varied within wide limits by varying deposition conditions such as power and pressure.

6 ACKNOWLEDGMENTS

This work was funded by the National Science Foundation (NSF, DMII Division) under Grant No. DMI-9713715. The authors would further like to thank the following people from Procter & Gamble Company for their contributions in this work: Mr. Paul France, for his invaluable suggestions and comments, Ms. Cindy Obringer and Ms. Saswati Datta, for doing the particle size distribution and specific surface area measurements and describing the method used, and Mr. Arseniy Radomyselskiy for setting up and tuning the laser interferometer.

References

1.   K. Upadhya and T. C. Tiearney, *JOM*, 1989, **41**, 6.
2.   A. Anders, S. Anders, I. G. Brown and I. C. Ivanov, 'Materials Research Society Symposium Proceedings', 1995, Vol. 388, Materials Research Society, Pittsburgh, PA, p. 215.
3.   H. Schonhorn, 'Polymer Surface', Wiley, New York, 1978, Chapter 10, p. 213.
4.   M. Hudis, 'Techniques and Applications of Plasma Chemistry', Wiley, New York, 1974, Chapter 3.

5.   W. J. van Ooij and K. D. Conners, 'ACS Symposium Proceedings Series', 1998, in press.

6.   A. Sabata, W. J. van Ooij and H. K. Yasuda, *Surf. Interface Anal.*, 1993, **20**, 845.

7.   K. D. Conners, W. J. van Ooij, S. J. Clarson and A. Sabata, *J. Appl. Pol. Sci: Polymer Symposia*, 1994, **54**, 167.

8.   N. Inagaki, S. Tasaka and H. Abe, *J. of Appl. Poly. Sci.*, 1992, **46**, 595.

9.   D. S. Sato and K. M. Toshihiro, *Polym. Prepr.*, 1997, **38**, 993.

10.   G. Nutsch and B. Dzur, *Elektrowaerme International*, Edition B: 1995, **53**, B201.

11.   'Handbook of Conducting Polymers', T. A. Skotheim, ed., Marcel Dekker, New York, 1986, Vol. 2, p. 281.

12.   W. J. van Ooij, S. Eufinger and T. H. Ridgway, *Plasmas & Polymers*, 1996, **1**, 229.

13.   G. Socrates, 'Infrared Characteristic Group Frequencies', Wiley, New York, 1994.

14.   C. J. Pouchert, 'The Aldrich Library of Infrared Spectra', Edition III, Aldrich Chemical Company, 1981, p. 1565.

15.   G. B. Street, T. C. Clarke, M. K. Krounbi, K. Lee, P. P. Victor, J. C. Scott and G. Weiser, *Molecular Crystals & Liquid Crystals*, 1982, **83**, 252, in 'Proc. of the Int. Conf. on Low-Dimens. Conduct., Pt. D', Boulder, CO, USA, Aug 9–14, 1981.

16.   H. Kobayashi, A. T. Bell and M. Shen, *Macromol.*, 1974, **7**, 277.

17.   W. J. van Ooij, S. Eufinger and S. Guo, *Plasma Chem. Plasma Proc.*, 1997, **17**, 123.

18.   W. J. van Ooij, S. Eufinger and T. H. Ridgway, *Plasma and Polymers*, 1996, **1**, 231.

EFFECT OF WATER PLASMA ON SILICA SURFACES: SYNTHESIS, CHARACTERIZATION AND APPLICATIONS

N.A. Alcantar, E.S. Aydil and J. Israelachvili*

Department of Chemical Engineering
University of California
Santa Barbara, CA 93106 USA

1 ABSTRACT

We have characterized silicon dioxide surfaces prepared by plasma enhanced chemical vapor deposition (PECVD) before and after water plasma treatment. We found that the water plasma treatment not only hydroxylated the silica surfaces, but also significantly reduced their roughness without affecting the thickness of the deposited films. These activated surfaces could then be rendered biologically compatible by reacting them with polyethylene glycol. The optimal water plasma parameters for creating smooth, chemically reactive silicon oxide films were investigated. The surface modification and polymer reaction were monitored by attenuated total reflection Fourier transform infrared (ATR-FTIR) spectroscopy, ellipsometry, atomic force microscopy (AFM) and contact angle measurements. Such surfaces are also being developed as more versatile and reusable substrates for surface force measurements using the SFA and AFM techniques.

2 INTRODUCTION

The name silica refers to a large class of materials with the general formula SiO_2 or $SiO_2 \cdot xH_2O$. Silica is used as an abbreviation for "silicon dioxide" in all its crystalline, amorphous, hydrated and hydroxylated forms. Silica is composed of interlinked SiO_4 tetrahedra. At the surface, the silica structure terminates in siloxane groups ($\equiv Si-O-Si\equiv$) or silanol groups ($\equiv Si-OH$). Siloxanes consist of two silicon atoms joined by an oxygen, whereas silanols can be classified as isolated or geminal depending on the number of hydroxyl groups attached to the silicon atom (one for isolated, two for geminal), and associated silanols, where two hydroxyl groups are attached to different silicon atoms close enough that they can form a hydrogen bond.[1]

The silica films for this work were produced by plasma-enhanced chemical vapor deposition (PECVD). This deposition technique forms amorphous silica and provides versatility since SiO_2 can be deposited on arbitrary shapes and many different surfaces including metals and plastics.

The surface of amorphous silica is randomly structured, so that the silanol groups are not ordered on the surface. In spite of this, we found that complete surface coverage of hydroxyl groups can be achieved by exposure of the surface to water plasma under optimized conditions without significantly roughening the surface.[2] It has been reported that fully hydroxylated silica has 4.6 ± 0.5 OH groups/nm^2 at the surface.[3,4] We were also able to reduce the roughness to *ca.* 10 Å from as high as *ca.* 60 Å under judiciously selected plasma treatment conditions. We refer to this hydroxylation as "activation" and the resulting surface as "activated", since the hydroxyl groups can then be reacted with other molecules such as polyethylene glycol to modify the silica surface.

Our goal is to make these silica coated surfaces biocompatible by attaching one end of a polyethylene glycol (PEG) chain to the "activated" silica film through a –SiOC– ester linkage. The underlying silica surface is then biologically compatible, since it is well known that PEG covered surfaces enhance protein rejection, nonimmunogenecity and nonantigenecity.[5–7] Silica was selected because it can easily be produced and activated, and it is nontoxic at low concentrations.[8,9]

3 EXPERIMENTAL METHODS

3.1 Deposition of Silica Surfaces

Thin silica films were deposited by reacting SiH_4 / Ar with O_2 in a plasma-enhanced chemical vapor deposition (PECVD) reactor.[8,13] PECVD has certain advantages over other silica production methods such as low deposition temperature, better control of stoichiometry and high purity of the films. In addition, thin films are easier to make by PECVD. The PECVD reactor, shown in Figure 1, utilizes a helical resonator plasma source operating at 13.56 MHz to sustain a low pressure (25 mTorr) high density plasma in a 2"-diameter quartz tube which is mounted on the top port of a six-way stainless steel cross. Oxygen gas is introduced from the top of the tube and flows through the plasma. Silane is fed a few centimeters above the sample through a gas injection ring that surrounds the sample stage. The gases are pumped out by a 300 L/s turbomolecular pump and the pressure in the reactor is controlled independently from the flow using a throttling valve, pressure transducer, and valve controller. In addition, the experimental setup has *in situ* ATR-FTIR spectroscopy and *in situ* spectroscopic ellipsometry capabilities to study various species on the film surface and the film thickness as the deposition proceeds. Lastly, the substrate temperature can be controlled via a heated copper electrode.

Deposition reactions have been described in detail by Han, *et al.*[8] In the experiments reported in this study, the system pressure was 25 mTorr and the total gas flow rate was maintained at 50 sccm by varying the flow rates of SiH_4 and O_2. The deposition rate was varied between 14 and 100 Å/min by adjusting the ratio of SiH_4/O_2 flow rate. The substrate temperature and radio frequency (rf) power were 250 °C and 80 W, respectively. We used silicon wafers as substrates immediately after they had been dipped in HF (49%) for 15 seconds and rinsed with deionized water for about a minute. This process removed the native silicon oxide film on the substrates' surface.

3.2 Water Plasma Activation

After deposition, the silicon dioxide films were water plasma treated to hydroxylate their surface (*i.e.* convert surface siloxane groups to silanol groups). Water plasma exposure was done in a different reactor than the one shown in Figure 1. Water plasma was created by a radio frequency (13.56 MHz) electric field between two parallel electrodes, which ionizes and dissociates the water vapor flowing between them. Water vapor was evaporated from a bulb filled with liquid water and fed into the reactor through a needle valve. The effects of plasma pressure, power and exposure times on the surface roughness were investigated. During these treatments, the substrate temperature was kept at approximately 20 to 25 °C.

Several methods were used to characterize the deposited silica films before and after water plasma treatment, which will be briefly described next.

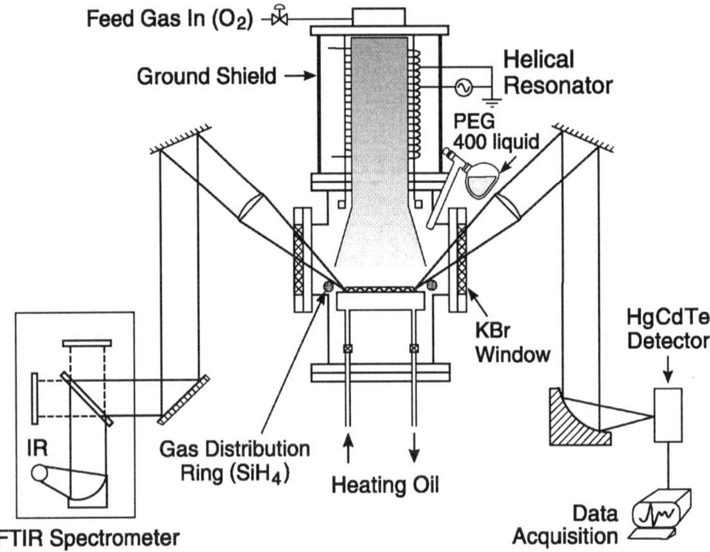

Figure 1 *Plasma enhanced chemical vapor deposition reactor and the in situ attenuated total reflection FTIR spectroscopy apparatus.*

3.3 Roughness and Thickness Measurements

A Nanoscope III AFM (Digital Instruments, Goleta, CA) was used to image and measure the roughness of the silica films using silicon nitride tips. The magnitude of the total force exerted on the samples varied from 10^{-7} to 10^{-6} N. The scan rate was 1 Hz over 1 μm^2 scan size. The images were recorded in contact mode using height acquisition data. Variable Angle Spectroscopic Ellipsometry, V.A.S.E. (Woollam, Co.), was used to measure film thickness.

3.4 Attenuated Total Reflection Fourier Transform Infrared Spectroscopy (ATR-FTIR)

Attenuated total reflection Fourier transform infrared spectroscopy is particularly well suited for studying surface absorbates on non-absorbing substrates with relatively high index of refraction such as silica. We used this technique to monitor the *in situ* absorbance and vibration frequencies of the silica constituents during deposition and water plasma treatment. In ATR-FTIR spectroscopy, infrared output from a FTIR spectrometer (Nicolet Magna 550) is focused at normal incidence onto one of the beveled faces of a GaAs (Ge or Si) crystal and transverses the sample undergoing multiple internal reflections (Figure 1). Detailed descriptions of this technique have been published.[8,10–12]

3.5　Contact Angle Measurements

We determined the degree of wettability of the silica films before and after water plasma treatment by measuring the contact angle of a water droplet placed upon them. The measurements were carried out at 24 ± 1 °C in a sealed chamber at *ca.* 100% relative humidity. The contact angles were recorded using a CCD camera and analysed by NIH Image (USA National Institutes of Health software by Wayne Rasband, ver. 1.60) to assure precision in our measurement.

4　EXPERIMENTAL OBSERVATIONS AND ANALYSIS

4.1　Analysis of Surfaces Before Water Plasma Treatment

Deposition rate, temperature, radio frequency (rf) power, and pressure are the most predominant parameters during plasma deposition of silica to obtain a uniform and reproducible film. For instance, they have a notable influence on porosity, hardness, refractive index, stoichiometry, roughness, water absorption, and density, which are among the most important film properties. Since our special concern was to produce regular and reproducible films, we kept the substrate temperature, rf power and total pressure identical for all the film depositions to maintain the reproducibility of their surface properties. We varied only the deposition rate in order to study its influence on the surface roughness of the films. Deposition rate depends on the ratio of SiH_4/O_2 gas flow rates.[13] The SiH_4/O_2 ratio values that were used for the silica depositions ranged from 0.002 to 0.2, which led to deposition rates between 14 and 100 Å/min.

　Approximately 5000 Å thick silica films were deposited at different deposition rates on silicon wafers. Table 1 shows the thickness and root mean square (RMS) roughness values obtained from ellipsometry and AFM, respectively. The deposition rates were calculated by dividing the final thickness by the deposition time. The deposition rate as a function of time is constant.

Table 1　*Thickness and RMS Roughness Values for Different Deposition Rate Silica Films.*

Thickness (Å)	Dep. time (min.)	Dep. Rate (Å/min.)	Roughness RMS (nm)
3860 ± 85	277	14	4.8 ± 0.5
5040 ± 85	360	14	5.0 ± 0.4
5030 ± 20	152	33	5.5 ± 0.4
4950 ± 20	95	52	5.4 ± 0.3
4520 ± 30	62	73	5.9 ± 0.4
4830 ± 20	61	79	5.6 ± 0.5
5070 ± 30	61	83	5.9 ± 0.5

We did not observe any strong correlation between thickness, deposition rate, and roughness of the films, although the roughness slightly increased with increasing deposition rate. It can be concluded that the roughness of silica films produced by PECVD was about 60 Å at high temperature and low pressure-high density plasma.

4.2   Effect of Water Plasma Conditions on Surface Properties of Silica Films

We found that water plasma treatments originally planned only to surface hydroxylate the silica surfaces also had a remarkable smoothing effect.

4.2.1 Effect of low pressure and low power water plasma.   Samples from each deposition batch listed in Table 1 were exposed to a low pressure (20 mTorr) and low rf power (12W) water plasma for 5 min. Subsequently, we measured their roughness and thickness. We found an unexpected reduction in the RMS roughness of the silica films after water plasma treatment, as shown in Figure 2.

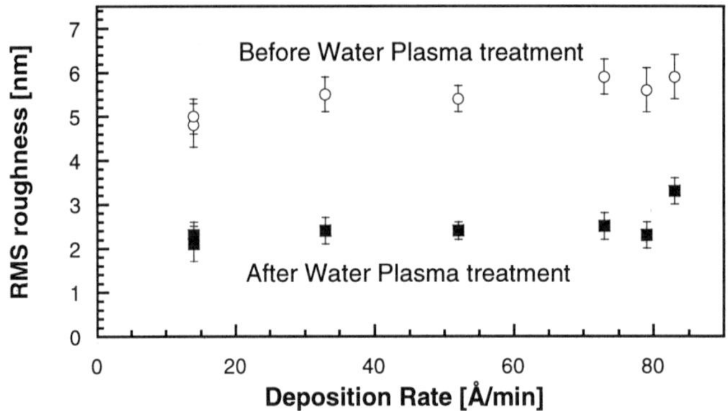

Figure 2   *Effect of water plasma activation on the RMS roughness of PECVD silica films.*

Moreover, although one might expect material to be etched away during this treatment, no substantial change in the film thickness was observed after water plasma treatments (Figure 3). We conclude that while the water plasma smooths the surface, the film is not etched. We also confirmed this by *in situ* ATR-FTIR measurements during water plasma exposure of silica deposited on GaAs crystals. These results will be discussed shortly.

4.2.2 Effect of exposure time of water plasma.   We exposed several PECVD silica films to water plasma for different periods of time at low pressure (20 mTorr) and low power (12W). The surface roughness decreased with increasing plasma exposure time, but the greatest effect was observed within the first five minutes. We determined that exposure to water plasma for 15 minutes resulted in the smoothest silica surfaces (Figure 4). Again, within measurement error, ellipsometry results before and after water plasma showed no net change in film thickness. We could obtain surfaces as smooth as 6 to 10 Å RMS roughness over 1 μm^2 areas.

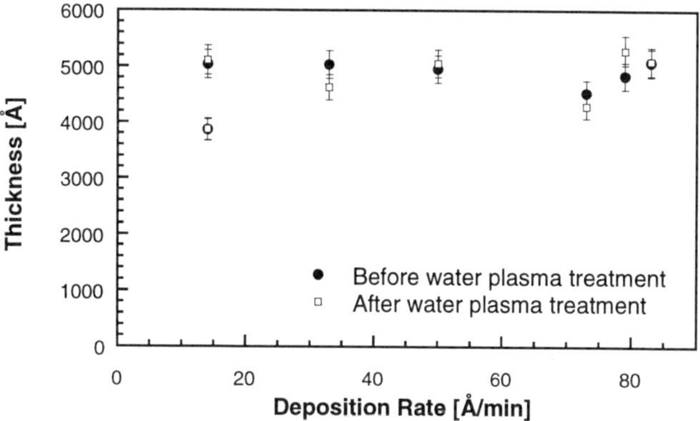

Figure 3   *Effect of water plasma activation on the thickness of PECVD silica films.*

4.2.3 Effect of water vapor pressure and rf power of water plasma on silica films. We also investigated the effect of water vapor pressure and plasma rf power as a function of exposure time on the roughness of silica films in order to establish the optimal water plasma treatment conditions. In all cases, the RMS roughness of the silica surfaces was measured before and after the water plasma modification.

We found that the RMS roughness values were lower than those for as-deposited silica when the silica surfaces were treated using: (i) high water vapor pressure (200 mTorr) and high rf power (80 W), (ii) high water vapor pressure (200 mTorr) and low rf power (10 W), or (iii) low water vapor pressure (20 mTorr) and high rf power (80 W). However, the roughness under these three conditions was still slightly higher than the values for silica surfaces treated with water plasmas at low pressure (20 mTorr) and low power (10 W). When low power plasma is used, the films became smoother with exposure time. When using high power plasmas for the treatments, the films begin to get rougher after reaching a minimum roughness. We attributed this finding to severe ion bombardment, which occurs with high density and high power plasmas. Figure 5 shows a comparison between the change in the RMS roughness of the silica films as a function of exposure time to a low pressure-low power plasma, and high pressure-high power plasma.

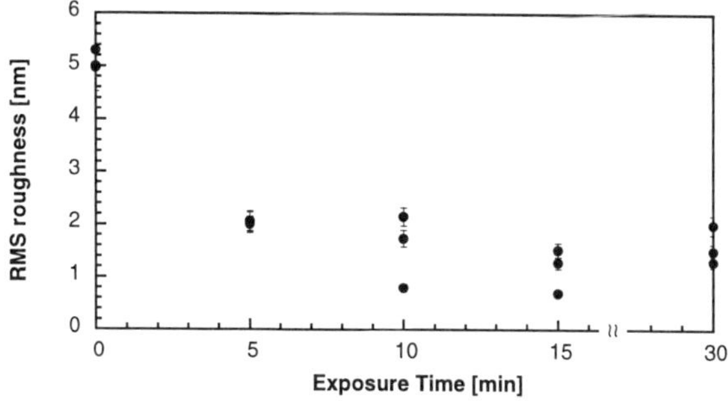

Figure 4   *Effect of water plasma exposure time on the RMS roughness of the PECVD silica films.*

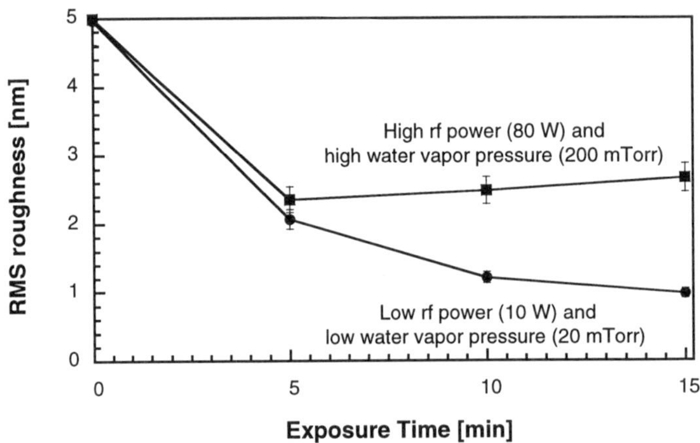

Figure 5   *Effect of rf power and water vapor pressure on the roughness of silica films deposited by PECVD.*

4.3   Determination of the Effect of Water Plasma on Silica Films by ATR-FTIR

Figure 6 shows typical ATR-FTIR spectra that were taken during the deposition of a thin film of silica (280 Å) at a deposition rate of 52 Å/min. The $\equiv$Si–O–Si$\equiv$ stretching frequencies of siloxanes are characterized by the sharp peak in the region of 1100–1000 cm^{-1} and the Si–O bending vibration at 800 cm^{-1}.[18] The associated silanol has a broad IR band located between 3700 and 3200 cm^{-1}, and the isolated silanol stretch band is a sharp peak at 3746 cm^{-1}.[1,13,18]

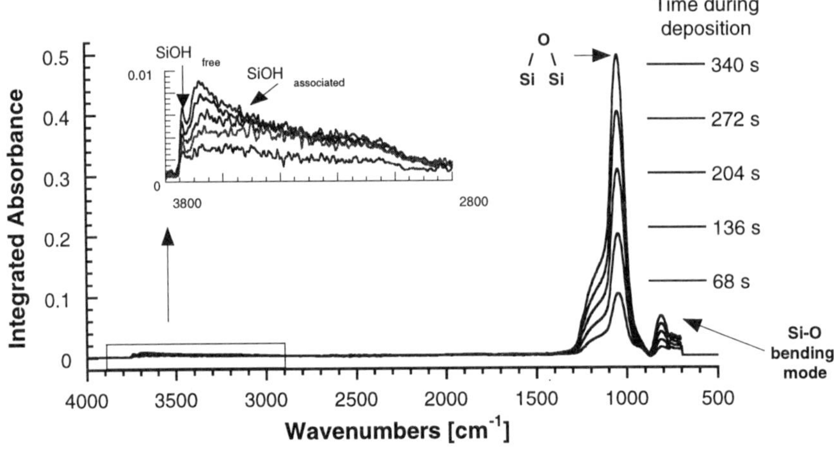

Figure 6   *ATR-FTIR Spectra during plasma enhance chemical vapor deposition (PECVD) of a silica film.*

The alteration of the silica surface was also monitored during water plasma activation using ATR-FTIR. The ATR-FTIR spectra in Figure 7 were taken in the so-called differential mode, where the spectra acquired during water plasma activation were divided by a spectrum of an untreated silica film. This procedure facilitates the detection of the changes on the surface due to water plasma activation. For example, an increase in the absorption frequency is observed due to the formation of chemical species on the surface. Similarly, a decrease in the absorption frequency is observed at the characteristic frequencies of species that have been removed from the surface.

As shown in Figure 7, several changes occurred when the surface was exposed to water plasma. First, the siloxane vibration band shifted. That is, the sharp peak located at 1020 cm^{-1} started disappearing during water plasma. However, an equivalent peak at higher wavenumbers (1080 cm^{-1}) started increasing at the same rate as the other disappeared. This finding proves that silica is not being etched away from the surface during water plasma, which agrees with the thickness measurements (Figure 3), but that the surface structure is being modified. Moreover, the change of absorbance with time was measured for these two peaks. We found that the change of disappearance of the peak at 1020 cm^{-1} is linearly proportional to the change of appearance of the peak at 1080 cm^{-1} as is shown in Figure 8. Therefore, water plasma does not etch silica, but modifies its surface physically and chemically by reducing the roughness and by shifting the absorption frequencies, respectively.

We also found the presence of two other peaks. One peak of medium intensity at 1440–1400 cm^{-1} could be assigned to in-plane OH bending, and another very broad peak at 3100–2400 cm^{-1} might be assigned to H-bonded OH stretch.[18,19] All alterations were likely produced by the interaction of H, O, OH and ions from the plasma with the silica surface. A close look at the vibration band for the free and associated silanol (3746–3200 cm^{-1}) showed that it reaches a constant value and saturates as a function of time, meaning that the silica surface has been fully hydroxylated. The exposure time that corresponds to the saturation of free silanols was about 8 minutes. Therefore, if we expose silica for about 15 minutes to water plasma, we get fully hydroxylated and smooth silicon dioxide films.

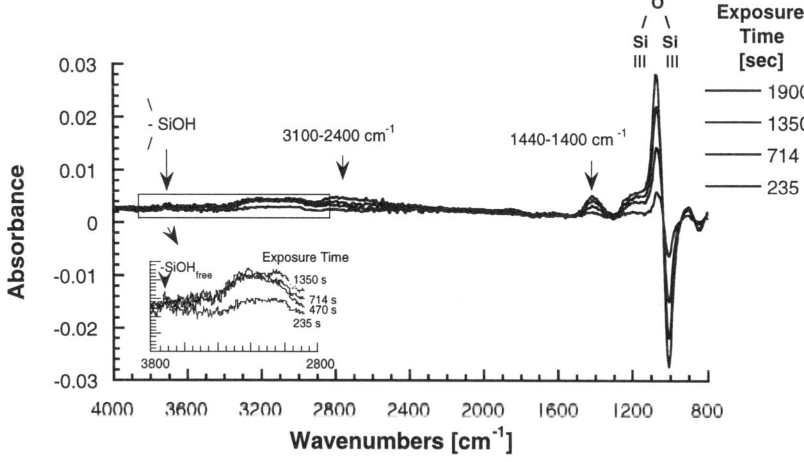

Figure 7　*ATR-FTIR of (PECVD) silica during water plasma activation.*

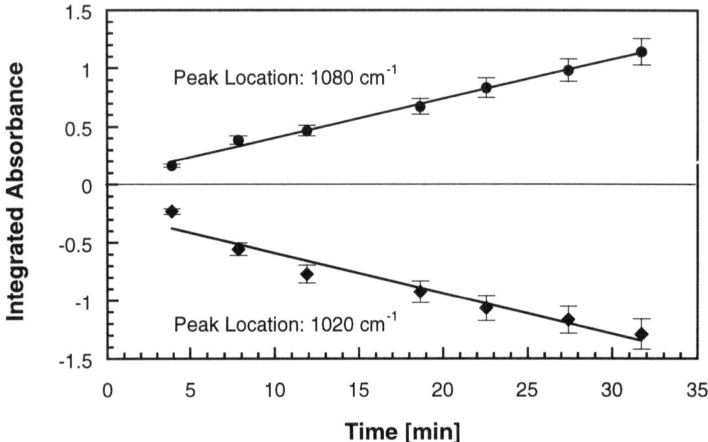

Figure 8   *Temporal behavior of* ≡Si–O–Si≡ *stretch during water plasma exposure of a silica film.*

4.4   Contact Angle Measurements

It is well known that the molecular species exposed at a surface can often be determined by measuring the contact angle of water, since this is a way to quantify the relative hydrophilicity of the surface. We measured the contact angles of water on PECVD silica and water plasma treated silica. Bare silica showed a high contact (58 ± 2°), while the contact angle decreased to 45 ± 1° after the surfaces were water plasma treated. In others words, water plasma treatment increased the wettability of the surfaces, since the number of hydrophilic hydroxyl groups increased relative to the hydrophobic siloxane groups.

5   APPLICATIONS

Silica can be used for creating biocompatible materials,[5] and as a replacement for mica as the substrate surface for measuring atomic and molecular interactions with the AFM and surface force apparatus (SFA).[6,14,15] Synthetic amorphous silica produced by PECVD can be deposited on optically polished silica disks. These disks are used as sample holders in the SFA technique.[17] The sample preparation method for using the SFA is complicated because it requires atomically smooth and very clean substrates. Mica has generally been used for this application because it is very smooth, inert, and provides particle clean surfaces for SFA experiments.[16] However, mica has some disadvantages. It should be replaced after each experiment because it deteriorates. In addition, its manipulation is tedious. The mica is also chemically inert and cannot be easily modified. We chose to replace mica with silica, because it has surface properties similar to mica. Furthermore, silica may be used for several experiments before it needs to be replaced, and silica films can be chemically modified to react with other molecules to resist protein adhesion and biological attack.[3,6,7] The studies of PEG grafted on silica films will be ready for publication shortly.

6 CONCLUSIONS

Physical and chemical changes are produced by treating plasma deposited silica films to water plasma. We have determined the optimal water plasma conditions to obtain the lowest surface RMS roughness values. The mechanism of this phenomenon is still unknown, but it can be concluded that the smoothing mechanism is not by etching, since removal of material has not been observed (*i.e.* there is no substantial change of thickness nor absorbance as a result of plasma treatment). We have found that exposing silica films for approximately 15 min to water plasma at low rf power (10 W) and low water vapor pressure (20 mTorr) produced smooth and hydroxylated silica films as shown by AFM and ATR-FTIR, respectively. Moreover, at high rf power or high water vapor pressure plasma conditions, the films were not as smooth as they were using the optimal water plasma conditions.

It was also determined that the wettability of the surfaces improved after they had been treated with water plasma. All of those facts about water plasma silica films are very important not only for their use in preparing biocompatible films, but also for using them in the surface forces apparatus (SFA), which has been limited to using mica-substrate surfaces.

7 ACKNOWLEDGMENTS

This work was supported by the MRL Program of the National Science Foundation under Award No. DMR-9123048. Authors would also like to acknowledge funding from an NSF-NYI award to E.S.A. (ECS-94-57758) and U.C. Biostar Program (97 21).

References

1. E. F. Vansant, P. Van Der Voort and K. C. Vrancken, 'Characterization and Chemical Modification of the Silica Surface', Elsevier Science, The Netherlands, 1995, Chapter 3, p. 59.

2. E. F. Vansant, P. Van Der Voort and K. C. Vrancken, 'Characterization and Chemical Modification of the Silica Surface', Elsevier Science, The Netherlands, 1995, Chapter 4, p. 81.

3. L. T. Zhuravlev, *Langmuir*, 1987, **3**, 316.

4. L. T. Zhuravlev, *Coll. Surf. A*, 1993, **74**, 71.

5. R. K. Iler, 'The Chemistry of Silica', John Wiley & Sons, New York, 1979, Chapter 6, p. 704.

6. N. A. Alcantar, J. N. Israelachvili, E. S. Aydil, "A Simple, Efficient and Flexible Method to Create Smooth PEG-coated Biocompatible Surfaces" (in preparation), 1998.

7. 'Poly(Ethylene Glycol) Chemistry, Biotechnical and Biomedical Applications', J. M. Harris, ed., Plenum Press, New York, 1992, Chapter 1, p. 1.

8. S. M. Han and E. S. Aydil, *J. Vac. Sci. Technol. A.*, 1996, **14**, 2062.

9. R. K. Iler, 'The Chemistry of Silica', John Wiley & Sons, New York, 1979, Chapter 7, p. 730.

10. S. C. Deshmukh and E. S. Aydil, *J. Vac. Sci. Technol. A.*, 1995, **13**, 2355.

11. E. S. Aydil, R. A. Gottscho, and Y. J. Chaubal, *Pure and Appl. Chem.*, 1994, **66**, 1381.
12. N. J. Harrick, 'Internal Reflection Spectroscopy', John Wiley & Sons, New York, 1967.
13. S. M. Han and E. S. Aydil, *Thin Solid Films*, 1996, **290**, 427.
14. J. N. Israelachvili, G. E. Adams, *J. Chem. Soc., Faraday Trans. I*, 1978, **74**, 975.
15. J. N. Israelachvili, *Proc. Natl. Acad. Sci. USA*, 1987, **84**, 4722.
16. J. N. Israelachvili, P. M. McGuiggan, *J. Mater. Res.*, 1990, **5**, 2223.
17. M. Ruths, 'Time-Dependent Interactions In Polymer And Liquid Crystal Systems', PhD Dissertation, University of California, Santa Barbara, April 1996.
18. D. Lin-Vien, N. B. Colthup, W. G. Fateley and J. G. Grasselli, 'The Handbook of Infrared and Raman Characteristic Frequencies of Organic Molecules', Academic Press, San Diego, 1991, Chapter 15, p. 258.
19. J. B. Lambert, H. F. Shurvell, D. Lightner and R. G. Cooks, 'Introduction to Organic Spectroscopy', Macmillan, New York, 1987, Chapter 7, p. 175.

SURFACE MODIFICATION OF EXTENDED WEAR CONTACT LENSES BY PLASMA-INDUCED POLYMERIZATION OF VINYL MONOMERS

Peter Chabrecek* and Dieter Lohmann

NOVARTIS-CIBA Vision
Advanced Research Unit
CH-4002 Basel, Switzerland

1 ABSTRACT

A new glow discharge downstream plasma process has been developed which enables deposition of a hydrophilic coating on siloxane hydrogel contact lenses showing the required properties. Argon is used as plasma gas and as carrier gas for the "afterglow-feed" of vinyl monomers. Radical polymerization of the vinyl monomers has been identified as the predominant reaction forming surface grafts of true vinyl polymer structure. The heavy fragmentation of organic species usually occurring under conventional plasma conditions can be largely avoided. The new process offers numerous opportunities for variations of coating composition, thickness, topography and degree of crosslinking as exemplified for poly-N-vinyl-2-pyrrolidone (PVP) grafts by XPS, ATR FTIR, AFM, dynamic contact angle and permeability measurements. PVP-coated contact lenses showed excellent permeability characteristics, stability, biocompatibility and on-eye performance.

2 INTRODUCTION

The development of novel soft contact lenses which are suitable for extended overnight wear is among the major goals in current contact lens research. In particular, continuous overnight lens wear for a period of 30 days (true extended wear = TEW) has recently become very attractive as an alternative to the use of daily wear disposable lenses (DWD). Only very few soft lens materials meet the extreme requirements of TEW applications with regard to gas permeability, wettability, biocompatibility and resistance to formation of irreversible deposits from tear components.

A hybrid-type soft contact lens composed of a highly gas permeable core and a highly wettable surface coating represents so far the most promising concept. Table 1 presents the basic requirements for hybrid-type TEW contact lenses.

Siloxane and perfluoropolyether-based materials, combined with hydrophilic segments, possess the high oxygen as well as ion (water) permeability required for extended wear contact lenses, but their surfaces are of pronounced hydrophobicity. Contact lenses made from these materials require modification of their surface properties to provide sufficient hydrophilicity, biocompatibility and ocular comfort.

Many surface treatment techniques for polymeric materials are known in the art.[1-2] Surface treatment based on a variety of gas plasma reactions has become an established

Table 1 *Basic Requirements for Hybrid-type Extended Wear Contact Lens.*

Lens Core	Lens Surface
Oxygen Permeability: > 80 Barrers*	Unhindered Oxygen Permeation
Water Uptake: > 20 % w/w	Contact Angle: < 60 °
Iondiff.Coeff.: > 1.5 x 10^{-6} mm^2/min.	Unaffected Ion Permeation
Elongation at Break: > 50 %	Coating Thickness: 20–100 nm
Light Transmissibility: > 90 %	Low Bacterial Adhesion
Shore A hardness: 25–35	High Deposition Resistance
E-modulus: 0.6–1.5 Mpa	Tear-Film Break Up Time: > 8 sec
	Autoclavability

$1\ Barrer = 10^{-11}\ cm^2 \bullet mL\ O_2/sec \bullet mL \bullet mmHg$

technology in the field of biomedical materials.[3,4] Among the most attractive features which made plasma processing the favored surface treatment technique in this area are:

- conformal, pinhole free films
- diverse surface chemistries
- firm adhesion and stability
- sterile, low levels of leachables
- applicable to a variety of substrates
- relatively easy to make
- mass production possible

In the contact lens market very few plasma-coated lens products have been developed and commercialized so far.[5–7] The reason for this may be that the majority of conventional plasma processes utilize the ionization and fragmentation of gases or evaporated organic molecules to form highly crosslinked polymeric deposit layers of unknown chemical composition, often having certain barrier effects for gases and liquids. The absence of a viable theory for predicting polymer structures represents the most significant handicap of this technique. Thus, one aim of our work was the development of a mild plasma process which allows the deposition of customized hydrophilic coatings of controlled polymer structure from diverse vinyl monomers, above all from *N*-vinyl-2-pyrrolidone. In addition, we probed whether poly-*N*-vinyl-2-pyrrolidone coatings, generated by this process, can provide suitable long-term wettability and biocompatibility to TEW contact lenses without affecting their superior permeability characteristics.

3 EXPERIMENTAL

3.1 Materials

The following polydimethylsiloxane/hydrogel extended wear contact lens materials, containing various contents of water, were used for surface modification: Alsacon 21 (AS 21), Glycon 17 (GL 17), Glycon 20 (GL 20), Betacon 23 (BT 23), Betacon 30 (BT 30), and Oxacon 15 (OTB 15).[8] The numbers refer to the actual water content (% w/w) of the various materials. Vinyl monomers used for plasma polymerization were as follows: *N*-vinyl-2-pyrrolidone (NVP), methyl methacrylate (MMA), *N,N*-dimethyl acrylamide (DMA), and 2-hydroxyethyl methacrylate (HEMA).

3.2 Plasma Experiments

Experiments on plasma-induced polymerization of the monomers were performed using a special type of plasma reactor for down-stream (often called after-glow) plasma processes (Figure 1b). The plasma reactor was developed in our laboratory in collaboration with ACR GmbH Niedereschach (Germany) in 1994. The reactor chamber is capacitively coupled to the RF power supply which operates at a fixed frequency of 27.13 MHz. In this reactor, the contact lens samples are placed outside the plasma discharge zone at variable distances. Discharge in the reactor is maintained by a non-polymerizing gas (Argon) fed directly into the electrode gap. Excited particles of plasma gas can escape from the discharge zone and diffuse to the deposition zone where a monomer gas is fed into the reactor and the lenses are located. Table 2 shows the basic plasma parameters used for down-stream plasma polymerization of NVP on contact lenses used.

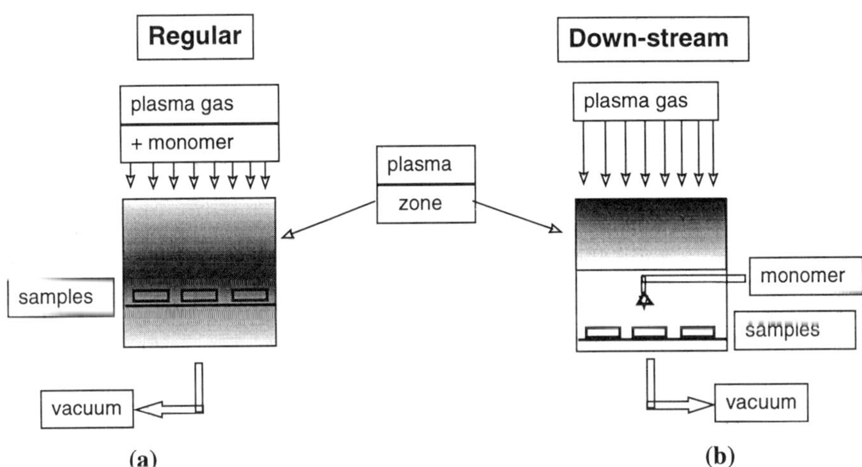

Figure 1 *Schematic representation of: (a) regular vs. (b) down-stream plasma conditions.*

Almost the same parameters were used for down-stream plasma polymerization of other monomers. Only the temperature of the monomer source was adjusted in order to keep the evaporation rate on an optimum level (see Tables 3 and 4).

3.3 Characterization Methods

Advancing and receding water contact angles of coated and non-coated lenses were determined with the dynamic Wilhelmy plate method using a Krüss K12 instrument (Krüss GmbH, Hamburg, Germany).

The water film break up time (WFBUT) on the lens surfaces was estimated by measuring the time interval from withdrawing the lenses from deionized water until observation of the first spot of rupture of the continuous water film.

Chemical structures of the deposited layers were determined by ATR-FTIR spectroscopy. A Bruker IFS-55 FTIR Spectrometer equipped with a liquid nitrogen cooled medium-bandpass mercury–cadmium–telluride (MCT) detector was used. A

Table 2 *Plasma Parameters for Down-stream Plasma Polymerization of NVP Studied in Most Detail.*

Ar flow into plasma zone	10 sccm
Ar flow through NVP source	10 sccm
Temperature of NVP source unit	40 °C
Pressure in plasma reactor	0.35 mbar
Plasma input power	45–150 W
On / off pulsed plasma frequency	1 μsec/3 μsec
Distance of lenses from plasma zone	10–20 cm
Time of plasma polymerization	5 minutes

germanium trapezoid crystal and an endface angle of 45° obtainable from Graseby Specac NIC was employed as an internal reflection element (IRE).

The oxygen permeability (Dk) of the lenses was determined at 35 °C in a wet cell using a Dk 1000 instrument from Applied Design and Development Co., Norcross, GA, USA. Thickness of the lenses in the areas being exposed for testing was determined by measuring with a Mitotoya micrometer VL-50.[8]

The ion permeability of the lenses was measured using the Ionoflux Technique.[8] This technique involves the use of a conductometer LF 2000/C with an electrode equipped with a temperature sensor (Wissenschaftlich-Technische Werkstätten GmbH, Germany). The donor chamber contained a salt solution, the receiving chamber contained 60 mL deionized water and a stir bar. The system was thermostated at 35 °C.

Topography and the thickness of the coatings were estimated by Scanning Force Microscopy (SFM) performed with a Nanoscope III (Digital Instruments, Santa Barbara, CA, USA) in Tapping Mode.

The adhesion and the cohesive integrity of the coatings were tested by subjecting coated lenses to autoclaving under standard sterilization conditions (121 °C, 30 minutes) and repeating the physico-chemical analyses.

Pre-clinical evaluation of the PVP-coated TEW lenses was accomplished by cytotoxicity testing using cell cultures and by assessment of overall on-eye performance using animal models (cats) first. The PVP-coated TEW contact lenses were finally clinically evaluated on eyes of human volunteers. The lenses were assessed in terms of their tear film break-up time (TFBUT), on-eye mobility, comfort, corneal staining as well as protein and lipid deposition.

4 RESULTS AND DISCUSSION

4.1 Plasma Polymerization

A conventional plasma process (see Figure 1a) is usually performed by evacuating a reactor and refilling with a low pressure gas. The gas is excited by electrical energy, usually in the radiofrequency (RF) range, such as 13.56 MHz. When the gas used to generate the plasma is polymerizable, molecules may polymerize either via the double bond or by formation of reactive species through fragmentation. Opening of a C=C bond requires less energy than dissociation of any C–C bond. The energy transferred to monomer molecules (primarily by inelastic electron impact) in glow discharge is so high that a polymerization by fragmentation becomes more important. If the reaction parameters are properly controlled, the reactive species combine to form a thin layer of a highly cross-linked solid deposit called plasma polymer.

Secondary effects of the plasma on the deposited layer (impact of electrons, ions, radicals, metastable species, and high-energy photons) lead to ablation and structural modification of the primary plasma polymer. Hence, only partially eroded cross-linked polymer layers, generated from highly fragmented species of used monomers, are obtained. The high cross-linking density in these conventional plasma polymers makes them quite brittle, which often precludes the possibility of their use as flexible coatings on soft, elastic substrates. In addition, the deposited organic films are of unpredictable structure and have complex, irregular compositions which usually deviate from that of the used monomers.

In contrast to the above mentioned conventional plasma polymerization, down-stream plasma polymerization (often called plasma-induced polymerization) is a process in which regular polymerization occurs in the presence of the plasma, but wherein the substrate (lenses) as well as the inlet for the monomer feed are located outside of the plasma zone (see Figure 1b). Radicals needed to initiate the polymerization process are generated in the plasma zone and transported into the polymerization zone below, where they interact with monomer molecules and lens surfaces. The collisions of radicals with monomer molecules lead to the formation of monomer radicals, their recombination and polymerization on lens surfaces.

The most important plasma parameters are the plasma power and the distance of the samples from the plasma zone. The optimal distance for NVP plasma-induced polymerization was found to be 10 cm. The plasma power should not exceed 150 W to avoid an erosion of sample surfaces. An increase of plasma power and the shortening of the distance between the plasma zone and the position of lenses increase the possibility of covalent binding and crosslinking of polymeric chains on the lens surfaces. As indicated from ATR FTIR spectra, a further increase of the plasma power or decreases in lens–plasma distance cause some structural modifications of the deposited PVP layers.

Table 3 summarizes some of the plasma parameters used for the down-stream plasma polymerization of NVP as well as water/air contact angle data determined on modified TEW contact lenses. Table 4 shows the most important plasma parameters and the resulting contact angles of DMA, HEMA and MMA down-stream plasma-modified TEW contact lenses. Because of their low vapor pressures at room temperature, the monomers used for polymerization had to be heated, except of MMA which was cooled to –60 °C.

The surfaces of all polymeric materials are modified very substantially when hydrophilic monomers are used for plasma-induced polymerization. The contact angle values decrease considerably, especially when DMA or NVP are used. The contact angles of all materials modified with HEMA were in the range of contact angles of poly-HEMA lenses.

Plasma-induced polymerization of methyl methacrylate turned out to be particularly difficult because of the very low boiling point of MMA. Although the deposit layers of PMMA on the above TEW lenses were very thin (approximately 10 nm) the ATR FTIR spectra clearly showed poly-MMA absorptions.

The excellent wettability and known biocompatibility of PVP coatings prompted us to optimize this process. The standard operating procedure for down-stream plasma-induced polymerization of NVP was established utilizing the plasma parameters described in Table 2 with plasma power 150 W and distance between lenses and plasma zone 10 cm. Using this procedure more TEW contact lenses were modified and characterized in more detail.

Table 3   *Down-Stream Plasma Parameters Used for PVP Grafting on Various Contact Lens Surfaces and Contact Angle Characteristics of Modified Lenses.**

Lens Cores/ Coatings	Plasma Power (W)	Distance from plasma zone (cm)	Contact angles Adv. / Rec. / Hyst.* (°)
AS 21	150	20	64.7 / 51.2 / 13.5
AS 21	150	10	50.3 / 45.2 / 5.1
GL 20	100	10	62.3 / 46.2 / 16.1
GL 17	150	10	58.8 / 40.4 / 18.4
GL 17	100	15	58.3 / 30.3 / 28.0
GL 17	60	25	60.2 / 21.1 / 39.1

** Contact angles of the unmodified TEW contact lenses are presented in Table 5; Adv. = advancing contact angle; Rec. = receding contact angle; Hyst. = hysteresis.*

Table 4   *Plasma Parameters for Down-Stream Plasma Polymerization of Various Vinyl Monomers on TEW Contact Lens Surfaces and Their Contact Angle Values.*

Material (Lens) Used	Monomer Used	Distance from Plasma Zone (cm)	Temp of Monomer Source (°C)	Plasma Power (W)	Contact Angle (°) Adv./Rec./Hyst.*
AS 21	DMA	15	30	150	61/40/21
BT 30	DMA				59/33/26
GL 17	DMA				57/40/17
GL 20	DMA				62/37/25
AS 21	HEMA	15	30	110	77/41/36
BT 30	HEMA				78/42/36
GL 17	HEMA				77/39/38
BT 23	HEMA				79/48/34
BT 23	MMA	15	−60	150	84/26/58

** Adv. = advancing contact angle; Rec. = receding contact angle; Hyst. = hysteresis.*

4.2 Characterization of PVP Coatings

4.2.1 The Wettability of PVP Coatings. The wettability of coated lenses was measured by means of the dynamic water/air contact angles. This method is one of the simplest techniques capable of directly evaluating the polymer–liquid interfacial free energy. Table 5 shows the water contact angles of PVP-modified lenses using the standard operating procedure described in Table 2 (plasma power 150 W, distance

Table 5 *Dynamic Contact Angles (Wilhelmy) and C.A. Hysteresis (°) of PVP-coated TEW Lenses.* *

Lens Cores / Coatings	Adv.	Rec.	Hyst.	WFBUT
AS 21	97	76	21	
AS 21 / PVP	50	45	5	
TF 40 (P-Hema)	78	33	44	6.5
TF 40 (P-Hema) / PVP	54	28	26	
OTB 15	101	56	45	
OTB 15 / PVP	70	55	15	
GL 17	111	86	25	
GL 17 / PVP	60	41	19	
GL 20	109	98	11	
GL 20 / PVP	70	51	19	
BT 23	92	64	28	
BT 23 / PVP	51	41	10	8.3
BT 30	96	78	18	
BT 30 / PVP	64	53	11	8.6

* *Adv. = advancing contact angle; Rec. = receding contact angle; Hyst. = hysteresis; WFBUT = water film break-up time.*

10 cm). All PVP-modified lenses show considerably lower contact angle values than the corresponding non-modified lenses, indicating the highly hydrophilic character of the coatings. Betacon lenses are *per se* a little more hydrophilic than Glycon lenses. A relatively low value for the contact angle hysteresis on coated lenses confirms the high stability and uniformity of the PVP coatings.

WFBUT on measured lenses is greatly affected by minor defects such as dust particles, which act as nucleation sites for rupture. For uniformly coated contact lenses, however, higher WFBUTs are achieved and a very slow shrinkage of the aqueous layer is observed, as opposed to the rapid and sudden rupture of the water film when an uncoated contact lens is tested. All PVP-coated Betacon lenses show excellent retention of films of water or saline, substantially in excess of poly-HEMA-based contact lenses (see WFBUT for PVP-coated B 23 and B 30 in comparison to TF 40 contact lens in Table 5). Thus, PVP coatings have a superior ability to retain aqueous films on substrates such as contact lenses.

Contact angles and WFBUT values remain unchanged after autoclaving of lenses with PVP coatings fabricated under optimized conditions. Even PVP-coated lenses which were subjected to repeated autoclaving cycles to test the resistance of the coatings to very harsh demands do not show substantial changes in contact angle values.

4.2.2 Chemical Composition and Structure of PVP Coatings. The chemical composition and structure of PVP-deposited films were analysed by ATR FTIR spectroscopy. Figure 2c provides a typical ATR FTIR spectrum of plasma-induced PVP on Betacon 23 TEW contact lenses. For comparison, the spectra of a linear PVP, polymerized by conventional radical polymerization, and of a plasma-polymerized PVP (NVP polymerized in plasma zone) obtained by Marchant[9] are given.

The ATR FTIR spectrum of the plasma-induced PVP coating (Figure 2c) obtained by the down-stream process described is very similar to the spectrum of a regular poly-NVP prepared by a conventional radical polymerization technique (Figure 2b). As the monomer does not pass through the zone of the highly reactive plasma gas, fragmentation of the monomer molecules can be largely avoided. With this process the structure of the polymer deposit can be controlled within a certain limit. Undesired surface erosion of susceptible substrates can be avoided and the formation of the polymer deposit is predominantly based on radical chain reactions. The polymerization mainly proceeds by opening the double bonds. The polymer chains, while exhibiting controlled crosslinking, are composed to a large extent of repeating units which are identical to the structure of the repeating units obtained by non-plasma radical polymerization of the polymerizable unsaturated compounds.

In comparison to the spectrum of PVP generated by Marchant (Figure 2a), the spectrum of the down-stream plasma-polymerized PVP does not show broadening and loss of resolution in the 1500–1000 cm^{-1} region that would be indicative of branching or ring opening reaction products. Not even the weak peak at 2150 cm^{-1} assigned to nitrile groups is present. Our ATR FTIR spectrum shows that plasma-induced PVP polymer contains a high content of unchanged poly-NVP chains. The main absorption peaks are: a strong C=O stretching absorption band at 1668 cm^{-1}, an absorption peak due to C–H stretching in the CH_2–C=O group (1423 cm^{-1}) and a strong C–N stretching absorption peak at 1286 cm^{-1}.

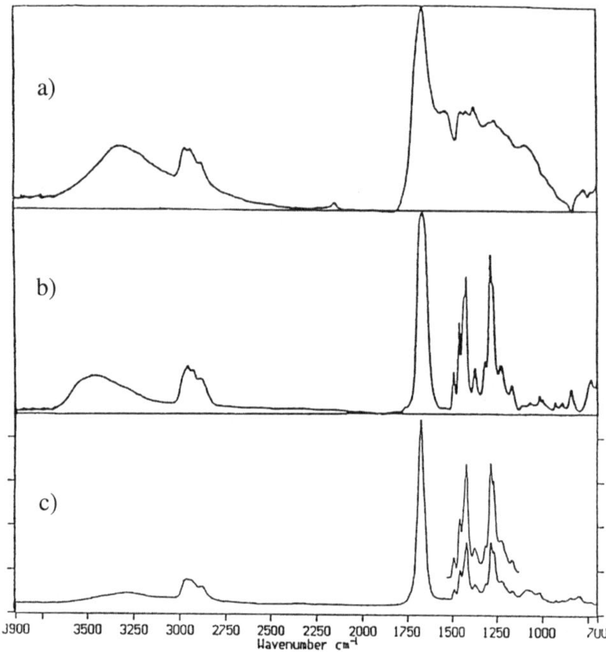

Figure 2   *ATR FTIR spectra from PVP films obtained the by: (a) in-glow plasma polymerization process, (b) conventional radical polymerization process, and (c) down-stream plasma-induced polymerization process.*

4.2.3 Ion Permeation.   It has been found that on-eye lens mobility is crucial for extended overnight wear of contact lenses.[10] Water permeability of the lenses indicated by ion permeability has been identified as one of the factors which control the on-eye mobility of a lens. Ion permeation measurements verify the idea of a critical ion diffusion rate required for on-eye mobility of contact lenses. Ion permeability of a contact lens depends to a large extent on the water content and the morphology of the bulk materials. From this point of view it was very interesting to evaluate the influence of the PVP coatings on the ion permeability of unmodified Alsacon, Betacon, Glycon and Oxacon lenses. The obtained values of relative diffusion constants before and after PVP modification (PVP coating thickness 60 nm) are illustrated in Figure 3.

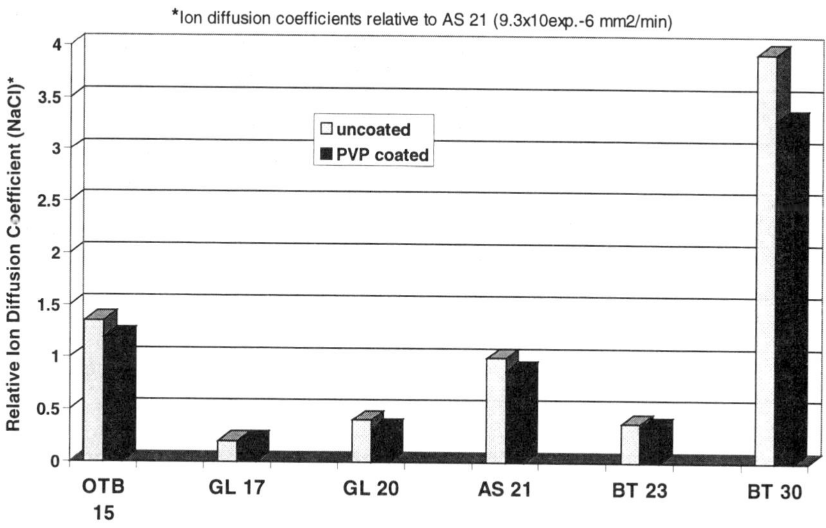

Figure 3 *Relative ion diffusivity of PVP-coated vs. non-coated TEW contact lenses.*

Ion permeabilities were lowered by the presence of a water-containing PVP-hydrogel layer on the surface of silicone-hydrogel lenses. Ion permeation was decreased by about 10 % after PVP-coating on almost all lenses in our plasma reactor. The extent of the decrease was larger for materials with high ion permeability than for the samples containing less water. The influence of the coating thickness has not yet been fully evaluated.

4.2.4 Oxygen Permeation. The influence of PVP-coatings on oxygen permeability of the same lenses is illustrated in Figure 4. As in the ion permeability study, the PVP-coating reduced this property – oxygen permeability – by about 10 %. Again, a more pronounced reduction was observed for materials with generally higher oxygen permeability than for the less oxygen-permeable materials. We can conclude from these the data that the influence of the PVP coating on the oxygen permeability is to some extent also dependent on the nature of the bulk material, which becomes particularly obvious with Glycon (GL) materials.

4.2.5 Stability Studies of PVP Coatings. The stability of the PVP coatings was studied by ATR FTIR spectroscopy before and after autoclaving of the lenses in phosphate buffered saline (PBS) at 121 °C for 30 minutes. While the characteristic peak intensity of weakly grafted PVP chains (1668 cm^{-1}) decreased considerably after autoclaving of the lenses, the peak intensity of strongly grafted PVP, generated by standard plasma operation procedure, remained practically unchanged after autoclaving (Figure 5).

4.2.6 Thickness and Topography. The thickness and topography of the plasma-induced PVP layers were determined by scaning force microscopy (SFM). One half of

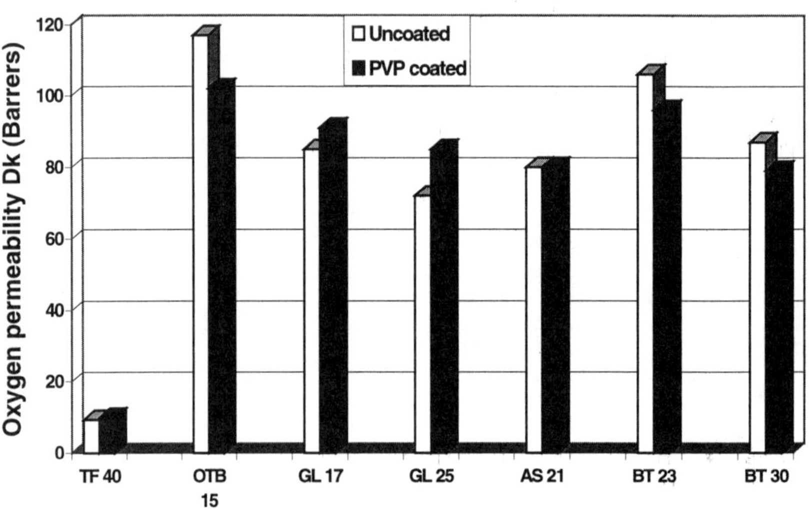

Figure 4 *Oxygen permeability of uncoated and PVP-coated TEW contact lenses.*

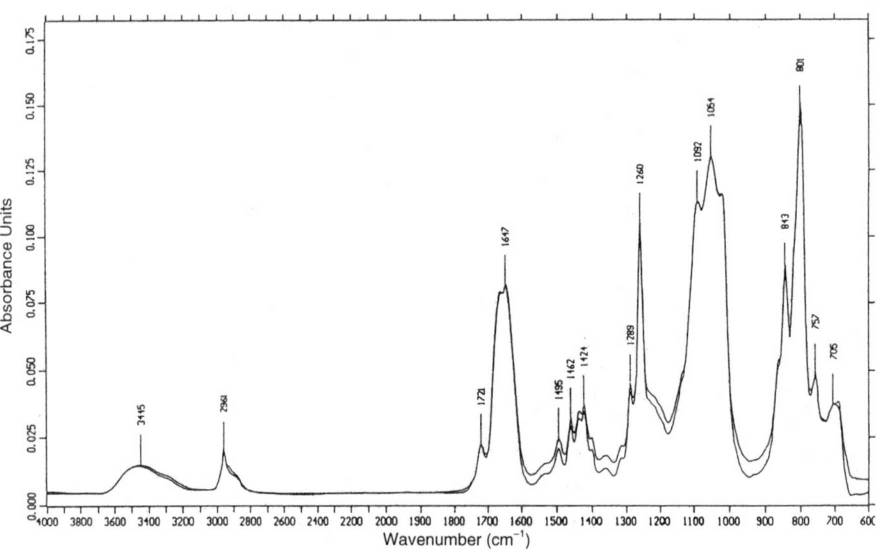

Figure 5   *ATR FTIR spectra of autoclaved and non-autoclaved down-stream plasma PVP-coated Alsacon 21 lenses.*

each type of lens was plasma-modified, whereas the second half was masked by covering it with a Teflon tape. After removing the tape, AFM measurements confirmed the half-coated status of all samples in both the dry and wet state of lenses by identifying the step between modified and non-modified halves of lenses (Figure 6b and 6d). The average thickness of the PVP layers determined on Betacon 30 lenses was found to be between 60 and 90 nm (Figure 6d). In some cases, if the generated layer was very thin, the presence of the coating was detectable only by a certain increase of surface roughness on the coated part. The minimum thickness of these coatings on lenses can be estimated at just below 50 nm.

SFM on dry and on wet coatings reveals the brain-type microstructures of the down-stream plasma-polymerized coatings. Figure 6c presents the surface topography of the PVP-modified Alsacon lens. Coatings from other monomers like DMA or HEMA showed similar topographies.

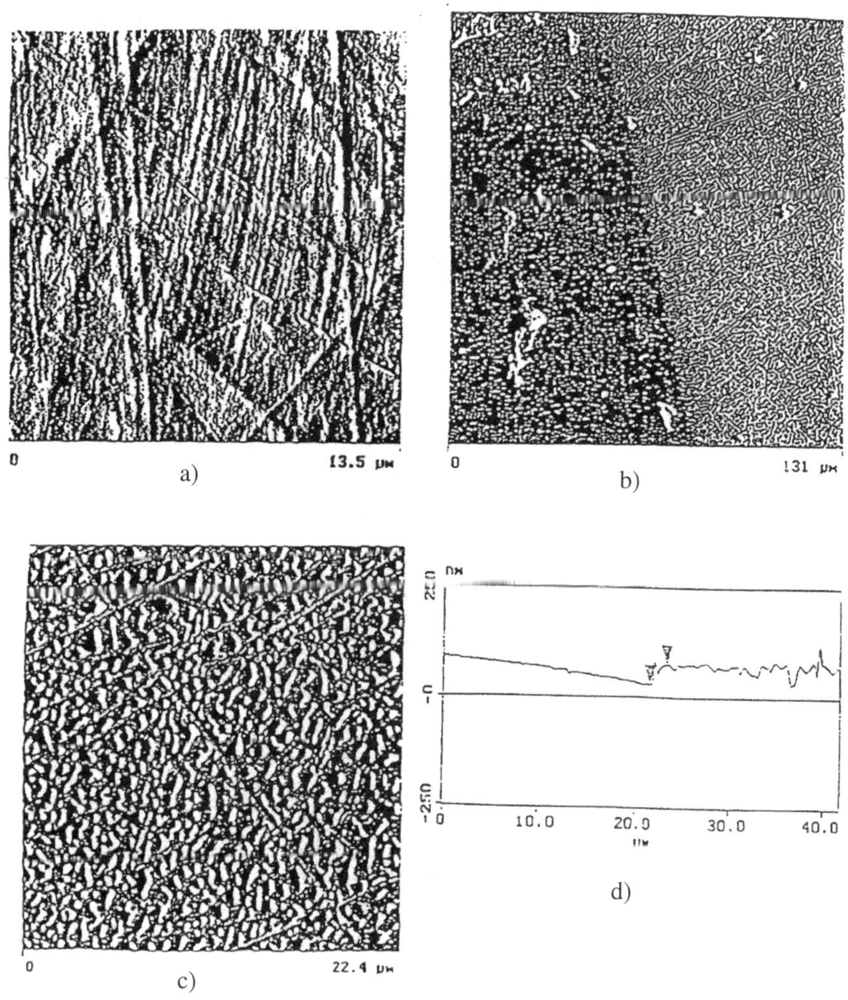

Figure 6 *SFM images of TEW Alsacon lenses: (a) non-modified, (b) half-coated with plasma-induced PVP, (c) whole coated with plasma-induced PVP, and (d) step between modified and non-modified areas.*

4.2.7 Clinical Evaluations. Clinical evaluations, performed at CRCERT Sydney, Australia, confirmed the biocompatible character of PVP coatings on all modified lenses. All lenses have shown favorable on-eye wettability and staining characteristics. As opposed to some coatings generated within a plasma zone, these coatings have not affected the essential on-eye mobility of the lenses. The coatings exhibited good permeability for oxygen, carbon dioxide, water and ions, had high wettability, stability towards tear fluid and resistance towards protein, mucin and lipid deposition as well as excellent comfort for patients during overnight extended wear. Detailed results will be published separately in the near future.

5 CONCLUSIONS

Down-stream plasma-grafted polymer coatings show a high potential with regard to applications on contact lenses, in particular for true extended wear applications. As exemplified for poly-*N*-vinyl-2-pyrrolidone the new down-stream plasma-polymerized coatings show firm adherence, good permeability, appropriate thermo-/hydrolytic stability and favorable biocompatibility properties. It is of particular importance for TEW applications that the new PVP-coated silicone hydrogel lenses maintain their favorable permeability characteristics and the essential on-eye mobility.

6 ACKNOWLEDGEMENTS

The authors are grateful to Paul Nicolson and Lynn Winterton of Ciba Vision Atlanta for their advice and multiple scientific communications. They also thank Miriam Bellus of Novartis Basel for the ATR FTIR investigations, Sergei Seiko of the University of Ulm, Germany, for the SFM measurements, and Jens Hoepken, Angelika Domschke, Thomas Hirt and Otto Kaiser of Ciba Vision Novartis Basel for their assistance and many valuable contributions to the project.

References

1. A. S. Hoffman, *Macromol. Symp.*, 1996, **101**, 443.
2. J. Jagur-Grodzinski, *Prog. Polym. Sci.*, 1992, **17**, 361.
3. N. Dilsiz and G. Akovali, *Polymer*, 1996, **37**, 333.
4. H. Yasuda, 'Plasma Polymerization', Academic, New York, 1985.
5. U.S. Pat. 4,143,949 (Mar. 13, 1979), Bausch & Lomb.
6. U.S. Pat. 4,312,575 (Jan. 26, 1982), H. K. Yasuda *et al.*
7. U.S. Pat. 4,409,258 (Nov. 10, 1983), Alcon Ltd.
8. U.S. Pat. 5,760,100 (Dec. 8, 1995), CIBA Vision Corp.
9. R. E. Marchant, D. Yu and C. Khoo, *J. Polym. Sci., Polym. Chem. Ed.*, 1989, **27**, 881.
10. A. Domschke, D. Lohmann, R. Baron, *Polym. Prepr.*, (ACS, Div. PMSE), 1997, **76**, 42.

SURFACE ORGANOMETALLIC CHEMISTRY ON METALS. CHEMICAL MODIFICATION OF PLATINUM CATALYST SURFACE BY REACTION WITH TETRABUTYLTIN; APPLICATION TO THE SELECTIVE DEHYDROGENATION OF ISOBUTANE TO ISOBUTENE

F.Z. Bentahar, F. Bayard, J.P. Candy* and J.M. Basset
LCOMS-CPE, 43 bd du 11 Novembre 1918, F-69616 Villeurbanne, France

B. Didillon
IFP, 1 et 4 av. de Bois Préau, F-92506 Rueil-Malmaison, France

1 INTRODUCTION

Supported Pt-based catalysts have been thoroughly investigated due to their tremendous importance in the petroleum and petrochemical industries. The conversion of hydrocarbons on supported platinum–tin bimetallic catalysts has been the subject of numerous studies.[1–13] Compared with catalysts containing only platinum, the bimetallic systems exhibit enhanced selectivity and stability. Two main effects (geometric and electronic) have been proposed in the literature to explain the catalytic behavior of these catalysts. The geometric effect states that there is a decrease in the number of Pt atoms in the ensemble constituting an active site.[14] The electronic effect proposes that there is a decrease in the strength of the bond between the adsorbed molecules and the surface metal atoms, caused by a change in the electronic properties of platinum.[15–17] Increasingly, authors describe experiments where the results can be explained by a geometric effect in which the number of contiguous Pt atoms is decreased by dilution with Sn atoms, and any electronic effects appear to play a minor role.[4,13,14,18,19] In such bimetallic Pt–Sn catalysts, the localization of tin atoms on the active surface became the key to catalytic performance. Kappenstein *et al.* compared PtSn/Al$_2$O$_3$ catalysts prepared by the classical co-impregnation procedure or via a [Pt(NH$_3$)$_4$][SnCl$_6$] complex precursor.[18] They found that the specific rate of dehydrogenation of *n*-butane into olefin at 420 °C is higher on the latter catalyst. They conclude that there are fewer contiguous Pt atoms on the sample prepared via the PtSn complex precursor. More recently, Stagg *et al.* prepared bimetallic PtSn/SiO$_2$ catalysts by various methods and measured their catalytic activity and selectivity for isobutane dehydrogenation at 500 °C.[13] They found that the best catalyst is prepared by impregnation with the complex PtCl$_2$(SnCl$_3$)$_2{}^{2-}$. These results demonstrate the prominent role of preparation procedure of bimetallic catalysts.

Among the various methods already used to add tin atoms, the surface organometallic chemistry on metal-route could be the most precise.[20] Previous studies have demonstrated that tetra *n*-butyl tin reacts selectively at 50 °C under hydrogen on reduced surface platinum atoms, leading to butane evolution and formation of a grafted organometallic species, Pt$_s$[Sn(*n*-C$_4$H$_9$)$_x$]$_y$.[21] This surface species undergoes complete hydrogenolysis at *ca.* 300 °C with formation of tin 'adatoms'. The formation of a Pt–Sn surface alloy can be accomplished by further treatment at 500 °C under hydrogen. It has been suggested that after treatment at 550 °C, tin atoms could be localized on the low coordinated surface atoms, *i.e.* edges and corners, in the case of silica-supported rhodium.[22] Notably, no studies have been done to describe the influence of metallic particle size on the initial surface reaction. This report describes the kinetics of the reaction between $Sn(n\text{-}C_4H_9)_4$ and various Pt/Al_2O_3 catalysts with increasing particle size.

2 EXPERIMENTAL

In this study, $\gamma c\text{-}Al_2O_3$ with a surface area of 215 m^2/g was used as support material. Pt/Al_2O_3 catalysts were prepared by dry impregnation using a toluene solution of platinum acetylacetonate, following the procedure already described.[23]

The dispersion (number of surface platinum atoms/total platinum atoms) of the sample was determined by chemisorption of hydrogen, oxygen and CO at 25 °C in a volumetric apparatus already described.[24] The stoichiometry of hydrogen, oxygen and CO adsorption (number of adsorbed hydrogen, oxygen atoms or CO molecules per surface platinum atom) are assumed to be 1.8 H/Pt_s, 1.0 O/Pt_s and 1.0 CO/Pt_s.[25,26] Prior to the adsorption measurement, the samples were reduced under flowing hydrogen at 350 °C for 3 h and then evacuated at the same temperature under vacuum (10^{-6} mbar) for 6 h.

The metal loading of mono and bimetallic samples was measured by elemental analysis. The metallic particle size was determined by electron microscopy (JEOL 100 CX). The average metallic particle size was correlated to dispersion, assuming that the metallic particles have a cubo-octahedral shape.[27]

The reaction between the *n*-heptane solution of $Sn(n\text{-}C_4H_9)_4$ and the monometallic catalysts was performed at room temperature in a closed glass reactor in the presence of one atmosphere of hydrogen following the procedure already described for supported rhodium catalysts.[28] Prior to the reaction, the desired amount of catalyst (2 g) was reduced at 450 °C under flowing dry hydrogen for 3 h. After cooling down to room temperature under hydrogen, the reduced catalyst was introduced under hydrogen into a Schlenk tube. Freshly distilled *n*-heptane (20 mL) and known amounts of tetradecane and isobutane (internal standards) were introduced in the Schlenk tube which was then closed under 1 atm of hydrogen. After 30 min of stirring, the desired amount of $Sn(n\text{-}C_4H_9)_4$ was added. The variation of $Sn(n\text{-}C_4H_9)_4$ concentration and the amount of butane evolved were then followed by gas chromatographic analysis of the liquid and gas phase respectively, carried out after increasing times, t, of reaction.

It has been verified that isobutane is never formed during the reaction of $Sn(n\text{-}C_4H_9)_4$ with alumina-supported platinum (the only product is butane) and that the equilibrium of isobutane and butane in both the gas and liquid phase is rapid. In this study, hydrogen used for the reduction of samples and for reaction in the closed reactor was fully deoxygenated and dehydrated by flowing through deoxo and zeolite traps. The *n*-heptane used as solvent was freshly distilled and kept under argon on pretreated molecular sieve, in order to avoid any trace amounts of water.

After reaction with $Sn(n\text{-}C_4H_9)_4$ at 25°C in *n*-heptane solution, the samples were extracted from solution by filtering. Excess $Sn(n\text{-}C_4H_9)_4$ was removed by washing with pure *n*-heptane. The samples were then treated under flowing hydrogen at increasing temperature (1 °C/min) up to 550 °C prior to the isobutane dehydrogenation reaction. Isobutane dehydrogenation was performed in a dynamic reactor working under atmospheric pressure at 550°C with a isobutane/hydrogen ratio of 1/1 (mol/mol). The conditions of the reaction have been described.[29]

If *m* is the amount of catalyst introduced (g), *Pt* is the platinum loading of the sample (w/w) (%), $f(iC_4^i)$ is the flow rate of isobutane (mol/s), and [A] is the sum of the products expressed as equivalent C_4, one can define:

- $[A] = [iC_4^=] + [nC_4] + [nC_4^=] + 3/4([C_3]+[C_3^=]) + 1/2([C_2]+[C_2^=]) + 1/4[C_1]$

- Conversion: $Conv.(\%) = 100 \cdot [A]/([A]+[iC_4])$

- Selectivity for isobutene: $Sel.(\%) = 100 \cdot [iC_4^=]/[A]$

- Activity: $r\ (mol/g/s) = Conv.f(iC_4^i)/m/Pt$

The flow rate of isobutane+hydrogen and the amount of catalyst were adjusted (in the range 40 to 200 mL/min and 20 to 50 mg) to obtain isobutane conversions of 5 to 10%.

Molecular modeling was performed using the 'SYBYL' software package from Tripos Associates,[30] running on an 'Indigo' computer from Silicon Graphics. We used the Tripos force field modified according to the Horner[31] determination for Sn–C distances.

3 RESULTS

3.1 Characterization of the Catalysts

In order to estimate the effect of metallic particle size on the kinetics and the stoichiometry of the surface reaction between the organotin compound and the platinum surface, various alumina-supported platinum catalysts were prepared and characterized. The amount of hydrogen, oxygen and CO adsorbed at 25 °C under 150, 50 and 30 mbar respectively on the various samples, and the corresponding dispersions (Pt/Pt_s) are reported in Table 1.

Table 1 *Metal Loading, Amount of H_2, O_2 and CO Adsorbed and Average Metallic Particle Diameter for the Two Alumina-supported Catalysts. For Each Value, the Corresponding Average Dispersion is Given in Parentheses. Disp. (Av.) is the Average Value of Dispersion from all of the Obtained Values.*

Sample	Pt (w) %	H₂ ads. Mmol/g	O₂ ads. Mmol/g	CO ads. Mmol/g	E.M. Nm	Disp.(av.)
Pt/Al-1	0.6	23 (0.82)	13 (0.84)	30 (0.97)	1.0 (0.90)	0.88
Pt/Al-2	3.2	52 (0.35)	36 (0.42)	74 (0.45)	2.3 (0.40)	0.41

Fairly good agreement between the three adsorbates is obtained. The average metallic particle size measured by electron microscopy and the corresponding dispersion of the samples, assuming a cubo-octahedral shape for the metallic particles, are reported in Table 1. There is a good agreement between the two methods.

3.2 Reaction of Sn(*n*-C₄H₉)₄ with Alumina-supported Platinum Catalysts at 25 °C

3.2.1 Stoichiometry of the Fixation of Sn(n-C₄H₉)₄. Reaction of Sn(*n*-C₄H₉)₄ with alumina-supported platinum catalysts in the presence of one atmosphere of hydrogen has been carried out at room temperature. Prior to the reaction, the samples are reduced at

450 °C under flowing dry hydrogen for 3 h. After cooling to room temperature under hydrogen, the samples are introduced under hydrogen, into the Schlenk tube and allowed to react with the *n*-heptane solution of $Sn(n\text{-}C_4H_9)_4$. Figures 1 and 2 indicate the variation of the concentration of $Sn(n\text{-}C_4H_9)_4$ and the amount of butane evolved with time for various initial concentrations of $Sn(n\text{-}C_4H_9)_4$ in the presence of Pt/Al-1 and Pt/Al-2.

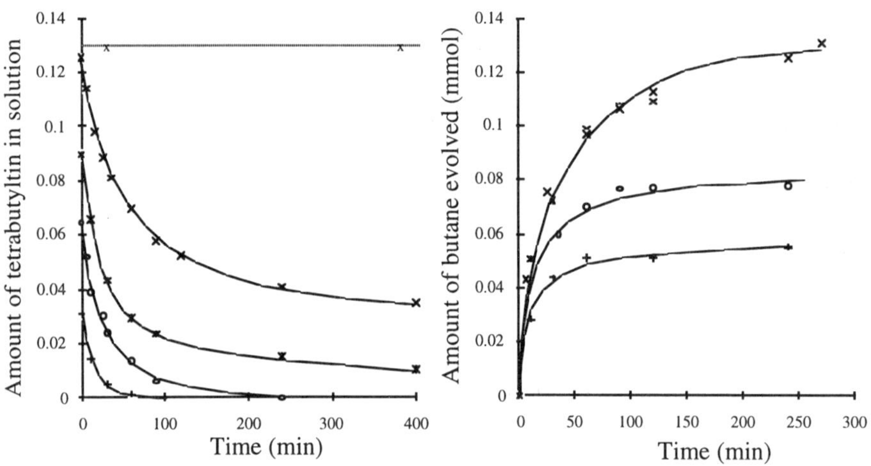

Figure 1 *Kinetics of $Sn(n\text{-}C_4H_9)_4$ reaction on Pt/Al-1 (full lines) and alumina (dotted line) with various initial concentration of $Sn(n\text{-}C_4H_9)_4$: $Sn_{int}/Pt_s = 0.5$ (+); 1 (o); 1.5 (*) and 2 (x).*

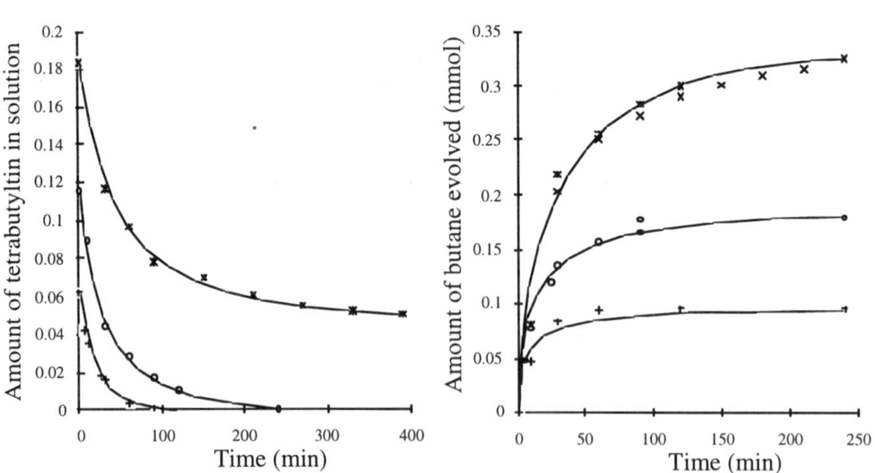

Figure 2 *Kinetics of $Sn(n\text{-}C_4H_9)_4$ reaction on Pt/Al-2 with various initial concentration of $Sn(n\text{-}C_4H_9)_4$: $Sn_{int}/Pt_s=0.5$ (+); 1 (o); 1.5 (*) and 2 (x).*

It seems that saturation of the surface is reached after *ca.* 5 hours of reaction. In Figure 3 we report the total amount of $Sn(n\text{-}C_4H_9)_4$ fixed by surface platinum atom (Sn_{fix}/Pt_s) after 24 hours of reaction as a function of the ratio $Sn(n\text{-}C_4H_9)_4$ introduced by surface platinum atom (Sn_{int}/Pt_s).

Clearly, there is a limit to the ratio Sn_{fix}/Pt_s which seems to depend on the dispersion of the samples. The value of Sn_{fix}/Pt_s is greater on the small particles (1.4) than on large ones (1.1). This is an unexpected result given that Ferretti *et al.*[28] found a direct relationship between the number of Sn_{fix} and the number of surface rhodium atom independent of dispersion, with a ratio Sn_{fix} by surface rhodium atoms of 0.8.

Blank experiments carried out on an alumina support without metallic platinum, but pretreated under the same conditions, did not give any detectable adsorption or reaction of $Sn(n\text{-}C_4H_9)_4$ (Figure 1, left).

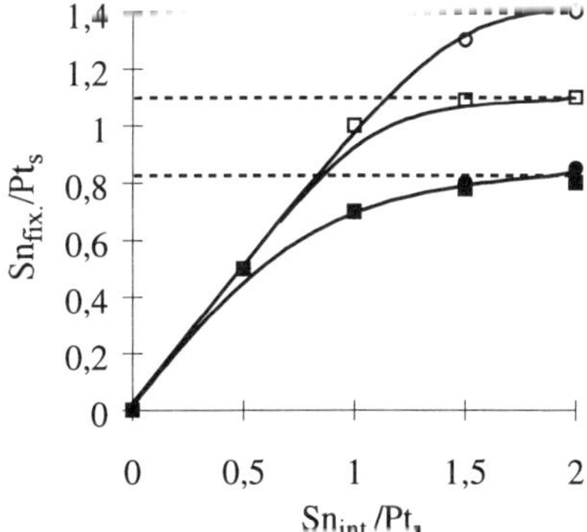

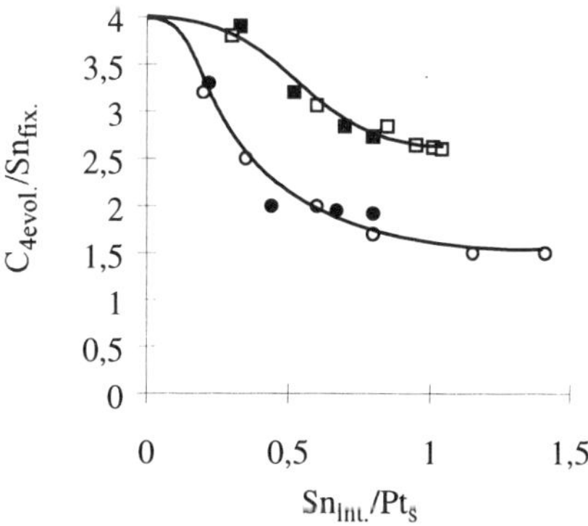

Figure 3 *Amount of Sn fixed (Sn_{fix}) on Pt/Al-1 (o) and Pt/Al-2 (□) treated at 450 °C under dry H_2 (open symbols) or after water addition (filled symbols).*

Figure 4 *Amount of butane evolved in function of surface platinum coverage. Symbols are the same as in Figure 3.*

Even given the results of these blank experiments, we cannot exclude for the case of alumina-supported platinum catalysts, that after treatment under dry hydrogen at 450 °C the Strong Metal–Support Interaction (SMSI) occurs, leading to an activation of the support (partial reduction) close to the metallic particle.[32–34] This activated support could react with $Sn(n\text{-}C_4H_9)_4$ even at room temperature, leading to formation of a grafted alumina surface complex which can be formulated as $Al\text{-}O\text{-}Sn(n\text{-}C_4H_9)_3$.[35] This phenomenon is related to the extent of interaction between the support and the metallic particle, and is much more pronounced in the presence of well dispersed samples.

The SMSI phenomenon was observed on alumina-supported catalysts treated under dry hydrogen at temperatures higher than 400 °C, and it has been demonstrated that the presence of water completely suppressed this phenomenon. Our experiments are performed using dry hydrogen, but this was not the case for the work of Ferretti *et al.*[28] In order to elucidate the possibility of such a process, the catalysts samples were treated with 50 µL/g (about 17 x 10^{20} molecules/g) of water before reaction with $Sn(n\text{-}C_4H_9)_4$. This amount of water was determined assuming that 5% of the surface hydroxyl groups of the alumina (about 15 x 10^{20} sites/g) are activated by hydrogen treatment at 450 °C. In Figure 3, we report the total amount of $Sn(n\text{-}C_4H_9)_4$ fixed by surface platinum atom

(Sn_{fix}/Pt_s) after 24 hours of reaction as a function of the $Sn(n\text{-}C_4H_9)_4$ introduced/surface platinum atom ratio (Sn_{fix}/Pt_s) for the two catalysts. Clearly, in this case, the limiting value is the same for the two catalysts and this value (0.8) is the same as that obtained with-supported rhodium.[28] It seems that the addition of water has completely inhibited the spillover phenomenon. Note that in this case, the amount of fixed tin does not depend on the dispersion of the samples.

3.2.2 Stoichiometry of Butane Evolution. Butane is the only one gas observed during the reaction of $Sn(n\text{-}C_4H_9)_4$ with alumina-supported platinum catalysts at 25 °C under atmospheric pressure of hydrogen.

For alumina-supported catalysts treated under dry hydrogen (without addition of water), the total amount of butane evolved by total amount of fixed $Sn(n\text{-}C_4H_9)_4$ (C_{4evol}/Sn_{fix}) as a function of surface platinum coverage (Sn_{fix}/Pt_s) is reported in Figure 4. It is clear that at low coverage (Sn_{fix}/Pt_s < 0.2), the number of butyl groups hydrogenolysed is close to 4 in both cases. This result is in agreement with the value of 4 butanes evolved per total amount of fixed Sn at low coverage, found by Ferretti *et al.* This can easily be explained if we assume that $Sn(n\text{-}C_4H_9)_4$ reacts quickly with the metallic surface, leading, at low coverage, to naked tin atoms 'chemisorbed' on the platinum surface (adatoms). At greater coverage, grafted organometallic fragments on the platinum surface $Pt_s[Sn(n\text{-}C_4H_9)_x]_y$ are formed with increasing 'x' values (decreasing values of C_{4evol}/Sn_{fix}). At full coverage, the value of C_{4evol}/Sn_{fix} is much higher on a poorly dispersed catalyst than on a well dispersed sample. We have seen before that the amount of $Sn(n\text{-}C_4H_9)_x$ grafted fragment is also greater on the more highly dispersed catalyst. We suggested in this case the formation of surface complexes grafted on the activated alumina surface. This alumina surface complexes can be formulated as $Al\text{--}O\text{--}Sn(n\text{-}C_4H_9)_3$[35] then on average, the amount of butyl groups remaining on the surface complexes by total amount of fixed Sn increases for the well dispersed catalyst.

If water is added to the catalyst after hydrogen treatment, we observe that the amount of butane evolved by grafted organotin fragment C_{4evol}/Sn_{fix} for increasing coverage (Sn_{fix}/Pt_s) follows almost exactly the corresponding curves obtained without water addition (Figure 4), but the values of Sn_{fix}/Pt_s are never greater than 0.8.

We can suggest that in all cases, $Sn(n\text{-}C_4H_9)_4$ reacts quickly on the metallic surface, leading to $Pt_s[Sn(n\text{-}C_4H_9)_x]_y$ surface complexes where $x = 0$ for y lower than 0.2 and, on average, $1 < x < 3$ for $0.2 < y < 0.8$, as in Scheme 1.

In the case of well dispersed samples treated under dry hydrogen and without addition of water, $Sn(n\text{-}C_4H_9)_4$ could also react slowly on the 'activated' alumina surface with formation of a $Al\text{--}O\text{--}Sn(n\text{-}C_4H_9)_3$ complex.[35] In this case, the value of y increases from 0.8 to 1.4 and the y value (C_4/Sn) can be as high as 2. The overall reaction could be schematically represented as in Scheme 2.

Let us now consider the effect of dispersion on the amount of remaining butyl fragments (x), on the platinum surface complexes $Pt_s[Sn(n\text{-}C_4H_9)_x]_y$ ($0 < y < 0.8$). Clearly, the x value at high coverage is greater in the case of a well dispersed sample. According to Van Hardeveld and Hartog,[27] a metallic particle could have a cubo-octahedral shape and in this case, the ratio of the number of metallic atoms located on the faces by the number of metallic atoms located on corner and edges increases with the size of the particle. The higher 'x' (C_4/Sn) value obtained at full coverage for the well dispersed catalyst (smaller particle size) could be explained if we assume that the hydrogenolysis of $Sn(n\text{-}C_4H_9)_4$ is deeper on metallic faces than on edges and corners, according to Scheme 3.

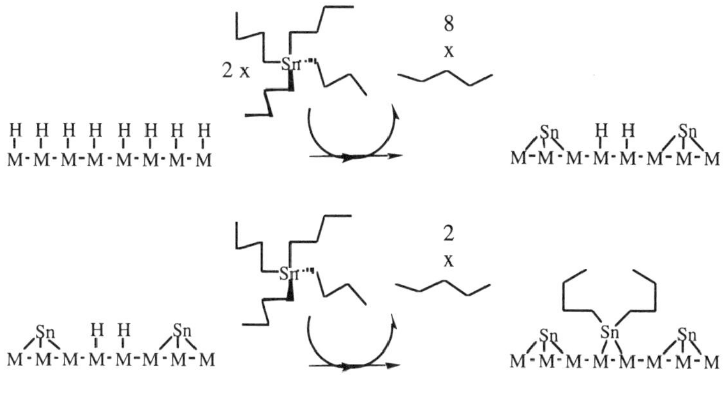

Scheme 1

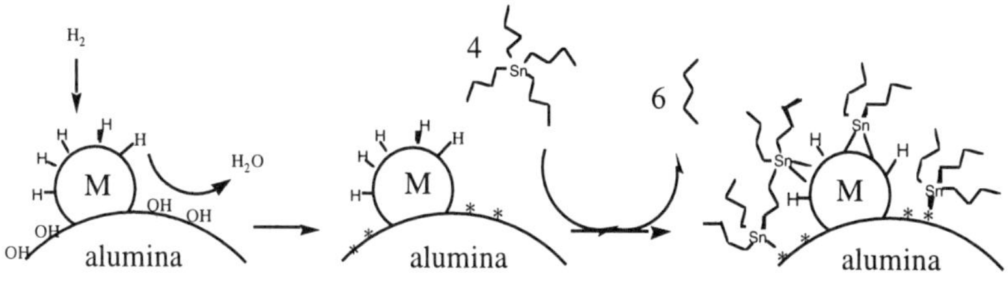

Scheme 2

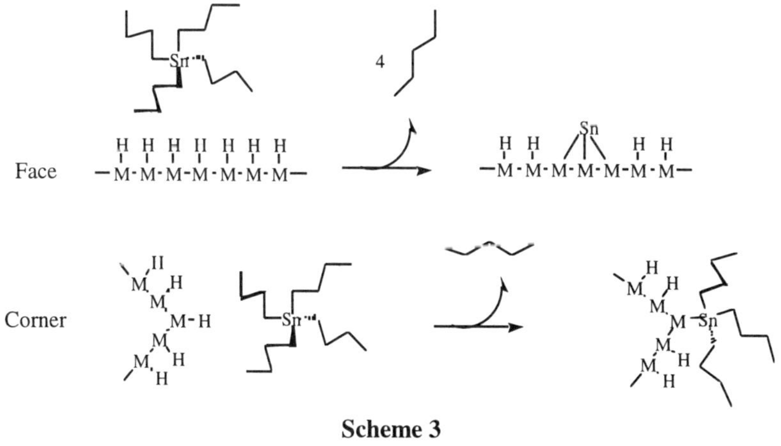

Scheme 3

In an attempt to verify this hypothesis, we use molecular modeling (Figures 5 and 6). We measure the C1-Pt and the C2-Pt distances for an organometallic fragment grafted on a platinum atom on the center of a [100] face (Figure 5) or on the corner (Figure 6). The conformation of the dihedral angle Pt–Sn–C1–C2 is fixed to zero degrees and we rotate the SnBu$_3$ around the Pt–Sn bond by 1 degree steps. As can be seen in Figures 5 and 6, the C2 atom comes 0.9 Å nearer than the C1, then the platinum surface is in a more favorable position to interact with C2 than C1. Moreover, as it can be seen in Figures 5 and 6, the shorter Pt-C2 distance is obtained on a face of a platinum particle (2.45 Å for the face and 3.84 Å for the coin position). We can then propose that the butyl groups of an organometallic fragment could be hydrogenolysed into butane faster when they are grafted on a face of a platinum particle than on a corner.

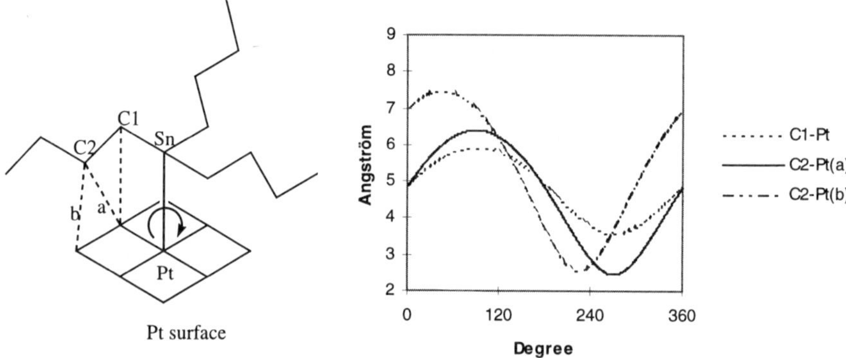

Figure 5 *Rotation of the SnBu₃ onto a (100) face of a Pt particle.*

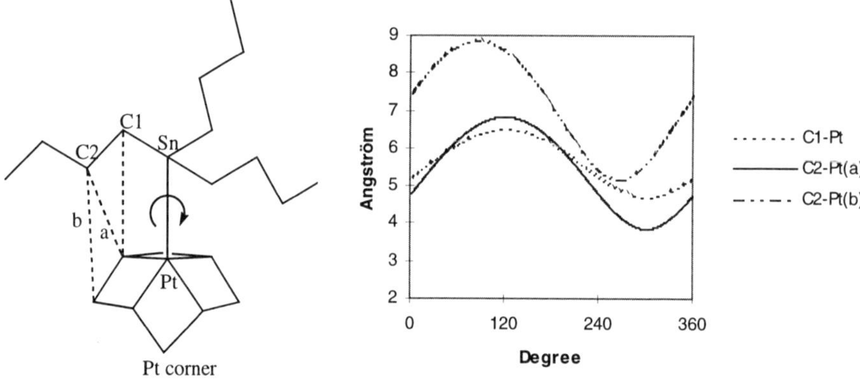

Figure 6 *Rotation of the SnBu₃ onto the corner of a Pt particle.*

3.3 Dehydrogenation of Isobutane into Isobutene

Both alumina-supported catalysts obtained after reaction at 25 °C under hydrogen of Pt/Al-1 and Pt/Al-2 with Sn(n-C₄H₉)₄ (Sn/Pt$_s$ = 0.7), completely lost the remaining butyl groups, when they are treated under flowing hydrogen at 550 °C prior to catalytic measurements. The aim of this part of the work is to demonstrate that bimetallic PtSn/Al₂O₃ catalysts obtained by the surface organometallic chemistry on metal route are active and selective in dehydrogenation of isobutane into isobutene.

In Figure 7 we report the catalytic activity and selectivity of the monometallic and of the corresponding bimetallic catalysts as a function of time on stream. With the monometallic and bimetallic catalysts, a decrease in activity is observed with time on stream. This effect is probably due to the formation of coke which readily occurs on the surface of platinum and leads to poisoning of the catalyst. With the monometallic catalyst, the initial selectivity for isobutene is low (less than 95 %) and slightly increases with time on stream. This effect could be also related to coke formation which plays the role of a selective poison.

When tin is present, the selectivity for isobutene increases in both cases as compared to pure platinum (to reach a value higher than 99 %). These results, which have already been observed in several other cases,[4,13,18] are remarkable and demonstrate the drastic effect of tin, especially when it is exclusively located on the platinum particle. The increase of selectivity for isobutene formation with the presence of tin could be explained simply by the 'site isolation' effect. It is generally admitted[4,13,14,18,19] that coke formation and hydrogenolysis processes occur on multimetallic sites of the platinum surface, while dehydrogenation reactions proceed on single Pt atoms. The presence of tin atoms regularly distributed on the platinum surface diminishes the occurrence of multimetallic sites as do copper atoms on nickel surfaces[36] or tin atoms on rhodium and nickel surfaces.[37,38] Increasing the amount of tin regularly distributed on the platinum surface strongly inhibits the hydrogenolysis process but slightly affects the dehydrogenation reaction, leading to a catalyst which is almost fully selective for isobutene.

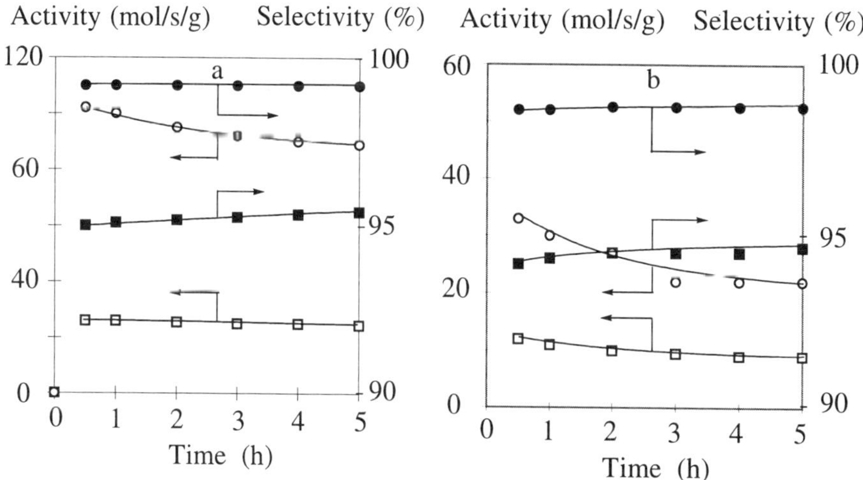

Figure 7 *Catalytic activity (open symbols) and selectivity for isobutene (solid symbols) in the dehydrogenation of isobutane at 550 °C on: (a) Pt/Al-1 (□) and PtSn/Al-1 (o), or (b) Pt/Al-2 (□) and PtSn/Al-2 (o).*

The effect of tin on catalytic activity is more complex. When Sn/Pt_s increases from 0 to 0.7, the catalytic activity based on the total number of Pt atoms increases drastically. The increase of activity with addition of tin (from $Sn/Pt_s = 0$ to 0.7) could be explained by the inhibition of coke formation (which poison the active surface). We can see in Figure 4 that the catalytic activity of the well dispersed sample PtSn/Al-1 is twice that of the less dispersed catalyst PtSn/Al-2. If we take into account the fact that the dispersion of the dispersed catalyst PtSn/Al-2. If we take into account the fact that the dispersion of the PtSn/Al-1 is twice that of PtSn/Al-2, the catalytic activities with respect to the number of surface atoms are the same for both PtSn catalysts. This result is in accordance with the observation that the catalytic activity for alkane dehydrogenation is structure insensitive.[14,39,40]

4 CONCLUSIONS

Two monometallic Pt/Al_2O_3 samples were prepared with dispersions of 85 and 50 %. The reaction of $Sn(n\text{-}C_4H_9)_4$ on the reduced Pt/Al_2O_3 samples leads to organometallic fragments grafted on the metallic surface $Pt_s[Sn(n\text{-}C_4H_9)_x]_y$ and to $Al\text{–}O\text{–}Sn(n\text{-}C_4H_9)_3$[35] surface complexes grafted on the activated support surface. The formation of this complex grafted on the support can be avoided by addition of traces of water. On reduced and water-treated Pt/Al_2O_3 samples, saturation of the metallic surface is achieved with 0.8 Sn_{fix}/Pt_s (y value of $Pt_s[Sn(n\text{-}C_4H_9)_x]_y$), without influence of the dispersion, but the amount of remaining butyl groups (x value) clearly depends on the dispersion of the sample. This value is greater on the well dispersed catalyst than on the poorly dispersed sample. This result is explained assuming that hydrogenolysis reaction of $Sn(n\text{-}C_4H_9)_4$ is deeper on faces of metallic particles than on corners or kinks. There is also deeper hydrogenolysis of butyl groups at low metallic surface coverage than at saturation. This is explained by self poisoning of the metallic surface by tin adatoms or by $Sn(n\text{-}C_4H_9)_x$ fragments.

The two bimetallic $PtSn/Al_2O_3$ (Sn/Pt_s = 0.7) catalysts prepared from the two monometallic samples were tested in the dehydrogenation of isobutane to isobutene. They are both more active and selective than the corresponding monometallic samples. These results are explained by the phenomenon of site isolation.

References

1.   E. Clippinger, S. Rafael and F. Mulaskey, *USA Patent 3,531,543*, 1970.
2.   E. O. Box, L.E. Drehman and F. Farha, *German Patent, 2,127,353*, 1970.
3.   I. B. Yarusov, E. V. Zatolokina, N. V. Shitova, A. S. Belyi and N. M. Ostrovski, *Catalysis Today*, 1992, **13**, 655.
4.   R. D. Cortright and J. A. Dumesic, *J. Catal.*, 1994, **148**, 771.
5.   R. Burch and L. A. Garla, *J. Catal.*, 1981, **71**, 360.
6.   G. Del Angel, F. Tzompantzi, R. Gomez, G. T. Baronnetti, S. R. De Miguel, O. A. Scelza and A. A. Castro, *React. Kinet. Catal. Lett.*, 1990, **42**, 67.
7.   Y. Fan, L. Lin, J. Zang and Z. Xu, in *Proceeding 10th Inter. Congr. Catal.*, Elsevier, Budapest, 1992, Vol. B, p. 2507.
8.   L. Wang, L. Lin, T. Zang and H. Cai, *React. Kinet. Catal. Lett.*, 1994, **52**, 107.
9.   J. Margitfalvi, M. Hegedüs and E. Talas, *J. Mol. Catal.*, 1989, **51**, 279.
10. I. B. Yarusov, E. V. Zatolokina, N. V. Shitova, A. S. Belyi and N. M. Ostrovskii, *Catalysis Today*, 1992, **13**, 655.
11. Y. Weishen, L. Liwu, F. Yining and Z. Jingling, *Catalysis Letters*, 1992, **12**, 267.
12. O. A. Barias, A. Holmen and E. A. Blekkan, *J. Catal.*, 1996, **158**, 1.

13. S. M. Stagg, C. A. Querini, W. E. Alvarez and D. E. Resasco, *J. Catal.*, 1997, **168**, 75.

14. P. Biloen, F. M. Dautzentberg and W. M. H. Sachtler, *J. Catal.*, 1977, **50**, 77.

15. J. Rodriguez and D.W. Goodman, *Acc. Chem. Res.*, 1995, **28**, 477.

16. M. W. Ruckman and M. W. Strongin, *Acc. Chem. Res.*, 1994, **27**, 250.

17. R. Burch, *J. Catal.*, 1981, **71**, 348.

18. C. Kappenstein, M. Saouabe, M. Guérin, P. Marecot, I. Uszkuruat and Z. Paal, *Catal. Letters*, 1995, **31**, 9.

19. W. M. H. Sachtler and R. A. Van Santen, *Adv. Catal.*, 1977, **26**, 69.

20. J. P. Candy, B. Didillon, E. L. Smith, T. Shay and J. M. Basset, *J. Mol. Catal.*, 1994, **86**, 179.

21. F. Humblot, B. Didillon, F. Le Peltier, J. P. Candy, J. Corker, O. Clause, F. Bayard and J. M. Basset, *J. Am. Chem. Soc.*, 1998, **120**, 137.

22. B. Coq, A. Grousot, T. Tazi, F. Figuéras and D. Salahub, *J. Am. Chem. Soc.*, 1991, **113**, 1485.

23. E. Merlen, P. Beccat, J. C. Bertolini, P. Delichère, N. Zanier and B. Didillon, *J. Catal.* 1996, **159**, 178.

24. J. P. Candy, A. El Mansour, O. A. Ferretti, G. Mabilon, J. P. Bournonville, J. M. Basset and G. Martino, *J. Catal.*, 1988, **112**, 201.

25. S. D. Jackson, J. Willis, G. D. McLellan, G. Webb, M. B. T. Keegan, R. B. Moyes, S. Simpson, P. B. Wells and R. Whyman, *J. Catal.*, 1993, **139**, 191.

26. J. P. Candy, P. Fouilloux and A. J. Renouprez, *J. Chem. Soc., Faraday Trans. I*, 1980, **76**, 616.

27. R. Van Hardeveld and F. Hartog, *Surf. Sci.*, 1969, **15**, 189.

28. O. A. Ferretti, C. Lucas, J. P. Candy, J. M. Basset, B. Didillon and F. Le Peltier, *J. Mol. Catal.*, 1995, **103**, 125.

29. F. Humblot, J. P. Candy, F. LePeltier, B. Didillon and J. M. Basset, *submitted for publication*.

30. *Tripos Associates,* St. Louis, MO, USA.

31. J. H. Horner and M. Newcomb, *Organometallics*, 1991, **10**, 1732.

32. J. B. F. Anderson and R. Burch, *Applied Catalysis.*, 1986, **25**, 173.

33. G. J. Den Otter and F. M. Dautzenberg, *J. Catal.*, 1978, **53**, 116.

34. C. C. A. Riley, P. Jonsen, P. Meehan, J. C. Frost and K. J. Packer, *Catalysis Today*, 1991, **9**, 121.

35. C. Nédez, A. Théolier, F. Lefebvre, A. Choplin, J. M. Basset and J. F. Joly, *J. Am. Chem. Soc.*, 1993, **115**, 722.

36. G. A. Martin, J. A. Dalmon, *J. Catal.*, 1980, **66**, 214.

37. M. Agnelli, J. P. Candy, J. M. Basset, J. P. Bournonville and O. A. Ferretti, *J. Catal.*, 1990, **121**, 236.

38. B. Didillon, J. P. Candy, A. El Mansour, C. Houtmann and J. M. Basset, *J. Mol. Catal.*, 1992, **74**, 43.

39. M. Guenin, M. Breysse, R. Frety, K. Tifouti, P. Marecot and J. Barbier, *J. Catal.*, 1987, **105**, 144.

40. D. W. Blakely and G. A. Somorjai, *J. Catal.*, 1976, **42**, 181.

CATALYTIC NO$_x$ REDUCTION WITH HYDROCARBONS OVER ALUMINA-SUPPORTED CATALYSTS

M.C. Kung, H.H. Kung,* K.A. Bethke, J.-Y. Yan and P.W. Park

Department of Chemical Engineering
Northwestern University
Evanston, IL 60208-3120 USA

1 INTRODUCTION

The demand for better air quality has been a driving force behind the development and implementation of various air-emission control technologies. In the urban area, a particular concern has been emission from transportation vehicles. Since the mid nineteen seventies, governments in industrial countries have set regulations on permissible emission levels of pollutant gases, such as hydrocarbons, nitrogen oxides (NO$_x$), and carbon monoxide. In recent years, these standards have been tightened. At the same time, there has been a strong push for higher fuel efficiencies for vehicles also. Since internal combustion engines that operate with an air-to-fuel ratio much higher than stoichiometric, also known as lean-burn engines, generally have higher fuel efficiencies, a significant volume of research in recent years has focused on the treatment of the exhaust gases from such engines. The typical conditions of these exhaust gases are shown in Table 1.

Table 1 *Typical Conditions of Engine Exhaust Gases.*

EngineType	Temp, °C	[NO$_x$], %	[O$_2$], %	Reductant for NO$_x$	Other
Diesel (heavy duty)	300–500	< 0.07	4–10	< 200 ppm HC	CO$_x$, H$_2$O particulate SO$_x$
Diesel (light duty)	200–350	< 0.07	4–10	< 200 ppm HC	CO$_x$, H$_2$O particulate SO$_x$
Gasoline (stoichiometric)	400–900	< 0.2	≈ 0	CO	CO$_x$, H$_2$O SO$_x$, H$_2$S
Gasoline (lean burn)	400–700	< 0.15	4–6	< 0.1 % HC	CO$_x$, H$_2$O SO$_x$

For these exhaust gases (except stoichiometric gasoline engine exhaust), a characteristic is the high partial pressure of oxygen relative to nitrogen oxides and hydrocarbons. While there are technologies to lower the partial pressure of hydrocarbons, there is still no satisfactory method to lower NO$_x$ to an acceptable level. This is because catalytic decomposition of NO$_x$ into dinitrogen and dioxygen is slow, although it is thermodynamically favorable, and catalytic NO$_x$ reduction with hydrocarbons in the exhaust stream has to compete with combustion of the hydrocarbons by the much higher partial pressure of oxygen. In addition, there are other requirements for a practical catalyst, including durability, efficiency over a wide range of operating conditions, and resistance to adverse effects due to water vapor and sulfur.

Because alumina is used in many industrial catalysts either as the catalytic material or as a stable support, it is of interest to test it for NO$_x$ reduction activity for the lean-burn exhaust. Recent studies have shown that alumina-supported samples can be hydrothermally stable.[1–8 and refs. therein] Here, we summarize the results of our study of using γ-Al$_2$O$_3$, with and without modification by Sn and Co oxides for NO$_x$ reduction.

2 EXPERIMENTAL

γ-Al$_2$O$_3$ was prepared from aluminum isopropoxide (99.99%, Aldrich Chemicals), which was first dissolved in 2-methylpentane-2,4-diol (Aldrich). After stirring for 4 h at 125 °C, H$_2$O was added. The white precipitate that was formed was aged at 80 °C for 15 h, filtered, and washed with 2-propanol, then dried in air at 105 °C overnight. The sample was then calcined in wet air using a published temperature schedule.[1] Its surface area was 250 m^2/g.

Co/γ-Al$_2$O$_3$ was prepared by incipient wetness impregnation of a Co acetate or Co nitrate solution. After impregnation, the solid was dried in an air oven overnight, and then calcined in air at 350 °C for 2 h and then at 800 °C for 2 h. Sn/γ-Al$_2$O$_3$ was also prepared by incipient wetness impregnation of an ethanol solution of SnCl$_2$ (Aldrich, 99.99%). After drying in an air oven at 100 °C overnight, the sample was calcined at 500 °C for 4 h and then at 800 °C for 2 h.

NO$_x$ reduction was carried out in a flow of 5 to 15% O$_2$, 0 to 10% H$_2$O, 0.1% NO or NO$_2$, and 0.1% propene, with the balance being He. The reaction was carried out in a quartz microreactor. The reaction products were analysed by on-line gas chromatograph with a TCD detector, sometimes in combination with a NO$_x$ analyser. In all the experiments reported here, no N$_2$O was detected in the product. Thus, the % NO conversion equaled the % N$_2$ yield.

3 RESULTS AND DISCUSSION

In NO$_x$ reduction with hydrocarbons, catalysts are compared with respect to the N$_2$ yield at various temperatures and partial pressures of the components in the reaction feed stream. Since the hydrocarbon is present in a limited amount in the feed, and it is a necessary reactant in the NO$_x$ reduction reaction, the N$_2$ yield can be understood as being determined by the ability of the catalyst to activate the hydrocarbon, and the competition between the reaction of hydrocarbon with NO *versus* O$_2$. It can be stated further that the temperature at which the maximum N$_2$ yield is obtained is determined by the former, and the maximum yield is determined primarily by the competitiveness factor, which is defined as the ratio of the rate of hydrocarbon conversion that leads to N$_2$ production to the total rate of hydrocarbon conversion.[1]

3.1 γ-Al$_2$O$_3$ Catalyst

Figure 1 shows N$_2$ yield as a function of temperature for NO reduction for two experiments of different conditions (curves A and B), and one experiment for reduction of NO$_2$ (curve C). For NO reduction, the N$_2$ yield was low below 450 °C. In the experiment with higher space velocity and oxygen and water partial pressures (curve A), the NO yield increased steadily with increasing temperature, reaching a maximum at around 600 °C. The hydrocarbon conversion paralleled this trend, being low below 450°C, but rising rapidly above 500 °C, reaching close to 100% above 550 °C. For the experiment with a slower space velocity and lower partial pressures of oxygen and water (curve B), the N$_2$ yield increases gradually with increasing temperature below about 510 °C. At around 510 °C, sustained oscillation in the N$_2$ yield and hydrocarbon conversion were observed. Above this temperature, the N$_2$ yield reached the maximum value, and the hydrocarbon conversion reached about 100%. As with other catalysts, the N$_2$ yield declined with further increase in temperature, while the hydrocarbon conversion remained at about 100%. The temperature at which oscillation was observed decreased to about 450 °C when a higher amount (1 g) of catalyst was used (Figure 2). Similar observation of oscillatory behavior has been reported by others.[3,9] Its origin is not understood, but is probably related to the coupling of the rates of production, reaction, and gas phase transport of some reactive intermediates and the heat transfer from one part of the catalyst bed to another.

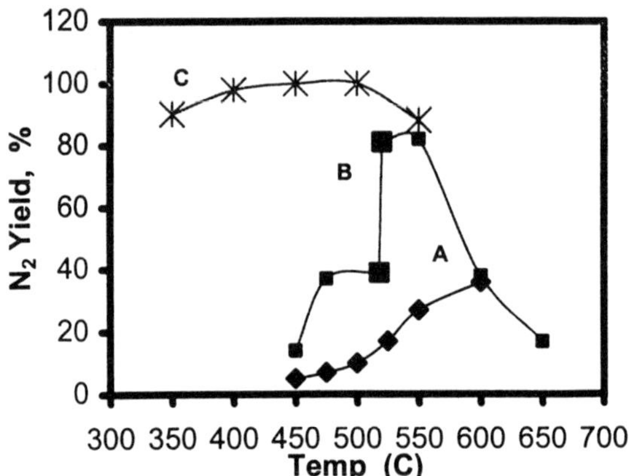

Figure 1 *N$_2$ yield in the reduction of NO or NO$_2$ over γ-Al$_2$O$_3$. Curve A ($\blacklozenge$): 0.2 g catalyst, 200 mL/min of 0.1% NO, 0.1% propene, 15% O$_2$, 10% H$_2$O, balance He; curve B ($\blacksquare$): 0.25 g catalyst, 100 mL/min of 0.1% NO, 0.1% propene, 6% O$_2$, balance He; curve C (*): 0.25 g catalyst, 100 mL/min of 0.1% NO$_2$, 0.1% propene, 5% O$_2$, balance He*

Unlike the reduction of NO, the reduction of NO$_2$ took place at much lower temperatures, and very high yields of N$_2$ were obtained (Figure 1, curve C). Simultaneously, the hydrocarbon conversions were also high. These data can be interpreted as follows. At the lower temperatures, neither NO nor O$_2$ is adsorbed dissociatively on γ-Al$_2$O$_3$. Therefore, there is little reaction between them and the hydrocarbon. That adsorption of NO or O$_2$ at these temperatures is associative can be deduced from the temperature-programmed

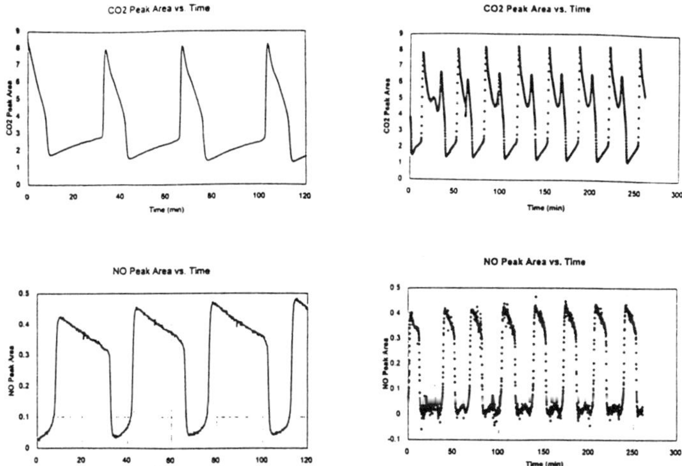

Figure 2 *Oscillation in N_2 and CO_2 yields in the propene reduction of NO over γ-Al$_2$O$_3$. 1 g catalyst, 100 mL/min of 0.1% NO, 0.1% propene, 6% O$_2$, balance He. Figures on the left at 451 °C, those on the right at 458 °C.*

desorption experiments (Figure 3). In this experiment, NO was first adsorbed at room temperature on γ-Al$_2$O$_3$, presumably at coordinatively unsaturated, surface Al ions (Lewis acid sites). Upon heating, molecular desorption occurred at *ca.* 400°C without any reaction. Because of the similar properties of NO and O$_2$, the latter is expected to behave similarly. Thus, γ-Al$_2$O$_3$ is quite inactive in oxidizing the hydrocarbon under these conditions. At higher temperatures, dissociative adsorption of these molecules becomes possible, and the alumina becomes much more active in oxidizing propene to oxygen-containing intermediates, which are much more active for NO reduction.[5,10] Thus, the alumina becomes active in the formation of N$_2$.

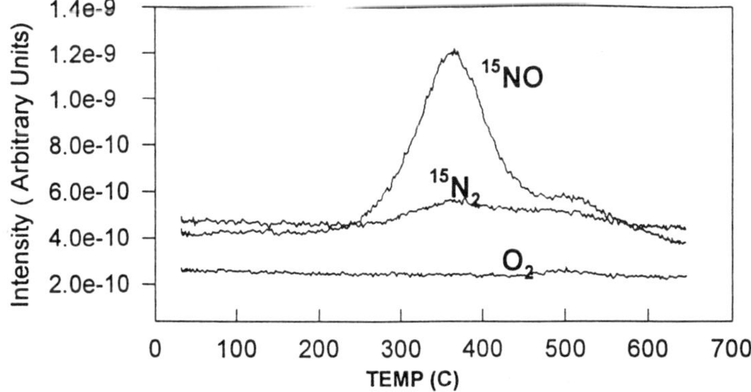

Figure 3 *Temperature-programmed desorption of ^{15}NO on γ-Al$_2$O$_3$. 0.1 g catalyst, He purge rate at 225 mL/min, heating rate 20 °C/min.*

When NO$_2$ is used instead of NO, since NO$_2$ is a stronger oxidizing agent, much more reactive, and is adsorbed much more strongly on γ-Al$_2$O$_3$ than O$_2$ or NO, it can react with propene on the alumina surface at low temperatures. Once activated by oxidation by NO_2, the hydrocarbon intermediate can react with NO, eventually resulting in the formation of N_2. The much stronger adsorption of NO_2 has been confirmed by Fourier transformed infrared spectroscopy, which showed the formation of surface nitrate and nitro species when γ-Al_2O_3 was exposed to NO_2 at room temperatures. These surface species could not be removed by heating *in vacuo* even at 450 °C.

These observations suggest that the low temperature NO reduction activities of γ-Al_2O_3 could be enhanced by introducing a component that is active for oxidizing propene at these temperatures. Cobalt oxide and tin oxide were chosen for this purpose.

3.2 Co/γ-Al$_2$O$_3$ Catalysts

The addition of 2 or 5 wt% Co ions to γ-Al_2O_3 enhanced the NO reduction activity significantly below 450 °C. As shown in Figure 4, 50% N_2 yield could be obtained at 400 °C with the 2 wt% Co sample. At the same time, the hydrocarbon conversion was also much higher. The NO reduction activity depends on the Co loading. As shown in Figure 4, the 5 wt% sample was less effective. In addition to Co loading, it has been shown that the catalytic activity also depends on the preparation method.[1]

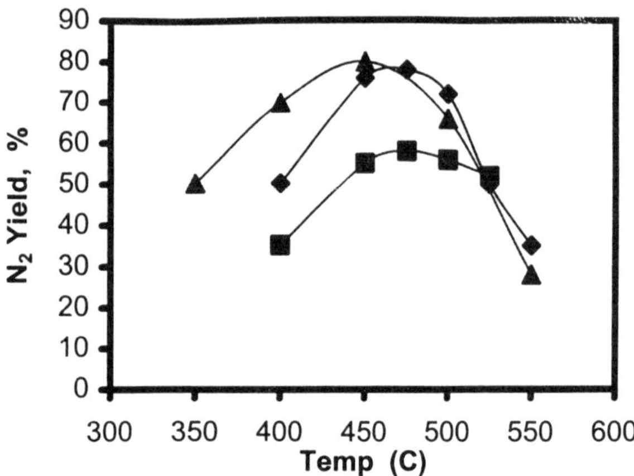

Figure 4 *N_2 yield in the reduction of NO or NO_2 by propene over Co/γ-Al$_2$O$_3$. 0.5 g catalyst, 100 mL/min of 0.1% NO or NO_2, 0.1% propene, 5% O_2, 1.7% H_2O, balance He. Curve A (♦): NO on 2 wt% Co sample; curve B (■): NO on 5 wt% Co sample; and curve C (▲): NO_2 on 2 wt% Co sample.*

It has been shown that Co ions are present in different forms on γ-Al_2O_3.[1] For these samples, the majority of them exist as Co^{2+} in $CoAl_2O_4$, the amount of which increases with increasing calcination temperature. Co ions in this form are present in the bulk solid, and are inactive for the catalytic reaction. Some Co ions are present as small Co_3O_4 crystallites on the alumina that can be identified by X-ray diffraction and as a peak at 380 nm in the UV–vis spectrum. The amount of this species decreases with increasing calcination temperature and decreasing Co loading. Co_3O_4 is known to be an active hydrocarbon combustion catalyst. It is much more easily reduced by hydrogen than other Co species. Its presence on the catalyst surface would promote the combustion of the propene reductant, thereby decreasing the competitiveness for the NO reduction reaction. The desirable Co species for the NO reduction reaction is proposed to be small cobaltoxo clusters containing octahedral Co ions, which are detected as a 480 nm peak in the UV–vis spectrum.[1] Desorption of coordinated water or oxygen from these species at elevated temperatures generates binding sites for the NO reduction reaction.

Interestingly, although γ-Al$_2$O$_3$ is much more active for the propene reduction of NO$_2$ than NO (Figure 1), the difference on Co/γ-Al$_2$O$_3$ is much less (Figure 3). In fact, γ-Al$_2$O$_3$ is more active for the reduction of NO$_2$ than Co/γ-Al$_2$O$_3$ at 350 °C. This is because the presence of Co ions not only catalyses the oxidation of propene by oxygen at the lower temperatures, it also catalyses the oxidation of propene by NO$_2$, while NO$_2$ is reduced to NO. This latter reaction depletes the NO$_2$ in the gas phase and minimizes the difference between a feed of NO$_2$ *versus* NO. A separate experiment to investigate the activity of the Co/γ-Al$_2$O$_3$ catalyst for NO oxidation to NO$_2$ and its reverse reaction showed that these two reactions occur at rates that are slow compared with the NO$_x$ reduction reactions.[1] Thus, they cannot account for the reason why Co/γ-Al$_2$O$_3$ is substantially less active than γ-Al$_2$O$_3$ in the reduction of NO$_2$.

Another interesting aspect is that the activity for NO reduction on Co/γ-Al$_2$O$_3$ depends on the nature of the hydrocarbon. Whereas substantial activity was observed using propene or propane as the reductant,[1] very low activity was observed using methane.[2] For example, the N$_2$ yield was less than 10% at 500 °C (Figure 5). It appears that for such an inert hydrocarbon as methane, the oxidation activity of Co is insufficient to activate it. The N$_2$ yield was found to increase substantially if a physical mixture of Co/γ-Al$_2$O$_3$ and H-zeolite (such as H-ZSM-5 or H-USY) was used (Figure 5).[2] These H-zeolites are very active for the reduction of NO$_2$ to N$_2$ by methane below about 550 °C, perhaps because the small pores in the zeolites concentrate the reactants methane and NO$_2$. Thus, an explanation of the synergistic effect of Co/γ-Al$_2$O$_3$ and H-zeolite is that NO is oxidized to NO$_2$ on Co/γ-Al$_2$O$_3$, which is then reduced by methane on the H-zeolite.

3.3 Sn/γ-Al$_2$O$_3$ Catalysts

Impregnation of Sn ions onto γ-Al$_2$O$_3$ also increases the low temperature NO reduction activity significantly.[11] Figure 6 shows the N$_2$ yield as a function of temperature for three catalysts containing 1, 5 and 10 wt% Sn. Even with a high space velocity (*i.e.* small amount of catalyst and high flow rate), over 25% N$_2$ yield was obtained at 400 °C for the 5 wt% sample. The N$_2$ yield increased with increasing temperature to reach a maximum of about 60% at *ca.* 475 °C. Under these temperatures, less than 10% N$_2$ yield was obtained over γ-Al$_2$O$_3$ (curve A, Figure 1).

The were no detectable X-ray diffraction peaks other than those of γ-Al$_2$O$_3$ for the 1 wt% Sn sample. For the 5 and 10 wt% samples, SnO$_2$ diffraction peaks were detected. Temperature-programmed reduction with 5% H$_2$/He showed a small, broad hydrogen consumption peak over 200–400 °C for the 1 wt% sample. For the 5 wt% sample, a broad reduction peak also appeared at 500–700 °C, which became prominent for the 10 wt% sample. This high temperature peak was assigned to reduction of large crystallites of SnO$_2$, and the broad, low temperature peak to finely dispersed SnO$_2$ on the surface of γ-Al$_2$O$_3$.

Bulk SnO$_2$ has been studied for this reaction.[12] Although it possesses NO reduction activity, the activity was quite low, probably because of low surface area of SnO$_2$. By comparison, either on per unit weight or unit volume basis, Sn/γ-Al$_2$O$_3$ is much more active. Since the bulk SnO$_2$ and Sn/γ-Al$_2$O$_3$ demonstrated about the same competitiveness factor at

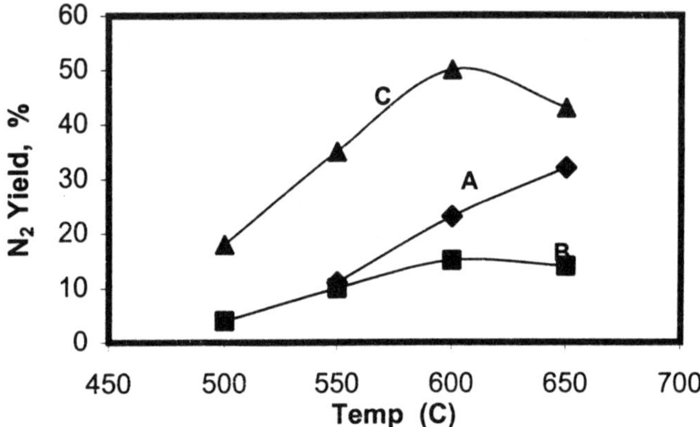

Figure 5 *N_2 yield in the reduction of NO by methane over Co/γ-Al$_2$O$_3$. 250 mL/min of 0.1% NO, 0.3% methane, 2% O$_2$, balance He. Curve A (♦) 0.5 g γ-Al$_2$O$_3$; curve B (■) 0.25 g Co/γ-Al$_2$O$_3$; curve C (▲) 0.25 g Co/γ-Al$_2$O$_3$ + 0.25 g H-USY zeolite.*

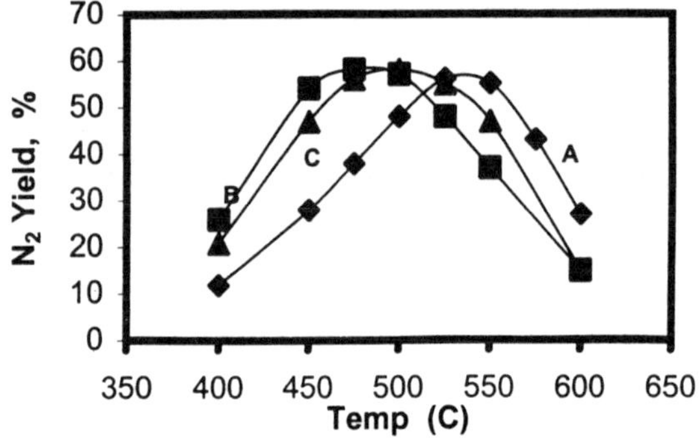

Figure 6 *N_2 yield in NO reduction by propene over Sn/γ-Al$_2$O$_3$ catalysts. 0.2 g catalyst, 0.1% NO, 0.1% C$_3$H$_6$, 15% O$_2$, 10% H$_2$O, balance He, total flow 200 cc/min. Curves A (♦), B (■), and C (▲) for 1, 5, and 10 wt% Sn/γ-Al$_2$O$_3$, respectively.*

comparable conversions of NO,[11] it is postulated that the role of γ-Al$_2$O$_3$ is to disperse the SnO$_2$ crystallites, thereby increasing their surface area substantially.[11]

4 CONCLUSIONS

γ-Al$_2$O$_3$ is active for NO reduction at high temperatures, but is quite inactive below 400 °C, probably because Al ions do not possess d-electrons and have poor reduction/oxidation ability. Thus, the catalyst cannot activate NO or O$_2$ at low temperatures. Introducing cobaltoxo species onto the surface of Al$_2$O$_3$ greatly enhances its NO reduction activities at the low temperatures by introducing activity for partial oxidation of the hydrocarbon reductant. Other species can be added in place of cobaltoxo clusters. For example, Sn/γ-Al$_2$O$_3$ is also an effective catalyst. Other examples include Ag/γ-Al$_2$O$_3$, Au/γ-Al$_2$O$_3$, and

Cu/γ-Al$_2$O$_3$. However, even Co/γ-Al$_2$O$_3$ is quite inactive for methane reduction of NO because methane is a very inert molecule. In this case, an H-zeolite is needed to enhance the activity.

5 ACKNOWLEDGEMENT

Financial support from the U.S. Department of Energy, Basic Energy Sciences, is gratefully acknowledged.

References

1.   J.-Y. Yan, M. C. Kung, W. M. H. Sachtler and H. H. Kung, *J. Catal.*, 1997, **172**, 178.
2.   J.-Y. Yan, H. H. Kung, W. M. H. Sachtler and M. C. Kung, *J. Catal.*, 1998, **175**, 294.
3.   K. A. Bethke and H. H. Kung, *J. Catal.*, 1997, **172**, 93.
4.   M. C. Kung, K. A. Bethke, J. Yan, J.-H. Lee and H. H. Kung, *Appl. Surf. Sci.*, 1997, **121/122**, 261.
5.   R. Burch, E. Halpin and J. A. Sullivan, *Appl. Catal. B*, 1998, **17**, 115.
6.   S. Sumiya, M. Saito, H. He, Q. C. Feng, N. Takezawa and K. Yashida, *Catal. Lett.*, 1998, **50**, 87.
7.   N. Okazaki, Y. Shiina, H. Itoh, A. Tada and M. Iwamoto, *Catal. Lett.*, 1997, **49**, 169.
8.   K. Shimizu, A. Satsuma and T. Hattori, *Appl. Catal. B*, 1998, **16**, 319.
9.   A. Obuchi, M. Nakamura, A. Ogata, K. Mizuno, A. Ohi and H. Ohuchi, *J. Chem. Soc., Chem. Commun.*, 1992, 1150.
10. K. A. Bethke, M. C. Kung, B. Yang, M. Shah, D. Alt, C. Li and H. H. Kung, *Catal. Today*, 1995, **26**, 169.
11. M. C. Kung, P. W. Park, D.-W. Kim and H. H. Kung, *J. Catal.*, submitted for publication.
12. Y. Teraoka, T. Harada, T. Iwasaki, T. Ikeda and S. Kagawa, *Chem. Lett.*, 1993, 77.

NOVELTIES OF HETEROPOLYACIDS SUPPORTED ON CLAYS: ROLE OF
WATER IN RATE ENHANCEMENT IN THE ETHERIFICATION AND
DEHYDRATION REACTIONS OF *t*-BUTANOL

G.D. Yadav* and N. Kirthivasan[†]

Chemical Engineering Division, University Department of Chemical Technology,
University of Mumbai, Matunga, Mumbai - 400 019, INDIA.
([†]Current address: Department of Chemistry, Michigan State University,
East Lansing, MI 48823 USA.)

1 INTRODUCTION

Clays of the bentonite and montmorillonite type show a remarkable increase in activity
when they are subjected to acid treatment. Such acid-treated clays were used decades ago
for catalytic cracking of hydrocarbons. K-10 montmorillonite and Filtrol are some of the
acid-treated clays that are commercially available. Heteropoly acids (HPA) supported on
K-10 montmorillonite are one of the few examples that exhibit synergism in the catalyst
industry. K-10, by itself a very good acidic catalyst, displays higher activity when an
HPA such as dodecatungstophosphoric acid (DTP), dodecatungstosilicic acid (DTS) or
dodecaphosphomolybdic acid (DPM) is supported onto it. We have previously reported
on the use of HPA/K-10 for the synthesis of methyl-*tert*-butyl ether and Bisphenol-A.[1,2]

Methyl-*tert*-butyl ether (MTBE) has been one of the fastest growing chemicals
worldwide during the past two decades, due to its tailor-made properties as a fuel
oxygenate in reformulated gasoline. The use of *t*-butanol in gasoline is also attractive
because its properties are close to those of oxygenate ethers and it is readily available as a
raw material for the synthesis of MTBE.

Dehydrations of aliphatic alcohols are among the most thoroughly investigated
reactions catalysed by metal oxides, which exhibit wide ranges of acid and base strength.
In the case of alcohol dehydration, the catalysts studied are sulphonated polymers such as
Amberlyst-15, acidic zeolites, clays and metal oxides such as γ-alumina.[3] Previous work
has shown that the Lewis acid sites (Al^{3+} ions) of γ-alumina are not involved in the
dehydration reaction.[4] The effects of additives, such as water, that can affect Brønsted
site behavior are thus of substantial interest.

In the course of our studies of HPA/K-10 as catalysts for the synthesis of MTBE
from *t*-butanol and methanol, as well as *t*-butanol dehydration, the role of water emerged
as a variable meriting detailed analysis. This paper presents these studies.

2 EXPERIMENTAL

t-Butanol, methanol, MTBE, 1,4-dioxane and all other chemicals were obtained from
well-known firms and were used without further purification. K-10 clay was obtained

from Fluka and Amberlyst-15 was obtained from Rohm and Haas. The various catalysts were prepared in the laboratory as described elsewhere.[1,2]

All experiments were conducted in a Parr autoclave of 100 mL capacity. The autoclave was equipped with a four blade pitched turbine impeller and could be set at varying speeds. The reactants and the particular catalyst were placed in the reactor and the temperature was raised to the desired value. An initial sample was then drawn and the reaction was monitored by periodic withdrawal of samples.

The samples were analysed by gas chromatography on a Porapak-Q column (2 m x 3.2 mm), with thermal conductivity detection. 1,4-Dioxane was used as an internal standard as well as a solvent. The quantitative analysis was done by comparison with standard synthetic mixtures.

3 RESULTS AND DISCUSSION

Table 1 shows that the best yields of MTBE were obtained with the clay-supported heteropoly acid, DTP/K-10, specifically, a 20 wt% loading of DTP on K-10 calcined at 285 °C as we have noted earlier.[1] Some aspects of HPA/K-10 catalyst have been discussed in our earlier communications. It was found that 20 wt% loading of DTP on K-10 calcined at 285 °C gave the best results as far as the catalyst activity is concerned.[1] Catalysts prepared under such conditions were used for study. All the reactions were performed under the absence of mass transfer resistance and under similar conditions unless otherwise specified.

Table 1   *Activities of the Catalysts used for the Synthesis of MTBE.**

Catalyst	% Conversion of t-BuOH	% Selectivity to MTBE
30 wt % DTP /silica, calcined at 650 °C	8	100
20 wt % DTP /K-10 clay, calcined at 285 °C	71	99
Cr^{2+} EXIL	44	95
Zr^{2+} EXIL	52	94
Sulfated zirconia	66	93
20 wt % DTS /K-10 clay, calcined at 285 °C	74	91
20 wt % DPM /K-10 clay, calcined at 285 °C	76	86
Amberlyst-15	74	86
Al^{3+} EXIL (25 meq Al^{3+}/g, dried at 120 °C)	25	84
30 wt % DTP /silica, calcined at 285 °C	55	76
Al^{3+} EXIL (25 meq Al^{3+} /g, calcined at 450 °C)	34	73
K-10 Montmorillonite	56	70
HZSM-5	77	52
Al^{3+} EXIL (5 meq Al^{3+}/g, calcined at 450 °C)	24	37
Na^+ Montmorillonite	21	33
20 wt % HPA /carbon, calcined at 285 °C	0	0

**Notes: The byproduct is isobutene. Reaction conditions: t-BuOH, 82.5 mmol; MeOH, 165 mmol; 1,4-dioxane, 34 g; catalyst loading, 2.2 % w/w of reaction mixture; reaction temperature, 85 °C; speed of agitation, 1000 rpm; reaction time, 6 h except for Amberlyst-15, for which it is 2 h; EXIL = Exchanged Interlayered Clays.*

3.1 Effect of Addition of Water in the Synthesis of MTBE

The effect of water addition was verified for higher concentrations of t-butanol. Our studies indicated that addition of small quantities of water affected only the dehydration of t-BuOH and not the etherification reaction. At higher concentrations of t-BuOH the competitive reaction of isobutylene with methanol would be reduced substantially. The mole ratio of water to methanol to t-butanol studied were 1 : 1 : 10 and 2.5 : 1 : 10. It was found that the rate of dehydration increased with the slight increase in water concentration. This was determined by the formation of isobutylene that was observed by GC. Figure 1 shows the plot of conversion with time for the t-butanol conversion with increasing water concentration for MTBE synthesis. Consequently, the dehydration of t-butanol was studied independently.

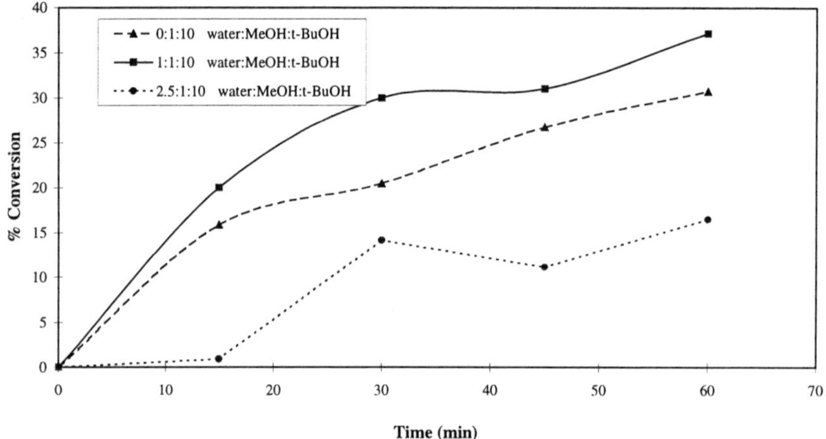

Figure 1   *Effect of water in the synthesis of MTBE.*

3.2 Dehydration of t-Butanol

The dehydration of t-butanol was studied using DTP/K-10 catalyst at 85 °C. A typical plot of conversion with time for t-butanol dehydration with varying concentrations of water is as shown in Figure 2. It is seen that as the concentration of water is increased the rate initially increases and then falls off. At a mole ratio of 1 : 4 of water : t-BuOH, there is an appreciable increase in the rate of dehydration. It can be surmised that a small amount of water increases the activity of the catalyst, whereas a further increase in the water content poisons the catalyst.

HPA/K-10 is calcined at 285 °C, a temperature sufficient to drive away all the water of crystallization and cause the lacunary phase of HPA to be lost. At temperatures above 300 °C, the HPA crystallites form aggregates and undergo a permanent decrease in surface area. However, at lower temperatures the permanent loss in activity is not observed. In the presence of aqueous media the HPA takes up the water to restore its lacunary phase, thereby rendering it active. The increase could be due to the formation of lacunary HPA formation that increases the turn-over number by opening itself up. A suitable model has been used in an attempt to analyse this phenomenon.

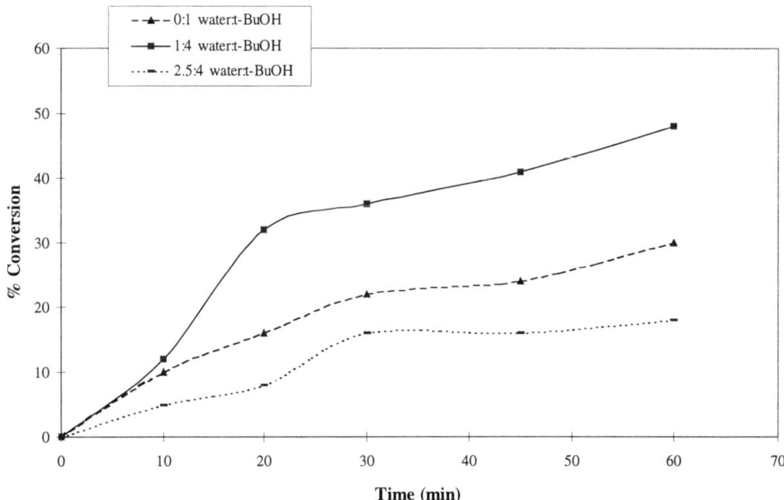

Catalyst Loading : 0.0225 g/cm3, Temperature : 85 deg C, Speed : 1000 rpm, Volume of reactrion mixture :
47 ml.

Figure 2   *Effect of water in the dehydration of t-butanol.*

4 ASPECTS OF HPA CATALYSIS

The crystal form of HPA depends on the amount of water of crystallization. Water molecules in the HPA crystal are arranged as protonated aquo-cations,[5] *e.g.* as $[H_3 \cdot 29H_2O]^{3+}$, in the face centered cubic lattice of $H_3PW_{12}O_{40} \cdot 29H_2O$. The water of crystallization can be eliminated on heating as easily as the water in zeolite, whereby the proton activity in HPA crystal is increased, resulting in higher activity in acid catalysed reactions. However, unlike the rigid anionic zeolite framework, the Keggin anions are still mobile even in dehydrated HPA crystal, and not only water but certain kinds of polar organic molecules can enter and leave the Keggin anions, being accompanied by a volume change of the HPA crystal. Nomiya *et al.*[6] have reported that this structural flexibility of HPA crystal should be taken into account when employing HPA as a heterogeneous catalyst. HPA may be a kind of mixed metal oxide in terms of chemical composition, but it must be strictly distinguished from ordinary mixed metal oxides in that HPA crystal is composed of Keggin anions which are arranged discretely. This structural feature of HPA is deeply associated with its unique properties and catalytic functions, in contrast to other mixed metal oxides and isopoly compounds such as paramolybdate.[7]

HPA used in aqueous or organic media retains its Keggin structure in the course of reactions. It has also been reported that the carbenium ions found in acid catalysed reactions are stabilized by heteropoly anion.[8]

The effect of pretreatment temperature on the activity of silica-supported $H_3PW_{12}O_{40}$ was remarkable because the maximum conversion was obtained at 150 °C with the 100 and 300 °C points exhibiting similarly low activities. This was explained in terms of remaining water of crystallization. $H_3PW_{12}O_{40} \cdot 25H_2O$ has an acid strength of $H_o > -3.0$ but it becomes more acidic ($H_o > -8.2$) when dehydrated by calcination at about 150 °C. It has been confirmed by IR spectroscopy that the Keggin structure of $H_3PW_{12}O_{40}$ is completely intact even after treatment at 300 °C. The reduction in catalytic activity on calcination was due to a decrease in the effective surface area of deposited crystallites of HPA through aggregation. It is also known that the unsupported $H_3PW_{12}O_{40}$ has a surface area of 8.3 m^2/g at 150 °C which reduces to 4.2 m^2/g at 300 °C, though the activity decreases more drastically. It has also been suggested by the same authors that for the liquid phase the reaction occurs not only on the surface of HPA but inside the crystallite of HPA supported on silica. Although HPA itself is really a protonic acid in nature, it may exhibit Lewis acid like catalytic function through the stabilizing effect of heteropoly anion on the benzyl cation as a reaction intermediate in the benzylation of benzene.[9] In this case $H_3PMo_{12}O_{40}$ supported on silica (15 % loading) showed maximum activity when calcined at 300 °C above which the HPA was found to decompose. For this case as well, the reaction is found to occur on the surface as well as within the crystallites of the supported HPA. $H_3SiW_{12}O_{40}$ and $H_3PW_{12}O_{40}$ do not decompose on the support in the range of 200–500 °C as was found by the same authors for acylation of *p*-xylene with benzyl chloride. The catalytic activity for the liquid phase dehydration of isopropanol at 80 °C with unsupported HPA is reported to correlate well with the total acid strength of the bulk HPA employed and the reaction proceeds in the interior of HPA crystals. On the contrary for vapor phase reaction at 150 °C, the conversions were lower. It has been explained that at 80 °C, a considerable amount of water is still included in the interior of HPA crystals in the form of aquo cations and this water can be exchanged with polar organic reactants, whereas at 150 °C only a meager amount of water can exist in the bulk HPA, rendering the crystal structure of HPA too rigid to accept the reactants throughout the bulk of the HPA.[10] Ogino and co-workers have reported that silica-supported hetero molybdic acids (5% w/w loadings) act as efficient catalysts for the vapor phase selective MTBE synthesis from isobutylene and methanol at 90–120 °C.[11] Higher activities of supported HPAs are ascribed to their stronger affinity allowing them to accept more polar organic substrates within their crystal bulk. Their higher ether selectivity may be due to milder acid strengths compared with heteropoly tungstic acids, which are less intimate with reactants, less susceptible to reduction by the reactants and possess stronger acidity than heteropoly molybdic acid, *i.e.* $H_3PMo_{12}O_{40}$/silica > $H_3SiMo_{12}O_{40}$/silica > $H_3PW_{12}O_{40}$/silica > $H_3SiW_{12}O_{40}$/silica.

Schwegler has presented a thorough analysis of heteropoly acids as catalysts in organic reactions.[12] According to their theory, in crystalline HPAs there are always a number of molecules of water of hydration present (normally 6 per Keggin unit). The protons are localized between two water molecules, forming $H_5O_2^+$ cations, or larger aggregates if more water molecules are present. So protons do not bind directly to the anion until the HPA is fully dehydrated, which is the case after heating to about 300 °C. At higher temperatures beyond 300 °C, two protons combine with an oxygen atom from the HPA and form water leaving an HPA molecule with oxygen vacancies and thus destabilizing it.

The positions of the protons in dehydrated HPAs are not well known and in all probability the protons are still quite mobile. Based on extended Hückel calculations, it is claimed that protonation would occur at the angular W–O–W bonds. At high temperatures other types of protons may be present in HPAs in the solid form. The HPAs formed are of the so-called lacunary type. They exist for both Keggin and Dawson types of structures of HPAs. The lacunary HPAs have different properties from the other complete HPAs in that (a) they are not really acidic, (b) they are not soluble in organic solvents (c) they have a buffering capacity, and (d) they are only stable in a fairly narrow pH range 3–7.

4.1 Pseudo-liquid Phase HPA Catalysis

The psuedo-liquid phase catalysis has characteristics of both dissolved and solid HPA. The psuedo-liquid phase is found under conditions intermediate to those for fully heterogeneous or homogeneous catalysis. The HPA is present as a solid, somctimes as crystals on a carrier, containing a certain number of molecules of water of hydration. If the catalyst is fully dehydrated the psuedo-liquid phase behavior is lost for most compounds. However, pyridine and ammonia are still taken up by the bulk of the catalyst even at 300 °C. Application of this typical behavior is possible in reactions where the reactant(s) and product(s) have different hydrophobicities. A reaction well suited to this phase of the HPA catalyst is dehydration of alcohols to alkenes. The acid sensitive product is obtained with an unusually high selectivity because the bulk of the catalyst is covered by hydrophobic alkene.[10,13–15] The dehydration of t-BuOH involves a hydrophilic t-BuOH molecule and a hydrophobic isobutylene molecule. Thus, the likelihood of adsorption of water on the sites is more than that of isobutylene.

Experiments on the effect of HPA loading on clay has shown that as the loading is increased the rate of MTBE formation increases. It should be recognized at monolayer coverage of HPA crystallites, the surface atoms are the heteroatoms (W, P) Si, Al and O and H^+ and OH ions. Whether lacunary HPAs are generated by the interaction of HPA on clay when calcined at 285 °C and whether they coordinate with surface cations of clay is a moot point. Since the catalysts were found to be stable on repeated use without any leaching of the HPA from the support, it appears that they are not simply physisorbed on the K-10 clay.

5 MODEL FOR t-BUTANOL DEHYDRATION

The foregoing literature review was presented in order to build a proper model for t-BuOH dehydration in the presence of DTP/K-10. Preliminary experiments had indicated that the amount of water initially present can enhance the rate of reaction, contrary to the usual behavior. Gates and coworkers have shown that the amount of water present in the reaction mixture of t-butanol and water acts as a diluent for a large concentration of water wherein the reaction is first order in t-butanol with a sulfonated polymer catalyst.[16–18] However, water acts as an inhibitor when present in the polymer phase. The role of water has thus to be ascertained for the present case.

The experiments conducted in the absence of any initial water did have some water at time t = 0 when the reaction commenced. The sampling at t = 0 showed the presence of water at 7.765 x 10^{-4} gmol/cm^3 at 85 °C. When water was added to the reaction mixture, the amount at zero time was 9.14 x 10^{-4} gmol/cm^3. It should be noted that dioxane was used as a solvent and hence the liquid phase was homogeneous phase with solid catalyst. The rate of reaction was found to increase from 9.64 x 10^{-4} to 2.46 x 10^{-3}

gmol·cm^{-3}·min^{-1} or from 0.0440 to 0.1123 gmol·gcat^{-1}·min^{-1} based on 2.19 x 10^{-2} gm/cm^3 of catalyst. When the amount of water was increased from 9.14 x 10^{-4} gmol/cm^3 to 1.14 x 10^{-3} gmol/cm^3, the rate was found to decrease from 2.46 x 10^{-3} to 5.46 x 10^{-4} gmol·cm^{-3}·min^{-1} or from 0.1123 to 0.0249 gmol·gcat^{-1}·min^{-1}. However, it should be noted that the rate was lower than when no initial water was added.

It appears that a certain minimum amount of water increases the total acidity of the catalyst by one of the following mechanisms:

- Since the catalyst was calcined at 285 °C, the water of crystallization was present in trace amounts, making the crystallite size of DTP/clay rigid. The reaction occurs only on the surface of DTP exposed to the *t*-BuOH molecules from solution which form *t*-BuOH$_2$$^+$ type of complex, because HPAs are known to be protonated acids.
- A small amount of water is required in the interior of the HPA crystal in the form of aquo cations, and this water can be exchanged with polar *t*-BuOH to form additional -BuOH$_2$$^+$ sites. In other words, when no water is present, not all protonated forms of HPA are available for acid catalysis, particularly from the interior of HPA (pseudo-liquid phase).
- When large quantities are taken, water gets adsorbed onto some sites thereby poisoning the catalysts for adsorption of *t*-BuOH.
- As has been suggested by some workers that Lewis acid centers in the Keggin structure are hydrogen bonded to water, generating additional Brønsted acidity, but still more water blocks the sites to poison them. The above hypotheses have a basis because for the synthesis of MTBE as well some quantity of water added initially has a positive effect on the rate of reaction, (Figure 3).

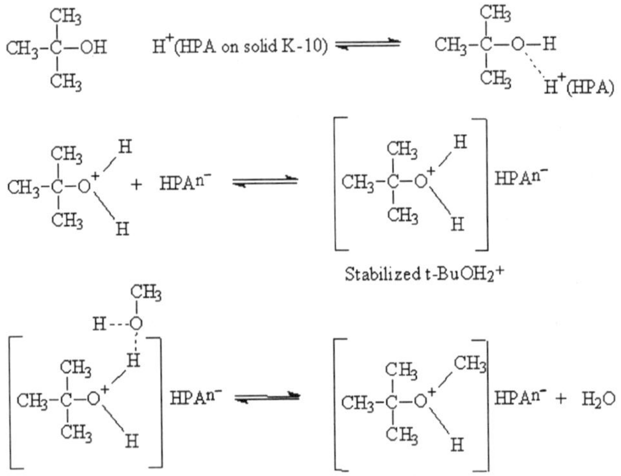

Figure 3   *Generation and stabilization of t-BuOH$_2$$^+$ with Keggin structure of HPA.*

To facilitate the discussion of the reaction kinetics to follow, we list in Table 2 the abbreviations and terms used.

Table 2   *Abbreviations and Terms used in the Discussion of Kinetics.*

A	=	methanol
B	=	*t*-butanol
C	=	concentration, $gmol/cm^3$ for liquid phase
C_A	=	concentration of methanol
C_B	=	concentration of *t*-butanol
C_t	=	total number of active sites
C_S	=	concentration of vacant sites
EXIL	=	exchanged interlayer
r_o	=	overall rate of reaction based on liquid phase volume, $gmol/cm^3/sec$
$k_1, k_2..._$	=	forward rate constants
$K_1, K_2..._$	=	adsorption equilibrium constant for reaction 1,2 ...
K_W	=	adsorption equilibrium constant for W
M	=	initial molar ratio of C_{AO}/C_{BO}
r_i	=	$-dc_i / dt$, rate of reaction species I
r_{BS}	=	rate of dehydration of *t*-butanol
S	=	vacant site
t	=	time
w	=	catalyst loading, g/cm^3 liquid phase
W	=	water
X_A	=	fractional conversion methanol $(C_{AO} - C_A)/ C_{AO}$
X_B	=	fractional conversion *t*-butanol $(C_{BO} - C_B)/ C_{BO}$

Let S_o be the original sites in the dehydrated catalyst and S_a be the additional sites generated due to the presence of trace water (on external addition), below its critical amount, beyond which poisoning of sites occurs due to adsorption of water. Let n be the number of centers which can be activated due to m moles of water resulting in p protonated sites.

$$n \ + \ mH_2O \ \rightleftharpoons \ p\,[\,H^+\,] = S_a \tag{1}$$

The exact stoichiometric amount of water required below the critical concentration cannot be ascertained *a priori*. In terms of site concentration,

$$Ct \ = \ C_{So} + C_{Sa} \tag{2}$$

where C_{So} = concentration of original active sites, C_{Sa} = concentration of additional sites generated due to traces of water.

The rate of dehydration of *t*-BuOH is given by:[19]

$$r_{BS} \ = \ \frac{k_B C_B Ct}{1 + K_B C_B + K_W C_W} \tag{3}$$

where $C_W = C_{wt} - C_{wa}$ = amount of water in excess over that required for generation of acid centers, and C_{wt} = total quantity of water required. Substituting for C_t from equation 2, equation 3 becomes

$$r_{BS} = \frac{k_B C_B C_{So} + k_B C_B C_{Sa}}{1 + K_B C_B + K_W C_W} \tag{4}$$

$$= \frac{k_B C_B C_{Sa}}{1 + K_B C_B + K_W C_W} + \frac{k_B C_B C_{So}}{1 + K_B C_B + K_W C_W} \tag{5}$$

$$= \frac{k_B C_B\, k_W C_{Wa} C_{So}}{1 + K_B C_B + K_W(C_{Wt} - C_{WC})} + \frac{k_B C_B C_{So}}{1 + K_B C_B + K_W(C_{Wt} - C_{WC})} \tag{6}$$

where C_{WC} = critical amount of water beyond which poisoning occurs. Note that in the above equation, it is assumed that $C_{Sa} = k_W C_{Wa} C_{So}$. The above equality may not be precisely valid but would give an idea about the effect of water.

Gathering terms we can rewrite equation 6 as

$$r_{BS} = \frac{k_B C_B\, [k_W C_{Wa} + 1]\, C_{So}}{1 + K_B C_B + K_W(C_{Wt} - C_{WC})} \tag{7}$$

For $C_{Wt} < C_{WC}$, the critical water concentration above which poisoning starts the above equation should not have any poisoning term and hence the rate of dehydration is

$$r_{BS} = \frac{-dC_B}{dt} = \frac{k_B C_B[k_W C_{Wa} + 1]}{1 + K_B C_B} \tag{8}$$

Note the constant C_{So} is merged with k_B without loss of generality.

Case I: At higher concentration of *t*-butanol, then $K_B C_B \gg 1$ and equation 8 simplifies to

$$\frac{-dC_B}{dt} = \frac{k_B\, [k_W C_{Wa} + 1]}{K_B} \tag{9}$$

The initial rate of reaction of *t*-butanol is given by

$$\left[\frac{-dC_B}{dt}\right]_i = \frac{k_B}{K_B} + \frac{k_B k_W C_{Wa}}{K_B} \tag{10}$$

or more simply, $r_i = (\text{constant 1}) + (\text{constant 2}) \cdot C_{Wa}$.

At two different concentrations it is possible to find k_B and k_W, or from the plot of initial rate against C_{Wa} these two constants could be found for experimental conditions where the rate had increased with water concentration.

Case II: For weak adsorption of *t*-butanol, then $K_BC_B \ll 1$ and equation 8 simplifies to

$$\frac{-dC_B}{dt} = [k_WC_{Wa} + 1]\, k_BC_B \tag{11}$$

The sites generated on addition of water are constant and do not vary with time, therefore equation 11 can be simplified by gathering constants to

$$\frac{-dC_B}{dt} = k_BC_B + k^lC_B \tag{12}$$

which is a typical first order equation. In terms of fractional conversion, upon integration of equation 12 we have:

$$-\ln(1-X_B) = [k_WC_{Wa} + 1]\, k_Bt \tag{13}$$

Thus, a plot of $-\ln(1-X_B)$ *vs.* time should give a slope of $[k_WC_{Wa} + 1]\, k_B$.

Case III: When water concentration is greater than critical, $C_{Wt} \gg C_{WC}$, then poisoning of sites occurs and the above equation can be further simplified to

$$r_{BS} = \frac{k_BC_BC_t}{1 + K_BC_B + K_WC_W} \tag{14}$$

if $K_WC_W \gg K_BC_B + 1$, then

$$r_{BS} = \frac{k_BC_BC_t}{K_WC_W} \tag{15}$$

On the other hand if $1 + K_WC_W \gg K_BC_B$, then inversion of equation 14 leads to

$$\frac{1}{r_{BS}} = \frac{K_W}{k_BC_WCt} + \frac{K_B}{k_BCt} \tag{16}$$

Thus, a plot of (initial rate)$^{-1}$ *vs.* C_{W0}/C_{B0} should be a straight line with intercept as (K_B/k_BC_t) and slope (K_W/K_BCt). Since k_B is already known and C_t is the catalyst loading, both K_B and K_w can be established.

6 VALIDATION OF THEORY

On the basis of the above model different plots were made. The plot of $-\ln(1-X_B)$ *vs.* time was found to be a straight line indicating the first order behavior of *t*-butanol dehydration (Figure 4). The plot of initial rate with water concentration showed a maximum at a concentration of 9.14×10^{-4} (Figure 5) while the plot of inverse of initial rate against C_{WO}/C_{BO} showed a reverse trend (Figure 6).

For a dilute water concentration (mole ratio of water to *t*-BuOH of 1 : 4), the data are well-fitted by a straight line passing through the origin (Figure 4). The slope is calculated as $[k_W C_{Wa} + 1]k_B$. Another experiment was conducted at a mole ratio of water to *t*-BuOH of 1 : 3, meaning thereby water concentration of generated water within was taken as t = 0, when no water was added. These two slopes are plotted in Figure 7. The intercept is $k_B = 4.3$ and slope is $k_B k_W$. Thus we obtain a value for k_W of 5.39×10^3 at a catalyst loading of 0.0219 g/cm^3.

The plot of (initial rate)$^{-1}$ *vs.* (C_{WO}/C_{BO}) shows that the rate increases from $C_{WO}/C_{BO} = 0.74$, goes through a maximum at $C_{WO}/C_{BO} = 0.52$ and declines drastically below $C_{WO}/C_{BO} \approx 0.5$. The poisoning is evident.

6.1 Conclusions on Role of Water on Dehydration Reaction

Water increases the rate of dehydration to a certain critical amount. Beyond that amount it poisons the activity of the catalyst. The cause of increase in the activity could be due to the formation of lacunary HPA that increases the turnover number of *t*-BuOH from the acidic sites, thereby increasing the rate and adsorption of water creating Brønsted acidity. Advanced spectroscopic techniques are required to conclusively prove the occurrence of such a phenomenon.

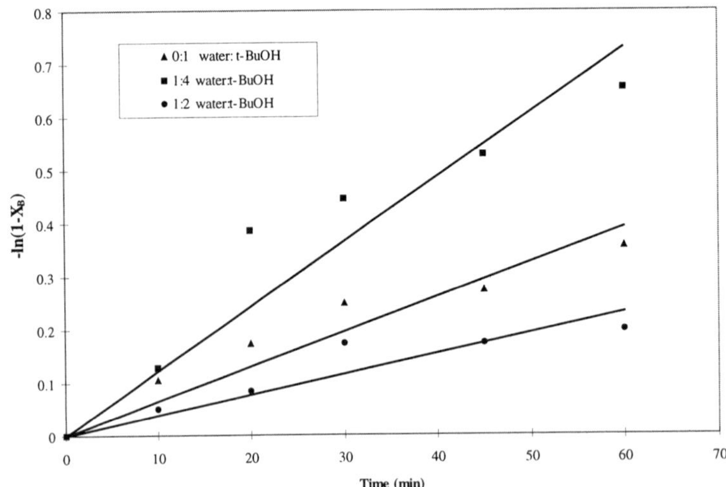

Figure 4   *Plot of –ln(1–X_B) vs. time.*

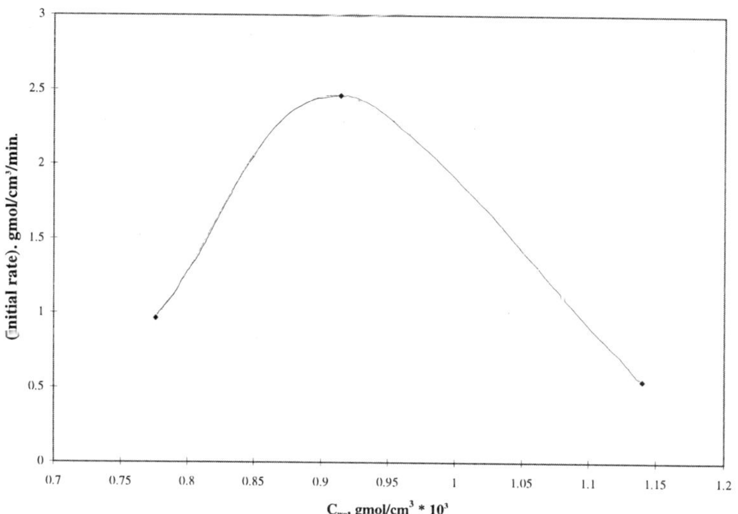

Figure 5   *Plot of (initial rate)$^{-1}$ vs. C_{wo}.*

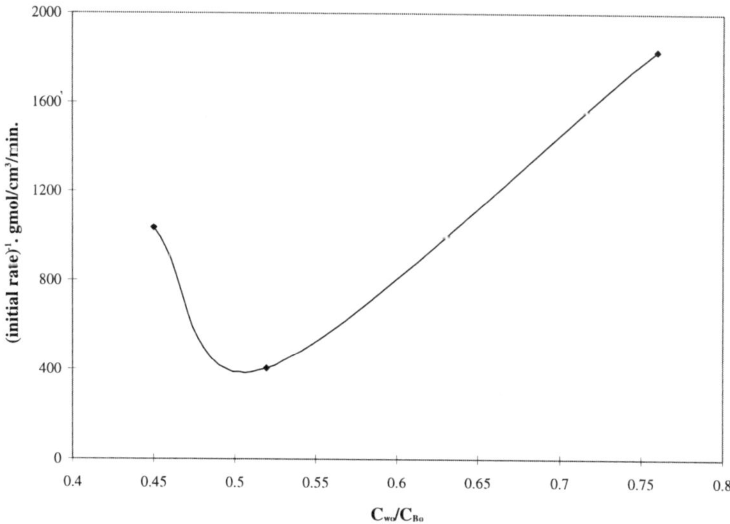

Figure 6   *Plot of (initial rate)$^{-1}$ vs. C_{wo}/C_{Bo}.*

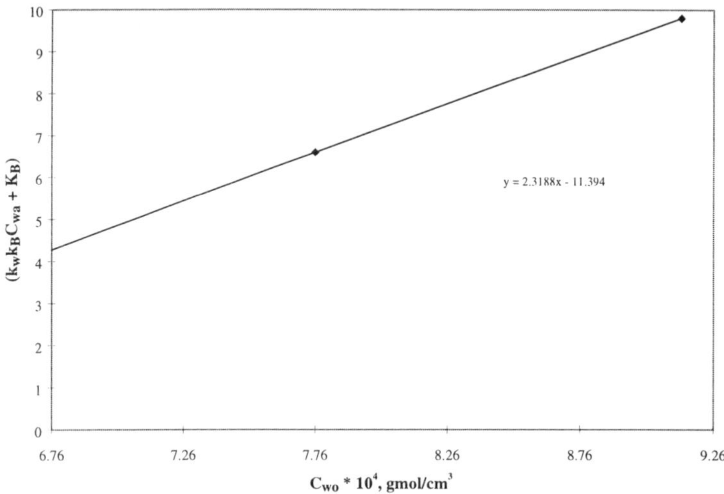

Figure 7. *Plot of $(k_w k_B C_{wa} + \kappa B)$ vs. C_{wo}.*

Further, more experimental data with different water concentration and mole ratio of water to *t*-BuOH and temperature effects on HPA/clay calcination must be collected.

7 EFFECT OF WATER ON ETHERIFICATION

Water concentration up to a critical value increases the Brønsted acidity of the solid catalyst, and hence the concentration of $BuOH_2^+$ increases. Thus, the rate of reaction of *t*-BuOH would also increase. An analogous equation could be written for it,[19] *i.e.*,

$$r_{BS} \ = \ r_0 \ = \ \frac{(k_W C_W + 1)(K_9 K_1 C_B C_A + k_2 K_1 C_B)Ct}{(1 + \Sigma K_i C_i + K_W(C_{Wt} - C_{WC}))} \tag{17}$$

where K_9 is the adsorption equilibrium constant for the reaction of methanol and adsorbed *t*-butanol. DTP/K-10 catalyst is hydrophilic and hence the numerator terms can be simplified to

$$r_{BS} \ = \ \frac{(k_W C_W + 1)(K_9 K_1 C_B C_A Ct)}{(1 + \Sigma K_i C_i + K_W(C_{Wt} - C_{WC}))} \tag{18}$$

For water concentration below the critical value,

$$r_{BS} \ = \ \frac{(k_W C_W + 1)(K_9 K_1 C_B C_A Ct)}{(1 + K_1 C_B)} \tag{19}$$

$$= \frac{k_W C_W\, K_9 K_1 C_B C_A Ct}{(1 + K_1 C_B)} + \frac{K_9 K_1 C_B C_A Ct}{(1 + K_1 C_B)} \qquad (20)$$

| Rate of reaction due to generation of additional Brønsted acidity | Rate of reaction in the absence of such a fact (no water) |

Furthermore, if $K_1 C_B > 1$, then equation 20 becomes

$$\begin{aligned}
r_{BS} &= k_W C_W\, K_9 C_A C_t + K_9 C_A C_t \\
&= (k_W C_W\, K_9 C_t + K_9 C_t) C_A \\
&= k_{ap} C_A \qquad (21)
\end{aligned}$$

For a fixed $C_W < C_{WC}$, equation 21 is a psuedo-first order equation, and

$$Ln(M) = k_{ap} t \qquad (22)$$

If, however, $K_1 C_B < 1$, then equation 20 becomes

$$r_{BS} = (k_W C_W + 1)\, K_9 K_1 C_B C_A Ct \qquad (23)$$

which is a typical second order reaction rate expression. When $M \neq 1$ the above equation on integration gives

$$\ln\{(M - X_B)/(M(1 - X_B))\} = (C_{AO} - C_{BO}) k_{ap} t \qquad (24)$$

but when $M = 1$ integration of equation 23 gives

$$\frac{1}{C_B} - \frac{1}{C_{BO}} = \frac{1}{C_{BO}} \cdot \frac{X_A}{(1 - X_A)} = k_{ap} t \qquad (25)$$

Thus, for $M \neq 1$ a plot of $\ln((M - X_B)/M(1 - X_B))$ *vs.* t will give a slope equal to $k_{ap}(C_{AO} - C_{BO})$, and for $M = 1$ a plot of $(X_A/(1 - X_A))$ *vs.* t will give a slope equal to $(C_{BO} k_{ap})$, from which k_{ap} can be found. Here

$$\begin{aligned}
k_{ap} &= (k_W C_W + 1)\, K_9 K_1 C_t \\
&= k_W C_W\, K_9 K_1 Ct + K_9 K_1 Ct \\
&= k_{ap-1} C_W + k_{ap-2} \qquad (26)
\end{aligned}$$

A plot of k_{ap} *vs.* C_W for $C_W < C_{WC}$ will give a slope equal to k_{ap-1} and intercept equal to k_{ap-2}. If catalytic sites are poisoned by excess water, equation 4 can be rewritten as

$$r_{BS} = \frac{K_9 K_1 C_B C_A Ct}{1 + K_1 C_B + K_W C_W} \qquad (27)$$

For $K_W C_W \gg (1 + K_1 C_B)$,

$$r_{BS} = \frac{K_9 K_1 Ct}{K_W} \cdot \frac{C_B C_A}{C_W} \tag{28}$$

This shows that for a fixed concentration of C_{BO} and C_{AO} at $t = 0$, the rate of reaction is inversely proportional to C_{WO}. Thus a plot of rate of reaction of t-BuOH against C_{WO} will be a straight line with negative slope.

8   CONCLUSIONS

In a single pot synthesis of MTBE from t-butanol and methanol wherein dodeca-tungstophosphoric acid on K-10 clay was used as catalyst, the amount of water was found to have a pronounced effect on the dehydration of t-butanol and also on the formation of MTBE. A small amount of water, up to a critical point, was found to increase the rate of dehydration and etherification reactions when studied independently. Any amount of water above the critical point gradually poisons the catalyst. The rate enhancement for etherification and dehydration reactions has been explained with a suitable model wherein the Eley–Rideal type of mechanism with adsorption of t-butanol and its reaction with methanol from the liquid phase is found to hold. When no methanol is present or when stoichiometric quantities of methanol are used, dehydration reaction contributes substantially as a parallel reaction to the overall rate of reaction of t-butanol. Furthermore, in presence of methanol the etherification reaction is much faster than the dehydration reaction. No free isobutylene is detected nor are its oligomers formed when the mole ratio of methanol to t-butanol is high.

9   ACKNOWLEDGMENTS

This work was supported under a research grant by CSIR, New Delhi. GDY acknowledges the Darbari Seth Endowment for the Chair and research support. NK thanks Professors J. Jackson and D. Miller for financial support, and J. Jackson for valuable suggestions.

References

1.   G. D. Yadav and N. Kirthivasan, *J. Chem. Soc., Chem. Commun.*, 1995, 203.
2.   G. D. Yadav and N. Kirthivasan, *Appl. Cat. A*, 1997, **154**, 23.
3.   B. C. Gates, 'Catalytic Chemistry', Wiley, New York, 1992.
4.   H. Knozinger, in 'The Functional Groups. The Chemistry of the Hydroxyl Group', S. Patai, ed., Wiley Interscience, New York, 1971, p. 642.
5.   A. F. Wells, 'Structural Inorganic Chemistry', 3rd edn., Clarendon Press, 1962.
6.   K. Nomiya, T. Ueno and M. Miwa, *Bull. Chem. Soc. Jpn.*, 1980, **53**, 827.
7.   M. T. Pope, 'Heteropoly and Isopoly Oxometallates', Springer-Verlag, 1983.

8.   Y. Izumi, K. Urabe and M. Onaka, 'Zeolites, Clays and Heteropoly Acids', VCH Publishers, Inc., 1992.
9.   Y. Izumi, N. Natsume, H. Takamine, I. Tamaoki and K. Urabe, *Bull. Chem. Soc. Jpn.*, 1989, **2**, 2159.
10.   M. Misono, T. Okuhara, T. Ichiki, T. Arai and Y. Kanda, *J. Am. Chem. Soc.*, 1987, **109**, 5535.
11.   S. Ogino, S. Iganishi and T. Matsuda, *Jpn. Pat. 1974*, 14909.
12.   M. A. Schwegler, 'Heteropolyacids as Catalysts in Organic Reactions', *Ph.D. Dissertation*, Technical University of Delft, Netherlands, 21 Oct 1991, p. 105.
13.   T. Baba and Y. Ono, *J. Mol. Catal.*, 1983, **37**, 327.
14.   T. Okuhara, T. Hashimoto, M. Misono, Y. Yoneda, H. Niiyama, Y. Saito and E. Echigoya, *Chem. Lett.*, 1983, 573.
15.   Y. Saito and H. Niiyama, *J. Catal.*, 1987, **106**, 329.
16.   B. C. Gates, J. S. Wisnouskas and H. W. Heath, Jr., *J. Catal.*, 1972, **24**, 320.
17.   B. C. Gates and L. N. Johanson, *J. Catal.*, 1969, **14**, 69.
18.   B. C. Gates and W. Rodriguez, *J. Catal.*, 1973, **31**, 27.
19.   N. Kirthivasan, *Ph.D. Dissertation*, University of Mumbai, 1995.

THEORETICAL STUDIES OF THE SURFACES OF MODIFIED AND UNMODIFIED FUMED SILICAS AND X/SiO$_2$ (X = Al$_2$O$_3$, TiO$_2$, GeO$_2$)

V.M. Gun'ko

Institute of Surface Chemistry
31 Prospect Nauki
Kiev, Ukraine 252022

1 INTRODUCTION

Dispersed individual and mixed oxides are widely used in industry. The properties of these oxides may be improved in many respects by modification of their surfaces with organic (OC) or organosilicon compounds (OSC).[1,2] Accomplishing these surface modifications is not trivial, as the reactions depend on many hard-to-control factors, such as the initial state of the surface, the amount of adsorbed water, the phase distribution in mixed oxides, surface topography, *etc.* And once formed, the bonds between the modifier and the oxide surface may or may not persist, depending on their polarity and polarizability, their reactivity towards common environmental factors such as water vapor, and the density of the surface coverage. Water plays the main role in these matters, as reactions occur mainly between the modifiers and surface hydroxyls (*i.e.* dissociatively adsorbed water) and the durability of the bonds depends on their hydrolysability. Additionally, water is a source of redox reagents such as •OH and a carrier of oxygen participating in degradation of the materials. Therefore, the study of surface reactions with the participation of water or hydroxyl groups is important for a deeper understanding of the above-mentioned issues.

Our focus in this work are surface active sites of metal oxides: MOH, $M_{(1)}O(H)M_{(2)}$, MOR, where M = Si, Al, Ti, Ge and R = OC, OSC. The principal goals are (1) to understand the interaction between these surface active sites and water (molecularly and dissociatively adsorbed, or upon its associative desorption, thermal decomposition of ≡SiOR groups), and (2) to understand the solvation of modified and unmodified oxides. The tools we employed in this theoretical study were the *ab initio* packages *Gaussian 94*[3] and *GAMESS*,[4] at different basis sets, and semi-empirical methods such as AM1,[5] PM3,[6] and NDDO[7] in a cluster approach.

2 THEORETICAL MODELING

According to experimental and theoretical studies,[8,9] the activation energy ($E^{\neq}$) of water desorption from silica or mixed metal oxides of silica with alumina, titania or germania, increases as follows: desorption of water molecules hydrogen-bonded ($E^{\neq}$ < 70 kJ/mol) < DAC of water ($E^{\neq}$ = 60–90 kJ/mol) < bridging OH + terminal OH → H$_2$O (90–150 kJ/mol) < two vicinal hydrogen-bonded or twin OH → H$_2$O (140–200 kJ/mol for silica) < two isolated OH → H$_2$O (200–250 kJ/mol for silica), where DAC is the electron-donor–acceptor complex. We assume that associative desorption of water starts in dense islands of OH groups or in contact between particles, and that such a reaction is limited by H$^+$ transfer at $E^* \approx$ 90–120 kJ/mol. The motion along the reaction coordinate in this reaction is linked only with a slight surface deformation and elongation of O–H upon H$^+$ transfer. For SiO$_2$/Al$_2$O$_3$ (SA), SiO$_2$/TiO$_2$ (ST), and titania, the subsidiary valence coordination numbers of metal (Al or Ti) and O atoms cause lower E^* in reactions with the participation of neighboring bridging (Brønsted acid sites, B-sites) and terminal hydroxyls in comparison with that for two ≡SiOH groups at the silica surface. Upon dehydroxylation of silica or SA, the potential energy surface (PES) sections along the reaction pathways have 2–3 maxima corresponding to reconstruction of the neighboring polyhedrons (side maxima of 60–70 kJ/mol for SA and up to 200 kJ/mol for silica) and H$^+$ transfer (the central highest maximum of 100 kJ/mol for SA and 240–270 kJ/mol for silica) between two OH groups upon the formation of a water molecule. The Si atoms are only 4-fold coordinated and bridging OH groups are absent on silica, so elimination of H$_2$O from two terminal groups involves formation of the defect sites ≡Si• and ≡SiO•, or strained rings =Si<$_O^O$>Si=. This requires higher energy than elimination of H$_2$O from bridging MO(H)M and terminal MOH with no formation of such strained bonds.

For dehydroxylation of silica, especially *via* reactions with the participation of isolated OH groups, the reaction pathway is longer than that for neighboring ≡MOH and ≡MO(H)M≡ groups at the surface of mixed oxides. The E^* value for elimination of H$_2$O from even geminal OH groups is high (250–270 kJ/mol by AM1 and 209 kJ/mol by MP4/6-31G**). However, the pathway is shorter than that for vicinal or isolated hydroxyls. Additionally, subsequent reconstruction of the surface can reduce the endothermic effect in this reaction because ≡Si–O groups are unstable. This is clear from comparison of the energy for the reactions

$$>Si(OH)_2 \; \rightarrow \; >Si{=}O + H_2O \qquad\qquad (270 \text{ kJ/mol})$$

and

$$2((HO)_4Si) \; \rightarrow \; (HO)_2Si{<}_O^O{>}Si(OH)_2 + 2H_2O \qquad (142 \text{ kJ/mol}).$$

The coordination number of M and O atoms at the oxide surface as well as the type of surface hydroxyls (Table 1) is of importance for the active site characteristics, *e.g.* their acidity or catalytic activity, which is maximum for B-sites such as ≡SiO(H)M≡ (M = Al or Ti). According to the calculations of the SA clusters, Al atoms tend toward the 4-fold coordinated state for individual AlO$_n$ polyhedrons in the silica matrices, as the bond length r_{Al-O} for two (r_{MO}, $r_{M\leftarrow O}$) from six bonds between AlO$_6$ and SiO$_4$ (or AlO$_4$ for the alumina cluster) units are longer than an ordinary Al–O bond in Al$_2$O$_3$ by 0.03–0.05 nm. This is observed presumably when the individual alumina phase is absent (*i.e.* there is only Al^{IV}). Similar results were obtained for ST, where Ti^{IV} atoms are observed in the silica matrix. These atoms M^{IV} correspond to Lewis acid sites, which are active in many catalytic reactions of OC and can promote dissociative adsorption of water.

The energy of dissociative adsorption of a water molecule (Equation 1) is higher than that of molecular adsorption, except for the case of pure SiO$_2$, in which the process

$$\equiv M\text{-}O\text{-}Si \equiv + \; H_2O \; \rightarrow \; \equiv M(OH)\text{-}O(H)\text{-}Si \equiv \qquad\qquad (1)$$

Table 1   *Parameters of Terminal and Bridging OH Groups (Ab Initio, Basis Set is* *LANL1DZ and* **6-31G(d)).*

Structure	q_M	$-q_O$	q_H	r_{OH} (nm)
SiOH*	2.192	1.015	0.467	0.09441
TiOH*	1.960	0.923	0.433	0.09406
GeOH*	2.144	1.002	0.457	0.09461
TiO(H)Ti*	1.772	1.110	0.449	0.09453
TiO(H)Si*	1.937$_{Ti}$ 2.240$_{Si}$	1.138	0.529	0.09491
GeO(H)Si*	2.121$_{Ge}$ 2.240$_{Si}$	1.006	0.456	0.09561
TiO(H)Al*	1.946$_{Ti}$ 2.107$_{Al}$	1.063	0.477	0.09526
AlO(H)Si*	2.032$_{Al}$ 2.240$_{Si}$	1.141	0.526	0.09543
SiOH**	1.488	0.848	0.476	0.09469
AlO(H)Si**	1.348$_{Al}$ 1.643$_{Si}$	0.946	0.530	0.09537
OH Ti<>Al* OH	1.317$_{Ti}$ 2.056$_{Al}$	1.143	0.455	0.09420

Table 2   *Energy of Bond Cleavage of Surface OH Groups.*

Reaction	ΔE_t (kJ/mol)	Method
$\equiv$TiOH $\rightarrow$ $\equiv$Ti$\bullet$ + $\bullet$OH	133	NDDO
$\equiv$TiOH $\rightarrow$ $\equiv$Ti$^+$ + OH$^-$	790	NDDO
$\equiv$TiOH $\rightarrow$ $\equiv$TiO$^-$ + H$^+$	1082	NDDO
Ti-O(H)-Ti $\rightarrow$ Ti-O$\bullet$-Ti + $\bullet$H	438	NDDO
Ti-O(H)-Ti $\rightarrow$ Ti-O$^-$-Ti + H$^+$	1019	NDDO
$\equiv$SiOH $\rightarrow$ $\equiv$Si$\bullet$ + $\bullet$OH	496	AM1
$\equiv$SiOH $\rightarrow$ $\equiv$Si$^+$ + OH$^-$	940	AM1
$\equiv$SiOH $\rightarrow$ $\equiv$SiO$\bullet$ + $\bullet$H	397	AM1
$\equiv$SiOH $\rightarrow$ $\equiv$SiO$^-$ + H$^+$	1299	AM1
$\equiv$AlOH $\rightarrow$ $\equiv$Al$\bullet$ + $\bullet$OH	360	AM1
$\equiv$AlOH $\rightarrow$ $\equiv$Al$^+$ + OH$^-$	869	AM1
$\equiv$AlOH $\rightarrow$ $\equiv$AlO$\bullet$ + $\bullet$H	483	AM1
$\equiv$AlOH $\rightarrow$ $\equiv$AlO$^-$ + H$^+$	1380	AM1
Al-O(H)-Si $\rightarrow$ Al-O$\bullet$-Si + $\bullet$H	487	AM1
Al-O(H)-Si $\rightarrow$ Al-O$^-$-Si + H$^+$	1122	AM1
Al-O(H)-Si $\rightarrow$ Al-O$^-$-Si + H$^+$	1113	NDDO

$\equiv$Si-O-Si$\equiv$ + H$_2$O $\rightarrow$ $\equiv$Si($\leftarrow$OH$^{\delta-}$)-O(H$^{\delta+}$)-Si$\equiv$ seems to be less likely than $\equiv$SiOH + H$_2$O $\rightarrow$ $\equiv$Si(H)O...HOH. However, dissociative adsorption of water is the most probable at the phase boundary of X/SiO$_2$. Formation of acidic $\equiv$M(OH)-O(H)-Si$\equiv$ groups at the interface leads to clustering of water near this bridge and the main part of adsorbed water is located at the phase boundary of X/SiO$_2$, where the hydrogen bond network can be strongly changed in a few statistical monolayers. The acid property of the $\equiv$M-O(H)-Si$\equiv$ groups correlates with the atomic charge value of H (Table 1, q_H), which is higher than that for terminal $\equiv$MOH groups. Also, the energy of H$^+$ removal is lower for B-sites (Table 2), but this energy is greater than for H$\bullet$ removal.

These results led us the conclusion that hydrolysis of the $\equiv$M-O-Si$\equiv$ bond occurs with the participation at least of two water molecules. The first creates the bridging and terminal OH groups, and the second attacks the Si atom and cleaves the bridge, as in Scheme 1.

$$\equiv\text{Si-O-M}\equiv \ + \ \text{H}_2\text{O} \ \rightarrow \ \equiv\text{Si-O(H}^{\delta+}\text{)-M(OH}^{\delta-}\text{)}\equiv$$

$$\equiv\text{Si-O(H}^{\delta+}\text{)-M(OH}^{\delta-}\text{)}\equiv \ + \ \text{H}_2\text{O} \rightarrow \ \equiv\text{SiOH} + \text{M(OH)}\equiv \ + \ \text{H}_2\text{O}$$

Scheme 1

The PES sections (Figure 1) demonstrate that the process corresponding to Equation 1 is more probable for $\equiv$Si-O-Al$\equiv$ than for $\equiv$Si-O-Si$\equiv$, but in all instances is exothermic. For similar reactions we typically see that the polarity of the active bonds is higher in the transition state (TS) than in the pre-reaction complex (hydrogen bonded or DAC). Therefore, the clustering of water molecules near this bridge can reduce E^* due to additional polarization of the bonds in the TS. It should be noted that transfer of two protons from two water molecules – from a molecule in DAC M$\leftarrow$OH$_2$ to the second H$_2$O (hydrogen-bonded), and then from the second H$_2$O to the O atom in Si-O-M – is not synchronous. Some displacement is observed in the phase of their motion and in the TS r_{OH1} = 0.1239 nm and r_{OH2} = 0.1113 nm.

It seems unlikely that dissociative adsorption of one molecule could occur without formation of the water cluster, especially for the $\equiv$Si-O-Si$\equiv$ bridge, which has an activation energy, E^*, of *ca.* 200 kJ/mol (Figure 1a). But for M = Al the second water molecule gives a slight change of E^*. For a pictorial rendition we used the bond length between the bridging oxygen atom and H from adsorbed H$_2$O (Figure 1a, solid lines) or between H from the first water molecule and O from the second (Figure 1a, dashed lines). The system motion corresponds to change in all the atom coordinates between the pre-reaction state and post-reaction complex, therefore the generalized reaction coordinate was used (Figure 1b). Interaction of the second water molecule with adsorbed (molecularly or dissociatively) H$_2$O gives additional stabilization of post-reaction complexes and the TS.

In the case of dissociation of a water molecule in the adsorbed cluster, a proton cannot form the bridging OH group but rather adds to another H$_2$O, thus forming H$_3^+$O or H$_5^+$O$_2$. This process is more probable for the boundary with liquid water. Using MNDO/H (a version of MNDO optimized for study of hydrogen bonds) we find the following:

$$\equiv\text{Si-O(H)-Al}\equiv \ \rightarrow \ \equiv\text{Si-O}^-\text{-Al}\equiv \ + \ \text{H}^+ \qquad\qquad \Delta E_t = 1122 \text{ kJ/mol}$$

$$\equiv\text{Si-O(H)-Al}\equiv\text{*5H}_2\text{O} \ \rightarrow \ \equiv\text{Si-O}^-\text{-Al}\equiv\text{*4H}_2\text{O*H}_3^+\text{O} \qquad \Delta E_t = 143 \text{ kJ/mol}$$

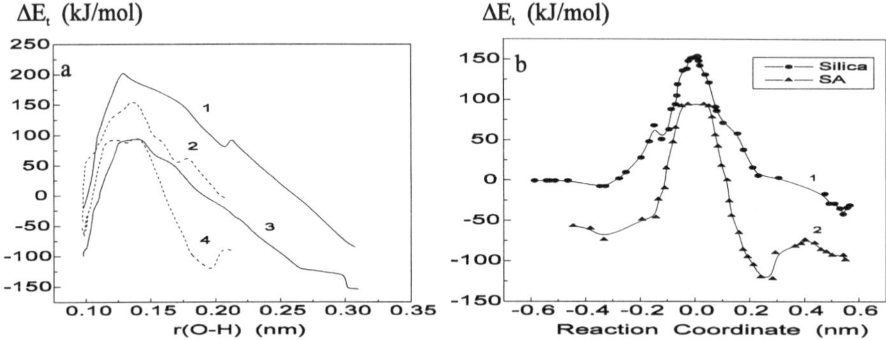

Figure 1   *Changes in the total energy (ΔE_t) as a function of (a) bond length (r_{oh}) and (b) reaction coordinate upon dissociative adsorption of H_2O onto Si-O-Si (A1, A2, and A1) and Al-O-Si (A3, A4, and B2) bridges, for participation of one molecule of water (A1 and A3), and participation of two molecules of water (A2, A4 and B).*

Consequently, the ion pair is not stabilized upon dissociation of the B-site in the presence of $5H_2O$, but ΔE_t is significantly lower than that without the adsorbed water cluster. With an increase of this cluster to $10H_2O$ or above, it is possible to stabilize the separated ion pair. That leads to a charging of the solid surface and the beginning of formation of the electrical double layer observed at the oxide/liquid water boundary.

The calculations of the solvation energy (E_s) by the AM1-SM1[10] method show that the difference in the E_s values upon solvation of the clusters $(O_3{}^*SiO)_3AlO(H)Si(OSiO_3{}^*)_3$ and $(O_3{}^*SiO)_3SiOSi(OSiO_3{}^*)_3$ (O^* is the O pseudo-atom) is small at -6 kJ/mol, but upon dissociation of the B-site, the difference in E_s values is -214 kJ/mol, with a contribution of -179 kJ/mol coming from the O^-. Notice that the E_s value for solvation of the charged cluster is relatively small as a result of the positive contributions of the boundary groups $SiO_3{}^*$ ($+70$ to $+80$ kJ/mol). Consequently, the main effect of solvation of the interfaces of AS is a result of interaction of water with the B-sites on their dissociation.

The modification of oxide with formation of $(-O-)_3Si-X$ bonds changes the electron-donor/electron-acceptor properties, and the polarizability of the surface. These changes make it possible to control the interaction between adsorbent and adsorbate.

The reaction shown in Equation 2, occurring at the silica surface, changes the properties of the active sites. For example, if X is any group besides F, its substitution for -OH in $\equiv$SiOH leads to an increase in E_{HOMO}, which can be localized on the $\equiv$SiX group. For example, when X is $-OCH_2NH_2$, $\Delta E_{HOMO} = +2.43$ eV relative to $\equiv$SiOH, which is consistent with the high electron-donor and proton-acceptor properties of such basic groups. However, the atomic charge q of the O atom in $\equiv$SiOH is the highest in the surface

$$\equiv SiOH + XY \rightarrow\ \equiv SiX + HY \tag{2}$$

groups, but q_N in $\equiv$SiOCH$_2$NH$_2$ is close to q_O (Table 3, q_{X1} and q_{X3}). Therefore, we can assume that bound groups having O or N atoms can strongly interact with polar groups, *e.g.* of the coupling agents that may be used for improvement of the cohesion and adhesion properties of coatings filled with metal oxides.

The $\equiv$SiOH groups can exhibit reactivity through not only O atoms but also H atoms in electrophilic substitution reactions (S_Ei) when the reactivity of $\equiv$SiOH is higher than that of $\equiv$SiOCH$_3$. Besides, $q_C < 0$ for X = -OCH$_3$ (Table 3, q_{X2}), meaning that S_Ei for $\equiv$SiOCH$_3$ groups is unlikely. In addition, E^* for the electrophilic substitution mechanism for the reaction $\equiv$SiOCH$_3$ + H$_2$O $\rightarrow$ $\equiv$SiOH + CH$_3$OH is higher than that for nucleophilic substitution on silicon, $S_Ni(Si)$. The probability of S_Ei reactions for other $\equiv$SiOCR groups can increase when R contains electron-acceptor atoms, such as O and N (Table 3, q_{X3} and q_{X4}) and in this case $q_C > 0.5$. However, nucleophilic substitution reactions $S_Ni(Si)$ for $\equiv$SiX groups are more probable than S_Ei as silicon has strong electron-acceptor properties ($q_{Si} \approx 1.5$) for all X, and 5-fold coordinated Si is more stable than 5-fold coordinated C. This causes Si-X (*e.g.* Si-OCR) bonds, which have high dipole moment, polarity and polarizability to be highly susceptible to hydrolysis. For example, the Si-O-Si bond in silica has a small normal component of the dipole moment (valence angle $\angle$SiOSi > 140°) and consequently has low reactivity and low hydrolysability. In contrast, $\angle$SiOC $\approx$ 124–129° for a typical alkoxysilane, whose dipole moment, μ, is high (Table 3) and near normal to the surface. Therefore, the Si-O-C bond can interact more strongly with polar molecules (*e.g.* H$_2$O) and its reactivity is higher especially in $S_Ni(Si)$ reactions. Specifically, we found that the activation energy of the $S_Ni(Si)$ reaction between Si-O-CH$_3$ and H$_2$O is lower than that for the same reaction between Si-O-Si and H$_2$O. The difference was found to be 90 kJ/mol by AM1, and 75 kJ/mol by MNDO/H. This explains the common experience that alkoxy groups are more hydrolysable than siloxy groups on oxide surfaces.

Hydrolysis of Si-X bonds at the silica surface is of importance for durability of the materials. According to the *ab initio* calculations (basis set LANL2DZ), the energy of hydrolysis of M-X (M = Al, Si, Ti, and Ge; X = OCH$_3$) depends slightly on M and equals -8, -1, 5, and -1 kJ/mol, respectively. The AM1 calculations on such a reaction for M = Si and X = NH$_2$, Cl, OCH$_3$, OPh, H, and CH$_3$ give $E^* = 36, 93, 111, 114, 225$, and 323 kJ/mol (relative to pre-reaction DAC Si$\leftarrow$OH$_2$) respectively. This gives ratios of rate constants for hydrolysis of: 6.3 x 10^7 : 1.7 x 10^3 : 3.0 x 10^2 : 1.7 x 10^2 : 2.0 x 10^{-12} : 1.0 x 10^{-25}, respectively, at 420 K. This result shows a very low probability of hydrolysis of Si-H and Si-C bonds, in accordance with their observed high stability and durability.

Oxide particles with silica, titania and alumina have strong local electrostatic surface fields, which can change the polarity parameters of adsorbed organic molecules. Using 169 point charges in the rutile titania lattice, and using q = 1.5 for the titanium nodes, and q = -0.75 for the oxygen nodes, we simulated the surface electrostatic field of rutile titania to understand its influence on the electronic structure of adsorbed organic molecules. For one small organic molecule we studied, H$_2$C=CCH$_2$OCH$_3$, we found that its dipole moment is doubled in the electrostatic field of the titania cluster. The atom charges increase too, and the change of enthalpy of formation, $\Delta\Delta H$, of this molecule relative to the free state is -75 kJ/mol, which is close to the known energy of the hydrogen bond between this molecule and the titania surface.

OC and OSC bound to pigment or filler particles can change not only adhesion properties of the coatings containing them, but also the light resistance of the coatings. In the case of methacryloyloxymethylenemethyl diethoxysilane (MADES) bound to silica and titania surfaces, electronic excitation gives different changes for E^α_{HOMO} and E^β_{HOMO} (for α and β electrons).[2] The spin density of the exciton (electron-hole pair) generated by UV irradiation is localized on the C=C bond of bound MADES, whereas on surfaces without MADES it is localized on the Ti-O or Si-O bonds. Therefore, such unsaturated bound

Table 3 *Calculated Parameters of Different X Groups Bound to the Si Atom in $(OH)_3SiX$ (6-31G(d) Basis Set).*

Structure	$-E_{HOMO}$ (eV)	E_{LUMO} (eV)	μ (D)	q_{Si}	$-q_{X1}$	q_{X2}	q_{X3}	q_{X4}	q_{X5}	$-q_O$	q_H
$Si(OH)_4$	13.49	4.73	0.02	1.488	0.848_O	0.476_H				0.848	0.476
$Si(OH)_4^*$	13.46	4.79	0.01	1.492	0.737_O	0.364_H				0.735	0.364
$R_4Si{\leftarrow}OH^-$	6.83	10.6	1.13	1.524	0.946_O	0.410_H				0.893	0.430
R_3SiF	13.87	4.57	2.09	1.571	0.487_F					0.842	0.482
R_3SiCl	12.54	4.48	2.43	1.393	0.359_{Cl}					0.829	0.485
R_3SiOCH_3	12.40	4.94	0.51	1.508	0.730_O	-0.159_C	0.171_H			0.847	0.475
$R_3SiOCH_3^*$	12.36	4.98	0.45	1.515	0.739_O	-0.003_C	0.120_H			0.735	0.362
R_3SiOCH_2OH	12.26	4.83	1.86	1.509	0.752_O	0.312_C	-0.739_O	0.472_H	0.177_H	0.843	0.478
$R_3SiOCHO$	12.89	4.64	1.33	1.516	0.683_O	0.569_C	-0.542_O	0.187_H		0.828	0.481
$R_3SiOCOOH$	13.16	4.74	0.99	1.522	0.693_O	1.010_C	-0.663_O	0.480_H	-0.62_O	0.826	0.482
$R_3SiOCH_2NH_2$	11.06	4.86	1.32	1.503	0.754_O	0.167_C	-0.843_N	0.169_H	0.369_H	0.844	0.476
$R_3SiOCHNH$	11.23	4.65	2.36	1.513	0.693_O	0.371_C	-0.628_N	0.182_H	0.333_H	0.837	0.482
R_3SiOCN	11.99	4.01	5.14	1.533	0.694_O	0.601_C	-0.428_N			0.831	0.490

*Note: R = OH. *Basis set is 6-31G(d,p); E_{HOMO} and E_{LUMO} are the energetic levels of the boundary molecular orbitals; μ is the dipole moment; X1, X2, X3, X4 and X5 are the first, second, third, fourth, and fifth atoms in bound groups; q_O and q_H are the atomic charges in OH groups.*

groups can increase the light resistance of coatings by short-circuiting redox reactions that would ordinarily lead to degradation of the organic polymers acting as binders in the coating. Such an effect is also seen with metal oxides that have been surface-modified with aromatic groups containing N atoms.

Silanes with reactive Si-X bonds (X = H, N, O, Cl, *etc.*) are widely used for modification of oxide surfaces. A silane containing multiple reactive Si-X bonds can retain some of these bonds after reaction with the surface, in which case these bonds may remain reactive and can by utilized for subsequent modification of the surface by organics or organometallics. Many reaction mechanisms may come into play in this "post-treatment". For example, hydrosilation occurs by H• transfer from a surface Si-H bond to the β-C in a C=C bond, but H$^+$ transfer is observed upon interaction of H$_2$O with Si-N, and either H$^-$ or H• transfer can occur in the thermal decomposition $\equiv$SiOCH$_3$ $\rightarrow$ $\equiv$Si-H + CH$_2$O.[11] Such changes in the reaction mechanism lead to the dependence of Si-X reactivity on different parameters of the electronic and spatial structure of these groups. For instance, the reactivity of Si-Hal bonds (Hal = F, Cl, Br, and I) increases as the atomic number of the halogen increases, due to an increase in polarizability and a decrease in strength of the Si-Hal bond. The reaction of Si-Hal with water occurs via the $S_{Ni}(Si)$ mechanism, so the reactivity of the Si-Hal bond should increase as the atomic charges of Si and Hal increase. However, q_{Hal} decreases with increasing n$_{Hal}$. Despite this decrease in q_{Hal} the LUMO (acceptor level localized on Si) energy drops with increasing n_{Hal}, which provides an increase of interaction of O (from H$_2$O) with Si in the DAC Si$\leftarrow$OH$_2$. Additionally, the increase in n$_{Hal}$ is accompanied by increasing polarizability of Si-Hal (Table 4, α). This influences favorable intermolecular interaction between OSC and polar surface groups in pre-reaction complexes, as dispersion interaction (which depends on polarizability) can contribute a significant portion to the interaction energy, especially for compounds with μ 0 and relatively low q(Hal) when the electrostatic component is small.

Table 4 *Parameters of Silicon Compounds with Different Si-X bonds (Ab Initio).*

Compound	α (a.u.)	μ (D)	q$_{Si}$	$-$q$_x$	$-E_{HOMO}$ eV	E_{LUMO} eV	Basis
SiH$_4$	22.6	0	0.334	0.084	9.59	1.40	B3LYP/6-31G(d,p)
SiF$_4$	17.0	0	1.373	0.343	11.92	0.18	B3LYP/6-31G(d)
SiCl$_4$	57.1	0	0.564	0.141	9.34	-1.38	B3LYP/6-31G(d)
SiCl$_4$	45.0	0	1.066	0.266	13.74	0.56	LANL2DZ
SiBr$_4$	72.0	0	0.668	0.167	12.09	-0.31	LANL2DZ
SiI$_4$	111.5	0	0.079	0.020	10.57	-0.97	LANL2DZ
H$_3$SiNH$_2$	28.5	1.41	0.551	0.729	6.85	0.63	B3LYP/6-31G(d,p)
H$_3$SiOH	24.4	1.46	0.637	0.613	8.21	0.17	B3LYP/6-31G(d,p)
H$_3$SiOCH$_3$	33.6	0.80	0.449	0.533	7.64	0.45	B3LYP/6-31G(d,p)
H$_3$SiF	16.6	1.55	0.955	0.501	13.28	3.33	B3LYP/6-31G(d,p)
H$_3$SiCl	26.6	1.81	0.500	0.277	8.63	-0.37	B3LYP/6-31G(d,p)
H$_3$SiBr	42.3	1.39	0.615	0.331	11.18	3.59	3-21G(d)
H$_3$SiI	62.3	1.53	0.512	0.271	9.98	2.85	LANL2DZ

Important information about the stability of the bond between modifier and surface can be obtained by the studying the thermal decomposition of surface groups. Theoretical calculations using quantum chemical methods and RRKM theory give kinetic parameters close to experimental data (Table 5), and provide a deeper understanding of the reaction mechanisms.[11] In the case of the transformations of $\equiv$SiOR groups, reactions leading to the appearance of $\equiv$SiOH groups are preferred, as other reaction paths give $E^{\neq}$ that is higher by at least 40 kJ/mol.

Table 5 *Theoretical and Experimental Parameters of Reactions at Silica Surface.*

$E^{\neq}_{obs}$ kJ/mol	$E^{\neq}_{calc}$ kJ/mol	$k_{o,obs}$ c^{-1}	$k_{o,calc}$ c^{-1}	Reaction
81 ± 8	90	–	–	$\equiv$SiOCH$_3$+H$_2$O $\rightarrow$ $\equiv$SiOH + CH$_3$OH
254 ± 18	246	$1.4^{+12}_{-1.3}\times10^{11}$	2.6×10^{11}	$\equiv$SiOCH$_3$ $\rightarrow$ $\equiv$SiH + CH$_2$O
188 ± 23	202	$4^{+196}_{-3.92}\times10^{11}$	8.0×10^{12}	$\equiv$SiOCHO $\rightarrow$ $\equiv$SiOH + CO
198 ± 17	230	$6^{+84}_{-5.6}\times10^{11}$	3.0×10^{12}	$\equiv$SiOC(O)CH$_3$ $\rightarrow$ $\equiv$SiOH + C(O)CH$_2$
249 ± 10	255	4×10^{15}	5.0×10^{14}	$\equiv$SiO(CH$_2$)$_3$CH$_3$ $\rightarrow$ $\equiv$SiOH + C$_4$H$_8$
140	160	2.2×10^{11}	1.2×10^{12}	$\equiv$SiOH + ClSi(CH$_3$)$_3\rightarrow$ $\equiv$SiOSi(CH$_3$)$_3$+HCl
–	75	–	–	$\equiv$SiOH + ClSiH$_3$ $\rightarrow^{NH3}\rightarrow$ $\equiv$SiOSiH$_3$ + HCl

3 CONCLUSIONS

Theoretical simulations of the structure of oxide surfaces, different active sites, and reactions of OC and OSC with oxides have been reported. Correlations between experimental data and calculated parameters provide additional information needed for improvement of oxide materials by their surface modification with these compounds.

References

1. V. M. Gun'ko, V. I. Zarko, R. Leboda, E. F. Voronin, E. Chibowski and E. M. Pakhlov, *Colloid. Surf. A*, 1998, **132**, 241.

2. V. M. Gun'ko, E. F. Voronin, V. I. Zarko, E. M. Pakhlov and A. A. Chuiko, *J. Adhesion Sci. Technol.*, 1997, **11**, 627.

3. *Gaussian 94, Revision E.1*, M. J. Frisch, G. W. Trucks, H. B. Schlegel, P. M. W. Gill, B. G. Johnson, M. A. Robb, J. R. Cheeseman, T. Keith, G. A. Petersson, J. A. Montgomery, K. Raghavachari, M. A. Al-Laham, V. G. Zakrzewski, J. V. Ortiz, J. B. Foresman, J. Cioslowski, B. B. Stefanov, A. Nanayakkara, M. Challacombe, C. Y. Peng, P. Y. Ayala, W. Chen, M. W. Wong, J. L. Andres, E. S. Replogle, R. Gomperts, R. L. Martin, D. J. Fox, J. S. Binkley, D. J. Defrees, J. Baker, J. P. Stewart, M. Head-Gordon, C. Gonzalez and J. A. Pople, Gaussian, Inc., Pittsburgh, PA, 1995.

4.   M. W. Schmidt, K. K. Baldridge, J. A. Boatz, S. T. Elbert, M. S. Gordon, J. J. Jensen, S. Koseki, N. Matsunaga, K. A. Nguyen, S. Su, T. L. Windus, M. Dupuis and J. A. Montgomery, *J. Comput. Chem.*, 1993, **14**, 1347.

5.   M. J. S. Dewar, E. G. Zoebisch, E. E. Healy and J. J. P. Stewart, *J. Am. Chem. Soc.*, 1985, **107**, 3902.

6.   J. J. P. Stewart, *J. Comp. Chem.*, 1989, **10**, 209.

7.   V. M. Gun'ko, V. I. Zarko, E. Chibowski, V. V. Dudnik, R. Leboda and V. A. Zaets, *J. Colloid Interface Sci.*, 1997, **188**, 39.

8.   V. M. Gun'ko, V. I. Zarko, B. A. Chuikov, V. V. Dudnik, Yu. G. Ptushinskii, E. F. Voronin, E. M. Pakhlov and A. A. Chuiko, *Int. J. Mass Spectrom. Ion Processes*, 1998, **172**, 161.

9.   (a) V. M. Gun'ko, E. F. Voronin, V. I. Zarko and E. M. Pakhlov, *Langmuir*, 1997, **13**, 250;   (b) V. M. Gun'ko, V. I. Zarko, V. V. Turov, E. F. Voronin, V. A. Tischenko and A. A. Chuiko, *Langmuir*, 1995, **11**, 2115;   (c) V. M. Gun'ko, V. I. Zarko, V. V. Turov, R. Leboda, E. Chibowski, L. Holysz, E. M. Pakhlov, E. F. Voronin, V. V. Dudnik and Yu. I. Gornikov, *J. Colloid Interface Sci.*, 1998, **198**, 141.

10.   J. Cramer and D. G. Truhlar, *J. Am. Chem. Soc.*, 1991, **113**, 8305.

11.   V. V. Brei, V. M. Gun'ko, V. V. Dudnik and A. A. Chuiko, *Langmuir*, 1992, **8**, 1968.

COMPOSITIONAL MAPPING OF CHEMICALLY MODIFIED GLASSY CARBON ELECTRODES WITH TAPPING-MODE SCANNING FORCE MICROSCOPY

G.K. Kiema, J.K. Kariuki and M.T. McDermott*

Department of Chemistry
University of Alberta
Edmonton, Alberta T6G 2G2 Canada

1 INTRODUCTION

Glassy carbon (GC) has enjoyed extensive usage as an electrode material due to its wide potential window, solvent compatibility and relatively low cost. As with many solid electrode materials, a number of schemes have been developed to ensure highly reactive and reproducible GC electrode surfaces. Procedures such as polishing, electrochemical pretreatment (ECP), vacuum heat treatment and laser activation have been extensively characterized in terms of their influence on electron transfer kinetics and background current.[1] Efforts to enhance the selectivity and sensitivity of GC surfaces have led to pathways for chemically modifying these interfaces. ECP was one of the first methods used to activate GC electrodes and effectively transforms compositionally the surface with a layer of graphitic oxide.[2-4] This procedure involves the application of one of several potential waveforms to the GC electrode in some suitable solvent and generally includes anodization of the electrode surface at potentials of 1.6 to 2.2 V *vs.* SCE. The resulting graphitic oxide layer formed on the surface can itself enhance electron transfer for some species (*e.g.* dopamine[5]) or can be further modified. Although ECP is a facile and flexible modification scheme, its effect on the morphology and surface architecture of GC has not been thoroughly explored.

Scanning probe microscopic (SPM) techniques have been successfully employed to characterize the two-dimensional structure of electrode surfaces. Despite its wide usage as an electrode material, however, GC has been the subject of only a handful of SPM studies.[6,7] This is likely due to the rough and ill-defined surface structure of GC compared with typical SPM substrates which are generally atomically flat. We present here a tapping-mode scanning force microscopy (TM-SFM) study of modified GC electrodes. This technique involves monitoring the amplitude of an oscillating cantilever as its integrated tip lightly "taps" the surface. TM-SFM is more suitable for probing relatively rough surfaces than the traditional contact mode SFM because the intermittent contact between the tip and sample in TM-SFM significantly reduces lateral forces experienced by the tip. We demonstrate here that TM-SFM is useful for tracking both morphological and compositional changes of GC surfaces induced by chemical modification. We have recently demonstrated that the phase shift of the oscillating cantilever relative to its driving waveform is sensitive to the adhesive interactions between the tip and sample.[8] Because adhesion depends on the chemical groups at the interface, phase contrast TM-SFM is sensitive to surface composition. In the present study, we use both topographic and phase contrast images to describe variations in GC electrodes induced by ECP and also probe the structure of covalently bound films to GC deposited by a recently reported technique.[9]

2 EXPERIMENTAL

2.1 Reagents

Eu^{3+}(aq) in 0.2 M $NaClO_4$ solution was prepared at 5 mM from $Eu(NO_3)_3 \cdot 5H_2O$ (Aldrich). 4-Diazo-*N,N*-diethylaniline (DDEA) and tetrabutylammoniumtetrafluoroborate (TAT) were obtained from Aldrich and used as received. Reagent grade acetonitrile (ACN) was used.

2.2 Electrode Preparation and Electrochemical Measurements

GC-20 (Tokai) electrodes were prepared by polishing with successive slurries of 1.0, 0.3 and 0.05 µm alumina in nanopure water on microcloth (Buehler). The GC electrodes were sonicated in nanopure water for 10 min between each polishing step. Polished GC electrodes were patterned by two methods. In some cases a freshly polished GC electrode was sprayed with small droplets of polystyrene (PS) dissolved in CCl_4. In most cases standard photoresist-based microfabrication techniques were utilized to create patterned surfaces. A freshly polished GC electrode was spin-coated with photoresist. A TEM grid was used as a mask through which regions of the photoresist were exposed to UV light and then developed. ECP was performed by poising a patterned electrode at +1.80 V *vs*. Ag/AgCl for 10 s to 120 s in either acidic electrolyte (1.0 M H_2SO_4) or basic solution (0.1 M NaOH). The patterned GC electrodes were then sonicated in acetone for *ca*. 10 min to remove the photoresist or PS. A three-electrode cell was used with a Ag/AgCl reference electrode and a Pt wire counter electrode. Cyclic voltammetry was performed at a scan rate of 0.2 V/s. All solutions were purged with N_2 gas for 5 min prior to electrochemical experiments. Attachment of *N,N*-diethylaniline was carried out in 5 mM DDEA in 0.1 M TAT/ACN solutions. A Ag/AgCl reference electrode (in saturated $LiClO_4$) was used in these modifications.

2.3 SFM Conditions

TM-SFM images were obtained in air using Nanoscope III (Digital Instruments, Santa Barbara CA). The Si cantilever was oscillated at *ca*. 300 kHz. Scanning was carried out with a constant oscillation amplitude. The scanning rate was between 0.5 and 1.0 Hz. The images presented here are representative of many images taken at different points on each sample. All topographic and phase contrast images presented in this study were collected simultaneously.

3 RESULTS AND DISCUSSION

The following sections describe our TM-SFM and electrochemical investigations of modified GC electrodes. In an effort to track the morphological and chemical alterations of GC surfaces with TM-SFM following chemical modification, it was necessary to compare directly the initial polished surface with the modified region in the same image. In most cases we used standard photoresist/mask patterning techniques to produce surfaces with segregated unmodified and modified regions. We also utilized small droplets of polystyrene (PS) which were nebulized from a carbon tetrachloride solution and sprayed on a polished GC surface to mask some regions. In all cases, the photoresist or the PS was dissolved from the GC substrate after modification and before imaging. We first present results for GC electrodes following ECP in basic and acidic solutions. We then describe the morphology of covalently bound aryl films on GC electrodes. We conclude by discussing the utility of TM-SFM for probing compositional changes of relatively rough electrode surfaces.

3.1 Electrochemical Pretreatment of GC in 0.1 M NaOH

The morphological changes induced by stepping the potential of a patterned GC electrode from 0 to 1.8 V *vs.* Ag/AgCl in 0.1 M NaOH for 2 min are illustrated in Figure 1A. Contained in the 100×100 μm topographic TM-SFM image are a series of depressions corresponding to the areas exposed to the pretreatment procedure by the pattern. From the cross-sectional profile the measured depth of the depressions is 330 nm for a 2 min oxidation. The plot in Figure 1B shows that the depth of the depressions produced by ECP in basic solutions is controllable and dependent on anodization time. These results indicate that significant alterations in the morphology of the GC surface accompany ECP in basic solutions likely as a result of electrochemically-induced etching. A mechanism proposed by Sherwood *et al.* for the oxidation of carbon fibers in basic solutions is shown in Scheme 1 and is consistent with our results.[10]

$$C_{(s)} + OH^- \rightarrow C_{(s)}OH_{(ads)} + e^-$$

$$4C_{(s)}OH_{(ads)} \rightarrow 4C + 2H_2O + O_2$$

Scheme 1

Compositional changes of GC surfaces induced by ECP in basic solutions were monitored with both TM-SFM and cyclic voltammetry (CV). Figure 2 contains 40×40 μm topographic (Figure 2A) and phase contrast (Figure 2B) images of a patterned GC electrode oxidized at 1.8 V in 0.1 M NaOH for 60 s. As mentioned above, the phase lag of the oscillating cantilever tapping the surface is sensitive to the interaction between the tip and sample. In addition, it has been shown that variations in sample mechanical properties as well as tip–sample adhesive differences can generate phase contrast.[11] Because the adhesion at a microscopic contact depends on interfacial chemical groups, any chemical alterations affected by ECP should be observed in phase images.

Depressions due to etching are again evident in Figure 2A, and are similar to those in Figure 1A. In Figure 2B a difference in phase lag is observed between the modified region inside the pattern and the original polished surface, a significant finding that implies a compositional difference between these two areas. A comparison with chemistry-induced phase contrast reported previously implies that the darker contrast observed at the polished region (outside the depressions) results from a greater adhesive interaction with the tip relative to that in the modified region (inside the depressions).[8] We believe that the Si–OH groups on the tip surface interact more strongly with the ubiquitous layer of polishing debris known to exist on GC electrodes than with the surface resulting from the ECP procedure. Although ill-defined, this layer is believed to consist of abraded carbon particles and contaminants originating from the polishing

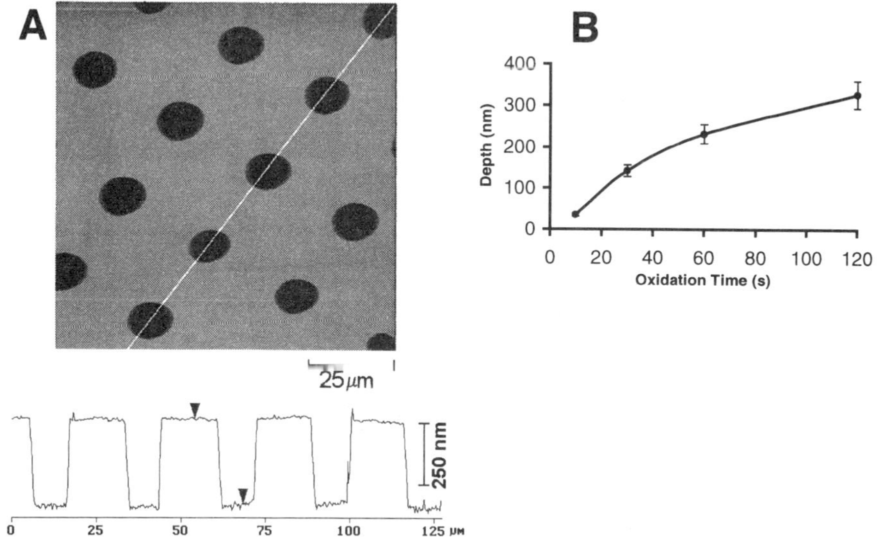

Figure 1　*(A) 100 × 100 μm TM-SFM image (z-scale = 500 nm) of the topography of a GC electrode following ECP at 1.8 V in 0.1 M NaOH for 2 min through a TEM pattern.　(B) Plot of depression depth vs. oxidation time for ECP at 1.8 V in 0.1 M NaOH.*

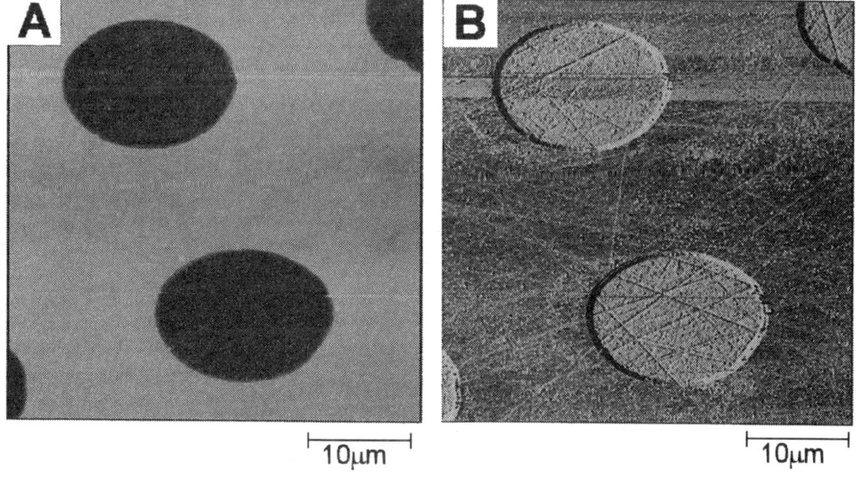

Figure 2　*40 × 40 μm TM-SFM images of a GC surface oxidized at 1.8 V in 0.1 M NaOH for 60 s through a TEM grid pattern.　(A) Topography (z-scale = 500 nm).　(B) Phase contrast (z-scale = 30 deg).*

slurries. Experiments indicate that this polishing layer is etched away in the first few seconds of base oxidation. As shown in Figure 3A, a 10 s ECP in 0.1 M NaOH through a PS mask (see the Experimental section) removes the top *ca.* 30 nm of the GC electrode. Because the resulting surface is free from the ill-defined film of polishing debris it is seemingly imaged at higher resolution as evidenced by the more pronounced polishing scratches. Interestingly, a minimal phase contrast is observed in Figure 3B between the polished and modified regions implying little chemical alterations are induced by a 10 s oxidation. However, this also argues a lack of mechanical differences between the two regions. This observation supports the conclusion that the phase contrast observed in Figure 2B is driven by ECP-induced chemical changes.

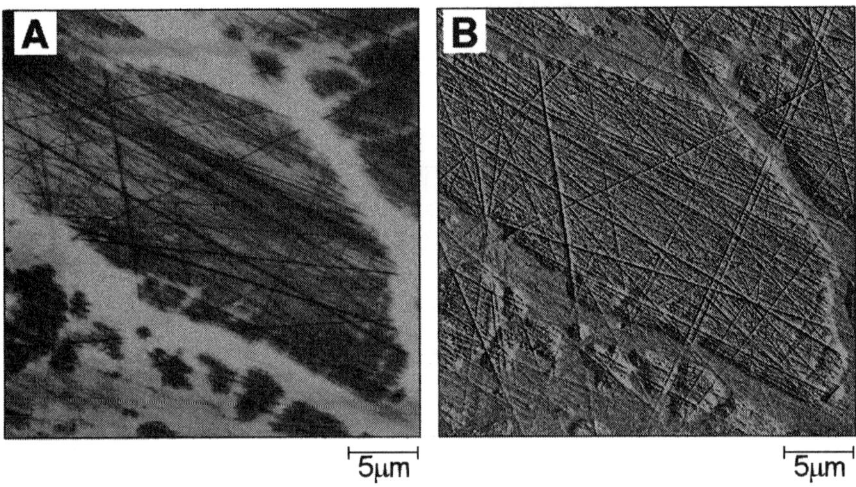

Figure 3   *30 × 30 µm TM-SFM images of a GC surface oxidized at 1.8 V in 0.1 M NaOH for 10 s through a polystyrene mask. (A) Topography (z-scale = 30 nm), (B) Phase contrast (z-scale = 20 deg).*

The electron transfer kinetics of several redox systems are sensitive to surface oxide species, particularly carbonyl groups, on GC electrodes. The aquated ions $Fe^{3+/2+}$ and $Eu^{3+/2+}$ are among these systems.[12] It has been shown that electron transfer, as measured by the peak separation in cyclic voltammograms (ΔE_p), is more facile to these systems at GC electrodes after ECP in acidic solutions. Table 1 shows that ΔE_p for $Eu^{3+/2+}$ decreases with oxidation time in basic solutions indicating that electron transfer for $Eu^{3+/2+}$ at GC electrodes is also facilitated by ECP in this media. Because of the known dependence of ΔE_p for $Eu^{3+/2+}$ on surface oxides, this correlation also implies that the chemistry of the GC surface is transformed during ECP in base via the deposition of a graphitic oxide.[3] As shown above, this chemical change can be tracked with phase contrast TM-SFM. Also listed in Table 1 is the observed phase contrast ($\Delta\Phi$) measured from images on patterned surfaces. The direction of the contrast is always consistent with that in Figure 2B where the phase lag on the polished region is greater than that at the oxidized region. As shown in Table 1, the compositional changes induced by oxidation in base as tracked by TM-SFM correlate with electron transfer to the $Eu^{3+/2+}$ redox system.

Table 1   *Voltammetric Peak Separations (ΔE_p) for $Eu^{3+/2+}$ and TM-SFM Phase Contrast ($\Delta\Phi$) for GC Electrodes Following ECP at 1.8 V in 0.1 M NaOH for Various Times.*

Oxidation Time (s)	ΔE_p (mV) for $Eu^{3+/2+}$	$\Delta\Phi$(deg)
Polished	265 ± 35	NA
10	259 ± 8	1.3–2.9
30	223 ± 17	5.0–7.5
60	184 ± 19	6.6–9.0

3.2 Electrochemical Pretreatment of GC in 1 M H$_2$SO$_4$

Figure 4 illustrates the effect of ECP in acidic solutions on the morphology and chemistry of GC. Figure 4A is the topographic image of a patterned GC substrate after oxidation in 1 M H$_2$SO$_4$ for 90 s. In contrast to Figure 2A, relatively little topographic changes are observed. Several rings are apparent likely corresponding to an enhanced oxidation at the boundary of the photoresist pattern. However, a significant contrast is observed in the phase image of Figure 4B. In this image the circular regions that were exposed through the patterned photoresist are easily observed, implying a notable compostional change. A chemical transformation is also supported by the correlation between the phase contrast and the electrochemical results in Table 2. Similar to the phase contrast for the base oxidized surfaces, the phase lag at the polished GC is greater (dark contrast) than that at the oxidized regions. This is again consistent with a greater adhesion between the tip and polished surface than between the tip and oxidized surface.

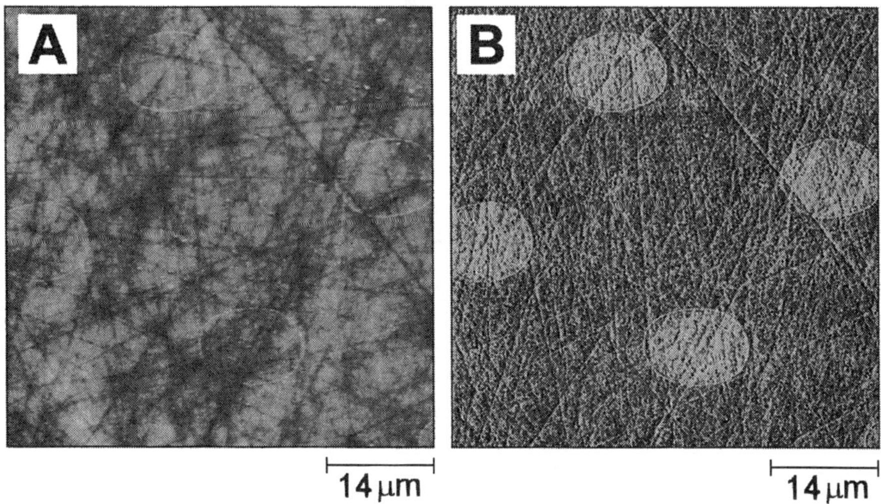

Figure 4   *55 ×55 µm TM SFM images of a GC surface oxidized at 1.8 V in 1 M H$_2$SO$_4$ for 90 s through a TEM grid pattern. (A) Topography (z-scale = 60 nm), (B) Phase contrast (z-scale = 20 deg).*

The exact chemical nature of this observation is presently under investigation. It is clear from Figure 4 that a morphological change in the GC surface does not accompany the change in surface chemistry induced by ECP in acidic solutions.

Table 2 *Voltammetric Peak Separations (ΔE_p) for $Eu^{3+/2+}$ and TM-SFM Phase Contrast ($\Delta\Phi$) for GC Electrodes Following ECP at 1.8 V in 1 M H_2SO_4 for Various Times.*

Oxidation Time (s)	ΔE_p (mV) for $Eu^{3+/2+}$	$\Delta\Phi$(deg)
Polished	265 ± 35	NA
30	130 ± 9	4.0–6.8
60	105 ± 5	–
90	80 ± 6	9.3–11

3.3 Reductive Attachment of *N,N*-Diethylaniline to GC Surfaces

A relatively new method to covalently bind functional groups to the surface of GC electrodes involves the electrochemical reduction of diazonium salts as illustrated in Scheme 2. It is thought that monolayers of aryl moieties are attached to GC electrodes via linear potential sweep or potential step in a solution of the appropriate diazonium salt.[9] Our interest in modified carbon electrodes prompted us to investigate the structure of these films with TM-SFM.

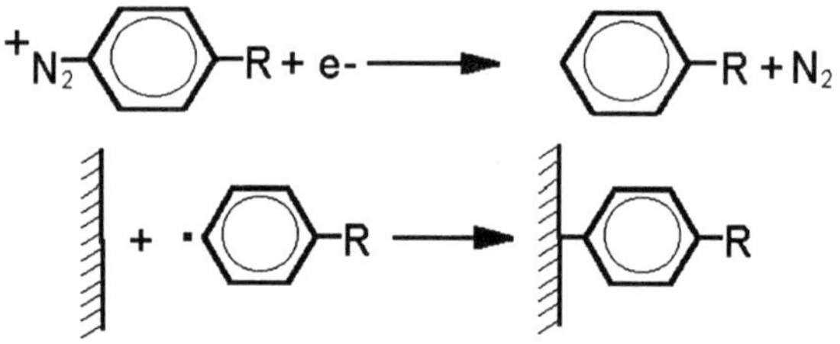

Scheme 2

Figure 5 contains images recorded over a boundary between a region modified with *N,N*-diethylaniline on the left side (R= –N(CH$_2$CH$_3$)$_2$) and the original polished surface (right side). A droplet of PS was used to create the boundary. The layer was deposited by stepping the potential of a GC electrode from 0 to –0.9 V in a 5 mM solution of 4-diazo-*N,N*-diethylaniline (DDEA) for 30 min. The boundary is apparent in Figure 5A due to the increased height of the film relative to the polished surface. A slight phase contrast is also observed in Figure 5B, highlighting chemical or mechanical differences. The film architecture appears to be discontinuous and comprised of closely spaced spherical structures. Surprisingly, the height of the layer measured from the polished regions is *ca.* 30 nm, which is much larger than expected for a single *N,N*-diethylaniline molecule.

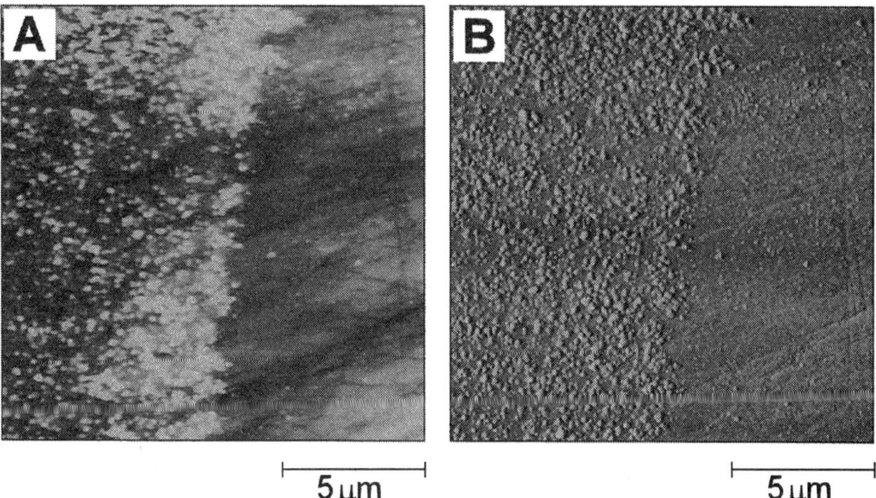

Figure 5 *15 × 15 μm TM-SFM images of the boundary between unmodified, polished GC (right side) and GC modified with N,N-diethylaniline. (A) Topography (z-scale = 60 nm). (B) Phase contrast (z-scale = 20 deg).*

In an effort to investigate further the anomalous height of the film in Figure 5, we examined *N,N*-diethylaniline films on highly oriented pyrolytic graphite (HOPG) surfaces. HOPG can be considered nearly single crystal graphite and exhibits an atomically flat basal plane as shown in the TM-SFM image in Figure 6A. Parts C and D of Figure 6 are TM-SFM images following *N,N*-diethylaniline deposition from a single linear voltage sweep between 0 and –0.9 V in a solution of 0.5 mM DDEA. As shown in the topography (Figure 6C) an array of segregated spherical structures are observed exhibiting heights of 5 to 15 nm. The phase contrast image (Figure 6D), however, reveals that these structures are not surrounded by purely basal plane graphite. Compared to the phase image of unmodified HOPG (Figure 6B) the areas surrounding the spheres appears textured. We interpret this textured pattern as being the initial *N,N*-diethylaniline monolayer. We believe that the higher spherical structures result from attachment of electrochemically produced aryl radicals to the already deposited monolayer. As depicted in Scheme 3, this polymerization can result in multilayers which can grow to yield the measured height in Figure 5A.

Scheme 3

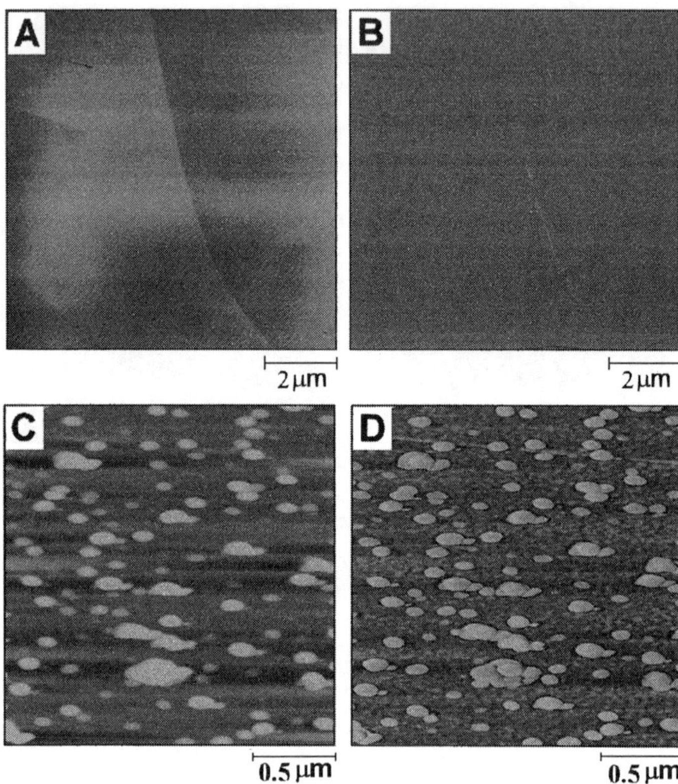

Figure 6   *(A) 10 × 10 μm topographic (z-scale = 10 nm) TM-SFM image of unmodified HOPG. (B) Corresponding phase contrast image (z-scale = 20 deg). (C) 2 × 2 μm topographic (z-scale = 20 nm) image of HOPG modified with a layer of N,N-diethylaniline. (D) Corresponding phase contrast image (z-scale = 20 deg).*

4 CONCLUSIONS

We have shown that TM-SFM can provide a great deal of information concerning interfacial transformations accompanying the chemical modification of electrode surface. The quality of the images on these relatively rough surfaces (*i.e.* not atomically flat) leads us to believe that TM-SFM with phase contrast can be utilized to track compositional changes in a variety of technologically relevant interfaces (*e.g.* polymer degradation, biomaterial compatibility). For GC electrodes, ECP in basic solutions results in both morphological and compositional alterations from the original polished surface.

The etching effect of anodization in aqueous NaOH opens pathways for the microfabrication of GC surfaces. These possibilities are currently being explored in our laboratory. The same pretreatment in acidic media produces only chemical changes. These solvent dependent variations should be considered when choosing an ECP procedure. Our experiments have also indicated that the deposition of covalently bound aryl groups may lead to multilayer formation although the generality of this proposal is still under investigation.

5 ACKNOWLEGMENTS

This work was supported by the Natural Sciences and Engineering Research Council of Canada (NSERC) and the Department of Chemistry, University of Alberta. We thank Dr. Glen Fitzpatrick of the Alberta Microelectronics Center (AMC) for the patterning of GC electrodes. We also thank Dr. Ken Westra of AMC for the gift of tapping mode cantilevers. The HOPG substrate was a gift from Dr. Author Moore of Advanced Ceramic Materials, Inc.

References

1. R. L. McCreery, 'Electroanalytical Chemistry', A. J. Bard, ed., Marcel Dekker, New York, 1991, Vol. 17, p. 221.
2. R. C. Engstrom, *Anal. Chem.*, 1982, **61**, 2310.
3. A. L. Beilby and A. Carlsson, *J. Electroanal. Chem.*, 1988, **248**, 283.
4. L. J. Kepley and A. J. Bard, *Anal. Chem.*, 1988, **60**, 1459.
5. D. M. Anjo, M. Kahr, M. M. Khodabakhsh, S. Nowinski and M. Wagner, *Anal. Chem.*, 1989, **61**, 2603.
6. See, M. T. McDermott, C. A. McDermott and R. L McCreery, *Anal. Chem.*, 1993, **65**, 937, and references therein.
7. P. Heiduschka, A. W. Munz and W. Gopel, *Electrochim. Acta*, 1994, **39**, 2207.
8. M. O. Finot and M. T. McDermott, *J. Am. Chem. Soc.*, 1997, **119**, 8564.
9. M. Delamar, R. Himi, J. Pinson and J. M. Saveant, *J. Am. Chem. Soc.*, 1992, **114**, 5883.
10. C. Kozlowski and P. M. Sherwood, *J. Chem. Soc., Faraday Trans.*, 1985, **81**, 2745.
11. A. J. Howard, R. R. Rye and J. E. Houston, *J. Appl. Phys.*, 1996, **79**, 1885.
12. C. A. McDermott, K. R. Kneten and R. L. McCreery, *J. Electrochem. Soc.*, 1993, **140**, 2593.

PERMEABILITY OF LAYERED POLYELECTROLYTE FILMS AS A FUNCTION OF pH

J.J. Harris and M.L. Bruening*

Department of Chemistry
Michigan State University
East Lansing, MI 48824 USA

1 INTRODUCTION

Layered polyelectrolyte films are attractive for sensing, separation, anti-corrosion, and adhesion applications because of their ease of synthesis and wide variability.[1,2] The feasibility of using these films will depend on their permeability and stability under relevant conditions. We report electrochemical studies of the permeability and stability of layered poly(allylamine hydrochloride)/poly(styrenesulfonate) (PAH/PSS) films in pH-3.2-, pH-6.3-, and pH-10-buffered solutions. Impedance spectroscopy shows that multilayer PAH/PSS films effectively passivate gold electrodes and that this passivation varies nonlinearly with the number of layers. The permeability of these films is much higher at pH-10 than at either pH 3 or pH 6. Higher permeabilities in basic solution result from a more open structure as well as a small amount of delamination.

Decher and coworkers demonstrated the synthesis of ultrathin organic films by sequential deposition of oppositely charged polyelectrolytes.[3] The process begins by immersing a charged surface (*e.g.* SiO_2, glass, Au modified with a charged monolayer) into a solution of an oppositely charged polyelectrolyte. A film forms due to electrostatic interactions between the substrate and the polyelectrolyte. Rinsing of the substrate with water and immersion in a solution of a second, oppositely charged polyelectrolyte yields another layer on the surface. Repetition of this process produces multilayers whose thickness depends on the number of depositions and the corresponding salt concentration of the polymer solutions.

Initial work on polyelectrolyte films employed poly(allylamine hydrochloride) and polysulfonates and sulfates,[4–8] while more recent studies incorporated molecules such as charged viruses,[9] proteins,[2] and DNA[10] into polyelectrolyte films. Ferguson used inorganic clays and poly(diallyldimethylammonium chloride) to form multilayer films,[11,12] while Mallouk showed that inorganic anions can be incorporated to create ionic-covalent multilayer films.[13–15] These studies demonstrate the versatility of electrostatic deposition as a method of film formation.

Prior investigations of polyelectrolyte films suggest a layered structural picture of these modified surfaces. XPS, radioactive labeling, and neutron reflectometry did not detect any incorporation of counter-ions in the films,[6,7,16] showing that each new layer compensates nearly all of the underlying charges. X-Ray reflectometry of deuterated films suggests significant overlap between layers.[6] The thickness of films strongly depends on the salt concentration in the deposition solution.[9] Even after film formation, immersion in solutions of high salt concentration increases the "dry" thickness of the film.[8,16]

Several groups monitored the growth of polyelectrolyte films using techniques such as small angle X-ray reflectometry,[3–5,7–10,17] UV/VIS spectroscopy,[3,18] neutron reflectometry,[7] radioactive labeling,[19,20] and optical ellipsometry.[18,21] To study transport, von Klitzing and Möhwald investigated the transport of neutral quenchers in polyelectrolyte films using time-dependent fluorescence.[22] Schlenoff and coworkers used electrochemical techniques to investigate electron-transfer through polyelectrolyte films containing redox-active layers.[23,24] To our knowledge, however, this study represents the first use of electrochemical techniques to investigate the permeability of charged species in polyelectrolyte films.

2 EXPERIMENTAL

2.1 Chemicals and Solutions

Poly(allylamine hydrochloride), 3-mercaptopropionoic acid, and sodium poly-(styrenesulfonate) were used as received from Aldrich. Solutions containing 1.0 M Na_2SO_4 were buffered with the following salts: 0.2 M CH_3COOH + 0.0036 M CH_3COONa (pH = 3.2), 0.025 M NaH_2PO_4 + 0.025 M Na_2HPO_4 (pH = 6.3), and 0.025 M Na_2CO_3 + 0.025 $NaHCO_3$ (pH = 10). Solutions for cyclic voltammetry contained 0.005 M $K_3Fe(CN)_6$ and AC impedance solutions contained 0.005 M $K_3Fe(CN)_6$ and 0.005 M $K_4Fe(CN)_6$ (Aldrich) in the above mentioned buffers. PSS solutions (15 mL total volume) were made with 55 mg of polymer, 1.5 mL of 0.01 M HCl and the supporting salt was 0.5 M $MnCl_2$ (pH = 2.3). PAH solutions (15 mL total volume) were made with 22 mg of polymer, 1.5 mL of 0.01 M HCl and the supporting salt was 0.5 M NaBr (pH = 2.5).

2.2 Film Preparation

Gold slides [200 nm of Au on 20 nm Ti on Si(100) wafers] were UV/O$_3$ cleaned for 15 minutes and immediately immersed in an ethanolic solution of 0.002 M mercaptopropionoic acid (MPA) for 30 minutes.[18] These slides were then rinsed with ethanol and Millipore water. To prepare one polyelectrolyte layer, the MPA-coated slide was immersed in PAH for 5 minutes, rinsed under a continuous stream of water for one minute, dipped in a PSS solution for two minutes and then rinsed for one minute. Additional layers were prepared similarly. Slides were dried with nitrogen only after the desired number of layers was deposited.

2.3 Instrumentation

Electrochemical measurements were made using a standard three-electrode cell containing a Ag/AgCl (3 M KCl) reference electrode and a platinum wire counter electrode. The working electrode was a gold slide sealed in a plastic holder with an exposed area of 0.1 cm^2. A CH-Instruments Electrochemical Analyzer was used to perform the impedance spectroscopy and cyclic voltammetry measurements. Impedance measurements were taken at frequencies ranging from 0.1 to 10^5 Hz using a 5 mV sinusoidal voltage centered around $E^{0\prime}$ of $Fe(CN)_6^{3-/4-}$. Charge transfer resistance (R_{ct})

values were obtained by fitting impedance data with a modified Randles circuit using CNLS regression (LEVM 7.0 by J. Ross McDonald, distributed by Solartron).[25]

Ellipsometric measurements were obtained with a M-44 Ellipsometer (J. A. Woollam Co., Inc.) using WVASE32 software. The thickness values for 1–5 layered films were determined using 44 wavelengths between 414 and 736.1 nm. Refractive indices of these films were assumed to be the same as those determined for a 10-layer film using ellipsometry. Films containing less than 5 layers are not sufficiently thick for refractive index determination or accurate in situ measurements. In situ ellipsometry was thus performed on 10-layer films. FTIR spectra were obtained using a Nicolet Magna-IR 560 containing a PIKE grazing angle (80°) attachment.

3 RESULTS AND DISCUSSION

3.1 Characterization of PAH/PSS Films

Ellipsometry of 'dry' films demonstrates uniform, stepwise growth of PAH/PSS films containing up to ten layers. (In this paper, we use the term layer to refer to a cycle of deposition of both PAH and PSS.) The thickness increase after the addition of each layer (with the exception of the initial layer) was 3.8 ± 0.2 nm. The value of 3.8 nm per layer is consistent with previous reports, which show single layer thickness values of between 2.27 and 9.0 nm.[3,4,9,10] *In situ* ellipsometry experiments show that a 10-layer film swells in water, but remains on the substrate in pH 3- and pH 6-buffered solutions. At pH 10, an additional swelling of *ca.* 15% occurs over a *ca.* 10 min immersion. Subsequently, a slow (10 min) *ca.* 15% decrease in thickness occurs that may be due to delamination. However, rinsing of a 5-layer film after immersion in pH-10 buffer can remove up to half of the film.

FTIR external reflection spectroscopy (FTIR-ERS) also confirms the stepwise growth of the PAH/PSS films. Additionally, FTIR-ERS spectra allow identification of functional groups that cannot be detected using UV/VIS spectroscopy.[3,18] Figure 1 shows the FTIR-ERS spectra of films containing from 1 to 5 layers. The sulfonic acid absorbances between 1216 and 1036 cm^{-1} dominate the spectrum of PAH/PSS films. The hump between 3200 and 2800 cm^{-1} is a combination of $-NH_3^+$ and CH_2 symmetric and asymmetric stretching vibrations. Benzyl–H stretches and $-NH_3^+$ deformations are visible between 1627 and 1410 cm^{-1}.[26] Absorbance increases linearly (Figure 1B) with the number of layers, confirming the ellipsometric data.

3.2 Permeability as a Function of pH

Electrochemical investigation of film permeabilities provides a sensitive probe of structural changes in PAH/PSS films. Cyclic voltammetry affords qualitative comparisons of the passivating abilities of different films based on peak currents and the shape of the voltammogram. Impedance spectroscopy is a complementary technique that often allows quantitative determination of kinetic and diffusion parameters. Impedance methods are also attractive because of the small sinusoidal potentials (5 mV) that are employed, as opposed to the wide potential window used with cyclic voltammetry.[27]

Peak currents (i_p) in cyclic voltammograms (CVs) of $Fe(CN)_6^{3-}$ vary with pH as given in Table 1. There is little difference between CVs in pH-3 and pH-6 solutions, but at pH-10, i_p increases by *ca.* 2000 % for electrodes coated with 4- and 5-layer films. A second immersion in pH-3 buffer does not reverse the increase in i_p that occurs in basic

Table 1 *Cyclic Voltammetric ($Fe(CN)_6^{3-}$) Peak Current Values ($\mu A/cm^2$) at pH-3, -6, -10, and pH-3.2 after pH-10.0 for a Gold Electrode Modified with PAH/PSS Films (scan rate: 0.1 V/s).*

Number of Layers	pH-3	pH-6	pH-10	pH-3 after pH-10
5	11 ± 2	10 ± 5	193 ± 110	240 ± 100
4	10 ± 1	7 ± 1	170 ± 40	160 ± 60
3	39 ± 15	30 ± 10	280 ± 40	260 ± 110
2	480 ± 210	440 ± 170	500 ± 70	720 ± 50
1	700 ± 40	670 ± 50	670 ± 70	830 ± 140

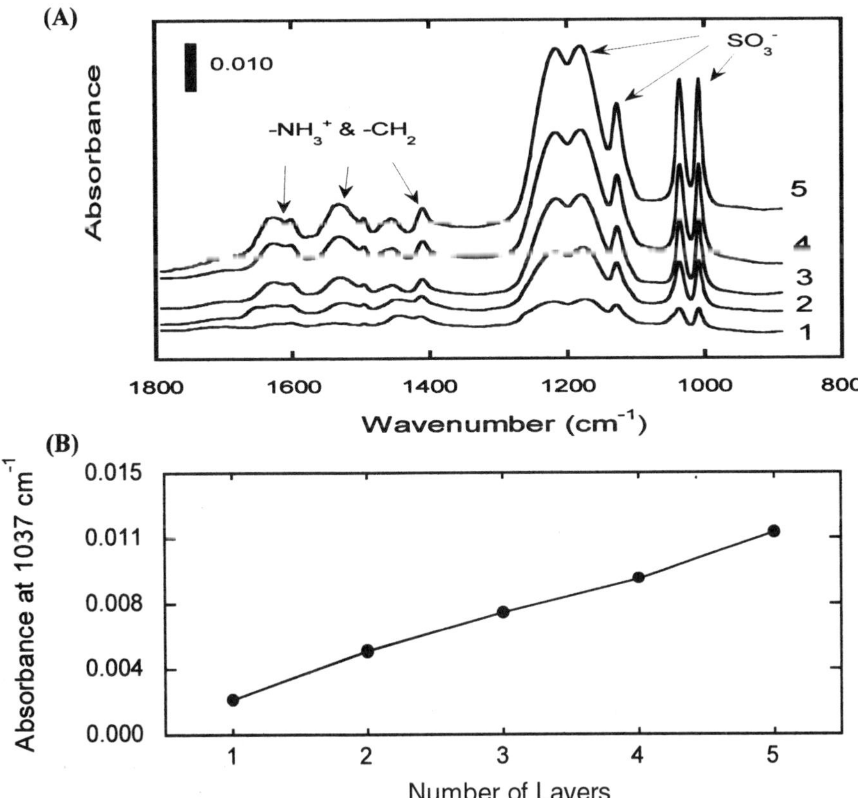

Figure 1 *FTIR-ERS spectra of polyelectrolyte films composed of 1–5 layers of PAH/PSS (A) and a plot of FTIR-ERS absorbances at 1037 cm^{-1} as a function of the number of layers in the film (B).*

solution. In fact, the CV of a 5-layer film-coated electrode measured in pH-3 buffer after a 12-minute immersion in pH-10 buffer shows a quasireversible voltammogram. These irreversible increases in peak current due to immersion in pH-10 buffer result from a combination of a change in film structure and delamination.

The shape of the CVs (Figure 2) also helps in understanding the processes occurring at the electrode–film interface. At pH-3 and -6 the CVs of a 5-layer film-coated electrode are broad and plateau-shaped, suggesting current originating from an array of microelectrodes (pinholes).[28,29] There is a barely distinguishable anodic peak seen in the pH-3 and -6 CVs which we attribute to small amounts of $Fe(CN)_6^{4-}$ absorption. At pH-10, the CVs show larger peak currents along with the plateau shape. The increase in the peak current values may be a consequence of larger pinholes or the formation of more densely packed pinholes. A small cathodic peak was visible in *ca.* 20 % of the experimental runs at pH-10 pointing to a transition from hemispherical diffusion to semi-infinite linear diffusion. At pH-3 after immersion in pH-10, the CV shows a quasi-reversible response (ΔE_p of 124 mV) with distinguishable anodic and cathodic peaks. This type of CV suggests overlap of the diffusion spheres at the electrode or semiinfinite linear diffusion.[30] Although the i_p values at pH-10 and pH-3 after pH-10 are sometimes similar in magnitude (Table 1), the change in shape of the CV shows variation in film structure or chemistry.

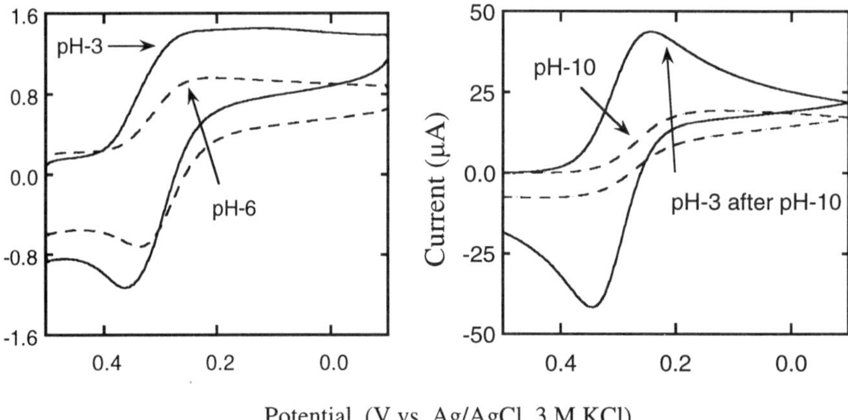

Figure 2   *Cyclic voltammograms ($Fe(CN)_6^{3-}$) for a gold electrode modified with a five layer PAH/PSS film. CVs were measured at pH-3, pH-6, (A) pH-10, and pH-3 after pH-10 (B) (scan rate: 0.1 V/sec).*

We employed impedance spectroscopy ($Fe(CN)_6^{3-/4-}$) to quantitatively investigate the permeability of polyelectrolyte films. To interpret impedance data, we used the modified Randles circuit shown in Figure 3E.[31] The modified Randles circuit contains film resistance and capacitance elements that are not included in a simple Randles model and yields an excellent fit to the data as demonstrated in Figure 3. Although the modified Randles circuit is superior to a Randles circuit in fitting the data, R_{ct} values are nearly the same for both models (within 25%). The circuit element of most interest to this work is R_{ct} because it relates directly to the accessibility of the underlying gold electrode.

Impedance data confirm the trends shown by cyclic voltammetry (Table 2). The difference between R_{ct} values measured in pH-3 and -6 solutions is small, but R_{ct} is much lower in pH-10 solutions and in pH-3 solutions after immersion in pH-10 buffer. The complex impedance plots for 5-layer films in pH-3 and pH-6 solution (Figure 3) show only a quarter circle with no visible Warburg line indicating complete kinetic control of the electron transfer process. The complex impedance plots for 5-layer films in pH-10

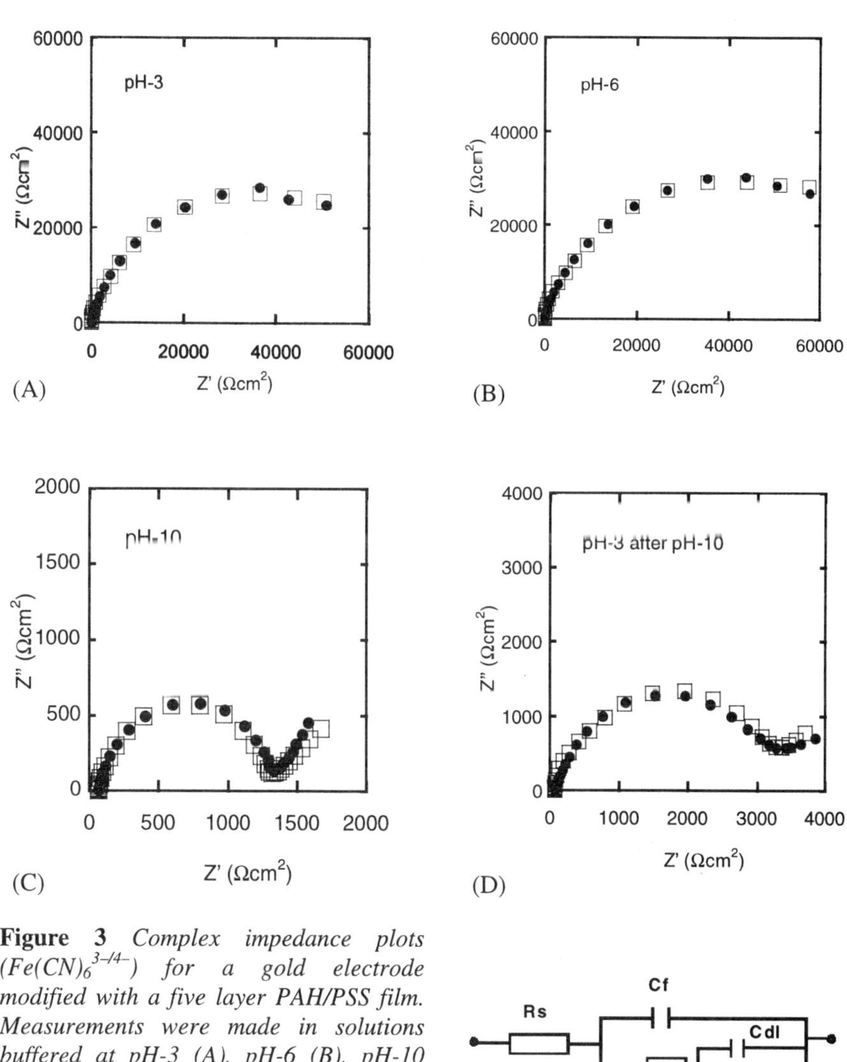

Figure 3 *Complex impedance plots $(Fe(CN)_6^{3-/4-})$ for a gold electrode modified with a five layer PAH/PSS film. Measurements were made in solutions buffered at pH-3 (A), pH-6 (B), pH-10 (C), and pH-3 (after pH-10) (D). Circles represent experimental data and squares are the fit to the data using the modified Randles equivalent circuit shown in (E)*

solution and pH-3 solution after immersion at pH-10 (Figure 3) are characterized by the appearance of small semicircles in the high frequency domain.

All of the electrochemical data show a definite increase in permeability upon immersion in pH-10 solution. At high pH values PAH should be deprotonated and Na^+ ions will enter the film to compensate the negative charge on the sulfonate groups. As a consequence of becoming deprotonated, PAH becomes more hydrophobic and the polymer likely rearranges to minimize its contact with the surrounding medium. The resulting PAH may assume a globular structure that disrupts the layered architecture. The new structure apparently contains a large number of pores that give rise to the dramatic increases in permeability at pH-10. When the films are reimmersed in a pH-3 buffer, the amines are again protonated, but the structure of the film apparently does not return to its initial state because permeability remains high. The high permeability at low pH after exposure to pH-10 buffer could be due to either delamination or a change in structure.

Table 2   *Charge Transfer Resistance Values (Ω-cm^2) at pH-3, -6, -10, and pH-3 after pH-10 for a Gold Electrode Modified with PAH/PSS Films*

Number of Layers	pH-3	pH-6	pH-10	pH-3 after pH-10
5	6.6 ± 2.2	6.5 ± 3.9	120 ± 50	290 ± 270
4	4.1 ± 0.8	10 ± 2	30 ± 20	90 ± 50
3	1.0 ± 0.3	2.3 ± 0.7	40 ± 20	60 ± 40
2	7.1 ± 2.9	10.5 ± 0.7	20 ± 10	4.5 ± 3.3
1	1.6 ± 0.6	5.5 ± 0.9	3.7 ± 3.6	1.4 ± 0.40

3.3 Permeability as a Function of the Number of Layers

Increasing the number of layers in the film should increase charge transfer resistance and decrease peak current values because pathways to the electrode surface will become fewer and longer. This is indeed the trend that we generally observe (Tables 1 and 2), although 4-layer films are sometimes more blocking than 5-layer films. The difference between 4- and 5-layer films is within the film to film variation. The most drastic changes in electrode blocking occur upon introduction of the third layer. For 1- and 2-layer films, CVs contain a quasireversible response. CVs of 3-layer film-coated electrodes, however, show plateau-shaped voltammograms at low pH values. Peak currents at an electrode covered with a 1-layer film differ by less than 20% from peak currents at a bare gold electrode. Addition of a second layer reduces i_p by less than 50%. The presence of a third layer, however, results in a > 10-fold reduction in peak current at low and moderate pH values.

Impedance spectroscopy also shows the dependence of permeability on the number of polyelectrolyte layers on the surface. For electrodes coated with 1- and 2-layer films, the Warburg diffusion element dominates the complex impedance plots. Addition of a third layer results in kinetic control of the reaction as seen by the appearance of only a semicircle in the complex impedance plot. There is a > 140-fold increase in charge transfer resistance between a two and three layer film as opposed to a > 5-fold increase between one and two layer films (low and moderate pH solutions).

The dependence of permeability on the number of layers suggests a different structure in the first two layers than in the rest of the film. Decher and coworkers previously showed that the first few polyelectrolyte layers can have a different thickness than later layers.[3,4,9,10] The first layers of polyelectrolyte bind only through a few sites.[6] With each subsequent layer the charge is multiplied until stepwise growth is achieved. This agrees with the electrochemical data, which suggest that there is greater access to the underlying gold electrode within the first two layers.

4 CONCLUSIONS

We have determined the permeability of layered PAH/PSS polyelectrolyte films to $Fe(CN)_6^{3-}$ in solutions of varying pH values. The permeability of layered PAH/PSS films decreases nonlinearly with film thickness. Film permeability is relatively independent of pH at low and moderate pH values, but increases drastically in basic solutions. *In situ* ellipsometric studies indicate that most of the film remains on the substrate in quiescent basic solutions, but electrochemical data show that structural changes occur at high pH and increase permeability.

References

1.	A. Ulman, 'An Introduction to Ultrathin Films, from Langmuir–Blodgett to Self-assembly', Academic Press, Boston, 1991.
2.	F. Caruso, K. Niikura, D. N. Furlong and Y. Okahata, *Langmuir*, 1997, **13**, 3427.
3.	G. Decher, J. D. Hong and J. Schmitt, *Thin Solid Films*, 1992, **210–211**, 831
4.	Y. Lvov, G. Decher and H. Möhwald, *Langmuir*, 1993, **9**, 481.
5.	Y. Lvov, F. Essler and G. Decher, *J. Phys. Chem.*, 1993, **97**, 13773.
6.	G. Decher, *Science*, 1997, **277**, 1232.
7.	J. Schmitt, T. Grünewald, G. Decher, P. S. Pershan, K. Kjaer and M. Lösche, *Macromolecules*, 1993, **26**, 7058.
8.	G. Decher, Y. Lvov and J. Schmitt, *Thin Sold Films*, 1994, **244**, 772.
9.	Y. Lvov, H. Haas, G. Decher, H. Möhwald, A. Mikhailov, B. Mtchedlishvily, E. Morgunova and B. Vainshtein, *Langmuir*, 1994, **10**, 4232.
10.	Y. Lvov, G. Decher and G. Sukhorukov, *Macromolecules*, 1993, **26**, 5396.
11.	E. R. Kleinfeld and G. S. Ferguson, *Science*, 1994, **265**, 370.
12.	E. R. Kleinfeld and G. S. Ferguson, *Chem. Mater.*, 1996, **8**, 1575.
13.	M. Fang, D. M. Kaschak, A. C. Sutorik and T. E. Mallouk, *J. Am. Chem. Soc.*, 1997, **119**, 12184.
14.	H.-N. Kim, S. W. Keller and T. E. Mallouk, *Chem. Mater.*, 1997, **9**, 1414.
15.	S. W. Keller, H.-N. Kim and T. E. Mallouk, *J. Am. Chem. Soc.*, 1994, **116**, 8817.
16.	G. B. Sukhorukov, J. Schmitt and G. Decher, *Ber. Bunsenges Phys. Chem.*, 1996, **100**, 948.
17.	Y. Lvov, G. Decher, H. Haas, H. Möhwald and A. Kalachev, *Physica B*, 1994, **198**, 89.
18.	F. Caruso, K. Niikura, D. N. Furlong and Y. Okahata, *Langmuir*, 1997, **13**, 3422.
19.	M. Li and J. B. Schlenoff, *Anal. Chem.*, 1994, **66**, 824.
20.	J. B. Schlenoff and M. Li, *Ber. Bunsenges Phys. Chem.*, 1996, **100**, 943.
21.	A. Tronin, Y. Lvov and C. Nicolini, *Colloid Polym. Sci.*, 1994, **272**, 1317.
22.	R. von Klitzing and H. Möhwald, *Macromolecules*, 1996, **29**, 6901.

23. J. B. Schlenoff, D. Laurent, H. Ly and J. Stepp, *Adv. Mater.*, 1998, **10**, 347.
24. D. Laurent and J. B. Schlenoff, *Langmuir*, 1997, **13**, 1552.
25. J. R. Macdonald, *Electrochimica Acta*, 1990, **35**, 1483.
26. N. B. Colthup, L. H. Daly and S. E. Wiberley, 'Introduction to Infrared and Raman Spectroscopy', Academic Press, San Diego, CA, 1990.
27. E. Sabatani and I. Rubinstein, *J. Phys. Chem.*, 1987, **91**, 6663.
28. V. P. Menon and C. R. Martin, *Anal. Chem.*, 1995, **67**, 1920.
29. C. Amatore, J. M. Savéant and D. Tessier, *J. Electroanal. Chem.*, 1983, **147**, 39.
30. O. Chailapakul and R. M. Crooks, *Langmuir*, 1995, **11**, 1329.
31. B. D. Cahan and C. T. Chen, *J. Electrochem. Soc.*, 1982, **129**, 700.

STABILIZATION OF ALUMINA SLURRIES IN THE PRESENCE OF OXIDIZERS FOR CHEMICAL MECHANICAL POLISHING

B.J. Palla,[†] M. Bielmann,[‡] R.K. Singh[‡] and D.O. Shah[†]*

[†]Department of Chemical Engineering, and
[‡]Department of Materials Science and Engineering
University of Florida
Gainesville, FL 32611 USA

1 BACKGROUND

In the microelectronics industry there has been an increasing need to achieve multilevel interconnections for increasingly complex, miniaturized devices and circuits. In order to fabricate high-performance multilevel devices, planarization of the chemical vapor deposited interlayer connection metals as well as the interlayer dielectrics is essential. Chemical mechanical polishing (CMP) is the preferred process by which thin films of metals and dielectric materials for multilayer integrated circuit manufacturing are planarized. CMP is a historically ancient technology, but there are significant differences in the present use of CMP in silicon integrated circuits manufacturing. Not only are the layers being removed by CMP very thin (usually less than 0.5 µm), but precise control of the remaining thickness, usually to within 0.01–0.05 µm, is also necessary while maintaining integrity of the underlying structures. Meeting these severe criteria has presented many new challenges to both scientists and engineers.[1]

1.1 The CMP Process

The CMP process, in general, involves rotating a polishing medium, referred to as a pad, against the silicon wafer while a polishing slurry is deposited between the pad and the wafer (Figure 1). The pad is used to provide support against the sample surface and to carry slurry between the sample surface and pad. Although there is some design variability in the equipment available to carry out this process, the wafer is normally attached to a shaft and lowered on top of the pad, which is affixed to a rotating table.[2]

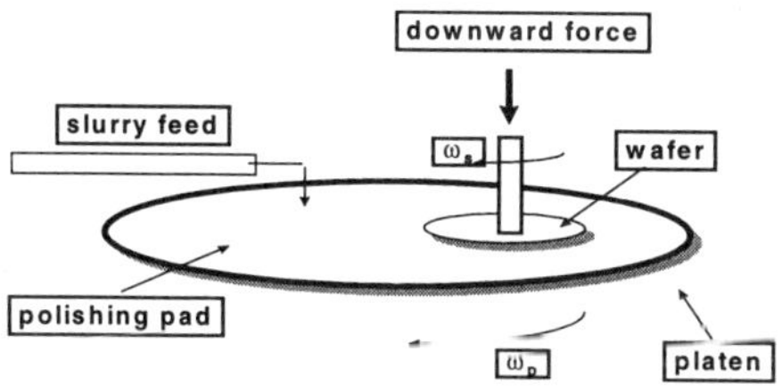

Figure 1 *Schematic of CMP process showing primary components and variables.*

There are therefore three main components of the CMP process: the wafer (or surface to be polished), the pad and the slurry. With these components come a large number of variables which can be manipulated in order to achieve the optimum polishing process for a given surface. These are listed in Table 1 and are classified according to component applicability.

The polishing slurry provides the means by which both chemical and mechanical action is used to remove and subsequently planarize the wafer surface. Mechanical action is accomplished by the use of an abrasive particle in the slurry. The type, particle size and concentration of abrasive material as well as slurry pH are all variables which affect the performance of the abrasive particles. The slurry achieves chemical action either by incorporating oxidizing agents or by incorporating additives which improve film dissolution, depending on the film being acted upon. It is both the chemical and mechanical action achieved by the slurry that provides global planarization of the wafer surface. Slurry components are chosen to optimize the output variables in the polishing process, which are given in Table 2.

Although the CMP process is already widely used in the microelectronics manufacturing industry, current operations rely heavily on empiricism and intensive research has only begun recently. The present work focuses on establishing processing–property correlations for slurry components in an effort to improve methods of slurry design for the CMP process.

Table 1   *Primary Input Variables in the CMP Process, Grouped According to Primary Components.*[1]

Slurry Variables	Pad Variables	Wafer Variables	Other Variables
Suspension stability	Fiber structure	Curvature	Temperature
Oxidizers	Pore size	Mounting	Pressure
Buffering agents	Compressibility	Film stack	Pad velocity
PH	Elastic modulus	Film stress	Wafer velocity
Slurry abrasive	Shear modulus	Film hardness	Frictional forces
Particle size	Hardness	Film microstructure	Pattern geometries
Concentration	Thickness	Cleaning sequence	Feature size
Other additives	Conditioning	Size	Pattern density
Surfactants	Ageing effects	Initial roughness	
Viscosity agents	Chemical durability		

Table 2   *Primary Output Variables in the CMP Process, Along with Desired Value.*[1]

Output Variable	Desired Value for Optimized Polishing
Polishing rate	Maximum
Planarization rate	Maximum
Surface roughness of polished surface	Minimum
Selectivity (ratio of surface film polishing rate to underlying film polishing rate)	Maximum
Adsorption of particles and chemicals to polished surface	Minimum with ease of removal

1.2 Tungsten Polishing by CMP

The use of tungsten as an interconnection metal has gained wide interest in the electronics industry. In this process, metallized vias or plugs are formed by a blanket tungsten deposition followed by a planarization step achieved by CMP. Via holes are etched through an interlayer dielectric (ILD) such as silica to interconnection lines or a silicon substrate etched below. Next, an adhesion layer such as TiN is formed over the ILD and into the via or plug, followed by the chemical vapor deposition of tungsten which fills the holes. Since this is a blanket deposition process, however, the tungsten is also deposited outside of the vias and plugs on the ILD surface, which must be subsequently removed by the CMP process. The tungsten removal process continues until the tungsten is completely removed from the ILD surface, leaving only the vias and plugs filled. Optimally, the top surface of the filled holes should be uniform and flat, and at the same level as the ILD surface. It is desired that the filled holes do not recess, which implies having a dip in the surface near the middle of the hole. Recessing causes a non-planar via layer to be formed which impairs the ability to print high resolution lines during photolithography steps which follow. It can also lead to voids and open circuits in the metal interconnections when formed above. An illustration of the formation of tungsten filled holes in the ILD layer utilizing CMP is given in Figure 2.[3,4]

CMP of tungsten is accomplished utilizing a slurry which includes an abrasive particle and a chemical reagent, typically an oxidizing agent. One slurry which has gained wide usage for tungsten CMP utilizes alumina (Al_2O_3) particles as the abrasive material in a harsh oxidizing environment. The demands placed on CMP performance require that the abrasive particles be of small size (typically less than 0.5 μm) and uniform size distribution. The presence of large particle agglomerates can lead to increased scratches and defects on the wafer surface. Oxidizing agents used in tungsten slurries include, but are not limited to, ferric nitrate, potassium ferricyanide, potassium dichromate, potassium iodate, potassium bromate and vanadium trioxide. A preferred oxidizing agent from these candidates is potassium ferricyanide, $K_3Fe(CN)_6$, which does not readily precipitate out of solution, cause corrosion, nor is it classified as a hazardous material. Potassium ferricyanide, when used in this type of tungsten CMP slurry, serves to rapidly oxidize the tungsten surface, while the alumina abrasive serves to remove the oxidized layer by mechanical action. The preferred pH for this type of slurry is in the 2–4 range. Higher pH values are known to be a cause of recessing in the filled holes after polishing, while lower pH values, although adequate for providing high removal rates, are hazardous and corrosive to the polishing apparatus. Both abrasive particles and oxidizing agent must be chosen in a concentration adequate to provide the desired effect without adding excess material to the slurry.[3]

2 MATERIALS

All materials were used as received without further purification. Water was de-ionized and filtered before use. Potassium ferricyanide was obtained from Fisher Scientific. AKP-50 (mean size between 0.1 and 0.3 μm) and AKP-30 (mean size between 0.3 and 0.5 μm) alumina particles were obtained from Sumitomo Co. γ-Alumina particles were supplied from Leco, Inc, with a listed mean particle size of 0.05 μm. Sodium dodecyl sulfate (SDS), specially pure, was obtained from BDH Chemicals, Ltd., cetylpyridinium

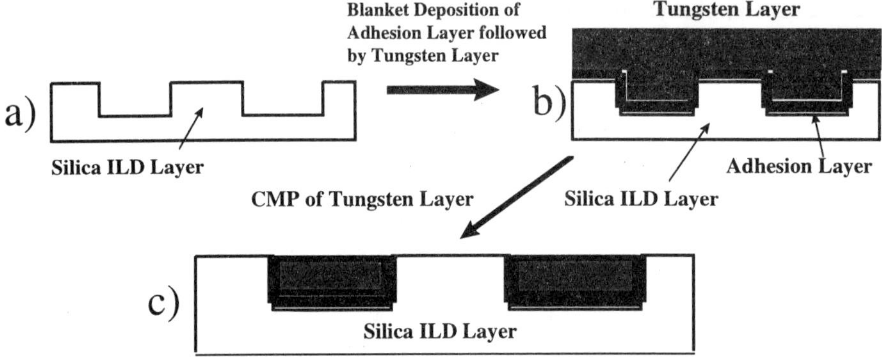

Figure 2   *Cross-sectional schematic of the tungsten CMP process: a) silica layer after etching, b) CVD blanket deposition of adhesion layer and tungsten layer, c) planarized tungsten surface after CMP.*

chloride (CPC) was obtained from Aldrich Chemical Co. and Triton X-100 was obtained from Sigma.

Polishing results were obtained on a Struer Rotopol 31 polisher using IC-1000/SUBA IV stacked pads from Rodel, a pressure of 6.74 psi, pad speed of 150 rpm and slurry flow of 0.1 L/min, with a wafer size of 1 in². AFM results were obtained from a Digital Instruments Nanoscope III, with a scan area of 1 μm. SEM results were obtained from a JEOL 200-CX scanning electron microscope. Particle size and zeta potential results, without sample dilution, were obtained from a Colloidal Dynamics Acoustosizer.

3 STABILIZERS FOR TUNGSTEN CMP SLURRIES

Alumina slurries for tungsten CMP have been shown to yield a high removal rate of tungsten. However, the slurries do not produce adequate planarization results when used without stabilizing agents. This should be explained by the fact that the alumina particles are not remaining dispersed without stabilizing agents present, and hence the particles are agglomerating. This phenomenon is investigated in this section.

3.1 Mechanism of Agglomeration of Alumina Particles

3.1.1 Introduction. Typically, alumina particles of the optimum particle size range for tungsten CMP applications at pH 2–4 do not agglomerate in water. This is because of the surface charge of the particles, which is sufficiently positive to provide electrostatic repulsion between the particles, preventing agglomeration. However, the addition of an oxidizing agent such as potassium ferricyanide to the alumina slurry can destabilize the particles by providing additional ions. The addition of salts such as potassium ferricyanide provides counter-ions of opposite charge that shield the electrostatic repulsion between particles. When this shielding is sufficient to overcome the repulsive barrier, agglomeration can occur. This effect is present in tungsten CMP slurries, in which potassium ferricyanide, typically used in concentrations near 0.1 M, acts as a salt and causes particulate agglomeration and a broadening of the particle size distribution.

3.1.2 Results. The particle size and zeta potential of alumina slurries at pH 4 with and without oxidizing agent are given in Table 3. These results were obtained from a Colloidal Dynamics Acoustosizer, with 5 wt. % AKP-50 (100–300 nm) alumina particles in the slurries. The results suggest that the ferricyanide ion ($[Fe(CN)_6]^{3-}$) present in high concentration in the slurry not only decreases the zeta potential to near zero, but actually reverses the zeta potential of the particles. Although the zeta potential of the particles with 0.1 M potassium ferricyanide present is about –30 mV, Table 3 shows that this charge is not sufficient to cause particle repulsion, so some agglomeration occurs. Not only is the average particle size in the slurry with 0.1 M oxidizer about 3 times larger than without oxidizer, but the size distribution limits given indicate a broader particle size distribution when oxidizer is present. The high limit with oxidizer is 1.68 μm, which indicates that there are a large number of particles in excess of 1 μm present in this slurry. The presence of particles larger than 1 μm are proposed to be responsible for scratching surfaces and increasing defect density. Both slurries have also been observed visually for settling rate comparison. These results are presented in Figure 3. These slurries were allowed to settle for 24 hours before being photographed. This figure clearly illustrates that the particles in the slurry with oxidizer have settled out of solution, while the particles in the slurry without oxidizer remain dispersed with little settling observed at the bottom of the cylinders. The settling, or destabilization, of the slurry with oxidizer is attributed to larger average particle size of these particles.

Table 3   *Particle Size and Charge of Alumina Particles With and Without Oxidizing Agent, Potassium Ferricyanide.*

Oxidizer Concentration (M)	Zeta Potential (mV)	Average particle size (μm)	68% size distribution limits (μm)
0	+83.8 ± 0.2	0.086 ± 0.002	(0.029 , 0.255) ± 0.005
0.001	–1.49 ± 1.0	0.671 ± 0.050	(0.172, 2.63) ± 0.30
0.1	–30.9 + 1.6	0.491 ± 0.048	(0.278, 1.68) ± 0.30

3.2 Initial Stabilization Investigation with Basic Surfactants

3.2.1 Motivation. It was determined that stabilization of the alumina particles in a tungsten CMP slurry could have significant effects on the polishing performance and handling requirements of the slurry. First, the use of a suitable stabilizing agent would eliminate the need for a two-part slurry system, which is commonly used in the tungsten CMP process. Second, the use of stabilizers could significantly effect the performance of the CMP process, including the removal rate, degree of planarization and selectivity. Finally, any stabilizers added to the slurry would have to be removed from the final product, which could pose a challenge for post-CMP cleaning. The use of additives such as surfactants in the slurry was determined as the most feasible route to stabilize the slurry.

3.2.2 Experimental Procedure. The first attempt to stabilize a tungsten CMP slurry investigated the effect of electrostatic forces between the surfactant molecules and

Figure 3   *Settling of alumina particles after addition of oxidizer, $K_3Fe(CN)_6$. Slurries are 1 wt. % AKP-50 alumina particles at pH 3,4, and 5 (left to right), allowed to settle for 24 hours before photograph.*

particles as a means of stabilization. One surfactant each was chosen from three main types of surfactants: cationic, anionic and nonionic. The slurry investigated was a 1 wt. % alumina slurry with AKP-50 (100-300 nm) particles in DI water. The surfactant concentration was chosen to be 10 mM, and the surfactant was dissolved in solution prior to addition of the oxidizer, potassium ferricyanide. After dissolution of potassium ferricyanide, the slurry pH was adjusted with nitric acid. The slurries were prepared in graduated cylinders in order to visualize stability over a long period of time. After considerable shaking, the cylinders were isolated until changes in particle level were minimal, after which photographs were taken. Common surfactants were chosen from each class, including cetylpyridinium chloride (CPC) as a cationic surfactant, sodium dodecyl sulfate (SDS) as an anionic surfactant and Triton X-100 as a nonionic surfactant.[5]

3.2.3 Results. Figure 4 is a photograph of the results with the three different surfactants added, and without surfactant added as a control, for (a) 0.001 M and (b) 0.1 M potassium ferricyanide. In (a) 0.001 M oxidizer, both CPC and SDS stabilize the suspension, while Triton X-100 shows little improvement over potassium ferricyanide alone. In (b) 0.1 M oxidizer, CPC remains stable to some extent, but particles are settled in large clumps. However, at this oxidizer concentration, both SDS and Triton X-100 are settled fully. These results imply that the cationic surfactant addition results in the most stabilization, but it is not sufficient to produce a stable suspension. The results can be explained partially by electrostatic repulsion effects caused by the charged surfactants. With only 0.001 M potassium ferricyanide added, acoustosizer results show from Table 3 that the charge on the alumina particles has been reduced to near 0, and therefore both SDS and CPC can adsorb on the particles without repulsion, form a bilayer, and promote dispersion by creating an electrostatic repulsion. The bilayers of CPC and SDS also form a tightly packed shield around the particles and may prevent attraction by physical repulsion of the bilayers. Triton X-100 cannot stabilize the suspension electrostatically due to its charge neutrality, and the presence of branched chains in its hydrocarbon backbone prevent the formation of bilayers packed tight enough to cause physical repulsion.   However, when 0.1 M potassium ferricyanide is added to the slurry, the particles exhibit a charge reversal as verified by acoustosizer results (zeta potential = −30.9 mV), due to the large amount of $[Fe(CN)_6]^{3-}$ ions in solution, and different results are observed.   The cationic surfactant (CPC) apparently continues to adsorb on the particles, as shown by the slight slurry stabilization, but a large amount of competitive adsorption is present, caused by the high ionic strength of the solution, and the surfactant alone cannot sufficiently prevent agglomeration.   The anionic surfactant (SDS) can no longer adsorb due to the electrostatic charge repulsion of its head group.

Figure 4   *Basic surfactant stabilization investigation in tungsten CMP slurries.   a) 0.001 M $K_3Fe(CN)_6$ in each slurry, b) 0.1 M $K_3Fe(CN)_6$ in each slurry. Slurries are 1 wt. % AKP-50 alumina at pH 3 (left) and pH 4 (right), with 10 mM labeled surfactant added prior to oxidizer.   Photographs were taken after 4 days of sedimentation.*

3.3 Use of Mixed Surfactant Systems

3.3.1 Motivation.   The use of surfactant systems as stabilizers for tungsten CMP slurries has been extended to mixed surfactant systems in an attempt to completely stabilize the slurry.   This is because mixed surfactants can have highly synergistic effects which lead to enhanced dispersion by the mixture which is not observed by the individual surfactants.   It was thought that the combination of an ionic surfactant which can induce electrostatic repulsion between particles along with a molecularly large nonionic surfactant which can enhance steric repulsion could produce the desired synergism necessary to stabilize the high ionic strength potassium ferricyanide slurries used for tungsten CMP.

3.3.2 Experimental Procedure. Different anionic, cationic and nonionic surfactants were mixed together in many combinations in order to compare results with expectations. The same settling observation procedure was used as with individual surfactants. The total surfactant concentration was constant at 0.10 M, with a 3:1 molar ratio of ionic:nonionic surfactant used in all experiments. The AKP-50 alumina concentration was kept constant at 1 wt. % and the potassium ferricyanide concentration was 0.1 M.

3.3.3 Results. Figure 5 is a photograph of two different combinations of surfactants, one using a cationic and nonionic surfactant and the other using an anionic and nonionic surfactant. These results clearly show that the combination of anionic/nonionic surfactants results in improved stability over the combination of cationic/nonionic surfactants, and these results were verified for many similar systems. This result is clearly an example of a synergistic effect when anionic and nonionic surfactants are used together. This is because the same surfactant types showed little or no stabilization when used alone, while the cationic surfactant did show some stabilization (see Figure 4b).

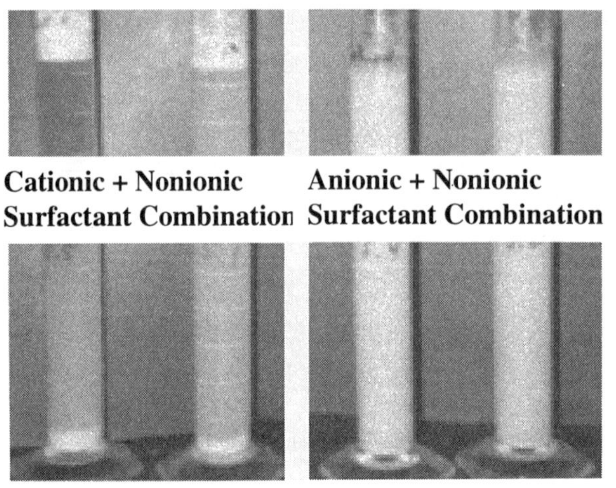

Figure 5   *Stabilization of slurries using mixed surfactant addition, as labeled. Slurries are 1 wt. % AKP-50 alumina at pH 3 (left) and pH 4 (right), with 10 mM total surfactant added in a 3:1 anionic:nonionic molar ratio. Photographs taken after 4 days of settling.*

Although the mechanism for stabilization in this complex system of surfactants, particles and oxidizing agent is not completely understood, a mechanism has been proposed for the stabilization of the suspension. The nonionic surfactants used in all cases contain large polar head groups that may adsorb to the alumina surface. These molecules are not considered large enough to promote steric stabilization when used alone, however, so the anionic surfactant is necessary for dispersion. The anionic surfactant can be attracted to the tail of the nonionic surfactant to form a mixed bilayer, which then causes repulsion from the large, tight packing of surfactant molecules around the particles. There may also be some electrostatic repulsion present due to the anionic head groups approaching one another. Although this mechanism may be oversimplified, it is thought that this is the general synergistic effect that causes dispersion of particles.

4 EFFECT OF SLURRY STABILIZATION ON TUNGSTEN CMP POLISHING PERFORMANCE

Although it is the subject of another publication by the same authors,[6] some of the results on the use of a stable slurry in the tungsten CMP process are presented here. In all cases, all variables were kept constant except the use of two different slurries. Both slurries contained 10 wt. % Leco 50 nm alumina particles and 0.1 M potassium ferricyanide at pH 4. One slurry contained no surfactant, while the other slurry contained a combination of anionic and nonionic surfactants, optimized for stability with respect to surfactant type, concentration and ratio. Results were obtained comparing the two slurries for polishing rate, surface roughness after polishing and alumina particle adsorption to the tungsten surface after polishing.

4.1 Polishing Rate

The rate of polishing of a tungsten wafer was determined by estimating metal thickness before and after polishing using a manual four-point sheet resistivity probe apparatus. Figure 6 shows a comparison of the polishing rates obtained for the two slurries. Results suggest that the use of stabilizing surfactant decreases the polishing rate of tungsten by about 30%. This difference can be attributed to a modification in the mechanical behavior of alumina particles caused by the surfactant. The surfactant decreases particle agglomeration, and hence decreases the effective particle size. The agglomerated particles present in slurries without surfactant increase the mechanical removal rate, and the sedimentation effect is expected to increase contact between the alumina particles and the tungsten surface.[7] Additionally, the surfactant could decrease the polishing rate due to a lubricating effect of an organic coating around the particles.

4.2 Surface Roughness After Polishing

The degree of local planarization is another key parameter in determining CMP process performance. The degree of local planarization can be measured by examining the surface roughness of the surface after polishing. The surface roughness of tungsten wafers after polishing with the same two slurries used for the polishing rate determination was obtained from atomic force microscopy (AFM). Figure 7(a) shows the surface morphology before polishing (rms roughness = 16 ± 3 nm), Figure 7(b) shows the surface after polishing with a slurry containing no surfactant (rms roughness = 2.65 ± 0.39 nm), and Figure 7(c) shows the surface after polishing with a slurry containing surfactant stabilizers (rms roughness = 1.57 ± 0.15 nm). The results show a marked decrease (about 40%) in surface roughness when the stabilized slurry was used for polishing. The improvement in planarization with the stable slurry can be explained by the modification of mechanical behavior of the alumina particles that also affects the polishing rate.

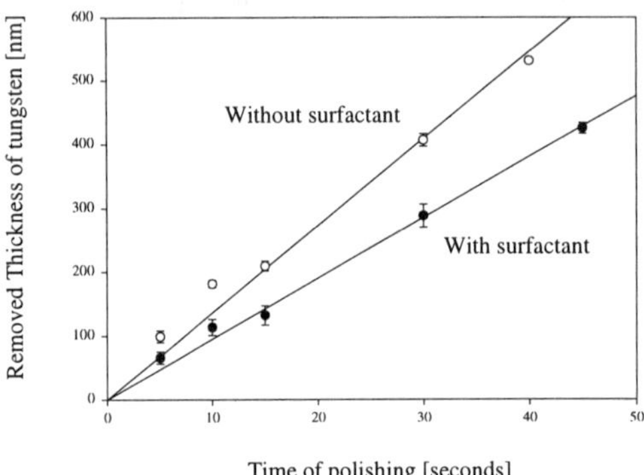

Removed Thickness of tungsten [nm]

Time of polishing [seconds]

Figure 6   *Effect of surfactant on removal (polishing) rate of tungsten. Slurries contain 5 wt. % Leco 50 nm alumina at pH 4 with 0.1 M K₃Fe(CN)₆ and optimized anionic and nonionic surfactant combination. Conditions: pressure=6.74 psi, pad speed=150 rpm, slurry flow=0.1 L/min on 1" x 1" sample.*

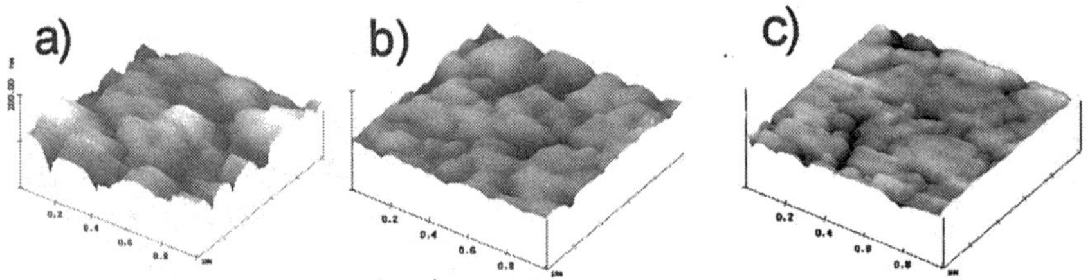

Figure 7   *AFM characterization of surface roughness of tungsten wafers. a) Before polishing, b) after polishing with slurry containing no surfactant, c) after polishing with slurry containing optimized anionic and nonionic surfactant combination. Each micrograph shown is 1 μm x 1 μm.*

4.3 Particulate Contamination of the Tungsten and Silica Surfaces

Particulate contamination of the tungsten and dielectric silica surface after CMP can lead to increased challenges in post-CMP cleaning. Addition of surface modifiers to the alumina particles can modify the tendency of the particles to attract to the polished surfaces.

Particulate contamination by the polishing process was estimated by conducting static dip tests in which tungsten and silica wafers were immersed for seconds in slurries containing larger alumina particles to facilitate viewing by SEM, and the wafers were rinsed with DI water. The number of residual particles was then determined from SEM micrographs of the wafers. Figure 8 shows the SEM pictures of both tungsten and silica surfaces, after immersion in the same two slurries used for polishing rate and surface roughness characterization. Figures 8 (a) and (b) show that addition of surfactant decreases the particulate contamination on a tungsten surface, and Figures 8 (c) and (d) show that the addition of surfactant also decreases contamination on a silica surface. These results can be explained through the van der Waals force of attraction between the particles and surfaces, since electrostatic effects are expected to be minimal in the slurries containing high concentrations of oxidizer. The Hamaker constant of tungsten is about 20×10^{-20} J,[8] while that of silica is 0.83×10^{-20} J.[9] Thus the van der Waals attractive forces are much larger for the tungsten surface than the silica surface, which explains the larger attraction of particles in both slurries to the tungsten surface. However, the adsorption of a surfactant layer in Figures 8 (b) and (d) creates a repulsive barrier between surfaces due to physical interaction of surfactant bilayers. This inhibits particles from approaching to close distances and being attracted by the strong van der Waals forces, thus limiting particulate contamination.

a)
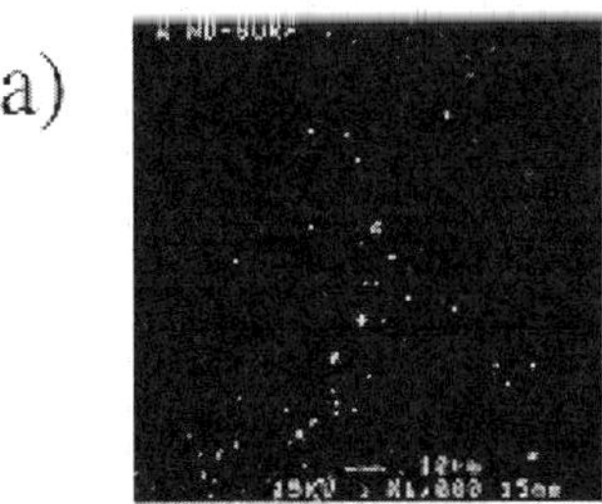

Alumina Particles on tungsten.
Slurry without surfactant

b)
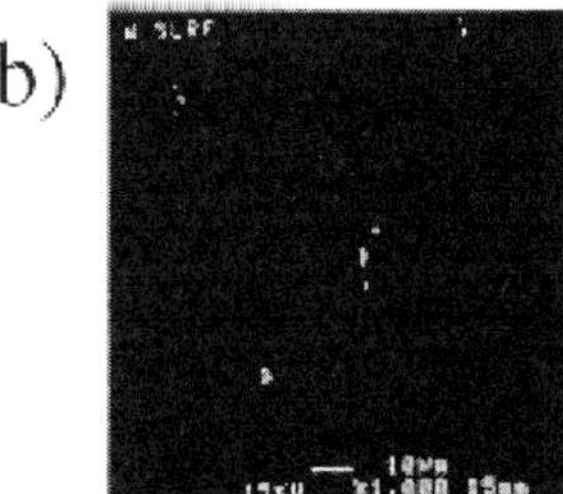

Alumina Particles on tungsten.
Slurry with surfactant

c)
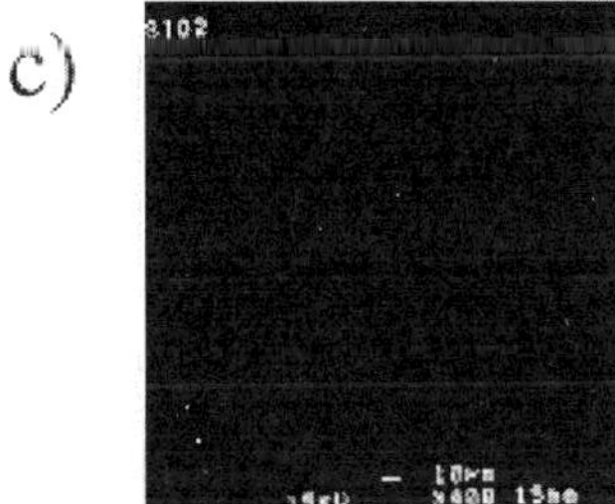

Alumina Particles on silica.
Slurry without surfactant

d)
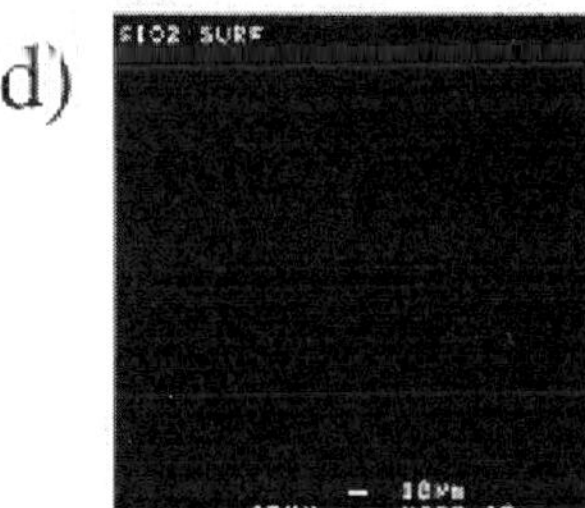

Alumina Particles on silica.
Slurry with surfactant

Figure 8   *Particulate contamination of tungsten and silica surfaces by dip testing. Slurries used are as labeled, all at pH 4 containing 5 wt. % AKP-30 (300–500 nm) alumina, 0.1 M $K_3Fe(CN)_6$ and optimized surfactant combination.*

5 CONCLUSIONS

The use of an appropriate slurry can be a significant factor in determining performance in the CMP process. In tungsten CMP, particle agglomeration caused by the typical high ionic strength of the slurry results in polishing performance that is less than optimum. The selection of surfactants which can stabilize a tungsten CMP slurry is a challenging task, but the use of a synergistic mixture of surfactants has been found to stabilize the slurry for a significant period of time. The results of using a stabilized slurry on CMP polishing performance is explained from the adsorption of an organic layer of surfactant on the abrasive particle surface as well as from the lack of particle agglomeration observed in the stable slurry. The use of a stable slurry compared to an agglomerated slurry results in decreased polishing rate, improved planarization and decreased particulate contamination of the polished surface. Therefore, the use of surfactant stabilized slurries can be a viable method for CMP of tungsten.

6 ACKNOWLEDGEMENTS

The authors would like to acknowledge the financial support of the Engineering Research Center (ERC) for Particle Science and Technology at the University of Florida. They also thank the other members of the CMP focus group, including Dr. Brij Moudgil, Dr. Chang-Won Park, Dr. Yakov Rabinowich, Joshua Adler, D. Kumar, and Uday Mahajan, for their invaluable suggestions.

References

1.   J. M. Stiegerwald, S. P. Murarka and R. J. Gutmann, 'Chemical Mechanical Planarization of Microelectronics Materials', John Wiley & Sons, NY, 1996, p. 1.
2.   W. C. O'Mara, 'Planarization by Chemical–Mechanical Polishing for Multilevel Metal Integrated Circuits', O'Mara & Associates, Palo Alto, p. 1.
3.   K. C. Cadien and D. A. Feller, *US Patent 5,516,346*, 1996.
4.   F. B. Kaufman, D. B. Thompson, R. E. Broadie, M. A. Jaso, W. L. Guthrie, D. J. Pearson and M. B. Small, *J. Electrochem. Soc.*, 1991, **138**, 3460.
5.   M. J. Rosen, 'Surfactants and Interfacial Phenomena', 2nd edition, John Wiley & Sons, NY, 1989, p. 7.
6.   M. Bielmann, U. Mahajan, R. K. Singh, B. J. Palla and D. O. Shah, *J. Electrochem. Soc.*, submitted for publication.
7.   E. A. Kneer, C. Raghunath, J. S. Jeon and S. Raghavan, *J. Electrochem. Soc.*, 1997, **144**, 3041.
8.   J. Visser, *J. Coll. Inter. Sci.*, 1972, **3**, 331.
9.   J. Israelachvili, 'Intermolecular and Surface Forces', Academic Press, London, 1992.

ASSESSMENT OF A THERMALLY REGENERATIVE BATTERY CONCEPT UTILIZING INTERCALATION PHENOMENA IN GRAPHITE

Pramod K. Sharma,* S.R. Narayanan and Gregory S. Hickey

Jet Propulsion Laboratory
California Institute of Technology
Pasadena, CA 91109 USA

1 ABSTRACT

Graphite is known to form intercalation compounds with halogens, such as bromine, and with alkali metals, such as potassium, cesium or lithium. The guest species enter the space between uniform layers of graphite atoms and form weak compounds at room temperature. The intercalation of such species in graphite is also associated with charge effects on the graphite as well as the intercalated species. One may conceptualize a battery based on this phenomenon in which graphite electrodes capable of undergoing intercalation reactions are immersed in a nonaqueous solution of a salt such as lithium bromide. The two electrodes are separated by an ion exchange membrane impermeable to bromine or Br_3^- anions. The intercalation of bromine into the graphite cathode leads to net power output.

Since intercalation of bromine in graphite is thermally reversible, the battery can be recharged by simply heating the cathode to a temperature sufficient to drive the bromine out of the graphite. In principal, the battery can also be recharged electrochemically. The critical step to verify this electro regenerative battery concept was identified as

$$CBr + e^- \Rightarrow Br^- + C$$

Experimentation was carried out using an electrochemical cell and bromine-intercalated graphite to verify that the above reaction step can occur electrochemically.

2 INTRODUCTION

Graphite is known to form intercalation compounds with halogens and alkali metals, in which the guest species enter the space between uniform layers of graphite atoms and form weak compounds at room temperature.[1-3] The guest species between the layers of graphite atoms often behaves as a two-dimensional fluid,[3] *i.e.* its mobility is restricted to two dimensions. A surface phenomenon of this type has certain selective and unique applications, *e.g.* in surface wetting and surface contamination transport processes.

We have previously reported on our studies of intercalation of bromine in graphite.[4] In this previous report, a graphite sample was brought into contact with bromine vapor in equilibrium with liquid bromine at room temperature. The bromine-treated sample was desorbed by stepwise heating to higher temperatures. This desorption experiment, the results of which are shown in Figure 1, demonstrated that a major part of the bromine is released upon heating the bromine-intercalated graphite to 120 °C.

It is known that the intercalation of halogens leads to a positive charge on the graphite while the halogen atoms assume a negative charge.[1] Similarly, the intercalation of alkali metals leads to a negative charge on the graphite while the alkali metals themselves assume a positive charge. These observations would suggest that coupling of two electrodes, one with intercalated halogens, and the other with intercalated alkali metals, has the potential to form a battery.

One may conceptualize a battery that takes advantage of this intercalation phenomenon. Such a battery would consist of a graphite anode and a thermally rechargeable graphite cathode, both immersed in a common body of electrolyte such as LiBr dissolved in a nonaqueous solvent such as diethyl carbonate, dimethyl carbonate, dioxane, propylene carbonate, ethylene carbonate and mixtures thereof. The two electrodes are separated by an ion exchange membrane impermeable to bromine and Br_3^- anions. A simplified sketch of the battery is provided in Figure 2. Some examples of materials suitable for the ion exchange membrane are provided in Figure 3.

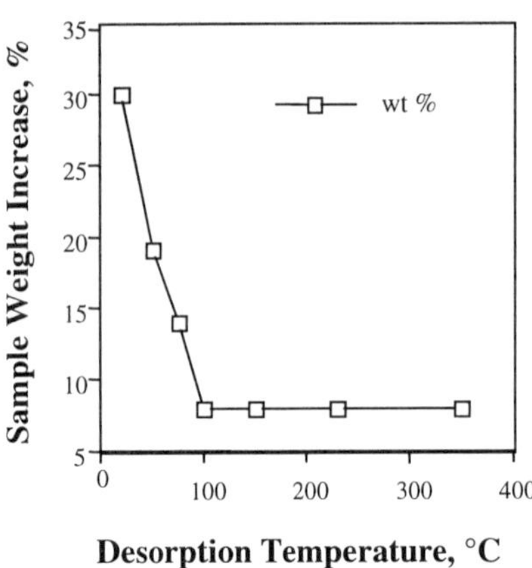

Figure 1 *Weight change of the bromine intercalated sample as a function of desorption temperature.*

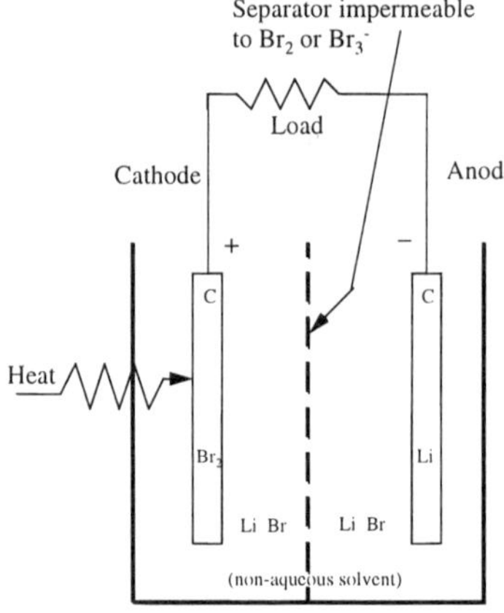

Figure 2 *Conceptual design of the battery based on bromine intercalation in graphite.*

It should be possible to regenerate such a battery either electrochemically (see below) or thermally. When the thermal option is used, the cathode alone needs to be heated. This may be achieved by using a configuration such as shown in Figure 4, which allows the two electrodes to be thermally isolated.

2.1 Electrochemical Reactions

In the fully charged state, there is free bromine in the compartment surrounding the cathode. In the presence of Br^- ions, bromine is complexed as Br_3^-, similar to the electrolyte in Zn/Br_3^- batteries. This is because Br_2 and Br^- cannot coexist, and react according to $Br_2 + Br^- \Rightarrow Br_3^-$.

I. PTFE-perfluoroalkoxysulfonic acid (NAFION®):

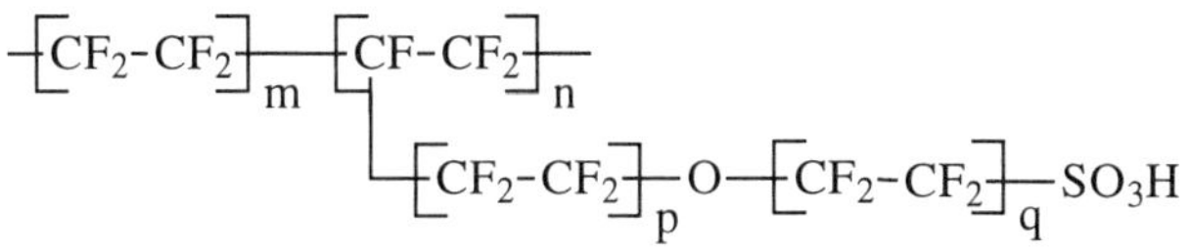

$$-\left[CF_2-CF_2\right]_m\left[CF-CF_2\right]_n$$

$$\left[CF_2-CF_2\right]_p-O-\left[CF_2-CF_2\right]_q-SO_3H$$

II. Polyhydrocarbon sulfonic acid:

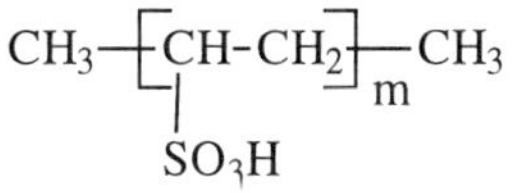

$$CH_3-\left[\underset{\underset{SO_3H}{|}}{CH}-CH_2\right]_m-CH_3$$

III. Polyaryloxyketone sulfonic acid:

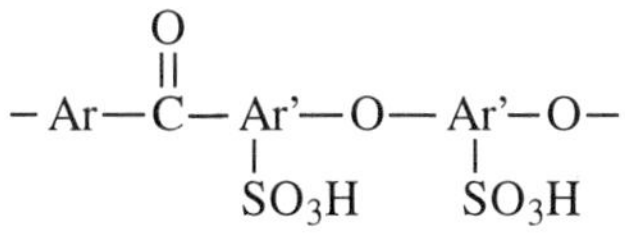

$$-Ar-\overset{\overset{O}{\|}}{\underset{\underset{SO_3H}{|}}{C}}-Ar'-O-\underset{\underset{SO_3H}{|}}{Ar'}-O-$$

Figure 3   *Examples of polymeric membrane separator materials and their structures.*

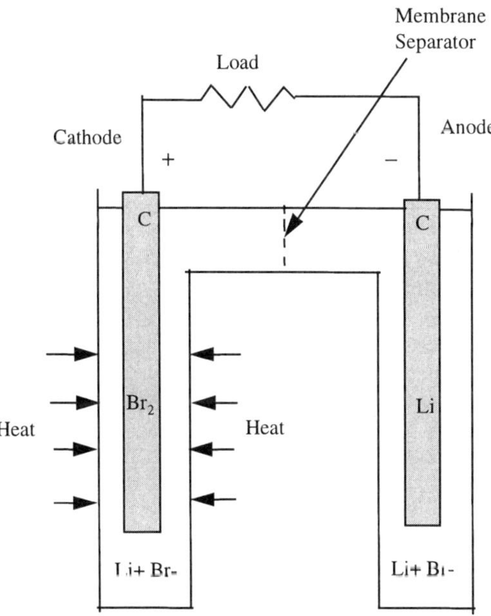

Figure 4   *Battery configuration allowing for selective heating of cathode compartment.*

During discharge, bromine enters the graphite cathode by intercalation while lithium comes out of the graphite anode. The Br^- generated inside the electrode reacts with Br_2 as above. Thus, the discharge processes may be represented as

$$\text{Anode:} \quad CLi \Rightarrow C + Li^+ + e^-$$

$$\text{Cathode:} \quad 2Br_2 + C + e^- \Rightarrow CBr + Br_3^-$$

Thus, at the end of discharge the cathode contains the intercalation compound CBr, which may be visualized as C^+Br^-, while the anode contains no guest species. These processes lead to an electrochemical power generation.

When the battery has been depleted, it can be recharged either thermally (by heating the cathode) or electrochemically. In the thermal option, the cathode can be heated using direct solar power, heat from a Radioactive Thermal Generator (RTG), or a suitable waste heat source. A temperature of 120 °C may easily be attained, for example, by concentrating solar radiation by a parabolic mirror. However it is achieved, heating of the cathode causes the intercalation compound to break up and bromine to come out of the cathode, thus reversing the intercalation process. This drives a current through an external circuit which connects the bromine electrode to the anode and will cause lithium ions to intercalate into the graphite anode forming CLi. The intercalation of Li in graphite anodes has been extensively studied[5] for use in lithium batteries. Thus, during recharging, the electrode processes may be represented as

$$\text{(heat)}$$
$$\text{Cathode:} \quad CBr + Br_3^- \quad \Rightarrow \quad 2Br_2 + C + e^-$$

$$\text{Anode:} \quad C + Li^+ + e^- \quad \Rightarrow \quad CLi$$

The ion exchange membrane (impermeable to Br_3^-) also serves as the electronic separator between the two electrodes. If the non-aqueous solvent is properly chosen, the battery can operate at temperatures as low as –60 °C during discharge.

2.2 Voltage and Specific Energy Calculation

In evaluating the practicality of a battery concept, the important parameters to consider are the cell voltage and the energy density. The expected cell voltage is easily calculated. For the bromine cathode, the voltage E_c is likely to be close to the standard reduction potential of the Br_2/Br^- couple which is 1.08 volts. For the lithium anode, the voltage E_a is close to –3.0 volts. Therefore, the cell voltage, $E_c - E_a$ is approaching

$$E_c - E_a = 1.08 - (-3.0) = 4.08 \text{ volts.}$$

The specific capacity obtained using Faraday's Law for a carbon–lithium anode and a stoichiometry of C_6Li is 372 mAh/g. For a carbon–bromine cathode, preliminary intercalation measurements[4] made with bromine vapor at room temperature (Figure 1) indicate a stoichiometry of $C_{32}Br$ and a resultant specific capacity of 69 mAh/g of carbon. As this is the smaller number, the cathode energy density will be controlling and, hence, the battery specific capacity will be limited to 69 mAh/g of carbon. Even limited by the bromine electrode, the specific energy (amp-hours X volts) is 280 mWh/g of carbon. It should be noted that when bromine intercalation is carried out at a higher vapor pressure, the cathode specific capacity (and hence specific energy) can be increased by a factor of 2 or more.

3 EXPERIMENTAL

That the battery can be thermally regenerated follows from the observation that intercalated bromine is lost upon heating the graphite cathode to 120°C. Assessing whether electrochemical regeneration is possible, however, was not as straightforward and required us to devise the test described below.

An aqueous solution of approximately 0.50 molar potassium sulfate (K_2SO_4) was prepared by dissolving 21.2 g of the salt in 300 mL of distilled water. This solution served as the electrolyte in the battery. We recognize that a rigorous test of this concept should be carried out with a non-aqueous electrolyte. However, in view of limited resources, a less-rigorous test was designed to be carried out in an aqueous medium electrolyte merely to verify the most important aspects of the concept.

The proposed test calls for using bromine-intercalated graphite as the cathode. But connecting electrical leads to bromine-containing graphite proved to be quite difficult due to bromine attack on the leads. This difficulty was overcome by using graphite paper as the electrode. Pieces of graphite paper were intercalated with bromine by contacting them with saturated bromine vapor at room temperature inside a sealed glass flask for a period in excess of 3 days. The bromine-intercalated paper was then removed from the flask and its top end was wrapped in a platinum foil which was pre-attached to an electrical cable. Another platinum wire was used as the anode in the cell. The reference electrode was a saturated Calomel electrode with a potential of 0.24 volts *vs.* Normal Hydrogen Electrode (NHE). See Figure 5 for the test schematics.

Tests were carried out for comparison purposes on a bromine-intercalated graphite electrode and a blank graphite electrode. In Test Configuration I, the cathode was the bromine-intercalated graphite paper and the anode was a platinum wire. In Test Configuration II, the cathode was a blank graphite paper and the anode was the same platinum wire. In both configurations the electrolyte was the 0.50 molar K_2SO_4 solution.

4 RESULTS

In Test Configuration I (with the bromine-intercalated graphite paper), the open circuit voltage was first measured and was found to be 0.834 volts. This value is close to the Br_2/Br^- standard reduction potential with the chosen reference electrode. The two electrodes were then connected and the Br_2 electrode was cathodically polarized with a current of about 10.5 micro-amps using an external power supply. About 20 minutes elapsed between connecting the electrodes and adjusting for the proper current. The potential of the bromine-intercalated graphite electrode was monitored as a function of time as the current was maintained at 10.5 micro-amps. The electrode potential *vs.* time profile obtained is shown in Figure 6. The total test duration was about one hour. However, the test was interrupted three times to measure the open circuit voltage, which was found to be close to 0.80 volts.

In Test Configuration II (with the blank graphite paper), the open circuit voltage was first measured and was found to be 0.132 volts. A cathodic current of about 10.5 micro-amps was then made to flow through the cell using the external power supply. The electrode potential as a function of time was monitored again as before. These measurements are also shown in Figure 6. We see that the electrode potential remains at the value of –0.400 volts, and is relatively independent of time. This voltage is basically determined by the H_2 evolution reaction at a pH of 5.0. This is the only reaction that can take place in the absence of bromine.

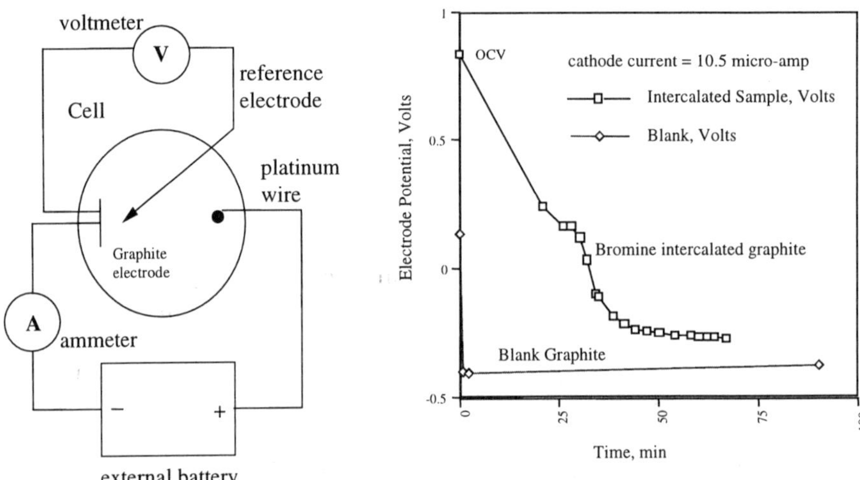

Figure 5   *Schematic of the apparatus used for the electrochemical measurements.*

Figure 6   *Comparison of the measurements for bromine-intercalated graphite (upper curve), and blank graphite (lower curve).*

In Test Configuration I, the expected reaction is

$$CBr + e^- \Rightarrow Br^- + C$$

The observed open circuit voltage and the electrode potential decay with time, as shown in Figure 6, are in agreement with a process described by the reaction shown above. Thus, it appears that chemically intercalated Br_2 can be de-intercalated electrochemically. The capacity of the Br_2 electrode could be dependent on the type of graphite used and can be higher, depending on operating conditions.

In the absence of Br_2 in graphite (Test Configuration II), the reduction reaction sustained is H_2 evolution. The observed electrode potential, which is relatively constant, is in agreement with this reaction. We also see in Figure 6 that the electrode potential of the bromine electrode approaches the value obtained for the blank graphite at long reaction times. This is consistent with de-intercalation of bromine from this electrode, and resultant behavior approaching that of blank graphite.

The remaining reaction to complete the cycle [$Br^- \Rightarrow 1/2Br_2 + e^-$] is a well studied electrochemical process and its demonstration is not warranted here.

5 CONCLUSIONS

The intercalation of bromine and alkali metals in graphite and resultant charge distributions suggest a novel battery concept. The concept allows thermal or electrochemical regeneration of the battery. The battery is expected to have a high voltage of about 4 volts and a reasonably good energy density. Limited proof-of-concept tests have been performed that support the conceptual battery. This type of a battery can find unique applications in space and on Earth.

6 ACKNOWLEDGMENTS

This research was carried out by the Jet Propulsion Laboratory, California Institute of Technology under a contract to the National Aeronautics and Space Administration.

References

1.   M. C. Robert, M. Oberlin and J. Mering, 'Lamellar Reactions in Graphitizable Carbons', in *Chemistry and Physics of Carbon,* Vol. 10, P. L. Walker and P. A. Thrower, eds., Marcel Dekker, New York, 1973, p.141.
2.   J. G. Hooley, 'Intercalation Isotherms on Natural and Pyrolytic Graphite', in *Chemistry and Physics of Carbon,* vol. 5, P. L. Walker, ed., Marcel Dekker, New York, 1969, pp. 321–372.
3.   E. D. Glandt, 'Statistical Thermodynamics of Two-Dimensional and Adsorbed Fluids', *Ph.D. Dissertation,* University of Pennsylvania, 1977.
4.   P. K. Sharma and G. S. Hickey, 'Kinetics of the Formation of Intercalation Compounds in Crystalline Graphite', in *Chemically Modified Surfaces: Recent Developments,* J. J. Pesek, M. T. Matyska and R. R. Abuelafiya, eds., The Royal Society of Chemistry, Cambridge, UK, 1996, pp. 76–88.
5.   N. Tahami, A. Satoh and T. Chsahi, *J. Electrochem. Soc.,* 1995, **142**, 371.

SURFACE MODIFICATION OF POLYSTYRENE USING MEDIATED ELECTROCHEMICAL OXIDATION

C.J. Tremlett and I. Mathieson*

Institute of Surface Science and Technology
Physics Department
Loughborough University
Loughborough, LE11 3TU, UK

1 INTRODUCTION

Polystyrene (PS) is a commodity plastic that is widely available and used for many applications. Its bulk properties make it ideal for use, for example, as containers, bottles, bags, *etc.* However, adhesion problems can be encountered when printing on it, or adhesively bonding it to other components. To overcome these adhesion problems, methods of surface modification have been developed which include acid etching, grit blasting, and flame, corona or plasma treatments. All these methods have distinct disadvantages. For example, 3-D objects cannot evenly be treated easily by flames or coronas, acid etching produces effluent and usage hazards, plasmas require non-ambient conditions, and grit blasting can be too aggressive and affect the aesthetics.

In this paper we report on the use of a new electrochemical method of surface modification, developed by Brewis, *et al.,*[1] that is capable of evenly treating 3-D objects. The technique was used to improve the adhesion of selected polyolefins by introducing oxygen-containing surface functionality.

The stability of the surface produced by mediated electrochemical oxidation (MEO) was studied and fully characterized using various surface techniques. The surface energy of polystyrene was increased without sacrificing the bulk properties of the polymer, and with short treatment times.

2 EXPERIMENTAL

Crystal grade polystyrene pellets, (Huntsman, UK), were melt pressed at 180 °C into finger shaped samples approximately 5 cm x 1 cm x 0.1 cm, between highly reflective and pure aluminium sheets (Ano-Coil, UK). They were treated for various lengths of time, using the MEO treatment, at two different temperatures, 25 and 50 °C.

In the electrochemical cell (Figure 1) the catholyte was 3.25 M HNO_3, and the anolyte was 3.5 M $AgNO_3$ in 3.25 M HNO_3 (Fisher Scientific, UK). Platinum electrodes were used and a constant current of 2 amps was applied. The whole arrangement was placed in a temperature-controlled ultrasonic bath. The oxidising agent, Ag (II), was created *in situ* at the anode by the oxidation of Ag (I) as $AgNO_3$; radicals that are produced oxidize organic material very readily. Silver acts as a mediator as it is readily reduced back to Ag(I) resulting in no net consumption. The nitric acid is also oxidized to form nitrous acid, but it too is readily reduced back by air. Samples were simply dipped into the Ag solution for the required length of time – 5, 15, 30, and 60 seconds. The polymer samples underwent a washing program of nitric acid (3.25 M) followed by at least 2 washes of deionized water.

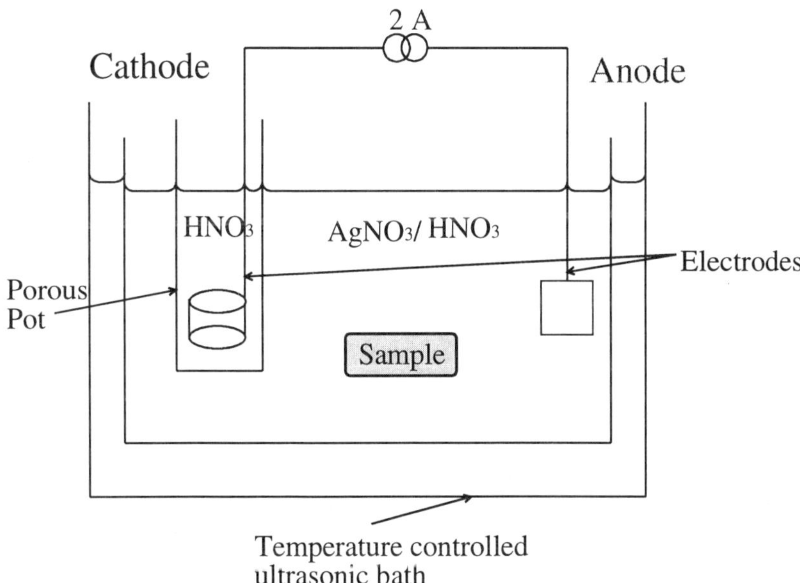

Figure 1 *Mediated Electrochemical Oxidation Reaction Cell.*

X-Ray Photoelectron Spectroscopy (XPS), was performed on an ESCALAB Mk I from VG Scientific, and was used to establish the degree of chemical modification from the electrochemical treatment on the polystyrene. A take off angle of 90° and anode conditions of 10 kV and 20 mA, with pass energies of 100 or 25 eV for broad scans and resolution scans, respectively, were used.

Advancing and receding water contact angles were measured to study changes in surface energy on a Krüss goniometer system, using water purified using a Millipore Milli-Q system (resistivity 18.2 MΩcm^2, surface tension 72.8 mJ m^{-2}). Surface energies were not calculated due to the difficulty of finding a suitable non-polar liquid. The most common liquids used for such measurements, *e.g.* bromonaphthalene and diiodomethane, dissolved the PS, and glycerol was too viscous for use in the syringe system.

Static secondary ion mass spectrometry (SSIMS) on a Cameca IMS 3F, was used to gain molecular fragment information from untreated and MEO treated PS. A positive 2° ion spectrum was taken with a 1° ion beam current $< 10^{-12}$ amp.

Vapour phase derivatization with trifluoroacetic anhydride (TFAA) (Aldrich, UK) at 10^{-4} atmospheres was used to selectively label the hydroxyl functionality produced by the MEO treatment. This aided XPS functional group quantification and identification, as carbon 1s peaks can be hard to resolve into the constituent portions.

Changes in surface topography were monitored using Atomic Force Microscopy (AFM), on a Burleigh SPM Aris 3000 in contact mode with silicon probes. The high surface static charging on PS was controlled by Au sputtering for 3 seconds. However, treated samples experienced fewer charging problems, so samples could be scanned before and after the sputter to ensure that no effect due to the sputtering was detected.

3 RESULTS AND DISCUSSION

The stability of the MEO treatment was monitored using XPS and contact angles. The concentration of surface oxygen on an untreated control PS sample was 0 atomic %. This remained the same throughout the trial, where the same storage conditions were used as for the treated samples. Advancing water contact angle for untreated PS was 92.5° and receding angle of 62° (± 2°, $n = 3$). Contact angles were subject to a maximum of ± 3.5° deviation where $n = 3$, the variation could be due to the roughness of the sample surface, or small local variations in functional group concentration. The MEO treatment and XPS assessment of surface oxygen concentration were subjected to a statistical analysis of variance (ANOVA) to compare the variances in the treatment and the XPS analysis. Four samples were treated for 15 seconds and each sample was analysed in three different areas. An F test compared the between-sample variance and the within-sample variance. The result concluded that there were no significant differences to a confidence limit of $P = 0.05$, and therefore all values fell within experimental error. All XPS results have an error of ± 0.5 atomic %.

As expected, the level of surface oxidation increased with treatment time, as shown by the XPS results in Figure 2. We should note that an optimum treatment level might not have been reached after 60 seconds; further experimentation to determine the optimum treatment time is ongoing. XPS established that at the temperatures and times under study, PS had a maximum of 13.7 atomic % oxygen. This was with a treatment time of 60 seconds at 25 °C. Water contact angles decreased with increasing oxygen concentration, indicating that the surface energy of the PS has been increased due to treatment, as shown in Figure 3.

**Effect of Treatment Time and Temperature on Mediated
Electrochemical Modification of Polystyrene**

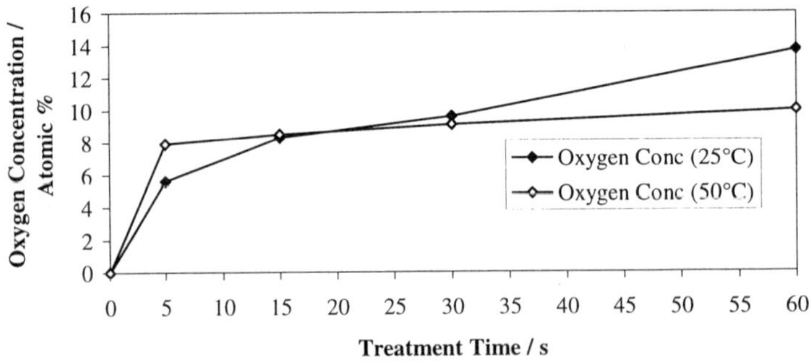

Figure 2   *Graph of the effect of treatment time and temperature on the MEO treatment of polystyrene.*

A certain amount of surface relaxation might have been expected and was evident from the XPS determined oxygen concentrations, Figure 4, and contact angles, Figure 5. The water contact angles increased with this observation, indicating a decrease in the surface energy of the PS. This occurred for all treatment times tested. However, the surface relaxation was not as great as that experienced with other polymer systems, *e.g.* polypropylene and poly(ethylene terephthalate).[2] This may be due to the restriction of movement around the hydrocarbon backbone chain caused by the large aromatic pendant group.

SSIMS analysis of untreated PS produced spectra with characteristic peaks, indicated by an asterisk (*) in Figure 6. MEO treated PS resulted in an extra peak at 105 mass units. This could possibly be assigned to the mass fragment of C_6H_5CO, implying that oxygenation is on the backbone and not the aromatic ring. This can be confirmed by higher resolution XPS.

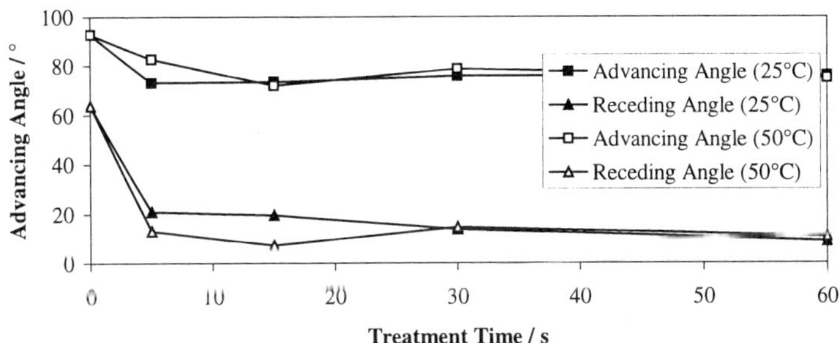

Figure 3   *Graph of the effect of MEO treatment on water contact angles.*

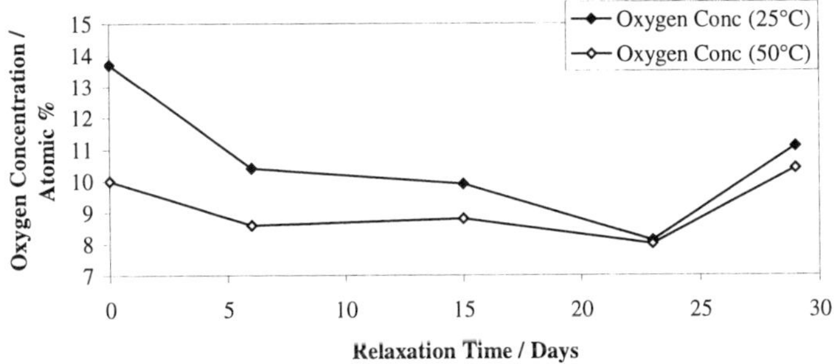

Figure 4   *Effect of surface relaxation on oxygen concentration.*

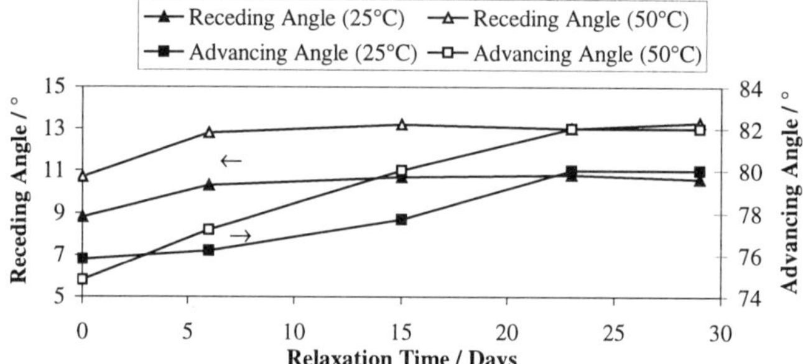

Figure 5　　*Effect of relaxation on contact angles.*

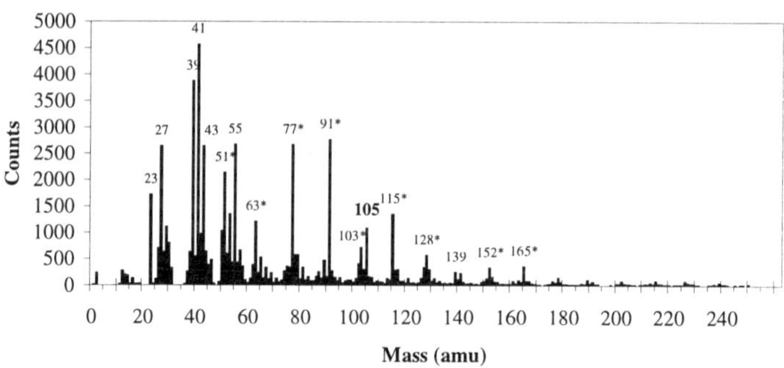

Figure 6　　*SSIMS spectrum of MEO treated polystyrene.*

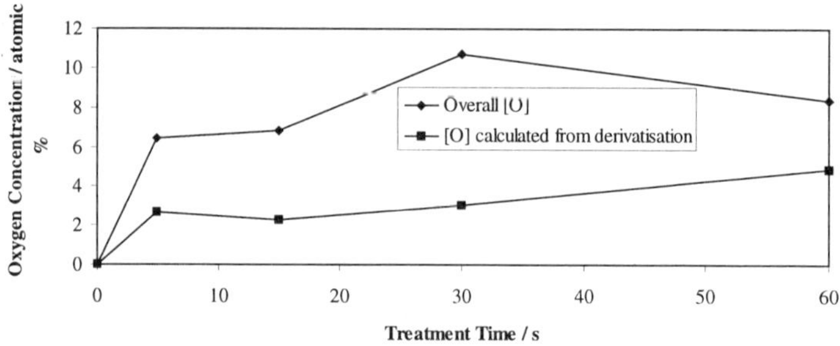

Figure 7　　*Proportion of oxygen concentration due to hydroxyl groups.*

Vapour phase derivatization followed by XPS revealed that about half of the oxygen on the surface was due to hydroxyl groups (Figure 7). Further derivatization studies with other selective labels are underway to identify the other oxygen-containing functionalities.

AFM revealed that the treatment did not greatly change the surface topography or average roughness (Figures 8 and 9). The original roughness, seen on the untreated PS, is due to the highly reflective aluminium sheeting (Ano-Coil, UK) against which the samples were pressed. Ageing did not introduce or change any features. The treatment itself affected the surface forces experienced by the AFM probe, making them more stable to scanning.

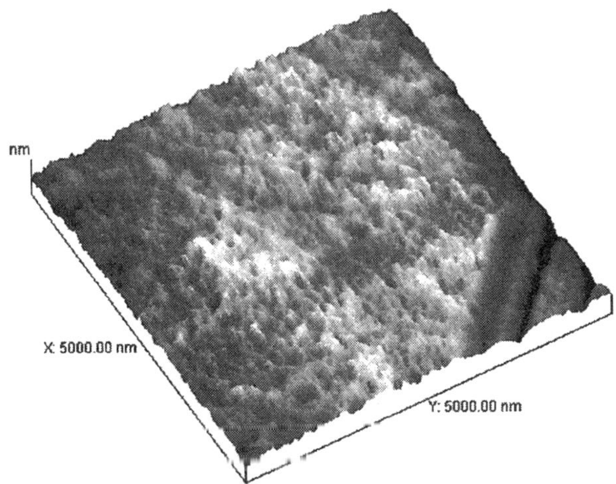

Figure 8   *AFM scan of untreated polystyrene.*

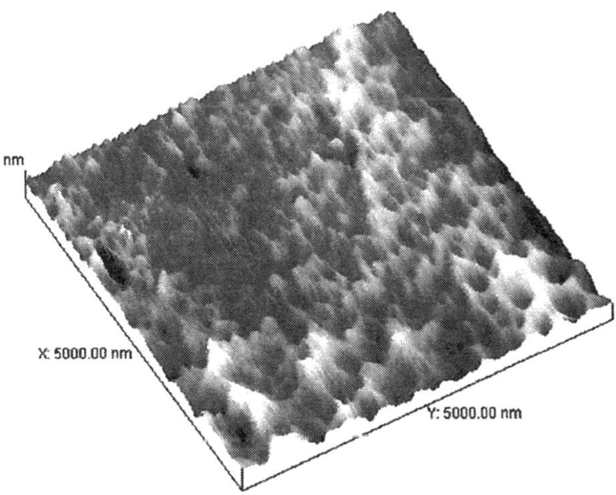

Figure 9   *AFM scan of MEO treated PS, 60 seconds at 50 °C.*

4 CONCLUSIONS

Mediated electrochemical oxidation of polystyrene is stable, quick, even and environmentally friendly. The aesthetics of the polymer are not affected in any way. The surface of polystyrene is oxidized to include hydroxyl groups. Further selective derivatization is required to identify the remaining functional group(s).

5 ACKNOWLEDGEMENTS

The authors wish to thank: M. D. Crapper, Physics Dept., Loughborough University, UK; D. M. Brewis, ISST, Physics Dept., Loughborough University, UK; R. H. Dahm, De Montfort University, UK; A. Dowell, Ano-Coil, UK; and D. Watson, Huntsman Chemical Co. Ltd., UK.

References

1. D. M. Brewis, R. H. Dahm and I. Mathieson, *J. Mat. Sci. (Letters)*, 1997, **16**, 93.
2. J. M. Hill, E. Karbashewski, A. Lin, M. Strobel and M. J. Walzak, 'Polymer Surface Modification: Relevance to Adhesion', K. L. Mittal, ed., VSP, Utrecht, Netherlands, 1996, Part 3, p. 273.

SPECTROSCOPIC AND ELECTROCHEMICAL PROPERTIES OF
$[(CN)_5Ru(CN)Ru(NH_3)_5]^-$ ANCHORED ON THIN FILM OF Ti(IV) OXIDE
DISPERSED ON THE SILICA GEL SURFACE

D. R. do Carmo[†], Y. Gushikem[‡] and D. W. Franco[†]

[†]Instituto de Química de São Carlos,-USP, C.P369,13560-970 São Carlos, SP Brazil
[‡]Instituto de Química, UNICAMP, C.P 6154,13083-970 Campinas, SP Brazil

1 INTRODUCTION

The reaction of the silica gel surface silanol groups with titanium (IV) alkoxide followed by subsequent hydrolysis yields a thin layer of Ti(IV) grafted on the surface of the supporting material.[1,2] Lewis acid sites of the highly dispersed titanium ions on the silica gel surface strongly interact[1] with anionic complex species such as $[Fe(CN)_6]^{-3}$ and $[Ru(edta)(H_2O)]^-$.

The present study originated from our interest in the chemistry of the chemisorbed ruthenium species,[3,4] due to its reactivity regarding the electron transfer process, and inertness to undergo substitution reactions. These properties make the ruthenium complex species extremely favorable to analytical studies.[5–7] As part of our efforts to understand the reactivity of immobilized species, the synthesis and properties of the heterobinuclear complex, $[(CN)_5Ru(CN)Ru(NH_3)_5]^-$ grafted on a thin film of Ti(IV) oxide, dispersed on the silica gel surface, are described. The influence of Zn(II) ions on the immobilized heterobinuclear complex properties is also discussed. Comparison between the properties of the immobilized species with their analogue complexes in solution is also presented.

2 EXPERIMENTAL

2.1 Reagents

All reagents and solvents were of analytical grade (Merck or Aldrich) and were used as purchased. Double distilled water was used throughout this work.

2.2 Silica Functionalization with Titanium (IV)

Silica gel having a specific surface area $S_{BET} = 500 \pm 76$ m^{2}g^{-1}, an average pore diameter of 0.6 nm, and a particle size of 0.023 ± 0.03 mm, was activated by heating at 423 K under vacuum (0.13 Pa) before use. About 70 g of this activated material was suspended in dry toluene (150 mL) and tetrabutyl orthotitanate monomer (24 mL) was added. The mixture was refluxed for 6 h under argon atmosphere (Equation 1).

$$n(\equiv SiOH) + Ti(OR)_4 \rightleftharpoons (\equiv SiO)_n Ti(OR)_{4-n} + n\,HOR \qquad (1)$$

The solid was filtered, washed with toluene (350 mL), then dried at 373 K under vacuum. The solid was hydrolysed and washed with demineralized water (Equation 2).

$$(\equiv SiO)_n\,Ti(OR)_{4-n} + (4{-}n\,)H_2O \rightleftharpoons (\equiv SiO)_n Ti(OH)_{4-n} + (4{-}n)ROH \qquad (2)$$

where $\equiv SiOH$ stands for the silanol groups and R= n-butyl. For the sake of brevity, the chemically modified silica $(\equiv SiO)_n Ti(OH)_{4-n}$ will be hereafter denoted as $\equiv TiOH$.

2.3 Measurements

2.3.1 Voltammetry Measurements. Electrochemical experiments were performed using a Princeton Applied Research (PAR) Model 273 potentiostat/galvanostat. The three electrode system used in the studies consisted of a modified carbon paste as the working electrode prepared by mixing the $\equiv TiOH$, analytical grade graphite and mineral oil), a saturated calomel electrode (SCE) as reference and a platinum wire as the auxiliary electrode. All measurements were carried out in an electrochemical cell filled with 1.0 M KCl supporting electrolyte solution at 298 K under argon atmosphere.

2.3.2 Electronic Spectra. The electronic spectra were recorded using a Hitachi U-3501 spectrophotometer. Since the refractive index of the carbon tetracloride and of the silica matrix are nearly the same, the electronic spectra of the immobilized complexes on the silica gel surface were obtained in CCl_4 suspension using a 1 mm pathlength quartz cell. Diffuse reflectance spectra of the bulk solid trinuclear complex were recorded between 350 and 1600 nm on a Guided Wave model 260 spectrophotometer, using a tungsten–halogen lamp as the radiation source, and detectors of Si and Ge.

2.3.3 Infrared Spectra. Infrared spectra were obtained using a self-supported pressed disk of the fine powder. Spectra between 2500 and 1500 cm^{-1} were recorded on a Bomen MB 102 FTIR spectrophotometer.

2.4 Preparations

2.4.1 Ruthenium Complexes. All manipulations of air-sensitive compounds were carried out under argon using standard techniques.[8] $[Ru(NH_3)_5OH_2](CF_3SO_3)_3$ and $[Ru(NH_3)_5Cl]Cl_2$ were prepared from $RuCl_3$ according to previously described methods.[9,10] The binuclear complex ion $[(CN)_5Ru^{II}CNRu^{III}(NH_3)_5]^-$ was generated in solution by reacting equimolar concentrations (10^{-3} M) of $K_4[Ru(CN)_6]$ and $[Ru(NH_3)_5Cl]Cl_2$ in aqueous solution at 60 °C, as shown in Equation 3.[11]

$$[Ru(CN)_6]^{-4} + [Ru(NH_3)_5Cl]^{+2} \rightleftharpoons [(CN)_5Ru(CN)Ru(NH_3)_5]^- + Cl^- \qquad (3)$$

The binuclear complex was isolated by precipitation adding methanol into the solution. Analysis, found (calcd): C, 13.70 (13.42); H, 3,43 (3.39); N, 24.9 (28.7). This elemental analysis suggests the formula: $K[(CN)_5Ru(CN)Ru(NH_3)_5]\cdot 3H_2O$.

On mixing $K[(CN)_5Ru(CN)Ru(NH_3)_5]^-$ and $Zn(CH_3COO)_2$ solutions in a 2:1 molar proportion, a deep blue solid precipitated (Equation 4). This light-sensitive, solid product, $Zn[(CN)_5Ru(CN)Ru(NH_3)_5]_2$, was filtered, then stored under vacuum in the dark. It is insoluble in H_2O, toluene, CCl_4, CH_3OH, CH_3CH_2OH and dioxane.

$$2[(CN)_5Ru(CN)Ru(NH_3)_5]^- + Zn(II)$$

$$\rightleftharpoons Zn[(CN)_5Ru(CN)Ru(NH_3)_5]_2 \qquad (4)$$

2.4.2 Formation of ≡Ti⁺[Ru(CN)₆]⁻⁴. The ≡Ti⁺[Ru(CN)₆]⁻⁴ surface complex was prepared as follows: 0.5 g of the solid (≡TiOH) was immersed in 25 mL of 10^{-3} M potassium hexacyanoruthenate(II) solution (pH *ca.* 3). This mixture was shaken for 30 minutes at room temperature and then filtered and washed with water. The following reactions are proposed to occur:

$$\equiv TiOH + H^+ \rightleftharpoons \equiv Ti^+ + H_2O \tag{5}$$

$$\equiv Ti^+ + [Ru(CN)_6]^{-4} \rightleftharpoons \equiv Ti^+[Ru(CN)_6]^{\,4} \tag{6}$$

In acid solution the surface of the material is positively charged due to formation of the Lewis acid site ≡Ti⁺, which reacts with $[Ru(CN)_6]^{-4}$ to form the surface species ≡Ti⁺[Ru(CN)₆]⁻⁴ (Equations 5, 6). At this stage, the surface color changes from white to yellow. This yellow solid was filtered, washed with water, dried, and stored under vacuum.

2.4.3 Formation of ≡Ti⁺[(CN)₅Ru(CN)Ru(NH₃)₅]⁻. The ≡Ti⁺[Ru(CN)₆]⁻⁴ complex was immersed in 25 mL of 10^{-3} M $[Ru(NH_3)_5OH_2]^{+3}$ aqueous solution and the mixture was stirred for 3 hours under argon atmosphere. The binuclear complex formation is indicated by the solid color change from yellow to deep blue:

$$\equiv Ti^+[Ru(CN)_6]^{-4} + [Ru(NH_3)_5OH_2]^{+3}$$

$$\rightleftharpoons \equiv Ti^+[(CN)_5Ru(CN)Ru(NH_3)_5]^- + H_2O \tag{7}$$

The resulting deep blue solid was filtered off, washed with 10^{-3} HClO₄ aqueous solution, dried under vacuum and stored under argon atmosphere in the absence of light. In the resulting complex (Equation 7), the cyanide group between the two ruthenium atoms, *i.e.* RuCNRu, is the bridging ligand.

2.4.4 Reaction of ≡Ti⁺[(CN)₅Ru(CN)Ru(NH₃)₅]⁻ with Zn(II). The complex shown in Equation 7, ≡Ti⁺[(CN)₅Ru(CN)Ru(NH₃)₅]⁻, was immersed in 50 mL of 10^{-1} M zinc acetate or zinc chloride solutions at pH *ca.* 4 and the mixture stirred for 30 min. The mixture was filtered and the solid washed with 10^{-3} M aqueous HClO₄ solution, dried and stored under vacuum.

3 DISCUSSION

3.1 Electronic Spectra

The main spectral characteristics of the compounds in this study are summarized in Table 1. The $[Ru(CN)_6]^{-4}$ ion in aqueous solution exhibits an absorbance maximum at λ_{max} 204 nm ($\varepsilon = 3.7 \times 10^4$ M⁻¹cm⁻¹) with a shoulder around 250 nm, where the former process is charge transfer in character (MLCT) and assigned[11] to the transition $^1A_{1g} \rightarrow {}^1T_{1u}$. For the immobilized species ≡Ti⁺[Ru(CN)₆]⁻⁴ this band is observed at 362 nm (with a shoulder at 390 nm) exhibiting a noticeable asymmetry with respect to the spectrum of the corresponding species $[Ru(CN)_6]^{-4}$ in solution. This asymmetry could be attributed to an increase in the d–d transition intensity, which is enveloped by the MLCT band, in the solid state.[12]

In a neutral aqueous solution, the $[(CN)_5Ru(CN)Ru(NH_3)_5]^-$ species exhibits one well defined absorption at λ_{max} 680 nm ($\varepsilon = 2.8 \times 10^3$ M^{-1}cm^{-1}) attributed to the metal to metal charge tranfer (MMCT) Ru(III)–Ru(II) process.   Another strong absorption, probably due to the hexacyano moiety, is observed below 300 nm.

The supported binuclear species ($\equiv$Ti$^+$[(CN)$_5$Ru(CN)Ru(NH$_3$)$_5$]$^-$) exhibits two absorptions: λ_{max} 369 nm and 642 nm (see Figure 1A).   These bands are attributed to MLCT and MMCT processes in the $[Ru(CN)_6]^{-4}$ and in the $[Ru^{II}CNRu^{III}]$ moieties, respectively.   These assigments are made by analogy with those described in the literature for the binuclear $[(CN)_5RuCNRu(NH_3)_5]^-$ ion in solution.[11,13,14]

Table 1   *Electronic, Vibrational and Electrochemical Parameters for the Mono and Binuclear Complexes Immobilized on the Matrix Surface and in Solution.*

Compounds	$E_{1/2}$ (mV)*		(ν_{CN})cm^{-1}	UV Visible (nm)
	$(E_{1/2})_a$	$(E_{1/2})_b$		
$\equiv$Ti$^+$[Ru(CN)$_6$]$^{-4}$		727	2058	362, 390
$\equiv$Ti$^+$[(CN)$_5$Ru(CN)Ru(NH$_3$)$_5$]$^-$	−254	729	2061	369,642
$\equiv$Ti$^+$[(CN)$_5$Ru(CN)Ru(NH$_3$)$_5$]/Zn	−225	1050	2101	366,578
[Ru(CN)$_6$]$^{-4}$		715	2048	204
[(CN)$_5$Ru(CN)Ru(NH$_3$)$_5$]$^-$	−254	720	2061	680
Zn[(CN)$_5$Ru(CN)Ru(NH$_3$)$_5$]$_2$	−225	1035	2101	620

Mid-point potential vs. SCE, T = 25° ±0.5 °C, in 1.0 M KCl, pH 3.1; uncertainty ± 10 mV

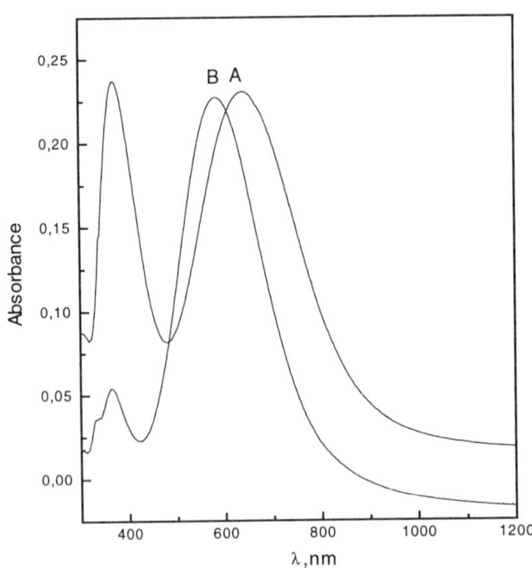

Figure 1   *Electronic spectra for: A) $\equiv$Ti$^+$[(CN)$_5$Ru(CN)Ru(NH$_3$)$_5$]$^-$,*
B) $\equiv$Ti$^+$[(CN)$_5$Ru(CN)Ru(NH$_3$)$_5$]$^-$/Zn.

The MMCT band of the heterobinuclear species undergoes a blue shift upon association to Zn(II). This shift is approximately 62 nm and reflects the stabilization of the highest occupied molecular orbital (HOMO), similar to that observed for other cyanide heterobinuclear complexes upon cyanide protonation.[15,16] As seen later, this MMCT blueshift follows the same trend of the Ru(III)/Ru(II) redox potential in the hexacyano fragment function of zinc coordination.

3.2 Infrared Spectra

Infrared spectra of the immobilized complexes on the silica gel surface were recorded between 2500 and 1500 cm^{-1} due to the strong absorption exhibited by the matrix outside this region.[17] The stretching vibrational modes, ν_{CN}, of the complexes are listed in Table 1.

In the IR spectrum of $\equiv$Ti$^+$[Ru(CN)$_6$]$^{-4}$ ν_{CN} is observed at 2058 cm^{-1}. This band is slightly shifted to higher frequency with respect to that observed for K$_4$[Ru(CN)$_6$]$^{-4}$ complex where ν_{CN} is observed at 2048 cm^{-1}. This shift occurs when the nitrogen atom of the CN ligand interacts with the surface Lewis acid site leading to $\equiv$Ti$^+$NC association.

Our observation is consistent with the well known fact that interaction of the CN group though a nitrogen atom with a Lewis acid site increases the CN stretching force constant.[18,19] The spectrum of the product of the reaction between $\equiv$Ti$^+$[Ru(CN)$_6$]$^{-4}$ and [Ru(NH$_3$)$_5$(H$_2$O)]$^{+3}$ (Equation 7) shows ν_{CN} at 2061 cm^{-1} (see Figure 2A). This ν_{CN} frequency may be shifted to higher energy due to the mechanical constraint on the CN motion imposed by the second metal center as a consequence of Ru–CN–Ru bridge formation.[20–23]

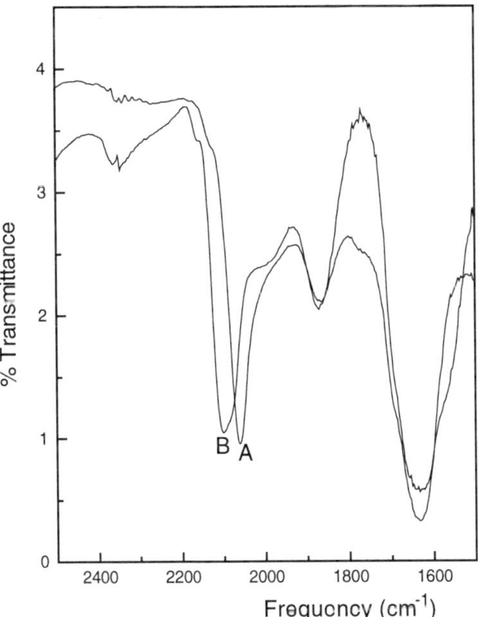

Figure 2   *Infrared spectra for: A) $\equiv$Ti$^+$[(CN)₅Ru(CN)Ru(NH₃)₅]⁻,
B) $\equiv$Ti$^+$[(CN)₅Ru(CN)Ru(NH₃)₅]⁻/Zn.*

When the deep blue immobilized complex reacts with Zn(II) the ν_{CN} frequency undergoes a 50 cm^{-1} blueshift (see Figure 2B). The observed ν_{CN} frequency at 2110 cm^{-1} for the immobilized complex agrees quite well with that observed for the complex in $Zn[(CN)_5Ru(CN)Ru(NH_3)_5]_2$ (see Table 1).

3.3 Electrochemistry

The ($E_{1/2}$) data for the complexes reported here are summarized in Table 1. The differential pulse polarograms (DPP) were recorded between –800 and 1100 mV since the titanium center is not electroactive in this potential range.[24]

The electrochemical behaviour of the $[Ru(CN)_6]^{-4/-3}$ species in solution or immobilized on the support, are quite similar. The redox potentials for the Ru(III)/Ru(II) couple in these hexacyano complexes are 715 and 729mV, respectively. For the binuclear complexes $K[(CN)_5Ru(CN)Ru(NH_3)_5]^{-/0}$ two well defined peaks are observed: $(E_{1/2})_a$ = –254 and $(E_{1/2})_b$ = 720 mV. For the corresponding supported species these values are $(E_{1/2})_a$ = –254 and $(E_{1/2})_b$ = 729 mV (see Figure 3A). These processes are assigned to the $(NH_3)_5\,Ru^{(III)/(II)}$ and $(CN)_5Ru^{(III)/(II)}$ couples in the binuclear complexes respectively, and are in agreement with reported values in the literature for the solution species.[25] The Ru(III) center in the $[Ru(NH_3)_5]^{3+}$ species moiety is 23 mV more difficult to reduce than the aquopentaammineruthenium species. The same trend is observed for the $[Ru(NH_3)_5]^{3+}$ species.

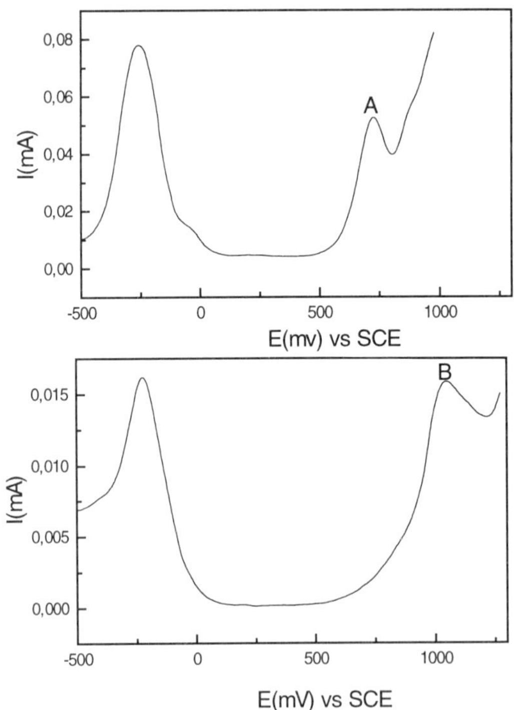

Figure 3   *Differential pulse polarogram (DPP) of: A) $\equiv Ti^+[(CN)_5Ru(CN)Ru(NH_3)_5]^-$,*
B) $\equiv Ti^+[(CN)_5Ru(CN)Ru(NH_3)_5]^-/Zn$.

As suggested by $(E_{1/2})$ data, the Ru(III)/Ru(II) couple behavior in the $[Ru(CN)_6]^{-4}$ ions is less sensitive than $[Ru(NH_3)_5]^{3+}$ to changes in the environment. The $(E_{1/2})_{Ru(III)/Ru(II)}$ in the hexacyano moiety exhibits almost the same values for the mono and heterobinuclear species, supported or not (see Table 1). Therefore, as far as the electrochemical data are concerned, no significant coupling between the two ruthenium centers could be detected in the heterobinuclear species. This is consistent with the absence of any intervalence band in the near infrared spectra of these species. However, upon reaction with zinc(II), the $(E_{1/2})_{Ru(III)/Ru(II)}$ in the $[Ru(CN)_6]^{-4}$ moiety is shifted (315 mV in solution and 321 mV when grafted, see Figure 3B). This ΔE shift is higher than that observed[15] for tri(poly)nuclear species $[(edta)Ru^{III}\text{-}(NC)\text{-}Ru^{II}\text{-}(CN)_4\text{-}(CN)\text{-}Ru^{III}(edta)]^{-6}$, where the redox potentials associated with the Ru(III)–Ru(II) couples show a displacement of 250 mV toward the most positive potentials.

This shift observed for the $(E_{1/2})_{Ru(III)/Ru(II)}$ couple in the hexacyano moiety in the $Zn[(CN)_5Ru(CN)Ru(NH_3)_5]_2$ and for the heterobinuclear supported species $\equiv Ti^+[(CN)_5Ru(CN)Ru(NH_3)_5]^-/Zn$ (*ca.* -21 kJ mol⁻¹) is likely to be a consequence of the electron withdrawing effects of the Lewis acid Zn(II) over one of the cyanide ligands. This effect is consistent with the λ_{max} shift observed in the electronic spectra of the heterobinuclear complexes upon association with Zn(II). In this case a blueshift of $62 \pm$ 2nm is observed on the MMCT (*ca.* -20 kJ mol⁻¹).

4 CONCLUSIONS

The spectroscopic and electrochemical properties of heterobinuclear complexes, immobilized on the silica gel surface modified with TiO_2 or in solution phase, are similar. The absence of significant matrix effects on the studied proportion suggests this material may be a good candidate as a support for a model of third generation catalysts.

5 ACKNOWLEDGEMENTS

The authors are very indebted to the Brazilian research foundations FAPESP, CAPES, FINEP and SINC/SHIMADZU (BRAZIL) for financial support and to Johnson Matthey for $RuCl_3$.

References

1. L.T. Kubota and Y. Gushikem, *Electrochim. Acta.*, 1992, **37**, 2477.
2. Y. Gushikem, C. R. M. Peixoto, U. P. R. Filho, L. T. Kubota and E. Stadler, *J. Colloid Interface Sci.*, 1996, **184**, 236.
3. M.T. Hoffmann, S. M. C. Neiva, M. R. Martins and D. W. Franco, in 'Chemically Modified Surfaces', H. A. Mottola and J. R. Steinmetz, eds., Elsevier, 1992.
4. M. C. Neiva, J. A. V. Santos, J. C. Moreira, Y. Gushikem, H. Vargas and D. W. Franco, *Langmuir*, 1993, **9**, 2982.
5. J. O'Shea, D. Leech, M. R. Smyth and J. G. Vos, *Talanta*, 1992, **143**, 39.
6. N. Oyama and F. C. Anson, *Anal. Chem.*, 1980, **52**, 1192.
7. A.L. Crumbliss, H. A. O. Hill and D. J. Page, *J. Electroanal. Chem.*, 1986, **327**, 206.

8.   D. F. Shriver, 'The Manipulation of Air-Sensitive Compounds', McGraw-Hill, New York, 1969, p. 51.

9.   E. Dixon, G. A. Lawrence, P. A. Lay and A. M. Sargeson, *Inorg. Chem.*, 1984, **23**, 294.

10. K. Heinz, 'Mixed Valence Ruthenium Ammine Dinitrile and Complexes', Ph.D. Dissertation, Stanford University, 1976.

11. B. Gray and N. A. Beach, *J. Am. Chem. Soc.*, 1963, **85**, 2922.

12. J. A. Olabe, H. O. Zerga and L. A. Gentil, *J. Chem. Soc., Dalton. Trans.*, 1987, 1297.

13. A. Vogler and J. Kisslinger, *J. Am. Chem. Soc.*, 1992, **104**, 2311.

14. G. Poulopoulo, Z.-W. Li, H. Taube, *Inorg. Chim. Acta.*, 1994, **225**, 173.

15. P. Forlano, F. D. Cukiernik, O. Poizat and J. A. Olabe, *J. Chem. Soc., Dalton Trans.*, 1997, 1595.

16. H. E. Toma and J. M. Malin, *Inorg. Chem.*, 1973, **12**, 1039.

17. G. D. Shukin and A. Apretava, *J. Pract. Spectrosc.*, 1988, **50**, 639.

18. J. F. Bertran, E. R. Ruhz and J. B. Pascual, *Spectrochim. Acta. Part A*, 1990, **46**, 167.

19. G. A. P. Zaldivar, Y. Gushikem and L. T. Kubota, *J. Electroanal. Chem.*, 1991, **318**, 247.

20. C. A. Bignozzi, R. Argazzi, J. R. Schoonover, K. C. Gordon, R. B. Dyer and F. Scandola, *Inorg. Chem.*, 1992, **31**, 5260.

21. K. D. Karlin, 'Progress in Inorganic Chemistry', Wiley and Sons, 1997, **45**, p. 28.

22. B. J. Coe, T. J. Meyer and P. S. White, *Inorg. Chem.*, 1995, **34**, 3600.

23. W. M. Laidaw and R. G. Denning, *J. Chem. Soc., Dalton Trans.*, 1994, 1987.

24. U. R. Evans, 'The Corrosion and Oxidation of Metals', Arnold, London, 1960, p. 143.

25. S. Siddiqui, W. W. Henderson and R. E. Sherped, *Inorg. Chem.*, 1987, **26**, 3101.

PHOTOPHYSICAL STUDY OF DONOR–ACCEPTOR-SUBSTITUTED STILBENES ADSORBED ON SILICA–TITANIA SURFACES

A. Eremenko,[†] N. Smirnova,[†] O. Rusina,[†] K. Rechthaler,[‡] G. Koehler,[‡] V. Ogenko,[†] and A. Chuiko[†]

[†]Institute of Surface Chemistry, National Ukrainian Academy of Sciences, Prospekt Nauki 31, UA-252022 Kiev, Ukraine

[‡]Institut für Theoretische Chemie und Strahlenchemie, Universität Wien, Althanstrasse 14, A-1090 Wien, Austria

1 INTRODUCTION

Semiconductor materials in bulk, powder, colloid and film forms have been widely used as photocatalysts for the photodegradation of organic pollutants.[1,2] An important problem with such materials is the limitation of their photoactivity to the UV region of the electromagnetic spectrum. The photosensitization of stable, large-bandgap semiconductors using organic dyes is a convenient method for extending the photoresponse into the near ultraviolet and visible region and has been widely employed in many applications.[3] Of particular interest are materials prepared from TiO_2 – due to its high refractive index, good chemical stability, high electrical resistance and interesting catalytic properties Mixed silica–titania compositions are also of significant technological importance. Silica glasses with a few mol % TiO_2 are used as catalysts and catalytic support materials. Such ultra-fine particles possess a highly porous morphology which is extremely convenient for modification with redox couples and sensitizers. In photocatalytic conversions, the interfacial electron transfer between adsorbates and surface results either in oxidation or reduction of a redox-active organic species. The rate for the charge injection from an excited singlet state of an organic dye is considered to be fast, and charge transfer occurs in less than 1–2 ns.[4] Organic dyes, *e.g.* oxazines, erythrosin B, eosin, $Ru(bpy)_3^{2+}$ and derivatives, have been extensively used in the sensitization of large bandgap semiconductors like TiO_2 due to their strong absorption in the visible region and ability to interact with semiconductor surfaces.[5–7]

Less is known about the application of push–pull stilbenes in photosensitization. The photophysical properties of stilbenes have attracted considerable interest. Several derivatives are applied as optical brighteners and laser dyes. The parent compound *trans*-stilbene undergoes *cis–trans* isomerization on excitation, a very well studied adiabatic photoreaction that proceeds via a so-called phantom-state. This conformation corresponds to a maximum on the ground state potential hypersurface. This pathway provides an effective deactivation funnel.[8,9] Donor–acceptor-substituted stilbenes, called push–pull-stilbenes, were studied with respect to charge separation in the excited state, as the dipole moment increases strongly. In earlier investigations of push–pull stilbenes it was proposed that the main deactivation process should be the twisting about the central double bond.[10] Furthermore, twisting around the single bonds was deduced from experimental results to form a twisted intramolecular charge transfer (TICT) state,[11–15] thus quenching the fluorescence quantum yield, whereas stilbenes with hindered internal rotation should not show this quenching behavior. However, it was shown on one push–pull-stilbene that preventing internal rotation is not a necessary condition for obtaining highly fluorescent compounds.[16] Application of donor–acceptor-substituted stilbenes on the surface of Ti-containing compositions is underlined, as it should provide the sensitization of heterogeneous photocatalysts to near ultraviolet and visible light. Adsorbed stilbenes in heterogeneous media are new, interesting photochemical systems providing the ability to vary fluorescence properties. There are some aspects of the photophysical properties of push–pull-stilbenes in heterogeneous media that would be of interest to study. Among these are the peculiarities of intramolecular charge-transfer processes in constrained and rigid media, the affect on the flexibility and reactivity of the photoactive species, the effect of type and concentration of active adsorption centers, and the polarity of the surface. Push–pull-stilbenes can be used as a new type of sensitive luminescence probe to study the surface chemistry of various adsorbents and catalysts, as well as visible light-sensitizers for titania photocatalysts.

Introduction of push–pull-stilbenes into heterogeneous media can be performed using several different methods, including adsorption from solution on porous bulk oxides, and incorporation during sol–gel transfer – when oxide matrices protect stilbenes from contact with the surroundings.

In this work we study the effect of silica and silica–titania surfaces (in the powder form) on the fluorescence and lifetimes of adsorbed push–pull-stilbenes. Two push–pull-stilbenes in particular, *E,E*-1-(4-Cyanophenyl)-4-(4-*N,N*-dimethylaminophenyl)-1,3-buta-diene, $C_{19}H_{18}N_2$ (**S1**) and *E*-9-(4-Cyanostyryl)-2,3,6,7-tetrahydro-1H,5H-pyrido[3,2,1-i,j]-quinoline, $C_{21}H_{20}N_2$, (**S2**) were adsorbed on the surface of silicas modified with 0.1, 0.3 and 1.0 wt% of TiO_2.

2 EXPERIMENTAL

Compounds **S1**, mp 259.5–262.5 °C and **S2**, mp 150.5–152.0 °C, were prepared as previously described.[8,16] The structures of both stilbenes are shown in Scheme 1:

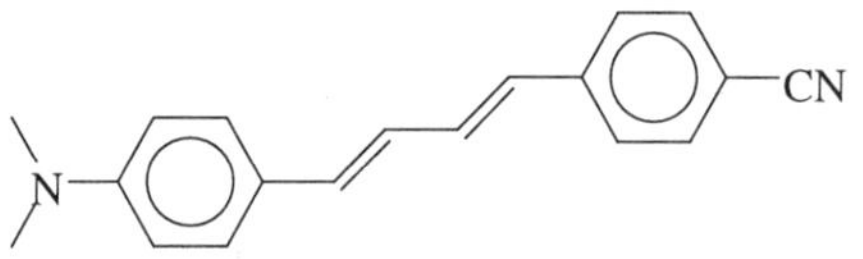

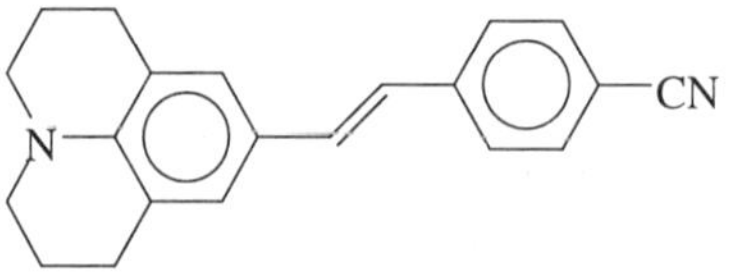

Scheme 1

The stilbenes were adsorbed on the surface of a series of porous silicates: silica Merck 100, and silica Merck 100 modified with 0.1, 0.3 and 1.0 wt% of TiO_2. To prepare silica–titania, tetraisopropyltitanuim in cyclohexane was added to the dried silica gel (T_{act} = 500 °C), then a catalytic amount of water was added slowly.[17] The sample was stirred at room temperature, and a large excess of water was added to ensure complete reaction of the alkoxide. The resulting silica–titania solid after heating at 500 °C contains an amorphous phase of titania as determined by X-ray powder diffraction.

Before adsorption of S1 and S2, the silica and silica titania samples were pretreated at 300 °C. Adsorption of S1 and S2 on these carriers was performed from butyl ether solutions, adsorption value $a = 10^{-7}$–10^{-6} mol g^{-1}. The solvent was removed from the samples with adsorbed stilbenes at 150 °C in air and under vacuum conditions. Samples are indicated as S1,SiO$_2$ and S2,SiO$_2$. Silica probes modified with 0.1, 0.3 and 1 wt% of TiO$_2$ are indicated as S1,SiO$_2$/TiO$_2$ and S2,SiO$_2$/TiO$_2$.

Emission and excitation spectra were studied for the samples in air and under vacuum to compare the effect of oxygen quenching and water vapor addition on the arbitrary intensity and lifetimes of the adsorbed species. Fluorescence and excitation spectra of the adsorbed S1 and S2 were recorded in the temperature range between 20 and 80 °C with a Perkin Elmer LS 50 B fluorimeter. Lifetimes of the adsorbed stilbenes were also registered. The apparatus and deconvolution procedure, based on the method of single photon counting, has been described elsewhere.[18,19]

3 RESULTS AND DISCUSSION

The affinity of silica and silica–titania samples for these stilbenes is very high. Polar push–pull stilbenes are easily adsorbed onto the OH-groups of SiO$_2$ and TiO$_2$–SiO$_2$ surfaces due to a hydrogen-bonding interaction. The degree of association between oxides and stilbenes is high. Adsorbed species could not be removed from either silica or silica–titania at room temperature with pentane and butyl ether; the molecules remained on the surfaces.

3.1 S1,SiO$_2$ and S2,SiO$_2$

Fluorescence and excitation spectra of the system under vacuum conditions are similar to the spectra in polar solutions: fluorescence maxima 540–560 nm, excitation maxima 390 nm (Figures 1 and 2).

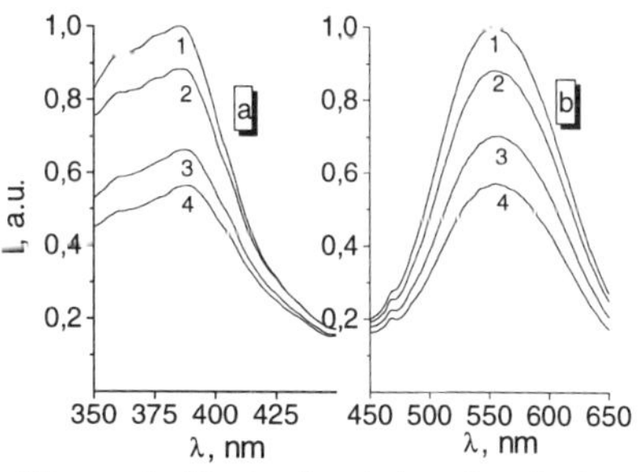

Figure 1 *Excitation (a) and emission (b) spectra of **S1** on SiO$_2$ under vacuum conditions at 20 °C (1), 40 °C (2), 60 °C (3) and 80 °C (4).*

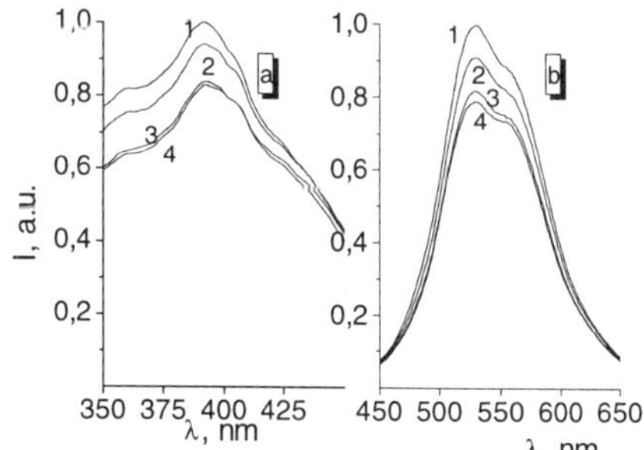

Figure 2 *Excitation (a) and emission (b) spectra of **S2** on SiO$_2$ under vacuum conditions at 20 °C (1), 40 °C (2), 60 °C (3) and 80 °C (4).*

When the temperature increases, the intensity of emission decreases. Lifetimes of S1 and S2 calculated using a biexponential fit are composed of short- and longtime components. After contact of the samples with air, the fluorescence properties of adsorbed species S1 and S2 change, the intensity decreasing in both cases. Spectral maxima of both fluorescence and excitation become smoother and less structured than under vacuum conditions. The λ_{max} of the S1,SiO$_2$ remains almost unchanged over the wide temperature range studied (20–80 °C) (Figure 3). In contrast to S1,SiO$_2$, fluorescence spectra of S2,SiO$_2$ changes dramatically. All spectra are blue-shifted after contact with air and they are very similar to that in hexane solution (Figure 4): the fluorescence maximum is located near 460–470 nm.

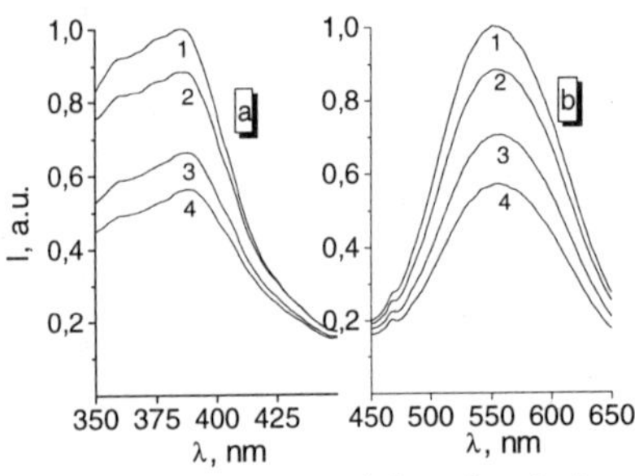

Figure 3 *Excitation (a) and emission (b) spectra of **S1,SiO$_2$** in air at 20 °C (1), 40 °C (2), 60 °C (3) and 80 °C (4).*

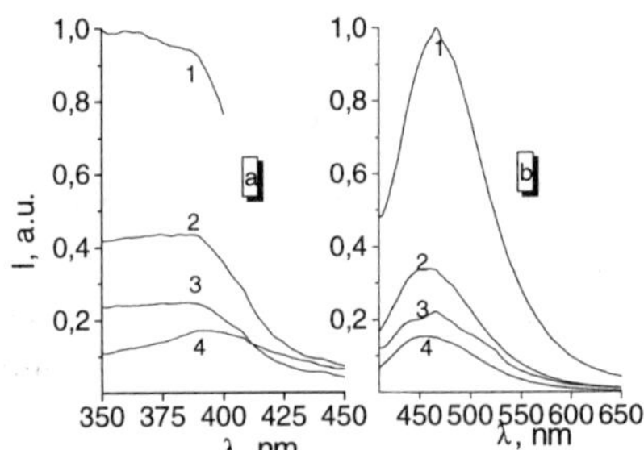

Figure 4 *Excitation (a) and emission (b) spectra of **S2,SiO$_2$** in air at 20 °C (1), 40 °C (2), 60 °C (3) and 80 °C (4).*

It is interesting to note that during the first 5 minutes after contact with air, the fluorescence maximum of S2 remains in the same position as in the vacuum prepared sample, even at 80 °C. The spectrum shows a blue shift after the material cools, and a second heating at 80 °C does not change its position. This post-adsorption heating step removes hydrogen bonded water on the surface used by stilbene for adsorption, thereby forcing probe migration on the surface into different adsorption sites. Probably a small amount of solvent remains in the pores of silica of S2,SiO$_2$ even after vacuum treatment. Repeated heating in air removes this residual solvent which can explain the blue shift of the fluorescence spectra. The results of the fluorescence spectra and lifetime measurements are given in Table 1.

3.2 S1,SiO$_2$/TiO$_2$ (0.1–1 % TiO$_2$) and S2,SiO$_2$/TiO$_2$ (0.1–1 % TiO$_2$)

Fluorescence spectra of both stilbenes on the surfaces of the TiO$_2$-containing samples in the vacuum cells are similar to those on the pure SiO$_2$ surface (similar to Figures 1 and 2). Emission intensity decreases with concentration of TiO$_2$ and becomes very low with 1.0 wt% TiO$_2$. Emission maxima of both stilbenes are blue-shifted after the stilbenes contact air, and coincide with those in the nonpolar solvent hexane (Figures 5 and 6). Fluorescence intensities of both stilbenes in air demonstrate a detectable dependence on the amount of Ti.

The results of the fluorescence lifetime measurements are very important for the understanding of the photosensitization process. As shown in Table 1, the fluorescence of the stilbenes both on silica and silica–titania surfaces exhibits a two-exponential decay. There is one component with a longer lifetime and the second one with a short τ, close to that of stilbene alone in solution. For instance, S2 $\tau_f = 0.46$ ns in hexane and 0.96 ns in ethanol[16] (compare 0.43 ns on silica and 1.18 ns on silica–titania surface). Both short and long-lived components could be attributed to adsorbed stilbene. The difference in fluorescence characteristics with the surface position of the stilbenes indicates a partitioning of stilbenes among different adsorption sites that possess different interaction characteristics with the

Table 1   *Spectral Characteristics of Adsorbed Stilbenes and Lifetimes of Adsorbed Species λ_F^{max} Denotes the Maximum of Fluorescence Emission, λ_{Exc}^{max} the Maximum of Fluorescence Excitation.*

Sample, conditions	Stilbene, a / mol g^{-1}	λ_f^{max} / nm	λ_{exc}^{max} / nm	$\tau_{f,1}$ / ns	$\tau_{f,2}$ / ns
SiO$_2$, vacuum	S1, » 10^{-7}	560 (shoulder at 460)	390	0.17 (73%) (450)	4.04 (27%) (450)
SiO$_2$, air (after vacuum)	S1	560 (shoulder at 460)	390	0.17 (79%) (450); 0.70 (81%) (560)	3.29 (21%) (450); 2.45 (19%) (560)
SiO$_2$/TiO$_2$ (1% TiO$_2$), air	S1	460	390	0.80 (56%)	2.92 (44%)
SiO$_2$, vacuum	S2, » 10^{-7}	525 (shoulder at 560)	395 (shoulders 360 and 420)	0.43 (450) (45%)	3.80 (450) (55%)
SiO$_2$, air (after vacuum)	S2	during first 5 minutes in air – 560, then – 450	390, wide	0.95 (58%)	4.57 (42%)
SiO$_2$/TiO$_2$ (1% TiO$_2$), air	S2	460	350 – 390	1.18 (49%) (450) 0.78 (47%) (505) 0.78 (74%) (555)	7.84 (51%) (450) 7.18 (53%) (505) 10.13 (26%) (555)

fluorophore. The 450 nm decay characteristics, using 390 nm excitation, indicate a concentration dependent partitioning of the adsorbed stilbenes with an increase in the contribution of a fast decay component ($\tau_1 = 0.17$ ns) in air. The biexponential decay profile analysis indicates that the amount of fast component in the decay is close to the long-lived one. The number of non-quenching sites that are occupied by S2 increases for the mixed oxide. In the presence of a higher concentration of titania in the system, the fluorescence of S2 shows a three-exponential decay with two long-lived components. The contribution of the fast decay

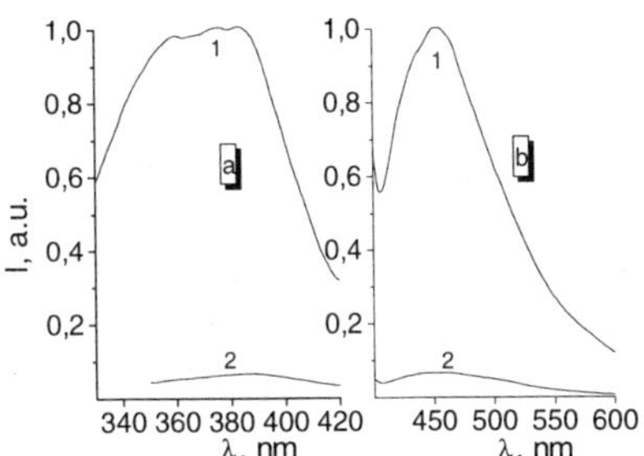

Figure 5 *Excitation (a) and emission (b) spectra of S1 on SiO$_2$/TiO$_2$ with (1) 0.1(1) and (2) 1 wt% of titania at 20 °C.*

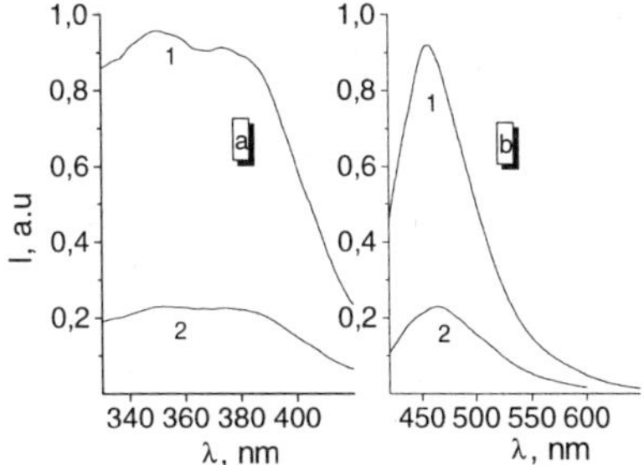

Figure 6 *Excitation (a) and emission (b) spectra of S2 on SiO$_2$/TiO$_2$ with (1) 0.1 and (2) 1 wt% of titania at 20 °C.*

component is larger in the presence of 1 % Ti ions indicating that a specific number of active sites exist on the surface which are occupied more favorably by S2 after solvent is removed. These facts can be explained in the context of the localization of titanium ions and stilbene molecules on the surface and inside the pores as well as their mutual orientation.

The nature of the active centers of silica and silica–titania samples is different. Silica gel is conventionally used as an inert support in most photochemical studies, and its surface has been well characterized in a previous work.[20] Photophysical studies with sensitive spectroscopic probes indicate that molecules adsorbed on silica surface sit in a polar environment. Silica gel surfaces also exhibit a heterogeneous distribution of adsorbed sites which leads to nonexponential decay of the excited molecule.[21-23] Since the adsorbed stilbenes on the surfaces are more rigid than that in solution, their adsorption will result in a decrease of internal conversion, leading to the longer lifetimes. Kamat[24] proposed that fluorescence quenching of sensitizer molecules adsorbed on the TiO_2 surface is due to electron injection from excited dye to the conduction band of titania. The fluorescence emission quenching on a pure TiO_2 surface should be attributed to electron transfer from the excited state of stilbene to the titania particles. Almost 100% of the emission of S1 and S2, as we observed, is quenched on the surface of pure TiO_2 with the appearance of a blue color of Ti^{3+} ions. The same is true on the silica–titania surfaces with a high concentration of TiO_2 (Si:Ti ratio is 1:3), when the blue color of the Ti^{3+} ions is registered in the presence of S2. The process of electron transfer (Equation 1) proceeds even in the absence of light. S1 is more stable in the presence of high concentrations of Ti, and a change of sample color is not visually observed.

$$\text{Stilbene} + Ti^{4+} \rightarrow \text{Stilbene}^{\cdot +} + Ti^{3+} \tag{1}$$

One could assume that degradation of molecules should result in the appearance of a dark color and desorption of decomposition products from the surface. Moreover, the products of decomposition could act as a quencher of the remaining stilbene on the surface. A decrease in fluorescence intensity has been observed in our experiments with S1 (S2), SiO_2/TiO_2 (1% TiO_2) (Figures 5 and 6). It can be assumed that the stilbene molecules are in direct contact with Ti on the surface of silica–titania. There is no evidence of photocatalytic or dark transformation of stilbenes on Ti-containing surfaces with low Ti concentration.

These facts, as well as the fully reversible temperature-dependent fluorescence intensity of adsorbed stilbenes, give evidence that in the case of a low concentration of Ti ions in the mixed system no transformation of stilbenes takes place. The nature of the active centers of silica–titania systems depends crucially on the titania concentration and on the localization of these particles in the porous structure of silica. In accordance with the sol–gel method synthesis used in this work, different structures, *e.g.* Ti-O-Si(OH), Si-O-Ti(OH), or Si-O(H)-Ti, can be formed on the silica surface. Silica–titania surfaces possess higher Brønsted activity, because water molecules can adsorb under dissociation on the titania surface at low temperatures.[25] Owing to Si-O-Ti bond hydrolysability and the possibility of a cleavage of these bonds, an amorphous silica with small particles of titania embedded in the pores of silica exists. Therefore, the active sites of titania (as TiOH, TiO(H)Ti, Ti^{3+}, *etc.*) are inaccessible for the adsorbed stilbenes at this concentration range. Titania in the mixed composition for $C_{TiO2} \approx 1$ wt% is not detected as an individual phase.

The change of the spectral position of the fluorescence maxima of adsorbed stilbenes in air may be connected with a change in local polarity around the adsorbed species due to the competitive adsorption of water molecules from air and removal of adsorbed stilbenes to places with lower polarity. One piece of evidence to support this view is the invariability of fluorescence spectra of S2 during the first minutes of air contact. Then adsorption of water occurs and adsorbed stilbenes are removed from their initial adsorption sites.

The possibility of the formation of intermolecular complexes between adsorbed stilbenes and oxygen and water molecules should also be taken into account.

4 CONCLUSIONS

- D–A-substituted stilbenes are sensitive fluorescence probes to estimate the polarity of the environment on the solid surfaces. Silica and silica–titania surfaces contain adsorption sites with different polarity of OH groups.
- Stilbene fluorescence is quenched effectively when the titania amount is 1 wt% in the mixed silica titania system. The quenching is attributed to electron injection from excited stilbene to titania.
- Stilbenes are chemically stable in the adsorbed state when the amount of titania is not larger than 1 wt% on the silica surface. Low sensitivity may be explained by the special position of the dopants in the silica matrix, when titania aggregates are localized in the pores of silica and are not directly accessible for interaction with adsorbed species. Stilbenes are involved in the process of electron transfer when the concentration of TiO_2 in the silica matrix is high enough to be distributed on the surface.
- The change of the spectral position of the fluorescence maxima of stilbenes on the SiO_2 surface may be connected with the change of local polarity around the adsorbed species due to competitive adsorption of water molecules from air and removal of adsorbed stilbenes to places with lower polarity

5 ACKNOWLEDGEMENTS

We are indebted to Martin Knapp, University of Vienna, for technical assistance. K. R. and G. K. thank the Fonds zur Förderung der Wissenschaftlichen Forschung in Österreich (project nr. P11880-CHE) for financial support.

References

1. I. Rosenberg, J. R. Brock and A. Heller, *J. Phys. Chem.*, 1992, **96**, 3423.
2. C. Kornman, D. Bahnemann and M. Hoffmann, *J. Phys. Chem.*, 1988, **92**, 5196.
3. P. V. Kamat, *Chem. Rev.*, 1993, **93**, 267.
4. R. W. Fessenden and P. V. Kamat, *Chem. Phys. Lett.*, 1986, **123**, 233.
5. D. Liu and P. V. Kamat, *J. Electrochem. Soc.*, 1995, **142**, 835.
6. J. Moser and M. Gratzel, *J. Am. Chem. Soc.*, 1984, **106**, 6557.
7. R. W. Fessenden and P. V. Kamat, *J. Phys. Chem.*, 1995, **99**, 99.
8. D. H. Waldeck, *Chem. Rev.*, 1991, **91**, 415; J. Saltiel and Y.-P. Sun in 'Photochromism – Molecules And Systems', H. Dürr and H. Bouas-Laurent, eds., Elsevier, Amsterdam, 1990, p. 64.
9. J. Michl and V. Bonacic-Koutecký, 'Electronic Aspects Of Organic Photochemistry', Wiley, New York, 1990, and references cited therein.

10. H. Gruen and H. Goerner, *Z. Naturforsch.*, 1983, **38a**, 928.

11. J.-F. Létard, R. Lapouyade and W. Rettig, *J. Am. Chem. Soc.*, 1993, **115**, 2441.

12. W. Rettig and R. Lapouyade, in 'Topics In Fluorescence Spectroscopy', J. Lakowicz, ed., Plenum Press, New York, 1994, vol. 4.

13. W. Rettig, W. Majenz, R. Lapouyade and G. Haucke, *J. Photochem. Photobiol.*, 1992, **A 62**, 415.

14. W. Rettig, W. Majenz, R. Herter, J.-F. Létard and R. Lapouyade, *Pure Appl. Chem.*, 1993, **65**, 1699.

15. R. Lapouyade, K. Czeschka, W. Majenz, W. Rettig, E. Gilabert and C.Rullière, *J. Phys. Chem.*, 1992, **96**, 9643.

16. K. Rechthaler and G. Köhler, *Chem. Phys. Lett.*, 1996, **250**, 152.

17. S. Ruetten, Ph.D. Dissertation, University of Notre Dame, 1996.

18. G. Köhler, *J. Photochem.*, 1986, **35**, 189.

19. D. V. O'Connor and D. Phillips, 'Time-correlated Single Photon Counting', Academic Press, New York, 1984.

20. J. K. Thomas, *Chem. Rev.*, 1993, **93**, 301.

21. R. Krasnansky, K. Koike and J. K. Thomas, *J. Phys. Chem.*, 1990, **94**, 4521.

22. A. Eremenko and A. Chuiko, *Res. Chem. Intermed.*, 1993, **19**, 375.

23. Y. Liu and W. R. Ware., *J. Phys. Chem.*, 1993, **97**, 5980.

24. P. V. Kamat, *J. Phys. Chem.*, 1989, **93**, 859.

25. R. L. Kurtz, *Surf. Sci.*, 1986, **177**, 526.

SYNTHESIS AND LIGAND METATHETICAL REACTIONS OF SURFACE-BOUND ORGANOMETALLIC INTERFACES

S.L. Bernasek,* J. Schwartz, A.B. Bocarsly, G. Lu, K. Purvis and S. Vanderkam

Department of Chemistry
Princeton University
Princeton, NJ 08544 USA

1 INTRODUCTION

Strong interactions between the surface of a metal and an organic overlayer can enhance the performance of a device requiring either surface passivation (corrosion prevention) or activation (microelectrode synthesis), and surface-bound organometallic complexes can be of use for interface synthesis between a metal and a dissimilar overlayer material.[1] Surface "packaging" of gold can be accomplished with alkanethiols but alkanethiols do not adsorb to many metals important in electronics or structural applications, in part because such metals are covered with a hydrophilic, often hydroxylated, native oxide film.[2] Adsorption of organics onto these native oxide surfaces *via* carboxylic[3,4] and hydroxamic acids,[1] and alkyl trichloro- or trialkoxysilanes,[1] however, is well-known.

We use ultrahigh vacuum (UHV) surface science to elucidate details of chemical synthesis of and bonding between an interfacial complex and a substrate surface. In fact, we use results obtained in UHV as a guide for the synthesis of materials under "near ambient" conditions. We have demonstrated the creation of covalently bound, functionalized overlayers on surface oxidized metals or oxides by protolytic attachment of labile organometallic reagents using surface –OH functionality. In particular, we have noted that metal alkoxide complexes react easily with –OH groups of a variety of substrates to give surface metal alkoxide species.

The development of metal alkoxides as interfacial complexes is an important conceptual and practical complement to conventional surface-modification technologies: by subsequent ligand metathesis, "divergent" synthesis of modified surfaces can be effected (Scheme 1a). Indeed, the surface alkoxide complex can be perceived as a versatile "molecular glue" that can be controllably modified to yield a desired, stable interface between a metal or metal oxide and an organic.

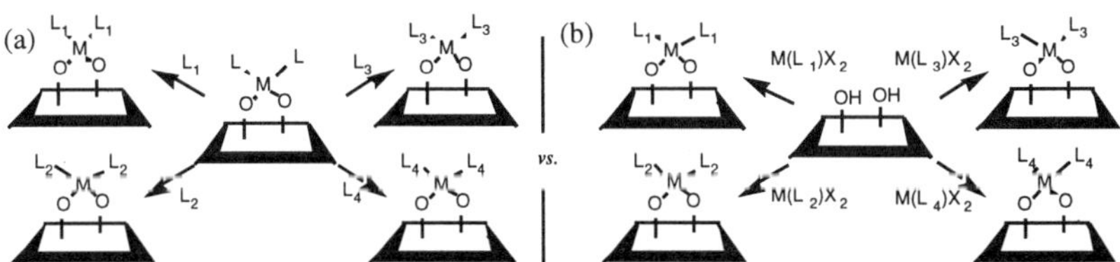

Schemes 1a and 1b: "Divergent" *vs.* "Linear" Synthesis

Utilizing readily available materials, it is possible to start with a common surface species and, through subsequent ligand replacement, prepare a variety of surface modified materials whose direct synthesis would be impossible or at least difficult to accomplish because of limitations in physical properties of the requisite precursors. Contrast this divergent route with "linear" ones (Scheme 1b) involving, for example, silanes, in which a new surface species must be prepared, often by tedious synthesis, for each modification of a given surface and in which functionality may be limited by virtue of synthetic incompatibilities with the silane "head group".

2 RESULTS AND DISCUSSION

2.1 Synthesis and Characterization of Surface Alkoxyzirconium Complexes in UHV

The surface hydroxyl group coverage on Al(110) can be controlled through cycles of adsorption of molecular water (170 K) and thermal desorption (300 K).[7] In turn, the thermal decomposition profile for the chemisorptive product of organometallic chemical vapor deposition (OMCVD) of tetra(*tert*-butoxy)zirconium (**1**) on hydroxylated Al(110) in UHV depends on the initial surface –OH group coverage.[8] Deposition of **1** on lightly hydroxylated Al(110) at 170 K, followed by desorption of the multilayer of **1**, gave [Al^{110}]–[O]–Zr(*tert*-butoxy)$_3$ (**Al^{110}-2**; XPS, C/Zr = 12.6 ± 0.5; expected C/Zr = 12) formed by protolytic loss of one alkoxy ligand. The FT-reflectance absorbance infrared spectroscopy (FT-RAIRS) spectrum of this overlayer exhibited a strong band at 2973 cm^{-1} (and 3 small bands at lower frequency) characteristic of ν_{C-H}. These and the other peaks observed[9,10] agree with those of **1**.[11] Desorption and thermolysis were probed by noting intensity changes (in particular for the strong band at 2973 cm^{-1}) which decreased with the same temperature dependence found for ligand thermolysis as measured by temperature-programmed desorption spectroscopy (TDS) and X-ray photoelectron spectroscopy (XPS). Thermolysis of **Al^{110}-2** (300–350 K) resulted in nearly total decomposition to ZrO$_2$ (by XPS, C/Zr ≈ 0); no intermediate zirconium species could be detected.

OMCVD of **1** onto –OH group-saturated Al(110) (170 K) followed by multilayer desorption (240–280 K) also gave **Al^{110}-2** (Scheme 2). XPS analysis based on Zr$_{3d}$ peak areas showed about 3–4 times as much complex chemisorption had occurred on the saturated *vs.* lightly hydroxylated Al(110) surfaces. Thermolysis at 350–450 K of –OH group saturated surface-bound **Al^{110}-2** did <u>not</u> result in total decomposition, but instead

Scheme 2: Deposition of 1 onto –OH Group Saturated Al(110)

gave a new surface complex, [Al^{110}]–[O]$_2$-(*tert*-butoxy)$_2$Zr, (**Al^{110}-3**) (XPS, C/Zr = 8.6 ± 0.5; expected C/Zr = 8), formed by protolytic loss of a second alkoxy ligand from **Al^{110}-2**; **Al^{110}-3** was stable to *ca.* 450K, above which it decomposed, leaving surface ZrO$_2$. It is important to note that TDS experiments can clearly differentiate between desorption from a multilayer *vs.* fragmentation–desorption of a chemisorbed species: Only desorption of the intact zirconium alkoxide gives rise to Zr-containing fragments in addition to *iso*-butyl ligand-group derived ones.

2.2 Metathesis Reactions of Surface Alkoxyzirconium Complexes with Phenols in UHV

tert-Butoxy ligands may simply be bulky, "insulating" groups. In contrast, phenoxides, because of π-delocalization, may facilitate efficient carrier injection between a surface-modified electrode and an organic overlayer in an electronic device. We found that Al^{110}-**2** and Al^{110}-**3** underwent smooth ligand exchange with a variety of phenols, HOC_6H_4X (X = H, CH_3, NO_2, OCH_3).[12] Metathesis of Al^{110}-**3** with a phenol was accomplished by cycles of exposure at 170 K followed by warming to 300 K. For phenol itself, FT-RAIRS analysis following one exposure cycle (0.6 L) showed the presence of both starting material and a new phenoxy complex. Repeating this procedure three more times resulted in complete loss of IR signals for the *tert*-butoxide ligand and an increase in intensity of those of the phenoxy ligand (Figure 1). That simple ligand metathesis and not some other degradation process occurred was demonstrated by XPS analysis of the reaction procedure. In particular, the XPS-determined atomic ratio of C to Zr measured prior to treatment with phenol was C/Zr = 8.2 ± 0.5 : 1 (expected: 8); after four cycles of exposure to phenol, C/Zr = 13.0 ± 0.5 (expected: 12), in good agreement with the stoichiometry, $[Al^{110}]$–$[O]_2$–$Zr(OC_6H_5)_2$ (Al^{110}-**4**; X = H) (Scheme 3). Further evidence for metathesis was obtained by analysis of the C_{1s} XPS; here, energy resolution is adequate to distinguish C bound directly to O ($BE[C_{1s}]$ = 287.1 eV) from C which is bound to C ($BE[C_{1s}]$ = 285.3 eV).[13,14] XPS analysis of the carbon region showed that the atomic ratio of $\underline{C}$–O *vs.* $\underline{C}$–C changed from 1:3.6 (expected: 1:3) before phenol exposure to 1:5.4 (expected: 1:5) following ligand metathesis. Qualitative compositional information was also obtained by TDS experiments in which both m/z = 39 ($C_3H_3^+$, derived from the benzene ring) and m/z = 57 ($C_4H_9^+$, characteristic of the *tert*-butoxy ligand) were monitored during thermal ramp: the appearance of $C_3H_3^+$ began at *ca.* 400 K (with the desorption peak at *ca.* 500 K); no signal at m/z = 57 was observed over the entire temperature range probed.

Ligand metathesis procedures were also performed using Al^{110}-**2**; exchange occurred more slowly than for Al^{110}-**3**. For the parent phenol, eleven cycles of dose (170 K) and removal (300 K) were performed; FT-RAIRS analysis showed that bands characteristic of the *tert*-butoxy ligand (2977 and 1197 cm^{-1}) were still (barely) observable; XPS analysis also showed that nearly complete exchange of phenoxide for *tert*-butoxide had occurred (C/Zr = 18.5 ± 0.5; expected: 18, for $[Al^{110}]$–$[O]$–$Zr(OC_6H_5)_3$) (Al^{110}-**5**; X = H) (Scheme 3). Two types of C were observed by XPS ($\underline{C}$–O *vs.* $\underline{C}$–C = 1:5.0; expected: 1:5) consistent with phenoxide ligation. Qualitative compositional information was also obtained by TDS experiments for Al^{110}-**5** as was done for Al^{110}-**4**. However, in contrast to the former case, both m/z = 39 ($C_3H_3^+$, derived from the benzene ring) and m/z = 57 ($C_4H_9^+$, characteristic of the *tert*-butoxy ligand) were observed during thermal ramp: the appearance of m/z − 39 began at *ca.* 400 K (with the desorption peak at *ca.* 500 K) and m/z = 57 was observed as a sharp peak at 350 K. Noteworthy is the observation that, except when the parent phenol was used with Al^{110}-**2**, incomplete exchange occurred, consistent with replacement of only 2 *tert*-butoxy ligands.

This difference in reactivity from **Al^{110}-3** might be due in part to relatively dense packing of ligand groups in [Al^{110}]–[O]–Zr(*tert*-butoxy)$_2$(phenoxy), which impedes further co-ordination of a phenolic molecule, necessary to complete metathesis. Comparatively, crowding in [Al^{110}]–[O]$_2$–Zr(*tert*-butoxy)(phenoxy) may not be so acute. No correlation between phenol acidity and rates of ligand exchange was noted. Rates seem to correlate with the melting point of the phenol, suggesting that perhaps ligand mobility on the surface was an important factor.

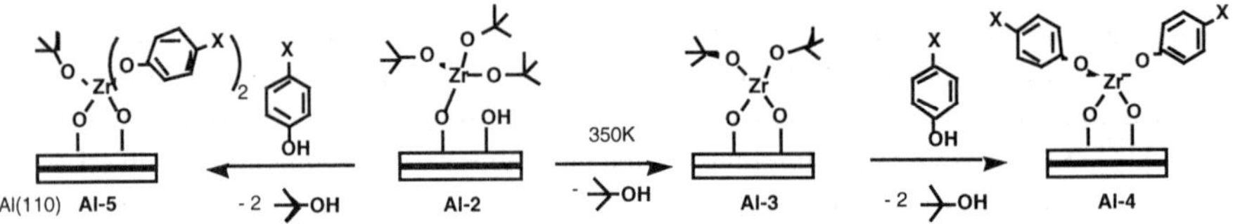

Scheme 3: Ligand Metathesis with Phenols

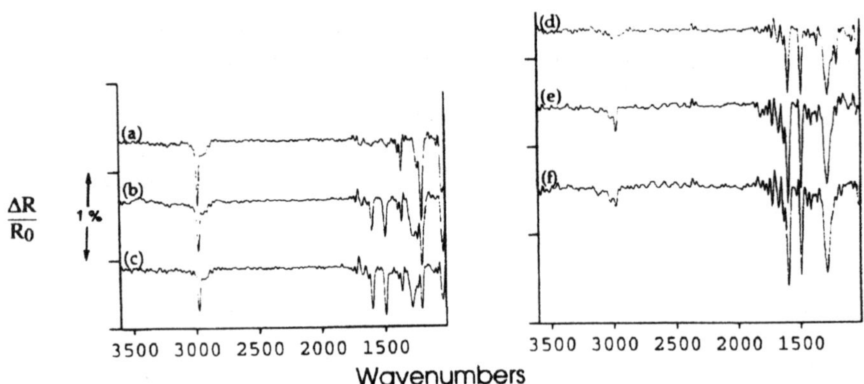

Figure 1   RAIRS *study of the reaction of Al^{110}-2 to give Al^{110}-5: (a) Al^{110}-2; (b) 1 cycle; (c) 2 cycles; (d) 5 cycles; (e) 9 cycles; (f) 11 cycles of phenol dose/desorb.*

2.3 Synthesis and Metathesis Reactions of Surface Alkoxyzirconium Complexes with Phenols Under "Near-ambient" Conditions

In a typical metathetical procedure,[12] a surface hydroxylated Al substrate[7] was treated with vapor of **1**. Protolytic deposition occurred with loss of *ca.* two alkoxide ligands to give product with an average stoichiometry, [Al]–[O]$_2$–Zr(*tert*-butoxy)$_2$(**Al-3**).[15] The substrate was then placed in an evacuable reaction chamber which was equipped with an inlet port attached to a small vial containing the phenol. Substrate processing conditions for exposure and evacuation depend on the volatility of the particular phenol. For the parent phenol, the reaction assembly was evacuated to approx-imately 10^{-2} torr for 30 minutes. The substrate was then subjected to four cycles of exposure (15 min with the reservoir opened to both the reaction assembly and the vacuum; 30 min with the reservoir opened to the reaction assembly, with no evacuation). Following the fourth cycle, the reservoir was isolated, and the chamber was evacuated for 1 hour. Specular reflectance

FTIR spectroscopy was used to monitor ligand metathesis; in particular, strong peaks for the *tert*-butoxy ligand (2973 and 1197 cm^{-1}) were reduced in intensity, and peaks corresponding to the phenolic ligands of [Al]–[O]$_2$–Zr(phenoxy)$_2$ (**Al-4**) grew with each exposure cycle.

2.4 Metathesis Reactions of Surface Alkoxyzirconium Complexes with Alkanecarboxylic Acids in UHV

Complete exchange of *tert*-butoxy ligands for carboxylates occurred for **Al**^{110}**-3** after several cycles of dosing (170 K) and warming (300 K), depending on the acid (Scheme 4).[16] Ligand metathesis was monitored by RAIRS by noting the disappearance of the band at 1200 cm^{-1}, characteristic of υ_{C-O} for the *tert*-butoxy ligands, and the increase in intensity of bands attributable to carboxylate υ_{sym} and $\upsilon_{antisym}$ at 1477 and 1600 cm^{-1}, respectively.[9,17] XPS analysis of the metathesis product (**Al**^{110}**-6a**; R = C$_3$H$_7$) established the stoichiometry of the overlayer to be close to [Al]–[O]$_2$–Zr(O$_2$CR)$_2$ (C/Zr = 9.8 ± 0.3, expected: 8). Further exposure to the acid gave [Al]–[O]–Zr(O$_2$CR)$_3$ (**Al**^{110}**-7a**; C/Zr = 14.3 ± 0.7, expected: 12); one surface O–Zr bond is apparently cleaved by the acid. It is interesting to note that although the same ratio, C/Zr = 12:1, is expected for both **Al**^{110}**-2** and **Al**^{110}**-7a**, resolution of the C$_{1s}$ spectrum, focusing on differences between alkoxide carbon (<u>C</u>–O) of **Al**^{110}**-2** and carboxylate carbon (<u>C</u>(=O)O) of **Al**^{110}**-7a**,[13] easily distinguishes between these ligand types. TDS analysis for the final surface species showed essentially no signal at m/z = 57 below 450 K, and confirms the absence of *tert*-butoxy groups. No carboxylate ligand desorption was noted for **Al**^{110}**-7a** or **Al**^{110}**-7b** (R = C$_7$H$_{15}$) which were stable to > 500 K.[8]

Only incomplete metathesis was observed for reaction between butanoic acid and **Al**^{110} **2**.[16] Even after 12 cycles of dose (170K) and warm (300K), bands characteristic of mixed alkoxide (1200 cm^{-1}) and carboxylate (1477; 1600 cm^{-1}) ligation were observed by RAIRS. Furthermore, TDS analysis for the resulting material (**Al**^{110}**-8**) showed a significant signal at m/z = 57 in the temperature range 300–400 K, characteristic of thermal decomposition of the *tert*-butoxy ligand.[8] As for **Al**^{110}**-2** and **Al**^{110}**-7a**, observed C/Zr ratios for **Al**^{110}**-2** and **Al**^{110}**-8** were nearly the same (*ca.* 12:1). Here, analysis of the deconvoluted C$_{1s}$ XPS signal as a function of butanoic acid dose cycle showed asymptotic replacement of alkoxide ligands by butanoate and its final stoichiometry to be [Al]–[O]–Zr(O$_2$CC$_3$H$_7$)$_2$(OC[CH$_3$]$_3$)$_1$. Incomplete ligand metathesis for **Al**^{110}**-2**, even after exhaustive cycles of dose and heat as noted for phenols, is attributed to steric factors in which the *tert*-butoxy ligands of **Al**^{110}**-2** are in a more congested environment than are those of **Al**^{110}**-3**. In this scheme, replacement of two of the three alkoxide ligands of **Al**^{110}**-2** leaves the third alkoxide group sufficiently sterically inaccessible that further metathesis is unlikely.

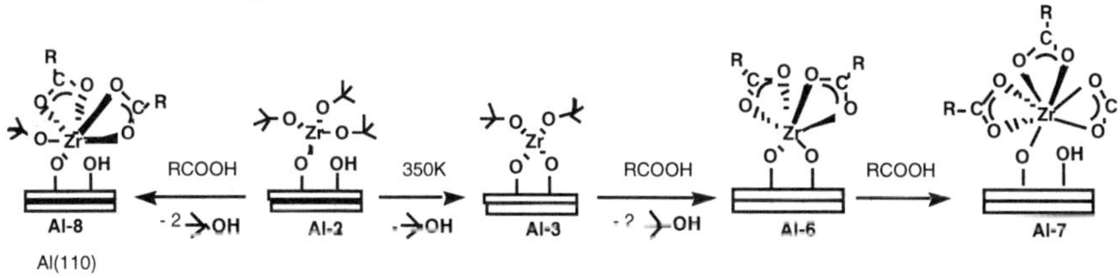

Scheme 4:   Metathesis in UHV to give Al-6, Al-7 and Al-8

2.5 Metathesis Reactions of Surface Alkoxyzirconium Complexes with Alkanecarboxylic Acids under "Near-ambient" Conditions

Passivating films for use in surface corrosion inhibition studies were prepared by metathesis reactions of **Al-3** with carboxylic acids; this gives surface-bound species of average stoichiometry [Al]–[O]$_2$–Zr(carboxylate)$_2$ (**Al-6**), determined by quartz crystal microbalance (QCM) measurements (Scheme 5).[19-21] These surface complexes are inaccessible by direct chemical vapor deposition routes because of low volatility of the tetracarboxylate complex precursor.[17] The reaction between carboxylic acids and metallic oxide surfaces, particularly of Al, has been studied as a means to attach organic films to oxides[2,4,5] but QCM measurements[9] show that simple alkanoic acids adsorb reversibly onto the oxidized Al surface. However, reaction of the carboxylic acid with **Al-3** enables irreversible subsequent adsorption of the alkanecarboxylic acid by formation of **Al-6**.[9] Infrared analysis of **Al-6** showed the alkane chains are not highly ordered,[22] and that the carboxylate ligands were η^2-ligated to Zr.[17] Perfluoroalkane carboxylic acids also react with **Al-3** to give the corresponding perfluoro-surface derivative.[23]

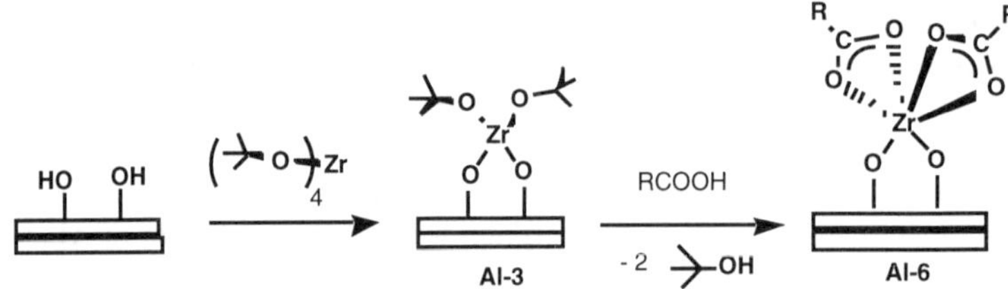

Scheme 5: Metathesis From Al-3 to Give Carboxylate Complex Al-6

2.6 Metathesis Reactions of Surface Alkoxyzirconium Complexes with Poly(ethylene-*co*-acrylic acids) under "Near-ambient" Conditions

Complex **1** can be used to form stable interfaces between poly(ethylene-*co*-acrylic acids) and a variety of metal oxides or surface oxidized substrates (oxidized Al, Fe, Cu, porous Si, and indium tin oxide [ITO]) *via* [metal]–[O]$_{4-n}$–Zr(*tert*-butoxy)$_n$ (**M-9**).[24] Both 5% by weight acrylic acid (PEAA$_5$) and 15% by weight acrylic acid (PEAA$_{15}$) co-polymers were studied for bonding to modified and unmodified substrates (see Scheme 6).

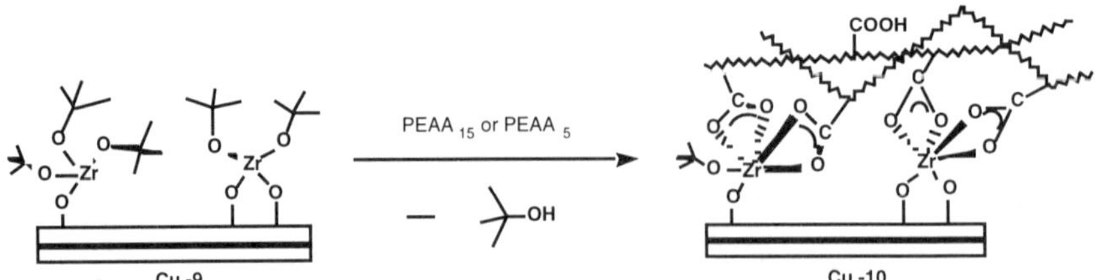

Scheme 6: Co-polymer Attachment

Polymers were applied to the substrates from solution, and both exhibited increased surface loading (by IR) onto the alkoxyzirconium-modified surface by formation of [metal]–[O]$_{4-n}$–Zr(*co*-polymer carboxylate)$_n$ (**M-10**) compared with the untreated surface. Enhanced bonding of the polymer *via* **M-9** was demonstrated by the stability of the deposited co-polymer layer to removal by washing. Differences between the polymer loading for substrates with and without Zr complex surface modification were most pronounced for the less acidic co-polymer (PEAA$_5$) and substrate surfaces of low inherent basicity. For Cu, even low acidity PEAA$_5$ adheres well as complex **Cu-10**.

3 CONCLUSIONS

Differences are observed between standard mechanical adhesion processing, using carboxylic acids or carboxylic acid copolymers, and that in which a metallic complex interface is first synthesized. The differences between these two cases may be understood in terms of the molecular interactions that enable contact between the generally hydrophobic polymer and the hydrophilic oxide substrate surface. In standard cases, this interaction is likely to involve either hydrogen bonding or proton transfer between the carboxylic acid group and a basic oxygen site of the surface;[9] the degree of proton transfer depends on substrate basicity.[25] These types of interactions are likely reversible,[9] since the critical proton is not removed from the ensemble interface. In contrast, for metallic alkoxide interfaces, reaction with copolymer carboxylate groups generates a strong covalent bond[9] between the substrate and the organic material. Since alcohol is readily lost from the ensemble, this carboxylate is formed irreversibly. Investigations using the metallic alkoxide complex interfaces to prepare robust ensembles of oxide surfaces and copolymers of very low net acidity are now underway.

4 ACKNOWLEDGEMENTS

The authors thank the National Science Foundation for support of this research.

References

1.   For example, see M. D. Porter, T. B. Bright, D. L. Allara and C. E. D. Chidsey, *J. Am. Chem. Soc.*, 1987, **109**, 3559.
2.   P. E. Laibinis, J. J. Hickman, M. S. Wrighton and G. M. Whitesides, *Science*, 1989, **245**, 845.
3.   For example, see A. Kumar, H. A. Biebuyck and G. M. Whitesides, *Langmuir*, 1994, **10**, 1498.
4.   D. L. Allara and R. G. Nuzzo, *Langmuir*, 1985, **1**, 45.
5.   J. P. Folkers, C. B. Gorman, P. E. Laibinis, S. Buchholz, G. M. Whitesides and R. G. Nuzzo, *Langmuir*, 1995, **11**, 813.
6.   Y. Xia, M. Mrksich, E. Kim and G. M. Whitesides, *J. Am. Chem. Soc.*, 1995, **117**, 9576.
7.   J. B. Miller, S. L. Bernasek and J. Schwartz, *Langmuir*, 1994, **10**, 2629.
8.   G. Lu, K. L. Purvis, J. Schwartz and S. Bernasek, *Langmuir*, 1997, **13**, 5791.

9.   Y. G. Aronoff, B. L. Chen, G., C. Seto, J. Schwartz and S. L. Bernasek, *J. Am. Chem. Soc.*, 1997, **119**, 259.

10. RAIRS spectra have a low frequency cut-off at *ca.* 1000 cm^{-1} because of the absorption of the UHV chamber CaF$_2$ windows.

11. C. T. Lynch, K. S. Mazdiyasni, J. S. Smith and W. J. Crawford, *Anal. Chem.*, 1964, **36**, 2332.

12. S. K. VanderKam, G. Lu, S. L. Bernasek and J. Schwartz, *J. Am. Chem. Soc.*, 1997, **119**, 11639.

13. U. Gelius, P. F. Heden, J. Hedman, B. J. Lindberg, R. Manne, R. Nordberg, C. Nordling and K. Siegbahn, *Phys. Scr.*, 1970, **2**, 70.

14. D. T. Clark, D. Kilcast and W. K. R. Musgrave, *J. Chem. Soc., Chem. Commun.*, 1971, 516.

15. J. B. Miller, S. L. Bernasek and J. Schwartz, *J. Am. Chem. Soc.*, 1995, **117**, 4037.

16. K. L. Purvis, G. Lu, J. Schwartz and S. L. Bernasek, *Langmuir*, 1998, **14**, 3720.

17. Compare with zirconium tetraacetate. For the η^2- carboxylate: 1540 ($v_{antisym}$); 1455 (v_{sym}): J. E. Tackett, *Appl. Spectrosc.*, 1989, **43**, 483.

18. These observations are consistent with estimates of bond energies: (alkoxide) C–O, 147 kJ/mol; (carboxylate) C–C, 310 kJ/mol; and O–Zr 776 kJ/mol. See, 'CRC Handbook of Chemistry and Physics,' D. R. Lide, ed., 1990–1991.

19. J. B. Miller and J. Schwartz, *Acta Chem. Scand.*, 1993, **47**, 292.

20. J. B. Miller and J. Schwartz, *Inorg. Chem.*, 1990, **29**, 4579.

21. J. B. Miller, J. Schwartz and S. L. Bernasek, *J. Am. Chem. Soc.*, 1993, **115**, 8239.

22. A. H. M. Sondag and M. C. Raas, *J. Chem. Phys.*, 1989, **91**, 4926.

23. B. Chen, S. K. VanderKam, and J. Schwartz, unpublished results.

24. S. K. VanderKam, A. B. Bocarsly, and J. Schwartz, *Chem. Mater.*, 1998, **10**, 685.

25. T. J. Gardner, C. D. Frisbie and M. S. Wrighton, *J. Am. Chem. Soc.*, 1995, **117**, 6927.

SILICA WITH COVALENTLY-BONDED BIS(DI-*N*-BUTYL) PHOSPHORTHIOTRIAMIDE: SYNTHESIS AND PROPERTIES

G. N. Zaitseva* and V. V. Strelko

Institute of Sorption and Problems of Endoecology
Academy of Sciences of Ukraine
Kiev, Ukraine 252680

1 INTRODUCTION

Silica with covalently bonded complexing compounds has attracted the attention of researchers who work with adsorbents for pre-concentration and separation of metal ions or with supported catalysts.[1,2] Phosphorthiotriamides contain the powerful metal-binding fragment P=S that shows affinity for metals with "soft" acidity,[3] which includes most toxic metal ions. Immobilization of phosphorthiotriamides can lead to selective adsorbent for solid-phase extraction of environmentally important metals such as mercury, lead and cadmium. In this paper a simple procedure for covalently bonding a phosphorthio-triamide to silica gel is presented, together with some analytical and physicochemical properties of this new material, $PTTA–SiO_2$.

2 EXPERIMENTAL PROCEDURE

2.1 General

Elemental analysis (C and N) of modified silicas was performed by the Dumas method. Sulfur content was determined by oxidation of modified silicas in air at 1400 °C with an AC-7432 sulfur microanalyser (USSR). Phosphorus content was determined by oxidation of modified silicas with bromine solution in acetic acid and transformation of phosphorus acid to molybdophosphate.[4]

Aqueous solutions of lead, zinc, cadmium and copper salts were prepared by dissolving exact amount of metals in nitric acid. Aqueous solutions of mercury, cobalt, nickel and iron(III) were prepared from the appropriate nitrates. The pH of these solutions was adjusted by addition of sodium hydroxide or nitric acid, and was measured with an EV-74 pH meter (USSR) before and after an adsorption experiment.

IR spectra of $PTTA–SiO_2$ were measured in the range of 1200–3800 cm^{-1} using a UR-10 spectrophotometer (Germany). $PTTA–SiO_2$, undiluted by KBr or other salts, was pressed (20MPa) into thin tablets for these IR experiments. UV–VIS spectra were measured in the 300–830 nm range by diffuse reflectance on a Specord M-40 (Germany). ^{31}P-NMR investigations were carried out on a Brucker CXP 200 spectrometer. Direct potentiometric pH titration was performed according to a published procedure.[5]

2.2 Preparation Of Adsorbent (PTTA–SiO$_2$)

Silica gel L (Chemapol, Czech Republic) with particle size 0.10–0.25 mm and surface area 300 m^2 g^{-1} was used as the matrix. It was washed with distilled water and dried at 450 °C for 8 h before modification. Aminosilylation of this silica was performed in dry toluene as described elsewhere.[1] Aminopropyl silica with 0.43 mmol·g^{-1} of bonded groups was obtained.

Trichlorophosphine sulfide (0.34 g, 2 x 10^{-3} mol) was added dropwise over a period of 2 h at room temperature to a slurry of the aminopropyl silica (10 g) and diisopropylethylamine (0.39 g) in 100 mL of dimethylacetamide. Di-n-butylamine (1.28 g, 1 x 10^{-2} mol) was then added and the resulting mixture was kept at 60 °C for 6 h. After being allowed to cool, the mixture was filtered to remove the modified silica, which was washed with acetone and methanol and finally dried at 20 °C with aspiration.

3 RESULTS AND DISCUSSION

Synthesis of bis(di-n-butyl)phosphorthiotriamide covalently bonded to silica gel (PTTA–SiO$_2$) was performed according to a surface assembling procedure[1] in two steps:

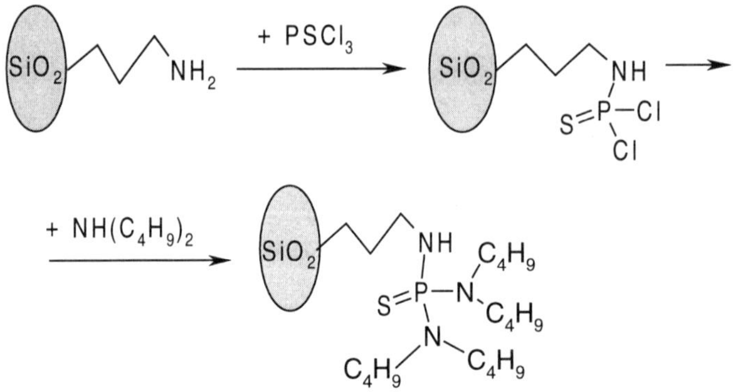

In the first step trichlorophosphine sulfide reacts with amino groups supported on the silica surface. A sterically blocked amine (diisopropylethylamine) is used to bind the hydrochloric acid liberated in this step to prevent quaternization and consequent deactivation of unreacted trichlorophosphine sulfide. It is important that the second and third chlorides in the SPCl$_3$ molecule are less reactive towards the amino groups than the first one, so single-point binding with surface can be expected. While the first step progresses well at room temperature, it is necessary to carry out the second step – amidation with 2 equivalents of di-n-butylamine – at elevated temperature and with an excess of the amine. The following mole-ratios of reaction components were found to be optimum: –NH$_2$: PSCl$_3$: HCl acceptor : secondary amine = 1 : 1 : 1.5 : 5.

Elemental analysis, IR, NMR, and electron spectroscopy were used to prove surface reaction. The concentration of bonded ligand, C_L, was calculated from multi-elemental analysis and from pH–potentiometric titration, as shown in Table 1.

Table 1 *Results of Elemental and pH–Potentiometric Analysis of PTTA–SiO$_2$ (C$_L$, mmol/g).*

Elemental analysis								pH–potentiometry
C,%	C$_L$	N,%	C$_L$	P,%	C$_L$	S,%	C$_L$	C$_L$
6.22	0.27	1.50	0.36	1.49	0.24	0.89	0.28	0.24

The IR spectrum of aminosilica shows C–H stretching bands at 2880 and 2910 cm^{-1} and C–H bending bands at 1380 and 1470 cm^{-1}. It also shows ν(NH$_2$) at 3308 and 3380 cm^{-1}, and δ(NH$_2$) at 1600 cm^{-1}. In the IR spectrum of PTTA–SiO$_2$, these latter bands due to the primary amino group are absent, but new bands at 1400–1415 cm^{-1} and at 1530 cm^{-1} (weak) appear. In the UV spectrum of PTTA–SiO$_2$ only an intense band at 29,920 cm^{-1} can be observed. The solid state MAS NMR ^{31}P spectrum of PTTA–SiO$_2$ exhibits a narrow resonance at 75 ppm, which is characteristic of a bent phosphorus atom in the SPN$_3$ moiety.[6]

Protolytic properties of PTTA silica were measured and equilibrium constants were calculated by non-linear least-squares methods under the model of "chemical reactions".[7] It was found that PTTA–SiO$_2$ exhibits weak acidic properties, with pK$_H$ = 5.2. The total content of the immobilized groups was determined from adsorption isotherms to be 0.24 mmol g^{-1}.

For determination of PTTA–SiO$_2$ adsorption characteristics, adsorption isotherms were built for different metals. The dependence of metal adsorption on solution pH was also studied under static conditions as described earlier.[5] The results are given in Figure 1, where we see that the affinity of PTTA–SiO$_2$ for metal ions decreases in the order:

$$Hg^{2+} > Pb^{2+} > Cu^{2+} = Cd^{2+} > Fe^{3+} > Zn^{2+} >> Co^{2+} > Ni^{2+}$$

Since adsorption of these metal ions on PTTA–SiO$_2$ is highly pH-dependent, with the optimum pH for adsorption differing by several pH units for many of them, PTTA–SiO$_2$ can be used as a selective adsorbent. For example, complete extraction of mercury ions from water solution is observed for pH higher then 2.0. Copper ions can be removed quantitatively at pH > 3.6, and iron ions at pH > 4.2.

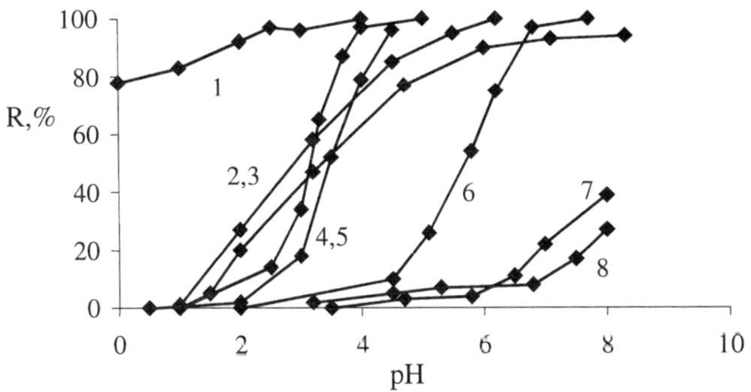

Figure 1 *Effect of pH on retention of some ions on PTTA–SiO$_2$: 1-Hg; 2-Pb; 3-Cd; 4-Cu; 5-Fe; 6-Zn; 7-Co, 8-Ni.*

The optimum condition for extraction of lead ions is pH > 5.7, and for zinc it is pH > 6.1. At pH = 4.0, 82% of lead ions and 74% of cadmium ions can be removed from water, and at pH > 5.8 extraction efficiency of cadmium ions plateaus at 90%. More than 95% of zinc ions remain in water under these conditions. In the pH range from 1.0 to 6.0, essentially all of the cobalt and nickel ions remain in solution.

The PTTA–SiO$_2$ prepared and studied here has a high capacity for adsorption of toxic metal ions due to the combination of the chemical affinity of the P=S group, and the high surface area of the silica support. Figure 2 illustrates adsorption isotherms of some metal ions on PTTA–SiO$_2$. The adsorption capacity for cadmium is 0.38 mmol/g, for lead 0.34 mmol/g and for copper 0.21 mmol/g. Thus, one gram of the PTTA–SiO$_2$ sorbent can quantitatively extract of 28.9 mg of lead, 19.1 mg of cadmium, and 5.1 mg of copper. In the best example of all, one gram of PTTA–SiO$_2$ sorbent can quantitatively remove up to 80 mg of mercury from water.

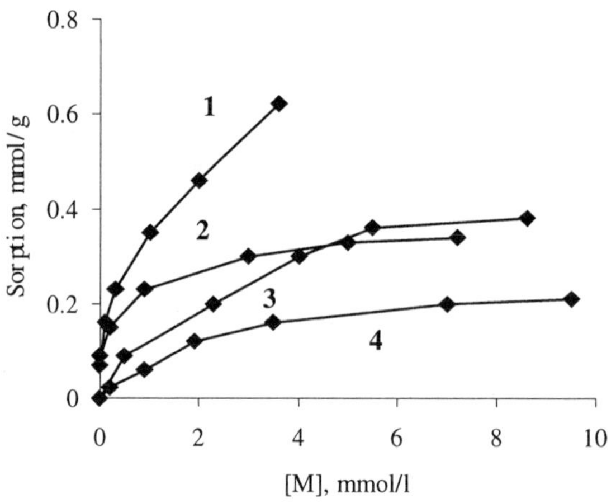

Figure 2   *Adsorption isotherms of: 1- Hg^{2+}, 2-Cd^{2+}, 3-Pb^{2+}, 4- Cu^{2+} on PTTA–SiO$_2$.*

References

1.   V. N. Zaitsev, 'Complexing Silicas: Syntheses, Structure of Bonded Layer and Surface Chemistry' (in Russian), Folio, Kharkov, 1997.
2.   Yu. V. Kholin and V. N. Zaitsev, 'Complexes of a Surface of Chemically Modified Silicas' (in Russian), Folio, Kharkov, 1997.
3.   W. E. Slinkard and D. W. Meek, *Inorg. Chem.*, 1969, **8**, 1811.
4.   V. N. Zaitsev, L. S. Vassilik, J. Evans and A. Brod, *Funct. Materials*, 1995, **2**, 33.
5.   G. N. Zaitseva and O. P. Ryabushko, *Ukrainskii. Khim. Zhurn.*, 1992, **58**, 965.
6.   S. R. Wade and G. R. Willey, *J. Inorg. and Nucl. Chem.*, 1981, **43**, 1465.
7.   Yu. V. Kholin, 'A Quantitative Physical-Chemical Analysis of Equilibrium on the Surface of Complexing Silicas' (in Russian), Oko, Kharkov, 1997.

COMBINATORIAL SYNTHESIS OF SILICA-SUPPORTED METAL BINDING AGENTS AND SENSORS

D. E. Bergbreiter,[†]* V. N. Zaitsev,[‡] E. Yu. Gorlova[‡] and A. Khodakovsky[‡]

[†]Department of Chemistry, Texas A&M University
College Station, TX 77843 USA
[‡]Chemistry Department, T. Shevchenko University
60 Vladimirskaya str. Kiev, Ukraine

1 INTRODUCTION

Metal-binding agents that are immobilized on appropriate inorganic supports are important in environmental chemistry,[1] in sensing chemistry[2] and in biochemistry.[3] However, in many respects, synthesis and the optimization of synthetic procedures leading to such binding agents is problematic.[4] For example, classical solution-state chemistry is typically used in surface modification and surface synthesis, but such classical approaches may not be optimized for reactions at a surface. Given the advantages of combinatorial approaches in drug discovery,[5] it is natural to expect that similar strategies would facilitate synthesis and development of immobilized binding agents. The results below validate this expectation

Our approach to the synthesis of supported metal binding agents combined features of our earlier work with silica bound hydroxamic acids[6] and prior combinatorial peptide synthesis.[7] We had two goals in mind. First, we wanted to evaluate various synthetic approaches to silica-bound metal binding agents/sensors that could reversibly bind soluble metal ions. Second, we wanted to test the colorimetric sensing ability of a group of structurally diverse hydroxamic acids.

To accomplish these goals, we prepared a library of surface-bound, metal-complexing hydroxamic acids, **1**, using silica-coated glass plates as a model for bulk silica. Within this library we varied the structure of the group linking the metal chelator

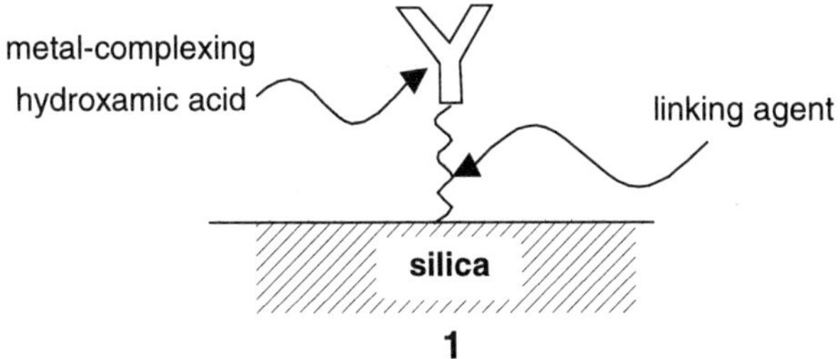

1

to the support, the structure of the hydroxylamine used to prepare the hydroxamic acid and the chemical procedures for the attachment of each species to the support. Qualitative colorimetric assays that were subsequently quantified with diffuse reflectance spectroscopy were used to evaluate the success of surface modification at each stage of the process. Metal binding was assayed using complexation of the hydroxamic acids to Fe(III). The use of common assay steps for all bound library members and intermediates helped minimize the time required to evaluate these synthetic procedures and to evaluate which of the product hydroxamic acids was best in a colorimetric assay for Fe(III).

2 EXPERIMENTAL

2.1 General Procedures

Dimethylformamide (DMF) solutions of the dicarboxylic acids or dicarboxylic acid derivatives (2 N) shown in Table 1 were used first to convert an aminopropylated silica surface into a carboxylic acid-containing surface. Then, the surfaces containing carboxylic acid groups were allowed to react with a 0.67 M DMF solution of a hydroxylamine hydrochloride in the presence of 1 N activating agent, typically carbonyl diimidazole (CDI), to prepare the surface-bound hydroxamic acids, **1**. Syntheses were carried out at 20 °C under atmospheric pressure.

Table 1 *Structures of Acylating Agents used to Modify Aminopropylated Silica.*

Carboxylic Acid Anhydrides				
2	3	4	5	6
Dicarboxylic Acid Chlorides				
7		8		9
Dicarboxylic Acids				
10		11		12

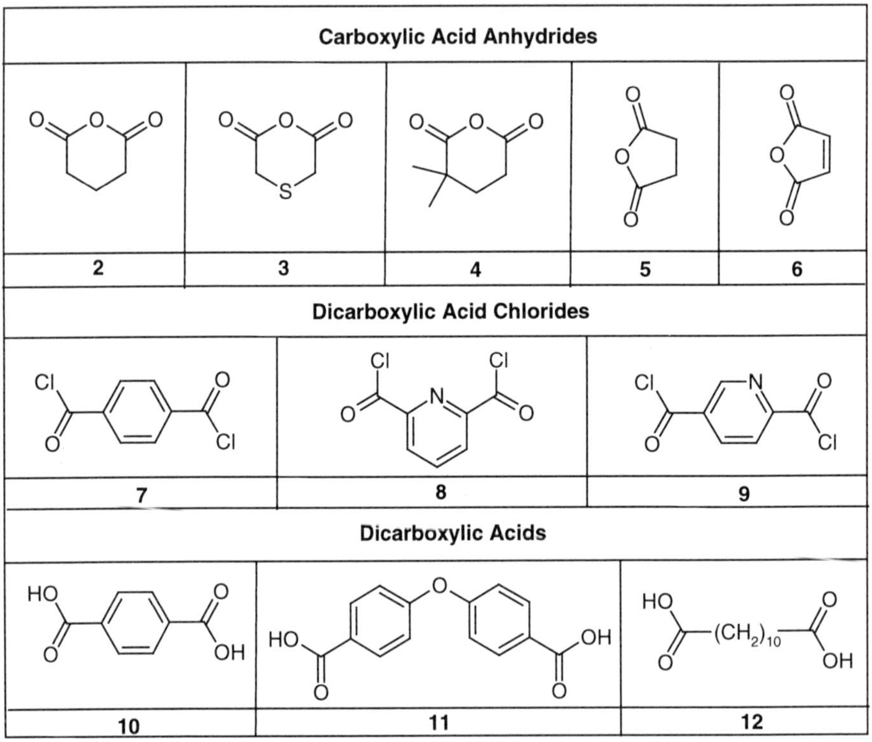

2.2 Detailed Synthetic Procedures for Library Members on Plates

All reactions were carried out on glass plates (100×50 mm) covered with silica. The entire plate was first aminopropylated with aminopropyltriethoxysilane. The concentration of aminogroups on the silica was 5×10^{-4} mol/g (2.6 μmol/m^2). The plates coated with aminopropylated silica were divided into 1 cm^2 sectors. Each sector contained *ca.* 0.01 g of aminopropyl silica (5×10^{-6} mol of amino groups). Before modification, the plates were dried at 120 °C. Individual sectors or spots on the plates were treated with an acylating agent from those listed in Table 1. A single plate thus contained multiple examples of immobilized carboxylic acids derived from the diacid derivatives in Table 1. Multiple examples of spots containing the same diacid derivative on individual or separate plates could also be prepared. The individual sectors on the plates each containing carboxylic acids derived from different acylating agents were all activated at the same time with CDI and allowed to react with hydroxylamine (**13**, **14** or **15**). Exposure of the hydroxamic acids formed in this reaction to an aqueous solution of ferric sulfate was accomplished by simply dipping the plates in an aqueous solution of ferric sulfate at pH 2.5.

$$NH_2OH$$

13　　　　　　**14**　　　　　　　　**15**

Estimation of the extent of reaction of the surface bound amine groups with the diacid derivatives and of the hydroxamic acids with Fe(III) was carried out qualitatively (visually) and semiquantitatively using diffuse reflectance spectroscopy (Figure 1). For example, to estimate the extent to which the amino groups of the support reacted with the dicarboxylic acid derivative, the intensity of reflected light from each sector of plate treated with the various dicarboxylic acid derivatives shown in Table 1 was analysed after the entire plate was treated with salicaldehyde. The intensity of the yellow light (due to formation of the salicaldehyde imine) decreased depending upon the extent to which the amines on the silica were consumed in the reaction with the diacid derivative. This extent of reaction could be estimated qualitatively by visually examining the plate after salicaldehyde treatment. The relative extent of reaction was also estimated in a semi-quantitative manner by diffuse reflectance spectroscopy using Equation 1. In this equation, I_{exp} is the average relative intensity of light from the experimental sector in question, I_0 is the average relative intensity of light from aminopropylated silica that was not modified at all.

$$W = (100\%)[(I_0 - I_{exp})/(I_0)] \qquad (1)$$

The binding of Fe(III) to hydroxamic acids on the plates was also estimated both qualitatively and semi-quantitatively. Since the goal of this work was to examine which complex gave the most intense colorimetric assay, the measured intensity of the light reflected from the matrix containing the Fe(III) complex was normalized based on the intensity of the light reflected from the colorless sector of bulk silica, I_0. The results of this assay using diffuse reflectance spectroscopy were in accord with the qualitative visual estimates of color intensity.

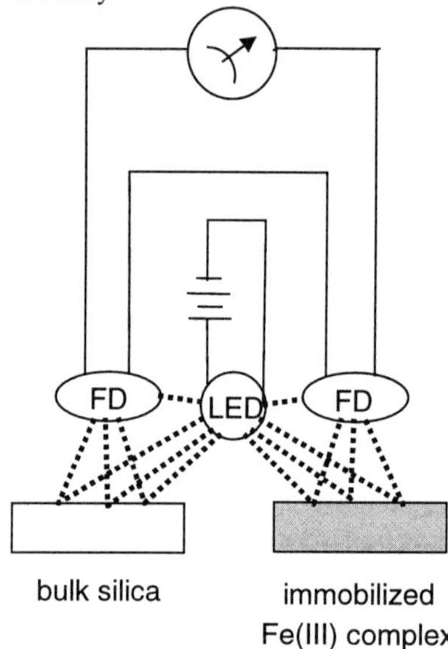

Figure 1   *Schematic diagram of the portable diffuse reflection visible spectrophoto-meter used (LED = light-emitting diode; FD = photodiode).*

2.3 Synthesis of Maleic Acid Immobilized on Bulk Silica

Bulk fumed silica (Aerosil 200, Degussa Corp.) was aminopropylated to form aminopropylaerosil containing 7.2×10^{-4} mol/g of amino groups. This aminopropylated silica sample (5 g) was then allowed to react with a 40 mL DMF solution of 1.76 g of maleic anhydride. The mixture was stirred at 40–60 °C for 4 h, then left at room temperature overnight. The silica was then washed with DMF and residual solvent was removed under vacuum. Spectral analysis (FT-IR, Figure 2) verified that acylation of the aminopropyl groups had occurred. An additional qualitative test involved treating this maleic anhydride-treated surface with salicaldehyde. The absence of a yellow color indicates that the maleic anhydride acylation was essentially quantitative as was true in the synthesis with silica-coated glass plates.

2.4 Synthesis of a Hydroxamic Acid on a Maleic Anhydride-modified Aerosil

Aminopropylated silica (4 g) that had been acylated with maleic anhydride was added to 36 mL of a DMF solution containing 0.96 g of CDI. After being stirred for 45 min at room temperature, this slurry was treated with 6 mL of a DMF solution containing excess hydroxylamine hydrochloride. The resulting mixture was stirred for 24 h. During the last hour of this treatment, the reaction temperature was increased to 70 °C to complete the reaction. Similar procedures were used with other acylated aminopropylated silicas to prepare other silica-bound hydroxamic acids. Typical IR spectra of the intermediates and products are shown in Figure 2.

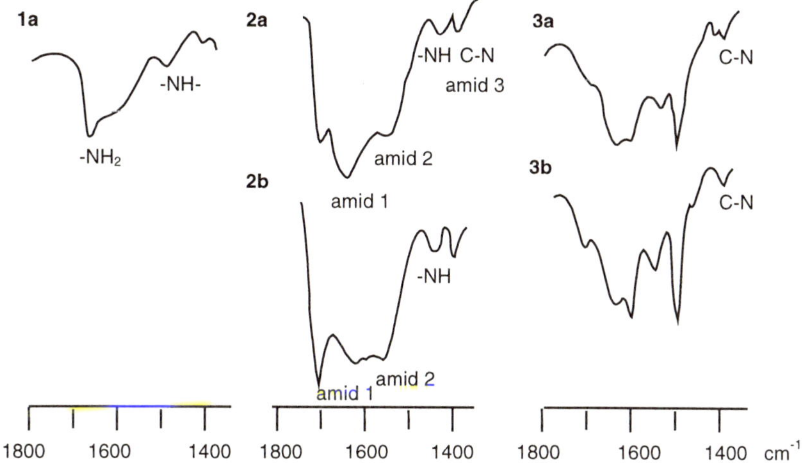

Figure 2 *IR spectra of modified silica: 1a) aminopropylated silica; 2a) maleic anhydride modified silica after conversion to a hydroxamic acid by reaction with 13; 2b) aminopropylated silica after maleic anhydride treatment; 3a) silica modified with 11 and then converted into a hydroxamic acid by reaction with 13; 3b) aminopropylated silica after modification with the dicarboxylic acid 11.*

3 RESULTS AND DISCUSSION

The first step in synthesis of the immobilized hydroxamic acids was coupling of a diacid or diacid derivative to the surface. The reactions shown in Equations 2–4 represent the three approaches used, corresponding to the three classes of acylating agents shown in Table 1. The specific reaction parameters that were varied to optimize the yield of each step were the structures of the acylating agent (see Table 1), and the ratios of the various reagents – initial aminopropyl groups, the acylating agent, and the activating agent (in the case of acylating agents **10–12**).

$$\{-(CH_2)_3NH_2 + \quad \text{(maleic anhydride, R)} \quad \longrightarrow \quad \{-(CH_2)_3NH-\overset{O}{C}-R-\overset{O}{C}-OH \quad (2)$$

$$\{-(CH_2)_3NH_2 + \quad Cl-\overset{O}{C}-R-\overset{O}{C}-Cl \quad \xrightarrow{H_2O} \quad \{-(CH_2)_3NH-\overset{O}{C}-R-\overset{O}{C}-OH \quad (3)$$

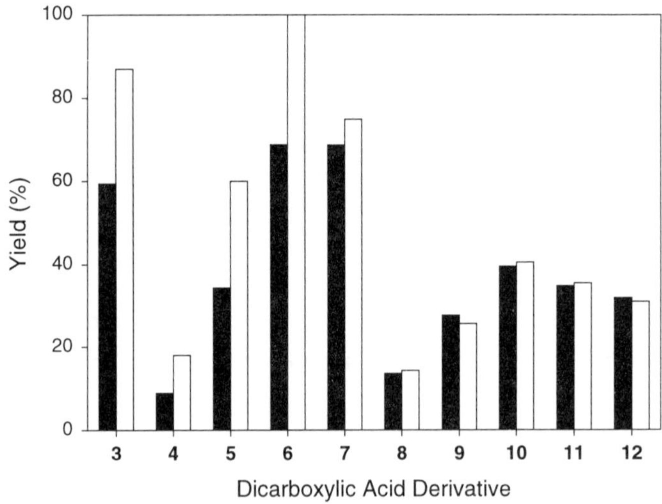

$$(4)$$

Figure 3 illustrates the extent of acylation using various acylating agents with varying ratios of reagents. As shown, the highest yield was attained using terephthaloyl dichloride. The lowest yield was obtained using the diacid chloride derived from 2,6-pyridinedicarboxylic acid and the dimethylated glutaric anhydride.

We also examined how the ratios of reagents and the choice of activating agents affected this synthesis of an immobilized hydroxamic acid. For example, when we varied the ratio of CDI to each of the diacid derivatives **10–12**, we found that an equimolar ratio of reagents was the best choice. When we examined another activating agent, dicyclohexylcarbodiimide (DCC), we found that CDI gave the best yields in acylation based on the extent of reaction of the surface-bound amine groups as determined by the salicaldehyde stain assay described above.

Once the acid groups were attached to the plate, the plate was treated again with CDI to activate the silica–bound carboxylic acids and then with hydroxylamine•HCl to form an immobilized hydroxamic acid. Of the three hydroxylamines we used, **13** formed a far more effective Fe(III) binding agent than did **14** or **15**, which was used in the iron-binding studies described below.

Figure 3   *The extent of acylation of aminopropylated silica with the various acylating agents in Table 1 based on a colorimetric assay using a salicaldehyde stain. Solid bars = 1:1 ratio of amino group to acylating agent, open bars = 1:2 ratio of amino group to acylating agent.*

After the synthesis of the immobilized hydroxamic acids was complete, a plate containing the hydroxamic acid library was placed in an aqueous solution of Fe(III) sulfate at pH = 2.5 for a couple of minutes. A reddish color developed, whose intensity was proportional to the concentration of hydroxamic acid. The intensity of this color was measured using the portable diffuse reflection spectrophotometer shown schematically in Figure 1.

The ability of the surface bound hydroxamic acids to bind Fe(III) was analysed on multiple samples prepared from silicas acylated with four of the anhydrides in Table 1. The relative binding ability of the hydroxamic acids derived from these four anhydride-treated aminopropylated silica samples was essentially consistent within each assay, although there were some sample-to-sample variations.

In these multiple assays, a group of four hydroxamic acids were prepared on a single plate. All the hydroxamic acids could then be assayed in a single Fe(III) treatment step. Multiple sets of four hydroxamic acids could also be treated with $Fe_2(SO_4)_3$ simultaneously. Visual examination of the color of various spots gave a useful qualitative sense of the relative binding ability of the four hydroxamic acids, but diffuse reflectance spectroscopy gave a more quantitative measure of the extent of reaction. The diffuse reflectance results are shown below in Table 2 with the intensity of reflected light for the Fe(III) complexes with each bonded hydroxamic acid being compared and the relative percent transformation (W,%) calculated as described above.

As shown in Table 2, the highest yields in formation of an Fe(III) derivative were obtained when glutaric anhydride or maleic anhydride was used as the acylating agent. Lower yields of an Fe(III) complex were obtained with hydroxamic acids derived from thioglutaric and dimethylglutaric anhydride-treated aminopropylated silicas. It is possible that these lowered amounts of binding of Fe(III) reflect the influence of an electron donor (the sulfur atom) or the steric effect (the two methyl groups).

Table 2   *Relative Extent of Reaction of Silica-bound Hydroxamic Acids with Fe(III) Sulfate as Determined by Diffuse Reflectance Spectroscopy.*

Anhydrides	Relative degree of transformation, W (%)								
	2	3	4	5	6	7	8	10	Average
1	54	61	45	61	54	76	68	74	62
4	15	17	16	17	16	22	24	29	20
5	33	33	33	33	41	59	56	69	45
6	69	65	47	60	60	83	65	100	69

The Fe(III) bound to these plates could be removed by washing in 1 N HCl or by treating the plate with an aqueous EDTA solution. Control experiments with plates that had not been treated with hydroxylamine (to form a hydroxamic acid) showed that hydroxamic acid formation was necessary in all cases for Fe(III) binding to occur.

To demonstrate the general utility of the procedures developed for library members on plates, we modified multigram quantities of bulk silica using the same chemistry (Sections 2.3 and 2.4). These studies showed that maleic anhydride was able quantitatively to acylate the aminopropyl groups on the bulk silica and that the resulting surface-bound carboxylic acid could be converted into an iron binding hydroxamic acid.

4 CONCLUSIONS

The optimization of a synthesis of silica-bound hydroxamic acids for binding of ferric sulfate was accomplished efficiently using common assay steps with a library of hydroxamic acids. By using this combinatorial approach, the effect of different spacer groups on the resulting hydroxamic acid could be readily assayed. Both qualitative visual assays of colorimetric stains and more quantitative diffuse reflectance spectroscopy experiments showed that higher yields were generally obtained in acylation using a two-fold excess of acylating reagent in the case of anhydrides and carboxylic acid chlorides. Since assays could be carried out on multiple samples visually or by diffuse reflectance spectroscopy, it was possible to use multiple samples to assay the reproducibility of the synthetic steps and to thereby confirm the conclusions about the relative acylating and iron-binding abilities of the various species studied. Finally, to confirm the results of library studies using multiple samples on plates, we successfully prepared larger samples of the same iron-binding hydroxamic acids on multigram quantities of bulk silica.

5 ACKNOWLEDGMENTS

Support of this work by the National Science Foundation (DEB, CHE-9707710) is gratefully acknowledged. VNZ also acknowledges a Fulbright Fellowship for support of a portion of his work on this project at Texas A&M University in 1995.

References

1. F. Silva and R. J. P Williams, 'The Biological Chemistry of the Elements', Oxford University Press, Oxford, U.K., 1991.

2. A. P. F. Turner, I. Karube and G. S. Wilson, 'Biosensors: Fundamentals and Applications', Oxford University Press, Oxford, 1989.

3. G. A. Robinson, in 'Methods of Immunological Analysis', R.F. Masseyeff, W. H. Albert and N. A. Staines, eds., VCH Verlagsgesellschaft, Weinheim, 1993.

4. V. N. Zaitsev, 'Complexing Silicas: Syntheses, Structure of Bonded Layer and Surface Chemistry' (in Russian), Folio, Kharkov, 1997.

5. A. Nefzi, J. M. Ostresh and R. A. Houghten, *Chem. Rev.*, 1997, **97**, 555.

6. V. N. Zaitsev, V. V. Skopenko, Y. V. Kholin, N. D. Donskaya and S. A. Mernyi, *Zh. Obsch. Khim.*, 1995, **65**, 529.

7. S. P. A. Fodor, J. L. Read, M. C. Pirrung, L. Stryer, A. T. Lu and D. Solas, *Science*, 1991, **251**, 767.

A NEW CLASS OF SILICA-BONDED ION EXCHANGERS

V.N. Zaitsev* and L.S. Vassilik

Chemistry Department, T. Shevchenko University
60 Vladimirskaya Str.
Kiev, Ukraine   252033

1 INTRODUCTION

Silica-based ion exchangers containing covalently bonded carboxylic acids, alkyl sulphonic acids, and quaternary ammonium salts have been described.[1,2] Silica gels with immobilized aminocarbonic acids are known to show selectivity towards transition metals. Aminophosphonic acids, because of specific electron donor properties, show great selectivity towards multiply charged s-, p- and f- ions, such as Al, Ga, La, Ce, Be, *etc.* Yet despite the promise they might have for (among others applications) rare metal ion separation, silica gels with covalently bonded aminophosphonic acids have not yet been reported.

In this research a simple one-step synthesis was developed to introduce a new class of hydrolytically stable silica-bonded ion exchangers with the general formula:

$$SiO_2 \sim\!\!\sim N \diagup\!\!\diagdown \begin{array}{l} PO_3H_2 \\ R \end{array}$$

wherein R = $-CH_3$, $-CH_2PO_3H_2$, $-C_2H_4(CH_2PO_3H_2)_2$, *etc.*[3] We report here the preparation and characterization of two examples from this class of silica-bound aminophosphonic acids, namely, amino-di(methylenephosphonic) acid, ADPA–SiO_2, and ethylenediamine-N',N'-di(methylenephosphonic) acid, EnDPA–SiO_2:

ADPA–SiO₂

EnDPA–SiO$_2$

2 EXPERIMENTAL

2.1   Instruments

IR spectra of silica-based ion exchangers were recorded on a Perkin-Elmer FT-IR spectrometer. Samples were prepared as thin films between KBr glasses. No dilution was used. Solid state MAS NMR spectra on ^{31}P and ^{13}C nuclei were acquired for wet and air-dried silica powders in the temperatures range 180–300 K on a Durham instrument (200 MHz). The contact time was varied from 0.1 to 20 ms.

2.2   Procedures

The concentration of bonded groups was inferred from the phosphorus content, which was determined spectrophotometrically by the following typical procedure:[4] modified silica (*ca.* 0.1 g) was weighed into a 50 mL volumetric flask, followed by addition of 0.6 mL of glacial acetic acid and 0.2 mL of bromine. The flask was heated on a sand bath until the color of the solution discharged. Then 5 mL of 25% H$_2$SO$_4$, and 10 mL of either 0.25% potassium vanadate or 5% ammonium vanadate solutions were added to the flask and diluted with water to volume. The solution was allowed to stand for 30 min to allow color to develop. The absorption was measured at $\lambda = 430$ nm, pathlength = 1 cm. A calibration plot was obtained using potassium dihydrophosphate solutions. Aminosilica was used as the control blank.   Linear regression produced the following algebraic expression relating the concentration of bonded groups, C_L (x 10^{-4} mol g^{-1}), to the optical density of the solution, A_{430}:

$$C_L = 1.13 \cdot A_{430} + 0.052 \quad \text{(correlation coefficient} = 0.998)$$

The concentration of immobilized groups was also independently determined by pH–potentiometric titration of the phosphonic acid groups. Since silica gel begins to react with alkali at pH > 8.5, titration was performed in 1 N KCl solution that diminishes the matrix dissolution rate. A sample of modified silica was placed into a vessel containing 25 mL of 1 N KCl solution, and 0.1 N sodium hydroxide was added dropwise. The pH of the solution was recorded continuously.

Metal binding properties of ion exchangers were examined potentiometrically by the following general procedure: modified silica (*ca.* 0.2 g) was placed in a glass vessel containing 25 mL of 1 M KCl, then a solution of metal salt was added until the molar ratio of metal : bonded ligand was 1:1. The slurry was mixed for 20–30 min, then titrated with 0.1 M NaOH. The stability of the metal complexes was determined using the model of "chemical reaction" together with non-linear least squares minimization.[5]

Elemental analysis (C,H,N) was performed in a service laboratory at London University according to the Dumas method.

2.3 Synthesis of the Chemically Modified Silicas

2.3.1 ADPA–SiO$_2$. Silica with immobilized amino-di(methylenephosphonic) acid groups was prepared according to the following reaction:

$$SiO_2 \sim NH_3^+Cl^- + H_3PO_3 + CH_2O \xrightarrow{110\text{-}125°C} SiO_2 \sim N\begin{cases} -PO_3H_2 \\ -PO_3H_2 \end{cases}$$

Thus, 20 g of protonated aminopropyl silica with 2.5×10^{-4} mol/g of bonded amino groups (C_L) was placed into a reaction vessel and treated with a solution of 0.98 g H$_3$PO$_3$ at 110–125 °C under mechanical mixing, after which 0.51 g of paraformaldehyde in an appropriate solvent was added dropwise over 1 h. After being allowed to cool, the modified silica was recovered by filtration, washed with dioxane in a Soxhlet apparatus and then with water, and finally dried in air at 110 °C.

2.3.2 EnDPA–SiO$_2$. This modified silica was prepared as described above using 6.3g of protonated ethylenediaminopropyl silica ($C_L = 5 \times 10^{-4}$ mol/g), 0.93 g of H$_3$PO$_3$ and 0.34 g of CH$_2$O in an appropriate solvent. A ninhydrin test for unreacted amino groups was used to judge completeness of reaction.[6] In most cases no color formation was detected and complete transformation of amino groups was assumed.

3 RESULTS AND DISCUSSION

Immobilization of aminophosphonic acids on silica was confirmed by elemental analysis, NMR and IR spectroscopy. Presented in Table 1 are the results of elemental analysis of ADPA–SiO$_2$ and EnDPA–SiO$_2$, and the bonded group concentration calculated from this data. Also shown in Table 1 are C_L values determined by an independent method – pH–potentiometric titration of the immobilized phosphonic acid groups.

Table 1 *Concentration of Bonded Groups (C_L, mmol g^{-1}) and Results of Chemical Analysis of Modified Silica.*

Silica	Method of analysis						
	Elemental analysis						pH titration
	C		H		N		P	
	%	C_L	%	C_L	%	C_L	C_L	C_L
ADPA–SiO$_2$	1.53	0.25	0.34	0.28	0.30	0.21	0.25	0.28
EnDPA–SiO$_2$	4.74	0.56	0.87	0.49	1.45	0.52	0.50	0.60

In the IR spectrum of ADPA–SiO$_2$ (Figure 1), an intense band at 1516 cm^{-1} that dominated the spectrum of the protonated aminosilica no longer appears, and a new, broad adsorption with a maximum at 3350 cm^{-1} appears. This new band remains unchanged under all conditions of temperature and vacuum treatment of the modified silica. We conclude that a zwitterionic structure of dry ADPA–SiO$_2$ is consistent with these data. If ADPA–SiO$_2$ is dried under vacuum at 110 °C with P$_2$O$_5$, an additional adsorption appears at 3689 cm^{-1}. We associate this adsorption with the stretching vibration of P–OH groups that becomes visible after the silica gel is thoroughly dehydrated. Unfortunately, IR is of little use in directly characterizing the phosphonic acid groups because the intense absorption due to Si–O–Si stretching in the 950–1200 cm^{-1} region obscures the P=O vibrations.

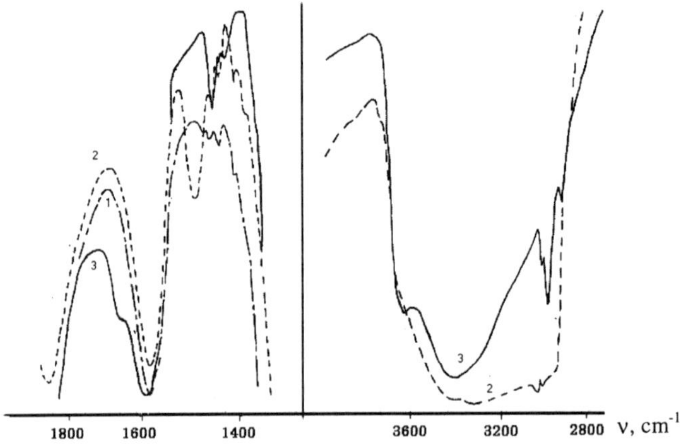

Figure 1 *IR spectra of SiO$_2$–NH$_2$ (1), SiO$_2$NH$_3$$^+Cl^-$ (2), and ADPA–SiO$_2$ (3).*

NMR spectra of ADPA–SiO$_2$ are relatively insensitive to the excitation methods. The static spectrum of ADPA–SiO$_2$ remained unchanged between room temperature and 180 K indicating that the line width in the 'magic angle' spinning spectrum results from a range of isotropic chemical shifts and not from motion of bonded groups. The solid state ^{13}C spectrum, Figure 2, shows two signals for the (presumably) equivalent carbon atoms

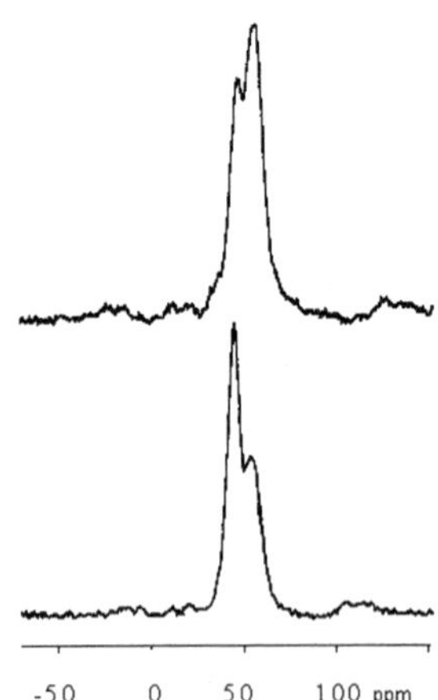

Figure 2 *Solid state ^{13}C-NMR spectra of monopotassium (upper) and dipotassium (lower) salts of ADPA–SiO$_2$.*

in the methylene links, namely at 44 and 53 ppm. The intensity ratio of these resonances changes with pH, so the signals were assigned to $-CH_2PO_3H^-$ and $-CH_2PO_3H_2$ groups, respectively.

The solid state ^{31}P-NMR spectra also show several chemical forms of bonded amino-di(methylenephosphonic) acid. In a broad multiplet centered at 10 ppm, three components can be distinguished corresponding to the grafted groups in different ionization environments. To verify the nature of signal splittings, ^{13}C spectra of ADPA–SiO$_2$ methylene groups were taken over a range of contact times. It can be seen from the spectra (Figure 3) that the two peaks differ quite markedly in their behaviour, reflecting differences in their environment relative to the nearest protons.

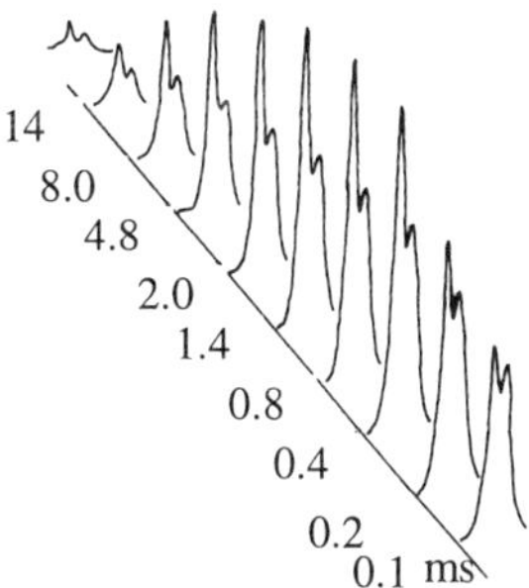

Figure 3 *Solid state ^{13}C-NMR spectra Of ADPA–SiO$_2$ for various excitation times.*

This result further supports the assignment of the ^{13}C-NMR signals at 44 and 53 ppm to methylene carbons of various branches of the amino-di(methylenephosphonic) acid. The observed effect can be explained by various chemical environments of the fragments in ADPA molecule, as in Figure 4. The splittings and significant signal broadening in the NMR spectra of the modified silica leads us to believe that there are strong hydrogen bonding interactions between grafted species and the silica matrix. This strong interaction prevents free rotation of grafted groups on the surface and causes their

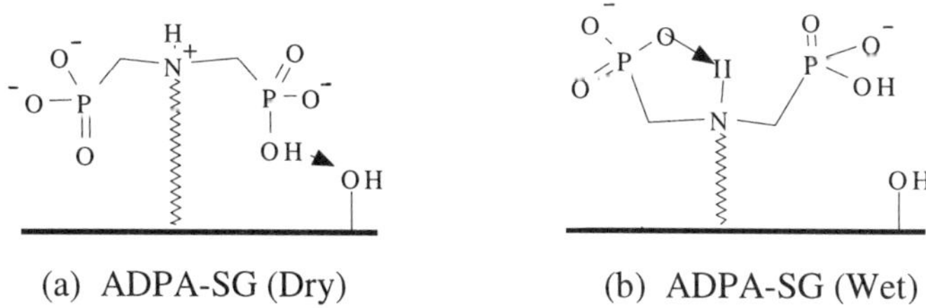

(a) ADPA-SG (Dry) (b) ADPA-SG (Wet)

Figure 4 *Structure of bonded layer on ADPA–SiO$_2$ (a) in air, and (b) in water.*

differentiation. This conclusion was confirmed by narrowing of ^{31}P NMR signals in static spectra of bonded aminophosphonic acids for the wet samples (Figure 5). Static ^{31}P-NMR spectra for hydrated ADPA– and EnDPA–SiO$_2$ also show very narrow resonances, with line shapes indicating motion nearly in the fast limit at room temperature. From this we conclude that hydration of the silica surface results in breaking of hydrogen bonds between surface bonded phosphonic acid groups and silanol groups, resulting in free rotation of the immobilized group and narrowing of the NMR signal (Figure 4, a → b).

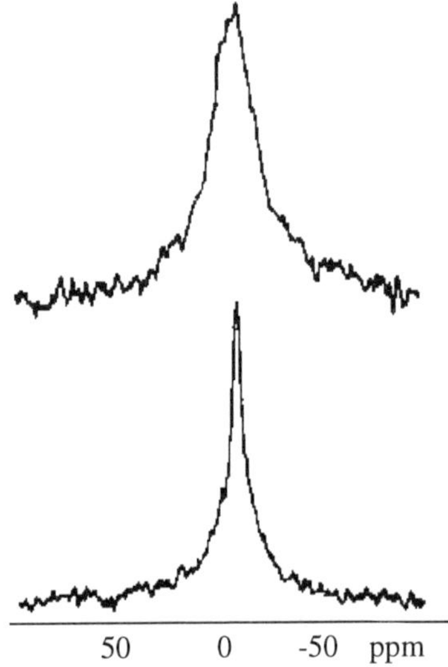

Figure 5   *Solid state ^{31}P-NMR spectra for (top) air-dried sample with MAS, and (bottom) wetted sample without MAS.*

The interaction of a proton with ADPA–SiO$_2$ and EnDPA–SiO$_2$ can be presented as a set of equations analogous to those of the native molecule in solution:[7]

$$
\begin{array}{c}
\text{C}_2\text{H}_5\text{-N}\overset{+}{\underset{}{\text{H}}}\begin{array}{l}\text{CH}_2\text{PO}_3\text{H}_2\\[2pt]\text{CH}_2\text{PO}_3\text{H}^-\end{array}\;\;\underset{\text{-H}^+}{\overset{\text{p}K_1>2}{\rightleftharpoons}}\;\;\text{C}_2\text{H}_5\text{-N}\overset{+}{\underset{}{\text{H}}}\begin{array}{l}\text{CH}_2\text{PO}_3\text{H}\\[2pt]\text{CH}_2\text{PO}_3\text{H}^-\end{array}\;\;\underset{\text{-H}^+}{\overset{\text{p}K_2=4\cdot68}{\rightleftharpoons}}\;\;\text{C}_2\text{H}_5\text{-N}\overset{+}{\underset{}{\text{H}}}\begin{array}{l}\text{CH}_2\text{PO}_3\text{H}\\[2pt]\text{CH}_2\text{PO}_3^{2-}\end{array}
$$

$$
\underset{\text{-H}^+}{\overset{\text{p}K_3=5\cdot92}{\rightleftharpoons}}\;\;\text{C}_2\text{H}_5\text{-N}\overset{+}{\underset{}{\text{H}}}\begin{array}{l}\text{CH}_2\text{PO}_3^{2-}\\[2pt]\text{CH}_2\text{PO}_3^{2-}\end{array}\;\;\underset{\text{-H}^+}{\overset{\text{p}K_4=11\cdot98}{\rightleftharpoons}}\;\;\text{C}_2\text{H}_5\text{-N}\begin{array}{l}\text{CH}_2\text{PO}_3^{2-}\\[2pt]\text{CH}_2\text{PO}_3^{2-}\end{array}
$$

Calculation of protonation constants for the fixed ligands was performed assuming that the acid–base equilibrium on the surface can be described by the same set of constants. We failed in calculation of pK$_1$. The second acidic constant was found to be nearly the same for immobilized and native analogue (4.53 ± 0.22 for ADPA–SiO$_2$, and 5.88 ± 0.15 for EnDPA–SiO$_2$). The third constant is significantly weaker for immobilized ligands than for native ones (8.39 ± 0.2 and 8.56 ± 0.2 for ADPA–SiO$_2$ and EnDPA–SiO$_2$, respectively). The final constant, pK$_4$, cannot be calculated since destruction of the silica matrix is observed at that pH.

ADPA–SiO$_2$ and EnDPA–SiO$_2$ are observed to adsorb some metal ions very strongly from aqueous solutions. Some selected results are shown in Figure 6.

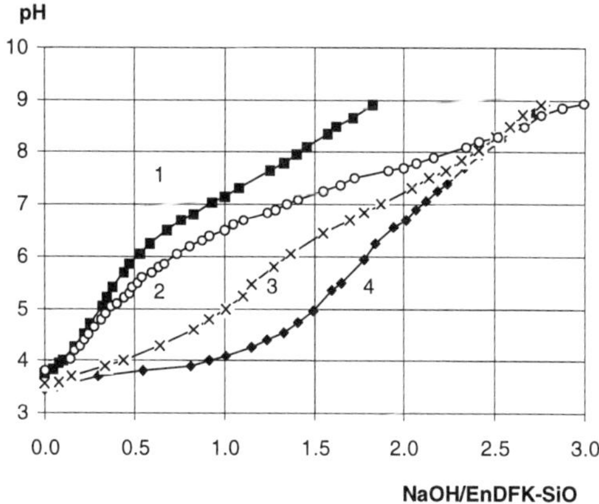

Figure 6 *Potentiometric titration curves for ENDPA–SiO$_2$ alone (1), and for ENDPA–SiO$_2$ complexes with the following metal ions: (2) Ca, (3) Cu, (4) La.*

We have determined that the adsorption affinity for the immobilized ligands in water decreases in the following order:

ADPA–SiO$_2$:   Eu $\approx$ Dy $\approx$ La $\approx$ Pb > Cu > Ni > Ca > Sr
EnDPA–SiO$_2$:   La $\approx$ Cu > Cd > Ca > Sr

Stability constants for numerous metal ions were calculated assuming that the metal ions react with fixed ligands according to scheme:

$$M^{2+} + H_2L^{2-} \leftrightarrow MH_2L \qquad LgK_1$$
$$M^{2+} + H_3L^{-} \leftrightarrow MH_3L^{+} \qquad LgK_2$$
$$MH_2L \leftrightarrow MHL^{-} + H^{+} \qquad LgK_3$$
$$MHL^{-} \leftrightarrow ML^{2-} + H^{+} \qquad LgK_4$$

The results for metal-ion adsorption experiments using ADPA–SiO$_2$ are summarized in Table 2. We conclude from these data that the complexes with ligands immobilized on silica are more stable than the corresponding complexes formed from the native (free) ligand in solution.

Table 2 *Equilibrium Constants for Metal Binding on ADPA–SiO$_2$.*

LgK	Metal ions						
	Ca^{2+}	Sr^{2+}	Ni^{2+}	Pb^{2+}	Cu^{2+}	La^{3+}	Eu^{3+}
LgK$_2$	2.85±0.17	2.84±0.1	4.46±0.14	4.41±0.16	3.14±0.3	4.83±0.3	4.45±0.5
LgK$_1$	5.75±0.16	4.87±0.09	7.51±0.13	8.95±0.4	8.5±0.4	8.87±0.4	9.05±0.4
LgK$_3$	−8.69±0.1	−9.32±0.12	−6.62±0.05	−5.45±0.14	−5.21±0.27	−7.58±0.15	−7.0±0.17
LgK$_4$	−	−	−9.42±0.14	−7.4±0.12	−8.1±0.2	−8.58±0.36	−7.1±0.21
S$_o^2$	1.3	1.0	0.3	1.2	1.8	2.0	1.25

4 SUMMARY

A new class of silica-based ion exchangers containing organophosphorus complexes covalently bonded to silica was described. Silicas with immobilized N-propylamino-di(methylenephosphonic) and ethylenediamino-N,N-di(methylenephosphonic) acids were obtained in high yield by surface assembling reactions. The new modified silicas were characterized by elemental analysis, FT-IR, ^{31}P- and ^{13}C-MAS NMR spectroscopy. Protolytic and metal binding properties of the silicas was investigated. High selectivity towards La^{3+} and Ca^{2+} was found.

References

1. B. B. Wheals, *J. Chromatography,* 1979, **177**, 263.
2. G. V. Kudryavtsev, S. Z. Bernadyuk and G. V. Lisichkin, *Usp. Khimii*, 1989, **18**, 684.
3. USSR Patent 1,613,129.
4. L. Mazor, 'Organic Analysis Methods', Mir, Moscow, 1986.
5. V. V. Skopenko, V. N. Zaitsev and Yu. V. Kholin, *Zhurn. Neorgan. Khimii*, 1987, **32**, 1626.
6. I. Taylor and A. G. Howard, *Anal. Chim. Acta*, 1993, **271**, 77.
7. R. P. Carter, R. L. Carrol and R. R. Irani, *Inorg. Chem.*, 1967, **6**, 939.

SILANE MODIFICATIONS TO CONTROL SURFACE INTERACTION FORCES AS MONITORED BY A NEW TECHNIQUE – EVANESCENT WAVE LIGHT SCATTERING; CHEMICAL MODIFICATION OF SOLID SURFACES FOR STUDIES OF FILTRATION SYSTEMS

Steve Truesdail, Aaron Clapp, Dinesh Shah* and Richard Dickinson

Department of Chemical Engineering & NSF Engineering Research Center for Particle Science and Technology
University of Florida
Gainesville, FL 32611 USA

1 INTRODUCTION

Although the relationship between surface–particle interactions and surface modifications may not be widely appreciated, it is nevertheless true that systematically modified surfaces are extremely valuable in the study of particle adhesion and association. The recent development of Evanescent Wave Light Scattering (EWLS) promises significant advancements in the study of surface–particle interaction forces, both for biological and non-biological colloidal particles.

The application of EWLS normally is limited to optically transparent surfaces to allow for evanescent wave propagation. We wished to explore the possibility that surface modification of the glass test surface used within the EWLS device could increase the range of possible particle interactions that can be studied with this technique. Thus, three surface treatments were evaluated within the EWLS system: 3-aminopropyltriethoxysilane (3-APTES), *N*-(2-aminoethyl)-3-aminopropyltrimethoxysilane (*N*-2-AE-3-APTMS) and a chemically coupled *N,N*-dimethylbutylamine (*N,N*-DMBA). These surface modifications were studied with respect to stability, charge and uniformity by AFM, FTIR and streaming potential. Preliminary EWLS studies of a system consisting of colloidal silica and 3-APTES-treated glass produced results that appeared to agree with observed particle interactions. The results we report here suggest that silane-based surface modification is a feasible means of expanding the range of particle forces that can be studied by EWLS. This will allow us to better understand interparticle forces present within filtration systems involving both biological and non-biological colloids.

Undesirable particle adhesion to surfaces can be linked to numerous biomedical, commercial and industrial problems. The better we understand the forces involved in such processes, the more successful we will be in engineering surfaces specifically to prevent or promote adhesion. EWLS is used to measure these forces within the range 10^{-14} to 10^{-11} N on particles 0.2 to 5.0 microns in diameter as a function of separation distance.

2 BACKGROUND

EWLS is based on the generation of an evanescent wave through total internal reflection of laser light within a prism optically coupled to a glass slide through the use of refractive index matching immersion oil. This wave propagates parallel to the glass slide surface and decays exponentially within the contacting aqueous media, as in Equations 1 and 2, where I(z) is the scatter intensity of a particle within the evanescent wave at separation distance z, I(0) is the intensity of scatter of a particle contacting the surface, n_i refers to the refractive index of the glass slide/prism and the surrounding media respectively, β represents the evanescent wave decay constant, λ_o is the wavelength of the incident laser and θ is the angle of the incident laser beam onto the prism.[1–3]

$$I(z) = I(0)e^{-\beta z} \tag{1}$$

$$\beta = \frac{4\pi}{\lambda_0}\left[n_1^{\,2}\sin^2\theta - n_2^{\,2}\right]^{1/2} \tag{2}$$

A three dimensional optical trap[4] is focused through a microscope objective lens to capture a single colloidal particle and control the separation distance between that particle and a glass (modified or unmodified) surface. This optical trap has an axial harmonic form in which the particle's displacement from the trap center results in variations of particle scatter intensity, a parameter monitored through the use of a photomultiplier tube. Through calibration of trap stiffness (γ) the displacement of the particle from the trap center yields the force between the colloidal particle and the solid surface as a function of relative trap position:

$$F_s = -F_t = \gamma(z_p - z_0) \tag{3}$$

$$\gamma = \frac{kT}{\sigma_z^{\,2}} \tag{4}$$

where F_s represents the force between the surfaces, F_t represents the force exerted by the optical trap, Z_p refers to the most probable particle position and Z_o is the trap center. The trap stiffness γ is defined as a function of temperature T, Boltzmann's constant k and the variance attributed to the displaced particle position in the optical trap σ_z.

3 EXPERIMENTAL

Colloidal silica particles with a 1.5 micron diameter were suspended in a 1×10^{-4} M or a 4×10^{-4} M KCl solution at a concentration of between 1,000 and 10,000 particles/mL at pH = 6.3. Approximately two to three hundred 10 micron spacer beads were then added to the suspension to prevent collapsing of the gap between the microscope slide and the cover slip. The edges of the cover slip were then sealed with silicon stopcock grease to prevent evaporation during analysis. The slide was then inverted and placed onto the microscope stage of the EWLS (Figure 1).

Total internal reflection of the He–Ne laser within the optically coupled glass slide and prism generates the evanescent wave. The particle captured with the 3-D trapping laser is then stepped into the evanescent wave through control of the objective lens focal point. The resulting scattered light is monitored with the photomultiplier tube (PMT).

A CCD camera mounted within the microscope eye piece (not shown in Figure 1) allowed for particle visualization and focusing of the optical trap. Once focused on the surface, the x–y motorized stage allowed for particle identification and capture with

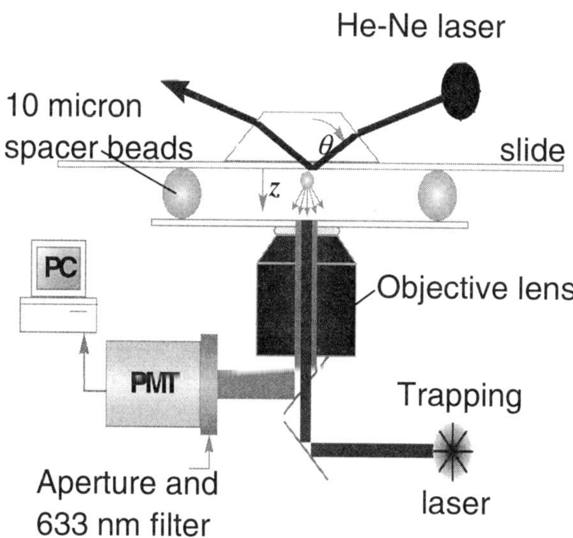

Figure 1 *Diagram of an inverted microscopy system and additional instrumentation as used for EWLS.*

the use of the three-dimensional optical trap. The particle could then be moved a large distance away from the slide surface and then gradually stepped inward towards that surface through automated control of the objective lens. The colloidal particle was generally stepped towards the surface at 20 nm intervals. The photomultiplier tube was then used to monitor the light scatter by the particle as it entered the evanescent wave and was acted upon by the approaching microscope slide surface forces.[5]

3.1 Silane Surface Preparation

The 3-APTES and *N*-2-AE-3-APTMS silanized glass surfaces were prepared through the use of a 95% ethanol/5% water solution to which the silane was added to reach a 2% final concentration. The glass slides were then dipped into the solution and removed after 2 min. These surfaces were then rinsed twice with ethanol and allowed to cure at room temperature for 24 h.[6,7]

The more complex *N,N*-DMBA surface coating involved three reactions as described by Suhara, *et al.*[8] First, acid-washed glass was exposed to 0.5 mL of 2,4,6,8-tetramethylcyclotetrasiloxane (TMCTS). The glass was then heated to 80 °C in a sealed container for 10 h, followed by an additional 20 h in an open container. The poly-TMCTS-coated surfaces were heated at reflux for 6 h in 500 mL of toluene containing 0.05 mg platinic acid, 0.05 mg tributylamine and 5 mL chloromethylstyrene, then rinsed thoroughly with toluene and allowed to dry under vacuum. In the final step, the glass was heated at 65 °C for 4 h in 500 mL of acetonitrile containing 50 mL *N,N*-DMBA. This product was then washed with water and dried under vacuum. This particular surface coating has been reported to have a zeta potential independent of pH (over the range of 2–10) and length of alkyl chain used in the final reaction (4 to 8 carbons).[8]

3.2 Surface Characterization

Streaming potential was used to monitor the apparent surface charge of the coated slide surfaces. Surface-coated materials were placed within the streaming potential sample cell and rinsed with distilled water (pH 6.3) for one hour at approximately 1 liter/min. Streaming potential measurements were then performed with a 1 x 10^{-4} M KCl solution to monitor the zeta potential of the various surfaces. The samples were then flushed with several liters of distilled water and periodically monitored with the streaming potential device to determine charge stability. Flow direction of the water used in the flushing procedure was periodically changed in order to ensure that dissimilar electrode environments did not develop in the proximity of the electrodes within the streaming potential cell and produce false readings. A fuller discussion of the device, procedure and accuracy has been published.[9]

The zeta potentials of the colloidal silica particles used in this study were determined with the use of a Brookhaven ZetaPlus electrophoresis instrument, model v3.21. Fourier Transform Infrared Spectroscopy (FTIR) was used to monitor the surface for the presence of nitrogen both before and after exposure to water. Tapping mode AFM was also used to evaluate the coating uniformity and roughness of a silanized glass surface.

4 RESULTS

Coating stability under water flow was monitored to aid in the preliminary evaluation of the surface modifications proposed for use within the EWLS apparatus. Glass samples modified by each of the three coating procedures outlined above were analysed with the use of a streaming potential device as a function of water elution volume. Figure 2 shows that all three surfaces started with positive zeta potentials. The *N,N*-DMBA chemically-coupled coating appeared to be the least stable of the three modifications. The 3-APTES and the *N*-2-AE-3-APTMS coatings appeared to achieve some degree of stability around –7 mV ± 3 mV at pH 6.3.

Preliminary EWLS studies were conducted with the various modified glass slides (after water exposure as shown in Figure 2) to evaluate the effect these coatings might have on the propagation of the evanescent wave. Both the washed 3-APTES and the *N*-2-AE-3-APTMS did not appear to affect the scatter properties of the probing colloidal particle. Overall, particle scatter at short distances between the colloidal silica and the modified glass microscope slide test surfaces (see Figure 3) was very similar to the scatter observed for the unmodified glass microscope slide surface. The three-step coupling reaction of *N,N*-DMBA, as reported by Suhara, *et al.*,[8] however, did not appear to provide the same uniform degree of scatter at all points tested on the surface.

Tapping mode AFM analysis of the 3-APTES modified surface after water exposure indicated a relatively uniform coating. The overall root mean square (rms) of the roughness on the coated surface was determined to be 0.58 nm with a z-range of 4.7 nm. The uncoated glass surface had a rms value of 0.28 with a z-value of 3 nm.

The EWLS was used both to qualitatively and quantitatively evaluate the surface forces between colloidal glass particles and a 3-APTES modified surface. Experimental observations and corresponding zeta potentials at pH 6.3 are shown in Table 1. The colloidal silica was also determined to have a zeta potential of –42 mV ±3 mV at 1 x 10^{-4}

M KCl and a zeta potential of $-36\,\text{mV} \pm 3\,\text{mV}$ at $4 \times 10^{-4}\,\text{M}$ KCl. In addition, the resulting force and energy curves generated with EWLS have been included as Figures 4 and 5.

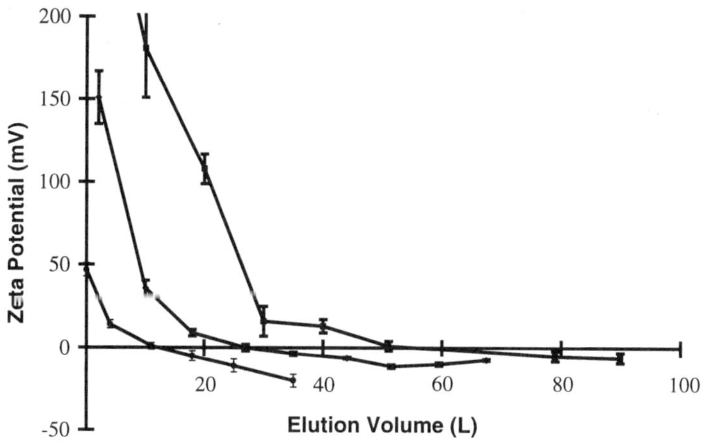

Figure 2 *Graph showing streaming potential results for three modified surfaces as a function of water elution volume. Elution and streaming potential runs were performed at pH 6.3. Curves from right to left correspond to 3-APTES, N-2-AE- 3-APTMS and the coupled N,N-DMBA.*

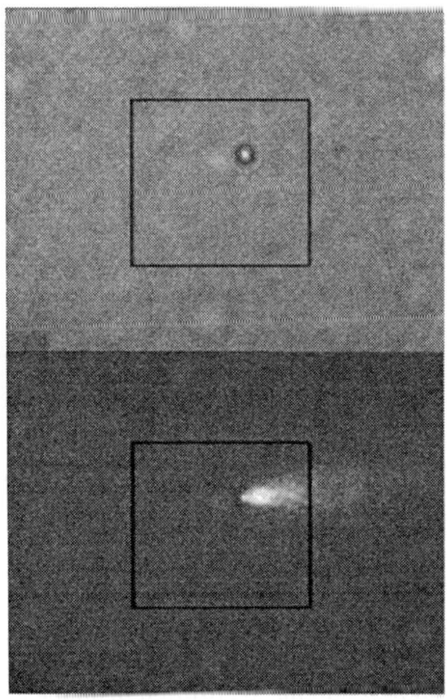

Figure 3 *Two CCD camera images showing particle scatter behavior for large separation distances between the colloidal silica particle and the 3-APTES glass surface (top) and for small separation distances (bottom).*

     Fundamental and Applied Aspects of Chemically Modified Surfaces

Table 1   *Summary of Observations on Colloidal Particle Interactions with EWLS Slide Surfaces.   Colloidal Adhesion and Brownian Motion Refer to the Observed Behavior of the Colloidal Silica.*

	[KCl]	**Surface Zeta Potential**	**Colloidal Adhesion**	**Brownian Motion**	**EWLS Minimum**
Unmodified Glass	1×10^{-4} M	-44 mV ±3 mV	All weakly Adhered	Slight	-3.25 kT
	4×10^{-4} M	-35 mV ±3 mV	All strongly adhered	None	N/A
3-APTES Glass	1×10^{-4} M	-7 mV ±3 mV	No particle adhesion	Substantial	-0.9 kT
	4×10^{-4} M	-5 mV ±3 mV	All very weakly adhered	Substantial	-1.5 kT

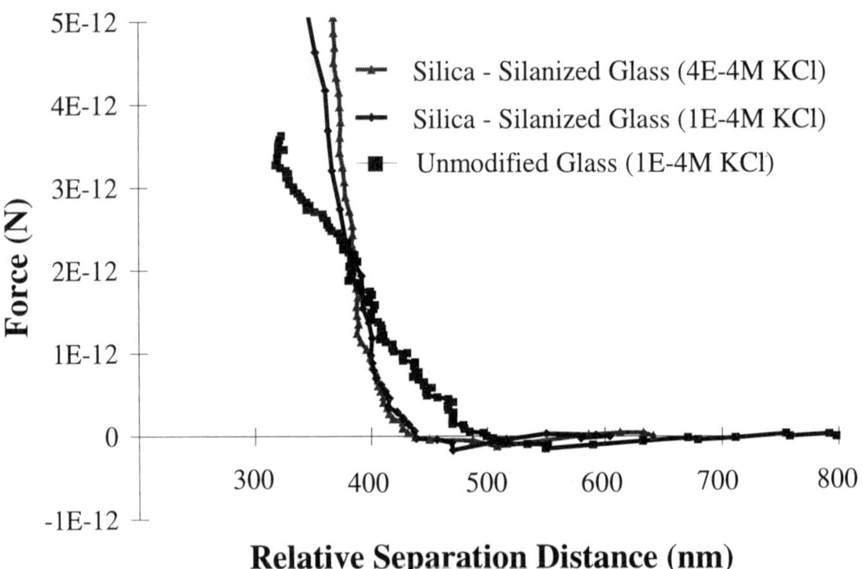

Figure 4   *Force curves for colloidal silica and 3-APTES / unmodified glass at pH 6.3 and varying electrolyte strengths. Curves represent EWLS data as calculated with equation (3) from the scattering signal detected by the PMT. Curves from right to left at a 350 nm separation distance correspond to the silanized glass and colloidal silica particle at 4×10^{-4} M KCl then at 1×10^{-4}M KCl. The final curve represents an unmodified glass and colloidal silica particle interaction at 1×10^{-4} M KCl.*

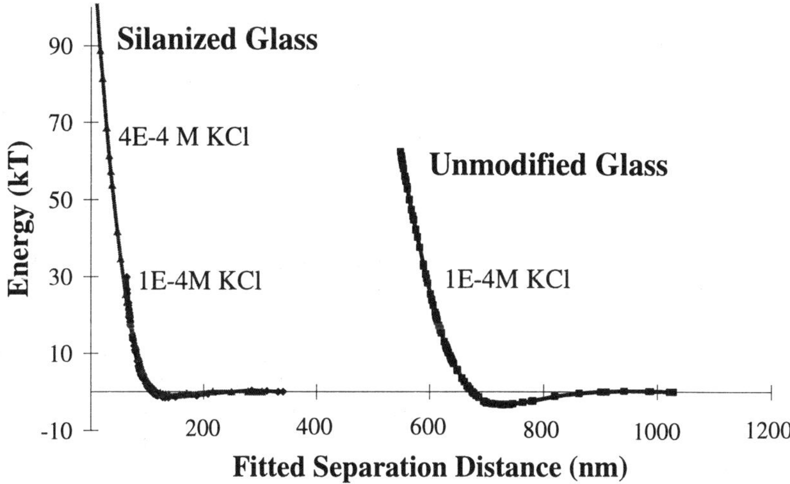

Figure 5 *Energy curves from integration of data in Figure 4. Left: silanized surfaces at [KCl] = 1 x 10^{-4} and 4 x 10^{-4} M (curves coincide). Right: unmodified glass at [KCl] = 4 x 10–4 M . Separation distance has been adjusted to correspond to the distances predicted with Wiese–Healy interaction theory.*[10]

5 DISCUSSION

EWLS was used to directly measure forces between a colloidal glass particle and a flat test surface within an aqueous environment. In the past this technique has successfully been used to measure surface forces for both biological and non-biological colloidal particles of small dimension (0.5 to 5 microns in diameter) with exceptional sensitivity (0.01 to 10 pN). The technique in these cases, however, was limited to an optically transparent unmodified glass surfaces.

In the present study, silanes were used to coat glass surfaces for EWLS analysis to determine their potential in producing wide ranges of uniform van der Waals, double layer repulsion and hydrophobic interactions. It is important that these surface modifications meet two criteria for successful use within the EWLS device: (1) coating thickness must be kept at a minimum to ensure sufficient penetration of the evanescent wave into the surrounding aqueous media, and to avoid interference with the evanescent wave decay constant; and (2) coatings must be stable in aqueous environments and uniformly applied to the glass substrate to prevent wide variations in surface properties from point to point.

Successful wave propagation with the 3-APTES and the *N*-2-AE-3APTMS appeared to suggest that the surface coatings on the glass slide were significantly thinner than the predicted evanescent wave penetration depth of 100 to 150 nm. The *N,N*-DMBA coating surface behaved the poorest of the three modified surfaces we studied. The thick 2,4,6,8-TMCTS surface-polymerized coating appeared to interfere with the transmittance of the evanescent wave and resulted in a non-uniform pattern of scattering at different positions along the slide surface. The apparent thickness of the coupled coating relative to that of the other two silanes also raised concerns as to the validity of our assumption that the coating would not significantly affect the refractive index of glass, which is used in the calculation of the evanescent wave decay constant. In addition, streaming potential studies of coating stability under water flow also suggested that the *N,N*-DMBA coating was a poor candidate for further analysis.

The 3-APTES and *N*-2-AE-3-APTMS coatings initially appeared to be unstable under water flow, the results shown in Figure 2 being attributed to loss of the electropositive nitrogen from the surfaces. These surfaces did appear to stabilize at a zeta potential of around –7 mV ± 4 mV. FTIR indicated that the nitrogen was still present on all coated surfaces after water exposure. Modifications to the *N,N*-DMBA reaction are currently underway to decrease the overall coating thickness and increase coating stability by replacing the TMCTS of the initial reaction with a triethoxysilane.

Surface non-uniformity within systems of particle interactions has been suggested as one of the reasons for disagreements between theory and experimental observations.[10–14] Heterogeneity in charge or roughness might significantly affect the reproducibility of force measurements as a function of slide position. AFM mapping of a washed 3-APTES modified glass slide appeared to show a relatively uniform surface with no large masses of coating present on the surface. The analysis indicated that the modified slides had a surface roughness with an rms value of 0.58 nm, with a z-range of 4.7 nm. The roughness of this modified surface was well below the rms values reported for flat surfaces being used in other AFM surface force studies (rms $\approx$ 3 nm).[15] It is therefore expected that this surface should be smooth enough to allow for reproducible surface force analysis, although surface roughness is expected to effect the interaction between the modified glass surface and the colloidal silica particle.[16]

EWLS was used to compare the interactions between glass surfaces – both a washed, 3-APTES modified and an unmodified microscope slide – and a colloidal silica particle. The unmodified glass surface appeared to retain all of the colloidal particles under solution conditions of 1 x 10^{-4} M KCl at pH = 6.3, where the zeta potential is –44 mV ± 3 mV. This result suggested that the energy potential curve between these two surfaces possessed an energy minimum of depth greater than 1–2 kT. (Brownian motion of colloidal particles are commonly assumed to have 1–2 kT of thermal energy.) The EWLS results indicated that this was in fact the case, with a measured minimum of –3.25 kT. At the higher salt concentration of 4 x 10^{-4} M KCl, where zeta potential = –36 mV ± 3 mV, all of the colloidal particles appeared to become strongly adhered to the unmodified glass surface, *i.e.* the particles showed no signs of Brownian motion. Under these conditions the adhesion force was stronger than the EWLS trapping force and the particles on the surface could not be captured for further study. The EWLS results for the washed 3-APTES surface (zeta potential = –7 mV ± 3 mV) at 1x10^{-4} M KCl showed a small minimum on the order of –0.9 kT. In this case, particle association with the modified glass surface would not be expected and was not observed. Increasing electrolyte concentration to 4 x 10^{-4} M KCl resulted in a measured energy minimum of –1.5 kT for the washed 3-APTES-coated glass slide (zeta potential = –5 mV ± 3 mV) and the colloidal silica particle (zeta potential = –36 mV ± 3 mV). This finding was accompanied by the observation that all of these colloidal particles were very weakly adhered to the surface, *i.e.* particles were still undergoing substantial Brownian motion. The generated interaction energy curves for all of these aforementioned systems have been included in Figure 5. The energy minima measured for the silanized surfaces have been magnified as Figure 6.

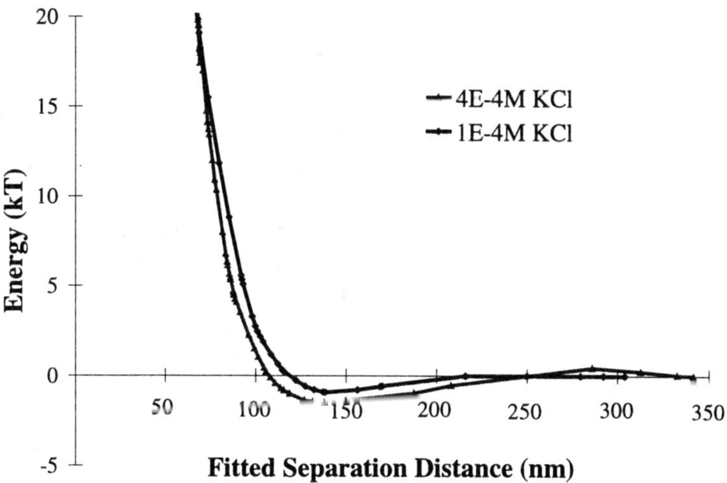

Figure 6 *Magnified section of the EWLS generated force curves shown in Figure 5, highlighting the differences in measured energy minimum for a colloidal silica particle and a 3-APTES modified glass surface at different electrolyte concentrations.*

6 SUMMARY

Surface silanization was successfully used to modify interaction forces between EWLS surfaces. The measured energy minima appeared to correlate well with the observed depositional behavior of the colloidal silica particles. The use of silanes to vary surface charge, hydrophobicity and structure should allow the EWLS technique to be used more widely in the study of surface forces, which doubtless will lead to the engineering of surfaces more suitable for both biological and inert particle applications.

7 ACKNOWLEDGMENTS

The authors would like to acknowledge the financial support of the Engineering Research Center (ERC) for Particle Science and Technology at the University of Florida, the National Science Foundation (NSF) grant # EEC-94-02989, and the Industrial Partners of the ERC. Additional appreciation is extended to Dr. Y. Rabinovich, Mr. Josh Adler and Mr. Alexander Patist for their assistance with the AFM characterization.

References

1. D. C. Prieve, F. Luo and F. Lanni, *J. Chem. Soc., Faraday Discuss.*, 1987, **83**, 297.
2. P. A. Temple, *Appl. Opt.*, 1981, **20**, 2656.
3. H. Chew, D. S. Wang and M. Kerker, *Appl. Opt.*, 1979, **18**, 2679.
4. A. Ashkin, *Biophys. J.*, 1992, **61**, 569.

5.   A. Clapp, A. Ruta, S. Truesdail and R. Dickinson, *Rev. Sci. Instrum.*, submitted for review, 1998.

6.   E. P. Plueddemann, 'Silane Coupling Agents', Plenum, NY, 1982.

7.   K. L. Mittal, 'Silanes and Other Coupling Agents', VSP, 1992.

8.   T. Suhara, H. Fukui and M. Yamaguchi, *Colloids Surf. A*, 1995, **95**, 29.

9.   S. E. Truesdail, G. Westermann-Clark and D. O. Shah, *J. Environ. Eng.*, accepted for publication, 1998.

10.   M. Elimelech, J. Gregory, X. Jia and R. A. Williams, 'Particle Deposition and Aggregation: Measurement, Modeling and Simulation', Butterworth, Oxford, 1995.

11.   J. N. Ryan and M. Elimelech, *Colloids Surf. A*, 1996, **107**, 1.

12.   W. J. C. Holt and D. Y. C. Chan, *Langmuir*, 1997, **13**, 1577.

13.   R. Kapur, J. Lilien, G. L. Picciolo and J. Black, *Bio. Mat. Res.*, 1996, **32**, 133.

14.   S. E. Truesdail, G. Lukasik, S. Farrah, D. O. Shah and R. Dickinson, *J. Colloid Interface Sci.*, in press, 1998.

15.   I. Larson, C. J. Drummond, D. Y. C. Chan and F. J. Grieser, *J. Am. Chem. Soc.*, 1993, **115**, 11885.

16.   L. Suresh and J. Y. Walz, *J. Colloid Interface Sci.*, 1997, **196**, 177.

STABILITY STUDY OF METAL OXIDE-COATED SAND

P. Kang, R. Tu and T. Van Reken

Department of Chemical Engineering
University of Florida
Gainesville, FL 32611 USA

1 INTRODUCTION

In recent years, a great deal of effort has been placed on the use of metallic hydroxide flocs and metallic hydroxide coatings to remove microorganisms from contaminated water. Metallic hydroxide flocs have been shown to be effective for the removal of bacteria and viruses from water following settling or filtration.[1] They have also been used to concentrate viruses from water as a part of virus detection procedures.[2,3]

Metallic hydroxide flocs were also used to coat sand and diatoms by using preformed flocs or by forming the flocs in solutions that are in contact with the media.[4–7] The treatment increases the media's ability to adsorb microorganisms from water. However, the coating procedure requires repeated treatments or long contact times. An alternate method of coating was developed by Farrah *et al.*, in which the metallic hydroxides were formed directly on the surface of the media.[8–10] With this method, a combination of metallic hydroxides can be deposited on the surface of the sand or other media. The combination of iron and aluminum hydroxides has been found to be more effective in removing microorganisms than the individual iron or aluminum hydroxide coating.[10] Surface–charge interaction is thought to be the primary contributor to the biofiltration process.

Industrial application of metallic hydroxide-coated sand requires that the coating display some degree of long term stability. Although the coating of iron and aluminum hydroxides by *in situ* precipitation method has been shown to be stable on a laboratory scale, we are interested in its long-term stability on a larger scale. For this purpose, a group of graduate and undergraduate students from the Advanced Separation Processes Thrust group of the National Science Foundation Engineering Research Center at the University of Florida designed and constructed a testbed column. This testbed filtration unit has the same features as commercial upflow moving bed filters (*e.g.* Dynasand, Parkson Corp., Ft. Lauderdale, FL). The features of the testbed column are described in detail in later sections. The leaching and attritioning of the coated sand as a function of time was studied by several methods, including inductively coupled plasma emission (ICP) elemental analysis, zeta potential measurement, and bacterial and viral adsorption properties.

2 APPARATUS AND PROCEDURE

2.1 Testbed Continuous Filtration Column

The dimensions of the testbed filtration column are given in Figure 1. The column consists of a water feed ring, a center tube, two triangular fins, and sampling ports. The feed ring consists of ¼ inch holes facing downward so that sand will not clog the holes. Because of this orientation, water initially enters the column in a downward flow, then flows upward through the sand in the column. The center tube is used to transport the sand to the top of the column. This is accomplished by injecting compressed air inside the center tube near its lower end. The air flow creates a suction that carries both the sand and water from the bottom of the column to the top. Once the sand is at the top, it is ejected from the center tube and falls back down the column. As the sand falls, water flows upward and removes the coatings that have come off the surface of the sand. The particulate-contaminated water is sent to a drain, while fresh water is fed to the column. The triangular fins are used to keep the center tube in position. They are made of plexiglass material. The sampling ports are made of ½ inch copper ball valves.

2.2 Experimental Conditions

The Parkson sand was obtained from Oglebay Norton Industrial (Brady, TX) and used as received. The sand was coated with a combination of iron and aluminum hydroxides by procedures outlined in Farrah *et al.* (Kona). Fifty lbs of metallic hydroxide-coated sand were then homogenized and placed in the testbed filter. Water flowrate was based on the superficial velocity (6.6 cm/min) established by the Dynasand pilot filter and set at 1.2 L/min. The pH of the tap water varies between pH 8.0 and 9.0. Air flowrate in the center tube was set at 0.4 L/min. The air flowrate was established empirically to match the sand turnover rate, 4–6 times/day, of the Dynasand pilot filter.

2.3 Sampling Procedure

For the stability of coating experiment, the column ran without the backwash section, *i.e.* no reject stream. For the first 8 days, a sample of approximately 1 L of sand was taken from the two lower sampling ports every 48 h. A 20 mesh sieve was used to rinse the sample. The sample was then homogenized to assure uniformity and 400 mL of the homogenized sample was stored in the refrigerator for later analysis, and the remaining sample was returned to the column. From day 9 to day 22, a sample was taken by the described procedure every 72 h. For the remaining days, a sample was taken every 500 h.

A primary method of determining the longevity of the treated sand is analysing the stability of the metals adsorbed onto the sand surface. This was done in our study by removing the metal coating from a portion of our sample by dissolving it in hot acid. The concentration of the resulting solution was then measured using ICP atomic emission. This number can then be converted into the more useful units of [mg (metal)/g (sand)].

2.4 Digestion and ICP Analysis

A primary method of determining the longevity of the treated sand is analysing the stability of the metals adsorbed onto the sand surface. This was done in our study by

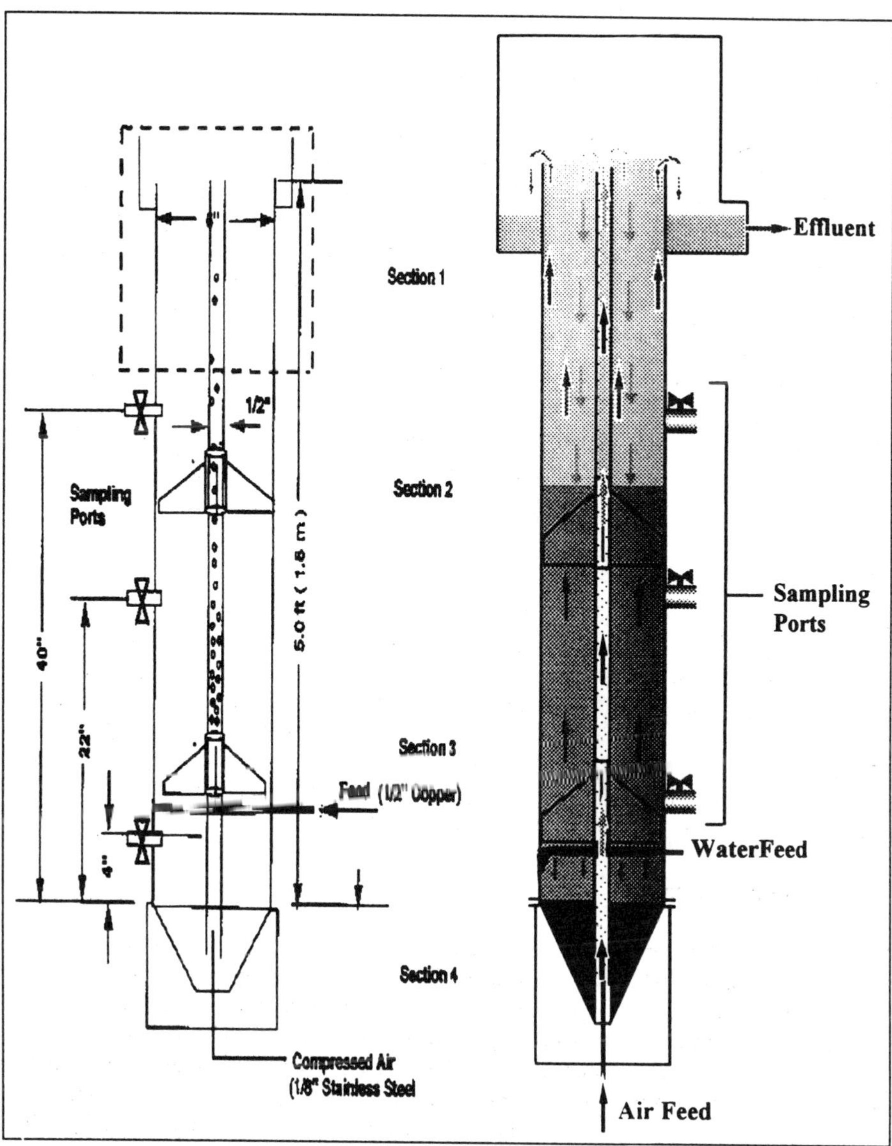

Figure 1 *Schematic diagrams of the testbed filtration column.*

removing the metal coating from the sand sample by dissolving it in hot acid. Three samples, weighing 20 g each from each sand sample, were added to 100 mL of an acid solution made up of 20% (by volume) concentration hydrochloric acid, 40% concentrated nitric acid, and 40% deionized water. The mixture was heated until the boiling point was almost reached. The hot mixture was filtered and the sand was rinsed several times with deionized water. The resulting solution was diluted to 1000 mL and measured using ICP atomic emission.

2.5 Preparation of Test Microbes

Batch cultures of *Escherichia coli* C3000 (ATCC 15597) were grown overnight to early stationary phase in Nutrient Broth (Sigma) at 35 °C with 120 rev min^{-1} shaking. Cells were harvested by centrifugation at 3000 × g for 10 min at 4 °C, resuspended in filter-sterilized MilliQ water to A_{550} = 0.40, and stained with 4',6-diamidino-2-phenyline-indole (DAPI) at a final concentration of 25 µg/mL for 30 min. Subsequently, cells were harvested and washed in 2 × volume of artificial ground water (AGW) adjusted to pH 7.0, with centrifugation at 7000 × g for 10 min at 4 °C, then resuspended in AGW (pH 7.0) for use in batch or column removal tests. Suspensions of bacteriophages (MS2, PRD-1) were prepared in AGW from refrigerated stocks kept in 30% tryptone soy broth (BBL) at pH 7.3. The microbial suspensions were kept at 4 °C until use.

2.6 Batch Removal Test

Batch removal tests with bacteria or phages were carried out using sand samples from the testbed column. In addition, unattritioned metallic oxide-coated and uncoated sands were used as controls. Initial bacterial concentrations were 10^5–10^6 PFU/mL. Five g of wet sand (rinsed with AGW) and 20 mL of suspension were used in the bacterial adsorption tests. Four g of sand from the testbed filter or dry control sand were used with 10 mL of suspension in the bacteriophage adsorption tests. The test microbes were contacted with the respective sands in 50 mL plastic centrifuge tubes affixed to a 70 cm diameter wheel that was rotated vertically at 30 revolutions/min at room temperature. The contact times were 2 h for the bacteria and 30 min for the phages. Blanks containing microbial suspension but no sand were run along with the other tubes. At the end of the contacting time, the sand in the tubes was allowed to settle for 10 min at room temperature. Then the supernatants were sampled. The phage samples were immediately diluted in 1 % tryptic soy broth (TSB). Bacterial samples and phage dilutions were stored at 4 °C until enumeration. All samples and controls were run in triplicate. The AGW has a pH of 7.0 and is buffered by bicarbonate.

2.7 Microbial Enumeration

Direct counts of bacterial samples were obtained within 48 h (preliminary experiments showed that cell losses were 5% or less after this length of storage) using epifluorescence microscopy (Apha *et al.*, 1992; Sherr *et al.*, 1993). The samples were not re-stained, hence only pre-stained cells were counted. This prevented interference from indigenous cells (background contamination or sloughed biomass). A Leitz microlab microscope, equipped with appropriate filter blocks for desired light excitation and emission, was used for visualization and counting. Samples were vortexed vigorously to mix well immediately before filtration. A 1.0 mL aliquot of sample was filtered through a sterile 0.20 µm pore size, black, nucleation track polycarbonate filter (Nuclepore, Costar Scientific Corp.), with a 0.45 µm pore size membrane filter (GN-6, Gelman) placed underneath to evenly distribute the vacuum. Filters were pre-wetted with filter-sterilized MilliQ water and a vacuum of 600 mm Hg applied. Three mL of filter-sterilized MilliQ water was added to the filter funnel immediately before applying vacuum to help disperse bacteria on the filter. Ten microscopic fields were counted per filter. Sample concentrations typically gave 25–50 cells per field, hence the total number of cells counted per sample exceeded the minimum range recommended by Kepner & Pratt (1994).

Bacteriophages were assayed within 24 h of sampling using the soft agar overlay method described by Snustad and Dean.[11] Host bacteria (*Escherichia coli* for MS2, *Salmonella typhimurium* for PRD1) were freshly grown in 30% TSB overnight at 37 °C without agitation. Serial dilutions of phage samples were added in 0.1 mL aliquots to tubes containing 0.2 mL host bacteria. Then, 5 mL soft agar (40–45 °C) was added to each tube, mixed, and poured while still warm onto plate count agar (DIFCO) containing 0.6 µg mL^{-1} crystal violet. The plates were cooled for 5 min, incubated overnight (6–8 h) at 37 °C, then plaques were counted.

2.8 Zeta Potential

The zeta potential of the sand was determined using a streaming potential apparatus constructed in our laboratory (Figure 2). The apparatus consisted of a flow-through cell which was packed with the sand being analysed, tanks to provide electrolyte (0.1 mM KCl) flow through the cell, manometers to measure the pressure drop across the cell, electrodes placed at the edges of the packed bed within the sample cell, and a voltmeter (Keithley) to measure the electrical potential across the electrodes. The sample cell was constructed from clear polycarbonate pipe, which allowed visual examination for air bubbles and uniformity of packing. Cell dimensions were 1.9 cm I.D. × 20.3 cm. The electrodes were 99% pure 18 gauge silver wire and 40 mesh size silver gauze (Newark Wire Cloth Co.) spot welded together so that the silver wire extended perpendicular from the center of the circular mesh. The mesh portion of an electrode was anodized in 0.1 mM HCl for 1 h using a copper cathode with a 5 mA current. The finished electrode was treated in 1 mM KCl overnight to stabilize the silver chloride coating produced in the anodizing process.

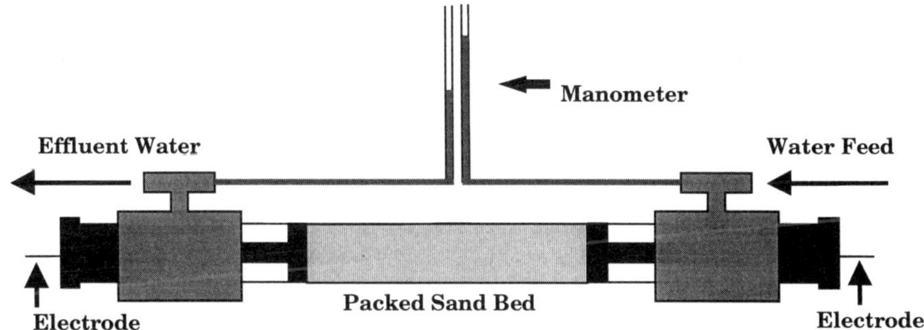

Figure 2 *Streaming Potential Apparatus.*

Electrodes were regenerated or replaced every two to three weeks to ensure proper performance. Sand samples were packed into the sample cell by the tap and fill method. The packed bed was flushed with CO_2 to remove all the trapped air, then 1–2 L of 0.1 mM KCl electrolyte solution was allowed to flow through the bed to allow stable voltage measurements. Measurements were performed within a 5 h period per sample to minimize the chance of biofilm lysis. The slope of the measured voltage *versus* pressure relationship along with the measured solution conductivity was used to compute the zeta potential according to the following equation:

$$\zeta = \frac{4\pi\mu K}{\varepsilon\varepsilon_0}\frac{\Delta E}{\Delta P} \qquad (1)$$

where μ = viscosity of water, ε = dielectric constant of water, ε_0 = dielectric constant of a vacuum, K = conductivity, ΔP = hydrostatic pressure change, and ΔE = voltage change. The apparatus provided reproducibility within an error of less than ±10% of the measured zeta potential at a 95% confidence level for single measurements. Standard deviation between samples was less than 3%.

3 RESULTS AND DISCUSSIONS

3.1 Metal Content

The amount of iron and aluminum coatings on the sand was analysed using an ICP spectrometer. The results of these analyses are shown in Figure 3. After the treatment and a superficial rinsing, the iron content was determined to be 1.61 mg/g of sand and the aluminum content was 0.811 mg/g of sand. These results were far above the values for untreated sand, which are less than 0.02 mg/g of sand for iron and aluminum. After the sand was loaded into the column, both metals showed an initial period of rapid attritioning. After three days in the column, there was approximately a 35% decrease in the amount of iron on the sand. There was a corresponding 25% decrease in aluminum coating during the same time period.

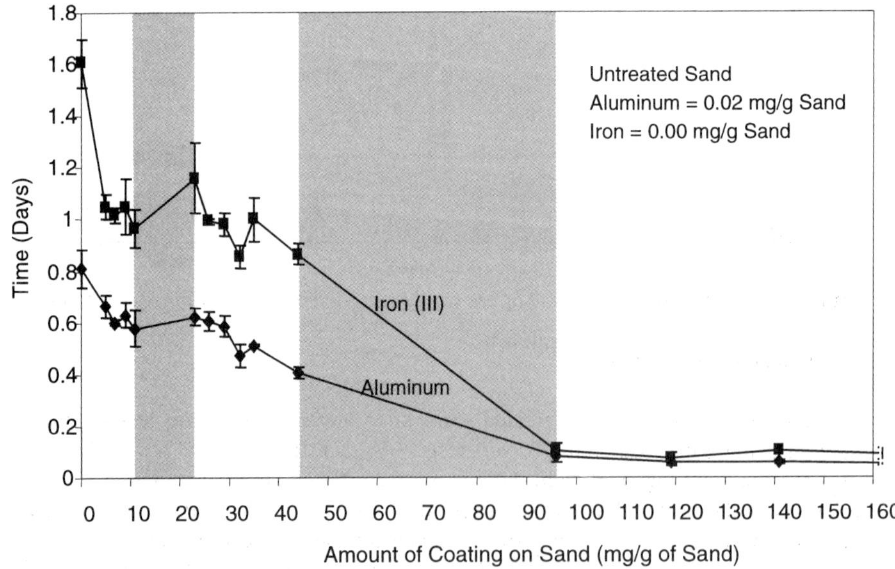

Figure 3 *Attritioning and leaching of the metal coating on sand as a function of time.*

After this initial decrease, the rate of decline becomes much more gradual. Over the next 40 days, both the iron and aluminum decreases only 10% further, and remained far above the baseline values. From day 11 to day 19, the column was not operational. From day 45 to day 86, the column was shut down again. During this down time, the sand in the column had to be removed twice in order to repair the center air tube. It is believed that these removals or handling of the sand caused the loss that was seen in Figure 3. The majority of the remaining metal content was removed during this second down period. After the second down time, the coating again remained stable, although at a concentration much closer to the baseline value.

3.2 Zeta Potential

The surface charge of particulate media, as measured by streaming potential, is another means of tracking the longevity of the treatment. The zeta potential of the coated sand was measured throughout the run (Figure 4) using the streaming potential instrument shown in Figure 2. Although the values never drop to the baseline value during the experiment, the reduction in zeta potential appears to follow the same trend as the other measured indicators. There is a rapid decrease in its value during the first few days, followed by a slower rate of decline. By the end of the run, the measured value of the surface potential falls to about –80 mV, which is still above the baseline value of –95 mV.

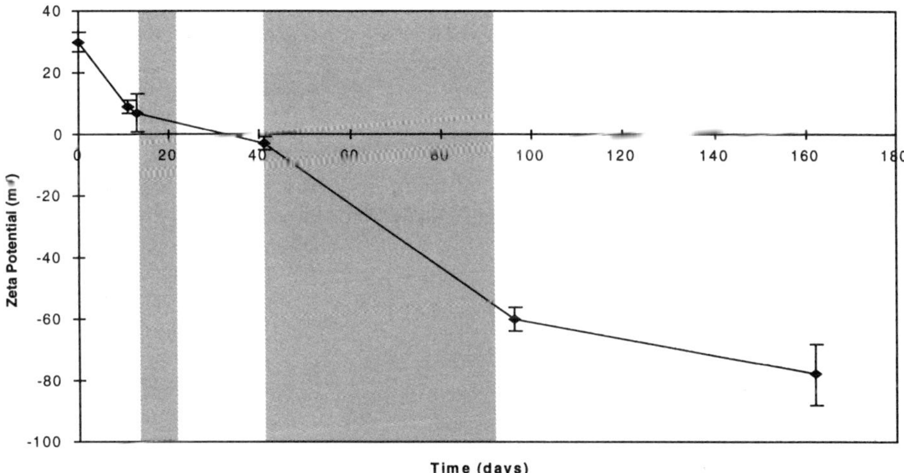

Figure 4 *Zeta potential at pH 7.0 of the sand from the column as a function of time.*

3.3 Bacterial Adsorption

The performance of the treated sand after attritioning in the test column was analysed using batch bacterial (*E. coli*) tests. At each time interval a sample from the untreated sand and the sand from the column were tested simultaneously to account for variations in the bacterial viability. These data are presented in Figure 5. There again appears to be a rapid decrease in performance during the first several days of exposure. A general downward progression in effectiveness continues over the whole of the run. This decrease in adsorption efficiency was correlated with the decrease in metal content and the increase in the electronegativity of the media. From day 140 to day 159, even though the adsorption efficiency is low, it is still above the value of the untreated sand, which indicates that the small amount of metal coating on the sand is still effective.

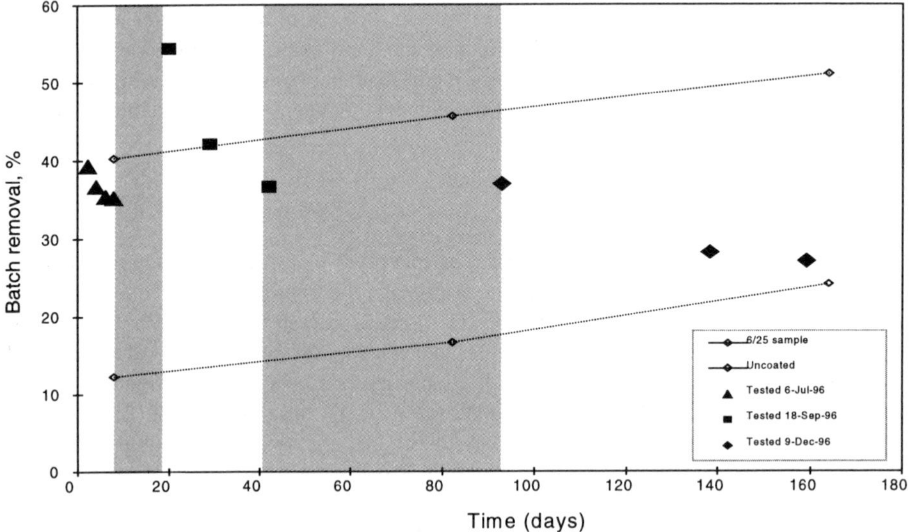

Figure 5 *Batch bacterial adsorption results for the sand as a function of time.*

3.4 Bacteriophage Adsorption

To evaluate the performance of the attritioned sand for bacteriophage removal, batch adsorption tests were conducted to measure the reduction of MS2 and PRD-1 bacteriophage. These data are shown in Figure 6. For the duration of the run, the metal

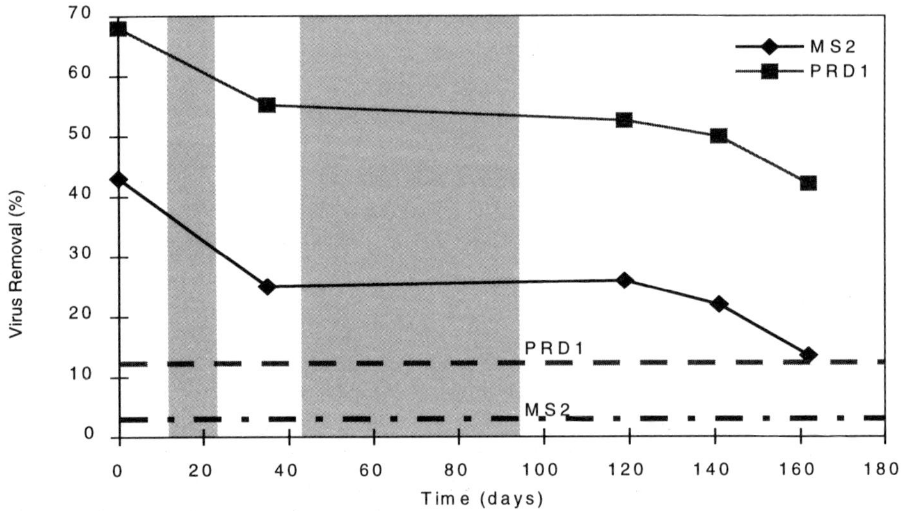

Figure 6 *Batch removal of bacteriophages as function of time.*

oxide-treated sand consistently removed several times more phage than untreated sand. In the batch test, the untreated sand removed about 4% and 12% of the MS2 and PRD-1 phage, respectively. The initial, unattritioned, treated sand far outperformed the untreated sand, removing about 43% of the MS2 and 68% of the PRD-1. In each case the removal efficiency decreased with time. The decrease in removal efficiency is more gradual than other indicators such as metal content analysis. These results suggest that for the removal of bacteriophages, only a small amount of coating on the sand is necessary. The values remained above the untreated baseline even at the end of the run, at 14 and 43%, respectively, for MS2 and PRD1.

4 CONCLUSIONS

In summary, the iron and aluminum hydroxide-coated sand is more effective than untreated sand in the removal of both bacterial and viral contaminants over a given time period, but the sand showed a gradual decrease in the amount of coating on the surface of the sand. Reduced concentrations of metal hydroxide on the surface can be correlated to the decreased filtration efficiency of micro-organisms (both bacteria and bacteriophages). This reduction in concentration asymptotically approaches the value of the untreated sand. A rapid initial rate of attrition during the first two days is followed by a more gradual degradation. The sand coating remained an effective filtration enhancement after at least 90 days in the continuous filter.

Industrial applications of this coating method require the stability of the metal hydroxide to fit the parameters of a given function. The coating process and cost can be evaluated on a case by case basis to determine the value of the filtration improvements for a given operation. The duration of the improved filtration properties may be appropriate for certain applications

5 ACKNOWLEDGEMENTS

The authors would like to acknowledge the financial support of the ERC at the University of Florida and the National Science Foundation (grant #EEC-94-02989). We would also like to thank all the professors on the interdisciplinary advanced separation team, including D.O. Shah, B. Koopman, S. Svoronos, H. El-Shall, S. Farrah, B. Moudgil, C. Park, and R. Dickinson. Also our thanks go out to the graduate students who have assisted greatly in analysis and experimental design, including S. Truesdail, G. Lucasik, Y. Li, P. Thatavarthy, R. Kalanraman, D. Davis, G. Zhan, J. Chen and J. Lorenzo.

References

1. G. Bitton, 'Wastewater Microbiology', Wiley-Liss, Inc., New York, 1994, p. 266.
2. P. Vilagines, B. Sarette and R. Vilagines, *Can. J. Microbiol.*, 1982, **28**, 783.
3. R. Walter, J. Durkop, B. Friedman, and H. J. Dobberkau, *Water Sci. Technology*, 1985, **17**, 139.
4   T. S. Brown, J. F. Malina, Jr., and B. D. Moore, *J. Am. Water Works Assoc.*, 1974, **66**, 735.

5.   Brown, T. S., J. F. Malina, Jr., and B. D. Moore, *J. Am. Water Works Assoc.*, 1974, **66**, 98.

6.   M. Edwards and M. M. Benjamin, *J. Wat. Pollut. Control fed.*, 1989, **61**, 1523.

7.   A. L. Mills, J. S. Herman, G. M. Hornberger and T. D. DeJesus, *Appl. Environ. Microbiol.*, 1994, **60**, 3300.

8.   S. R. Farrah and D. R. Preston, *Appl. Environ. Microbiol.*, 1985, **50**, 1502.

9.   S. R. Farrah, D. R. Preston, G. A. Toranzos, M. Girard, G. A. Erdos and V. Vasuhdivan, *Appl. Environ. Microbiol.*, 1991, **57**, 2502.

10.   J. Lukasik, S. R. Farrah, S. Truesdail and D. O. Shah, *Kona Powder and Particle, No 14*, 1996, p. 87.

11.   D. P. Snustad and D. S. Dean, 'Genetic Experiments with Bacterial Viruses', W. H. Freeman and Co., San Francisco, 1971, p. 65.

NEW USE OF CYANOSILANE COUPLING AGENT FOR DIRECT BINDING OF ANTIBODIES TO SILICA SUPPORTS. PHYSICO-CHEMICAL CHARACTERIZATION OF MOLECULARLY BIO-ENGINEERED LAYERS

S. Falipou,[†] J.M. Chovelon ,[†*] C. Martelet,[†] J. Margonari[‡] and D. Cathignol[‡]

[†]IFoS, CNRS/UMR 5621, Ecole Centrale de Lyon, BP163 69131 Ecully Cedex, France
[‡]INSERM, Unité 281, 151, Cours Albert Thomas 69424 Lyon Cedex 03, France

1 INTRODUCTION

Since the pioneering work of Nelson and Griffin,[1] the immobilisation of biological molecules on a variety of solid inorganic supports has become a widely used technology. A great number of biomedical applications, particularly medical diagnostics, involve binding of biospecies (antigen, antibody, enzyme) to various solid matrices. For example, in the case of biosensing applications it is necessary to graft the involved biospecies on silicon, polymers or metallic-based materials.[2,3] For immunoaffinity chromatography, functionalized beads are required. The methods used for functionalization are often tedious, rather expensive and can induce a denaturation of the biological species.[4] In addition, in all cases it is necessary to characterise the efficiency of the grafting processes through topographic studies (atomic force microscopy) or analytical investigation. For this purpose silica-based substrates like mica or glass slides with similar reactivity and with a well controlled surface state (low rugosity) have been used.[5,6]

Various chemical reagents can be used to modify surfaces in order to immobilise proteins. Silanes and thiols are among the most widely reagents used to fix enzymes, antibodies or other biologicals.[7] Thus, methods based on covalent binding to substrates have been developed.[8] The main difficulty of such methods is that they often need the use of an additional coupling agent. Only a few studies deal with a direct protocol for protein immobilisation.[9,10]

In this paper we present a simple protocol to fix biological species to silica-based surfaces by using a monofunctional silane reagent: 3-cyanopropyldimethylchlorosilane. This silane reagent was used without any further derivatization. The system allowed a strong – but not covalent – grafting of antibodies through their glycosylated regions (–OH groups) onto solid supports.[11] Glass slides were used first to show the efficiency of such a silanization process leading to a high rate of immobilised antibodies on the silica surfaces. Then the process was used to anchor antibodies to floating hollow silica microbeads that were able to recognise CD45 rat cells.

The originality of this study is that such hollow, low-density silica microbeads, have never before been used. This solid support was chosen as it could be used in ultrasound applications. Cavitation could be used for the destruction of cells *in vitro*:[12] gas bubbles in the suspension could collapse and damage cell membranes causing their death. Thus, acoustic cavitation could be used as an alternative medical therapy for cancer treatment. Prat *et al.*,[13] obtained subsequent hepatic lesions *in vivo* in rabbits by acoustic cavitation coupling the injection of artificial bubbles in the liver with an ultrasonic treatment.

2 EXPERIMENTAL SECTION

2.1 Reagents

3-Cyanopropyldimethylchlorosilane (purity 95%) was purchased from Aldrich. The following commercial analytical grade reagents were used: sulfuric acid and potassium dichromate from Merck, tetrahydrofuran from Sigma, xylene from Carlo Erba and 2-methylbutane from Aldrich. Low density hollow silica microbeads (SDT-60, 0.6 g/cm) were obtained from NewMet.

Purified sheep anti-mouse monoclonal antibody (SAM Ab), purified mouse anti-rat CD45RC clone (OX22) monoclonal antibody and biotin-conjugated mouse anti-rat CD45RC clone (OX22) monoclonal antibody were obtained from Pharmingen. FITC-conjugated F(ab')$_2$ Fragment of goat anti-mouse antibody was purchased from Dako. Ficolite was obtained from Dutscher.

Phosphate-buffered saline (PBS), bovine serum albumin (BSA) and Tween 20 were purchased from Sigma, carbonate/bicarbonate buffer was prepared by mixing Na_2CO_3 0.1 M and $NaHCO_3$ 0.1 M and adjusting the pH to 9.4. Barbital (Veronal) buffer was prepared by dissolving 4.12 g of sodic Veronal and 0.8 g of Veronal in 1L of distilled water and adjusting the pH to 8.6. Citrate buffer was prepared by mixing 1.78 g of Na_2HPO_4, 0.7 g of citric acid and 134 mL of distilled water.

2.2 Preparation of the Surface of the Samples

2.2.1 Pretreatment. The cleaning and hydroxylation of silica surfaces play an important role in the grafting process. Glass slides were degreased under ultrasonic stirring with acetone. Silica microbeads were not degreased before the hydroxylation treatment due to sample handling difficulties.

Glass slides and floating silica microbeads were treated with sulfochromic solution for 30 min at 70–80 °C to remove any coatings or contaminants and to create hydroxyl functions at the surface. Glass slides were washed with deionized water and dried by a pure nitrogen stream. The washing procedure used for microbeads is more critical. After the sulfochromic treatment, microbeads were allowed to float to the top of the suspension while the underlying fluid was drawn off. Then, deionized water was added to the microbeads which were once again allowed to float. This process was repeated until a complete elimination of the sulfochromic solution took place. Then the microbeads were partially dried in ambient air with gentle heating.

2.2.2 Silane activation. After the cleaning procedure, glass slides and microbeads were dried at 140 °C in vacuum (0.1 Pa) for 2 h in order to remove physisorbed water. For glass slides, pure silane was introduced under nitrogen atmosphere whereas for microbeads, a solution of 5 x 10^{-2} moles of silane in 2-methylbutane, which ensured a good wetting of the microbeads, was prepared prior to its introduction. In this latter case, solvent was removed in the following step under vacuum. The condensation reaction on the silica sample was performed in sealed vessels under vacuum for 48 h at 70 °C. The excess silane was eliminated by washing with THF, xylene and deionized water. For glass slides, substrates were blown with dry nitrogen, whereas for microbeads, they were allowed to dry in air at ambient temperature.

2.2.3 Immobilization of immunologicals species.   Antibodies were added onto silanized slides at concentrations ranging from a few ng/mL to several µg/mL.   They were allowed to adsorb for 1h at 37 °C and 1 night at 4 °C in a wet atmosphere.

The ELISA assay was done with mouse anti-rat CD45RC in concentrations ranging from a few µg/mL to several mg/mL in different buffers, using glass slides activated with the cyanosilane as substrate.   This assay was used first to test the validity of cyanosilane as coupling agent, and second to determine the optimum antibody level, coupled by varying the antibody concentration and the incubation buffer.   The most convenient buffer was found to be the carbonate buffer with a pH of 9.4.

2.2.4 Immobilization of cells onto silica microbeads.   Experiments and controls were done using silanized or non silanized substrates.   Different fixing modes were tested using monoclonal mouse anti-rat antibodies which recognise CD45 molecules expressed on rat lymphocytes.   Another biochemical link was tested with the insertion of an antibody between the complex cells/CD45 and the substrate, the SAM Ab.   In all cases, cells were added to the microbeads in a ratio higher than 5×10^5 cells/mg of microbeads.

The volume was 500 µL to 1mL, and the microbeads and cells were incubated for 2 h at 4 °C with agitation.   After incubation, centrifugation at 450–500 g allowed microbeads to float to the top of the suspension whereas non-fixed cells fell to the bottom.   Then, the effluent was drawn off.   Floating microbeads were observed under a microscope and photographs of the cell rosets were done.

2.3 Techniques of Characterization

2.3.1 Contact angle measurements on planar substrate.   Contact angle measurements were made by the sessile drop method on a Digidrop instrument (GBX Scientific Instruments).   Measurements were made on drops of pure water deposited at 22 °C.   The reported contact angles are the average of results obtained with at least 15 drops for each surface treatment.

2.3.2 Atomic force microscopy.   Atomic force microscopy (AFM) images were obtained in air using a Park Scientific Instrument (Autoprobe).   V-shaped cantilevers (PSI) with a normal spring constant of $0.05~\mathrm{Nm^{-1}}$ and a pyramidal Si_3N_4 tip were used in this study.   Either $100 \times 100~\mu m^2$ or $5 \times 5~\mu m^2$ scanners were used.   Images were taken in the contact or tapping modes.

2.3.3 Infrared spectroscopy on microbeads.   Methods previously described do not allow the characterization of microspherical substrates.   Infrared measurements were made by the "Service Central d'Analyses" of the CNRS on a Nicolet 20SXC instrument (diffuse reflection mode).   Results were given after accumulation of 2048 scans.

3 RESULTS AND DISCUSSION

3.1 Silanization Process

Considering the geometry of the silica beads, classical methods of characterization of the silanization cannot be applied.   Among these methods, contact angle measurements offer the better wettability test when pure fluids and smooth surfaces were used.   So, to evaluate the wettability after silanization on flat samples, glass slides were used.   Such silica-based samples would have a behaviour similar to microbeads.   For silane-treated and untreated glass slides, Table 1 summarises contact angle results at equilibrium.

Table 1   *Contact Angle Measurements on Glass Slides.*

Substrate	Liquid	Treatment	Angle (Degrees)	Standard Deviation	Number Of Measures
Glass slides	pure water	none	34	7	27
		sulfochromic at 70–80 °C	< 15	////	////
		blank	40	2	16
		silanization	71	2	27

Without any treatment, a value with an important standard deviation ($\theta = 34 \pm 7°$) was obtained showing the heterogeneity of the surface. Glass slides, when cleaned with hot sulfochromic solution, are almost totally wetted by water ($\theta < 15°$), demonstrating the efficiency of such a cleaning procedure. Surface hydrophilicity is due to the presence of a great number of hydroxyl groups on the surface, required for the silanization process. Great care must be taken with cleaned samples, as organic molecules can readily be adsorbed on these reactive surfaces, inducing an increase of the contact angle within a few minutes. In this condition, as soon as the treatment is finished, the samples are placed under vacuum.

After the silanization, contact angles increase up to around $71 \pm 2°$ proving the presence of silane. It is noteworthy that the surface treatment induces a more homogeneous surface (low standard deviation). To be sure that no contamination occurs, a sample treated according to this process but without the silanization step was tested. In this case the contact angle increased to 40°. This value was less than the one obtained with the silane but higher than the one obtained after sulfochromic treatment.

For characterizing the molecular organization of silane on glass slides, AFM techniques have been used. Substrate modifications after reaction were checked by measuring surface roughness variations. Table 2 shows the evolution of this surface roughness parameter (rms). After silanization, the roughness of the surface was found to be higher than the non treated one (rms = 0.7 nm compared with 0.3 nm), values which could be compared to those obtained by Perrin *et al.*,[14] who used the same chain length for the silane reagent.

Figure 1, obtained in tapping mode (1μm x 1 μm), shows that the silane layer was dense and rather homogeneous, except for a small number of pinhole defects. Using these defects, the thickness of one layer was determined to be 0.58 nm. This figure also shows areas with higher thicknesses, proving that in some cases aggregates can be formed. The origin of these aggregates can come from possible hydrogen bonding interactions between cyano moieties and silanols. These interactions are strong enough to resist different washing processes.

Table 2   *Surface Roughness Parameter Evaluation.*

Substrate	Roughness (rms)
non-treated glass slide	0.3 nm
silanized glass slide	0.7 nm
silanized glass slide + antibodies	>5 nm

Line	Height	Distance
[A]	1: 5.84 Å	1: 471 Å

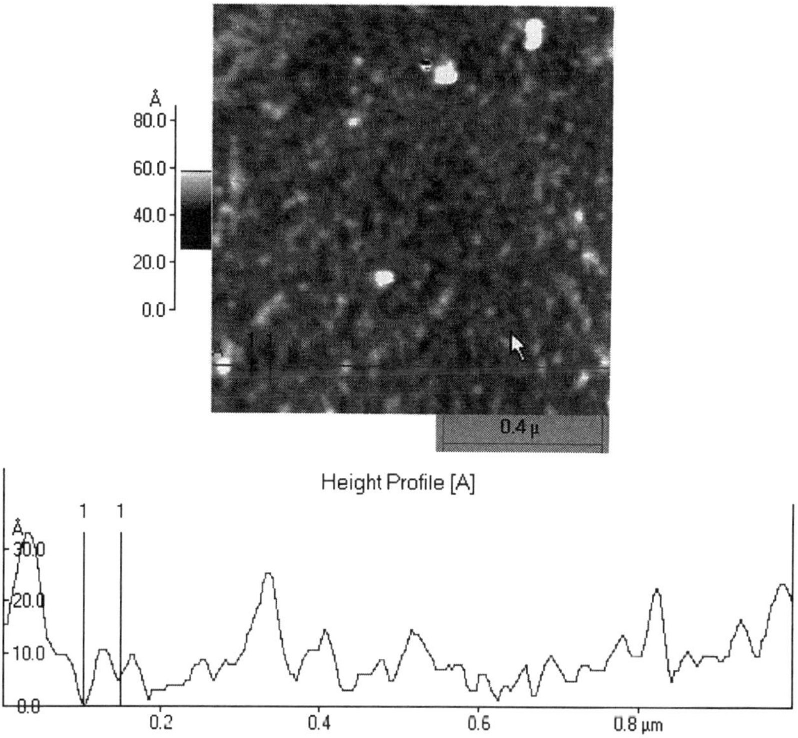

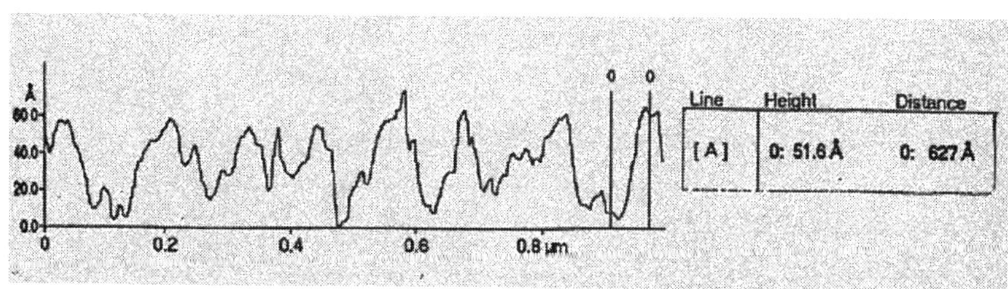

Figure 1 *AFM image and profile of silanized surface.*

Figure 2 *AFM profile of the surface after grafting of antibodies.*

In order to confirm the presence of silane on treated microbeads, diffuse reflection infrared spectroscopy was used. These spectra (not shown here) prove the presence of nitrile and alkyl functions at the surface of the microbeads, whereas no peaks were obtained with untreated microbeads. Under the infrared analysis conditions used, if only one monolayer was present on the surface, no peak would be visible, taking into account the sensitivity of the apparatus. In our case, the presence of nitrile and alkyl peaks confirmed that the silanization process occurred by forming a few aggregates. In this context, it seems that silanization effects could be considered similar for both substrates.

3.2 Immobilisation of Immunological Species

Figure 2 shows an AFM profile taken in tapping mode in air from antibodies deposited onto cyanosilane. The image shows aggregates with dimensions around 5 nm vertically. This suggests that lateral interactions occur between the adjacent IgG molecules. It was not surprising to find such lateral interactions, as this protein has some self-aggregating properties. The depth of this layer of deposited antibodies roughly matches the known size of one IgG molecule (5 x 15 nm),[4] confirming the horizontal orientation of the antibodies on the surface.

3.3 Cell Binding

Table 3 compares the results of different modes of cell immobilisation. A camera coupled to an optical microscope was used to collect the data. Results are expressed as " + " for subsequent immobilisation, " +/– " for a weak immobilisation and " – " for no immobilisation of cells on the substrate.

Table 3   *Results of the Cell Binding Assays.*

Substrate And Fixation Mode	Results	Molecular Assemblies
(1) silane activated microbeads with CD45 Ab	+/–	
(2) non-silanized microbeads sensitized cells	–	
(3) silane activated microbeads non sensitized cells	–	
(4) silane activated microbeads sensitized cells without SAM Ab	+/–	
(5) silane activated microbeads sensitized cells with SAM Ab	+	

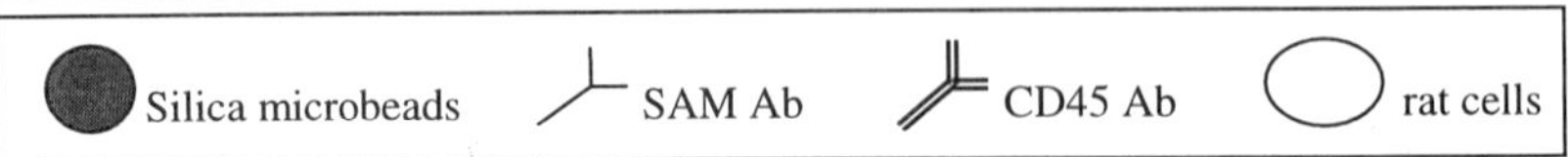

This table shows that only activated microbeads incubated with sensitized cells lead to a subsequent immobilisation of the cells. Figure 3 shows a photograph of cell aggregation around microbeads. In this case, microbeads grafted with SAM Ab acting as a spacer allows a better approach between cells sensitized with CD45 Ab and microbeads, which is in good agreement with the literature.[15] Moreover, the addition of SAM Ab, a coupling agent, fixed on substrates, and able to recognise mouse species on lymphocytes could minimize unexpected fixation or removal of antibodies to substrates.

In order to validate the following experimental procedure, blanks were done in order confirm that no fixation occurs when microbeads were not silanized and, in addition, that SAM Ab do not recognise CD45 molecules expressed on lymphocytes.

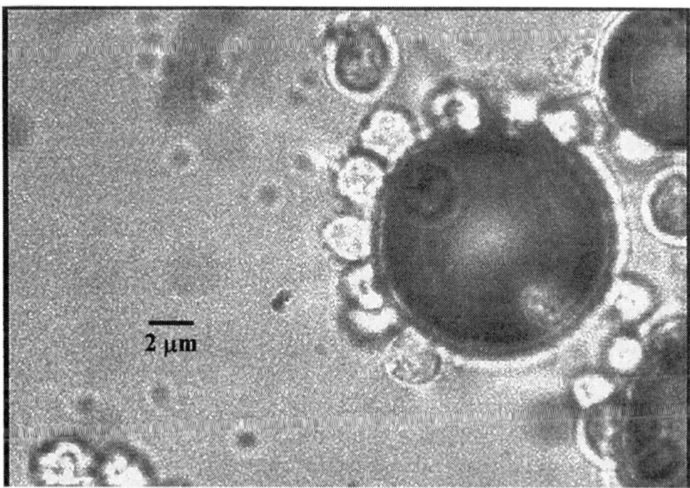

Figure 3 *Photograph of cells around silica microbeads.*

4 CONCLUSION

The application of a well controlled functionalization process has shown that activated hollow silica microbeads are able to strongly and selectively fix rat cells. In a further step, cell destruction could be obtained as microbeads would be destroyed under ultrasonic power. But the presence of mineral fragments which cannot be dispersed forbid a direct use in blood. Biocompatible supports allowing a destruction through acoustic cavitation have to be found. For this purpose we propose that functionalized liposomes are potential candidates.

References

1. J. M. Nelson, E. G. Griffin, *J. Am. Chem. Soc.*, 1916, **38**, 1109.
2. H. Maupas, *PhD Dissertation*, Ecole Centrale Lyon, 1995.
3. H. Maupas, C. Saby, C. Martelet, N. Jaffrezic-Renault, A. P. Soldatkin, M. H. Charles, T. Delair and B. Mandrand, *J. Electroanal. Chem.*, 1996, **406**, 53.

4.   E. Harlow and D. Lane, 'Anticorps: un manuel de laboratoire', Editions Pradel, Paris, 1991.

5.   S. Karraasch, M. Dolder, F. Schabert, J. Ramsden and A. Engel, *Biophys. J.*, 1993, **65**, 2437.

6.   M. E. Browning-Kelley, K. Wadu-Mesthrige, V. Hari, G. Y. Liu, *Langmuir*, 1997, **13**, 343.

7.   E. P. Plueddemann, 'Silane Coupling Agents', Plenum Press, New York, 1982.

8.   H. H. Weetall, 'Immobilized Enzymes, Antigens, Antibodies And Peptides. Preparation And Characterization', M. Dekker, Inc., New York, 1975.

9.   M. Chiong, S. Lavandero, R. Ramos, J.C Aguillón and A. Ferreira, *Anal. Biochem.*, 1991, **197(1)**, 47.

10. N. M. Pope, D. L. Kulcinski, A. Hardwick and Y. A. Chang, *Bioconjugate Chem.*, 1993, **4**, 166.

11. C. Saby, N. Jaffrezic-Renault, C. Martelet, B. Colin, M. H. Charles, T. Delair and B. Mandrand, *Sensors and Actuators*, 1993, **B15–16**, 458.

12. F. Prat, J. Y. Chapelon, B. Chauffert, T. Ponchon and D. Cathignol, *Cancer Research*, 1991, **51(11)**, 3024.

13. F. Prat, T. Ponchon, J. Y. Berger, J. Y. Chapelon, P. Gagnon and D. Cathignol, *Gastroenterology*, 1991, **100(5)**, 1345.

14. A. Perrin, V. Lanet and A. Theretz, *Langmuir*, 1997, **13**, 2557.

15. DYNAL, Technical Handbook (Second Edition).

Subject Index